Student Solutions Manual

for
Tussy and Gustafson's

Elementary Algebra

Third Edition

Kristy Hill
Hinds Community College

Alexander Lee
Hinds Community College

THOMSON

BROOKS/COLE

Australia • Canada • Mexico • Singapore • Spain • United Kingdom • United States

Printed in the United States of America
1 2 3 4 5 6 7   08 07 06 05 04

Printer: Darby Printing

0-534-38685-7

For more information about our products,
contact us at:
Thomson Learning Academic Resource Center
1-800-423-0563

For permission to use material from this text or
product, submit a request online at
http://www.thomsonrights.com.
Any additional questions about permissions can be
submitted by email to **thomsonrights@thomson.com.**

Cover Image: Kevin Tolman

Thomson Higher Education
10 Davis Drive
Belmont, CA 94002-3098
USA

Asia (including India)
Thomson Learning
5 Shenton Way
#01-01 UIC Building
Singapore 068808

Australia/New Zealand
Thomson Learning Australia
102 Dodds Street
Southbank, Victoria 3006
Australia

Canada
Thomson Nelson
1120 Birchmount Road
Toronto, Ontario M1K 5G4
Canada

UK/Europe/Middle East/Africa
Thomson Learning
High Holborn House
50–51 Bedford Road
London WC1R 4LR
United Kingdom

Latin America
Thomson Learning
Seneca, 53
Colonia Polanco
11560 Mexico
D.F. Mexico

Spain (including Portugal)
Thomson Paraninfo
Calle Magallanes, 25
28015 Madrid, Spain

TABLE OF CONTENTS

Chapter One   An Introduction to Algebra   Page

Section 1.1   ---   1
Section 1.2   ---   3
Section 1.3   ---   8
Section 1.4   ---   10
Section 1.5   ---   13
Section 1.6   ---   16
Section 1.7   ---   19
Section 1.8   ---   23
Key Concept   ---   27
Chapter Review   ---   28
Chapter Test   ---   33

Chapter Two   Equations, Inequalities, and Problem Solving   Page

Section 2.1   ---   36
Section 2.2   ---   40
Section 2.3   ---   45
Section 2.4   ---   48
Section 2.5   ---   54
Section 2.6   ---   61
Section 2.7   ---   68
Key Concept   ---   76
Chapter Review   ---   77
Chapter Test   ---   89
Cumulative Review ---   93

Chapter Three   Linear Equations and Inequalities in Two Variables   Page

Section 3.1   ---   99
Section 3.2   ---   101
Section 3.3   ---   109
Section 3.4   ---   115
Section 3.5   ---   119
Section 3.6   ---   125
Section 3.7   ---   131
Key Concept   ---   137
Chapter Review   ---   138
Chapter Test   ---   145
Cumulative Review ---   148

Chapter Four　　　　　　Exponents and Polynomials　　　　　Page

    Section 4.1 -- 153
    Section 4.2 -- 157
    Section 4.3 -- 163
    Section 4.4 -- 166
    Section 4.5 -- 171
    Section 4.6 -- 174
    Section 4.7 -- 177
    Section 4.8 -- 180
    Key Concept -- 186
    Chapter Review -- 187
    Chapter Test -- 194
    Cumulative Review -- 197

Chapter Five　　　　　　Factoring and Quadratic Equations　　　Page

    Section 5.1 -- 202
    Section 5.2 -- 205
    Section 5.3 -- 207
    Section 5.4 -- 210
    Section 5.5 -- 212
    Section 5.6 -- 214
    Section 5.7 -- 216
    Key Concept -- 221
    Chapter Review -- 222
    Chapter Test -- 225
    Cumulative Review -- 227

Chapter Six　　　　　　Rational Expressions and Equations　　　Page

    Section 6.1 -- 232
    Section 6.2 -- 237
    Section 6.3 -- 242
    Section 6.4 -- 246
    Section 6.5 -- 254
    Section 6.6 -- 261
    Section 6.7 -- 273
    Section 6.8 -- 280
    Key Concept -- 285
    Chapter Review -- 287
    Chapter Test -- 298
    Cumulative Review -- 304

Chapter Seven   Solving Systems of Equations and Inequalities   Page

    Section 7.1 --- 310
    Section 7.2 --- 315
    Section 7.3 --- 324
    Section 7.4 --- 335
    Section 7.5 --- 342
    Key Concept --- 347
    Chapter Review --- 349
    Chapter Test --- 358
    Cumulative Review --- 363

Chapter Eight   Radical Expressions and Equations   Page

    Section 8.1 --- 369
    Section 8.2 --- 373
    Section 8.3 --- 377
    Section 8.4 --- 380
    Section 8.5 --- 385
    Section 8.6 --- 391
    Key Concept --- 394
    Chapter Review --- 395
    Chapter Test --- 401
    Cumulative Review --- 404

Chapter Nine   Solving Quadratic Equations, Functions   Page

    Section 9.1 --- 412
    Section 9.2 --- 421
    Section 9.3 --- 430
    Section 9.4 --- 434
    Section 9.5 --- 442
    Section 9.6 --- 446
    Key Concept --- 451
    Chapter Review --- 453
    Chapter Test --- 464
    Cumulative Review --- 469

Appendix   Statistics   Page

    Statistics --- 480

PREFACE

Kristy Hill and Alexander Lee would like to receive comments from both students and instructors about the more detailed worked out problems of the Student Solutions Manual. Both of us have heard our own students ask the question about the Student Solutions Manual, "How did they get that answer?" We both decided to put in more detailed steps so the students would not have to ask the instructor that question again. We hope we have succeeded in producing a friendlier Student Solution Manual for you, the student.

Kristy Hill, Hinds Community College, Raymond Campus, krhill@hindscc.edu
Alexander Lee, Hinds Community College, Rankin Campus, ahlee@hindscc.edu

SECTION 1.1

VOCABULARY

1. The answer to an addition problem is called the **sum**. The answer to a subtraction problem is called the **difference**.

3. **Variables** are letters that stand for numbers.

5. An **equation** is a mathematical sentence that contains an = symbol.

7. The **horizontal** axis of a graph extends left and right and the **vertical** axis extends up and down.

CONCEPTS

9. equation

11. algebraic expression

13. algebraic expression

15. equation

17. a) multiplication and subtraction
    b) x

19. a) addition and subtraction
    b) m

21.

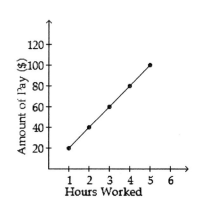

23. They determine that 15-year-old machinery is worth $35,000.

NOTATION

25. $5 \cdot 6$ or $5(6)$

27. $34 \cdot 75$ or $34(75)$

29. $4x$

31. $3rt$

33. lw

35. Prt

37. $2w$

39. xy

41. $\dfrac{32}{x}$

43. $\dfrac{90}{30}$

PRACTICE

45. The product of 8 and 2 is 16.

47. The difference of 11 and 9 is 2.

49. The product of 2 and x is 10.

51. The quotient of 66 and 11 is 6.

53. $p = 100 - d$

55. $7d = h$

57. $s = 3c$

59. $w = e + 1{,}200$

61. $p = r - 600$

63. $\dfrac{l}{4} = m$

65. $d = 360 + L$

Lunch time (minutes) L	School day (minutes) d
30	$d = 360 + \textbf{30}$ $= \textbf{390}$
40	$d = 360 + \textbf{40}$ $= \textbf{400}$
45	$d = 360 + \textbf{45}$ $= \textbf{405}$

67. $t = 1{,}500 - d$

Deductions d	Take-home pay t
200	$t = 1{,}500 - \textbf{200}$ $= \textbf{1,300}$
300	$t = 1{,}500 - \textbf{300}$ $= \textbf{1,200}$
400	$t = 1{,}500 - \textbf{400}$ $= \textbf{1,100}$

69. $d = \dfrac{e}{12}$

APPLICATIONS

71. $90° - \text{Angle } 1 = \text{Angle } 2$

Angle 1 (degrees)	Angle 2 (degrees)
0	**90**
30	**60**
45	**45**
60	**30**
90	**0**

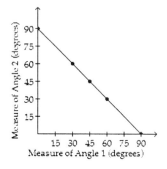

WRITING

73. Answers will vary.

75. Answers will vary.

CHALLENGE PROBLEMS

77. $s = t + 1$

s	t
10	$\textbf{10} + 1 = \textbf{11}$
18	19
$\textbf{34} - 1 = \textbf{33}$	34
47	48

SECTION 1.2

VOCABULARY

1. Numbers that have only 1 and themselves as factors, such as 23, 37, and 41, are called **prime** numbers.

3. The **numerator** of the fraction $\frac{3}{4}$ is 3, and the **denominator** is 4.

5. Two fractions that represent the same number, such as $\frac{1}{2}$ and $\frac{2}{4}$, are called **equivalent** fractions.

7. The **least or lowest** common denominator for a set of fractions is the smallest number each denominator will divide exactly.

CONCEPTS

9. $2 \cdot 2 \cdot 3 \cdot 5 = \mathbf{60}$

11. $\frac{4}{12} = \frac{1}{3}$

13. $\dfrac{\cancel{2} \cdot 2 \cdot \cancel{3}}{\cancel{2} \cdot \cancel{3} \cdot 5} = \dfrac{2}{5}$

15. a) To build up a fraction, we multiply it by $\boxed{1}$ in the form of $\frac{2}{2}$, $\frac{3}{3}$, or $\frac{4}{4}$, and so on.
    b) To simplify a fraction, we remove factors equal to $\boxed{1}$ in the form of $\frac{2}{2}$, $\frac{3}{3}$, or $\frac{4}{4}$, and so on.

17. a) 3 times
    b) 2 times

NOTATION

19. a) $\frac{5}{16}$

    b) $\frac{3}{3} = \boxed{1}$

    c) $\dfrac{5 \cdot 3}{16 \cdot 3} = \dfrac{15}{48}$

PRACTICE

21. 1, 2, 4, 5, 10, 20

23. 1, 2, 4, 7, 14, 28

25. $75 = 3 \cdot 5 \cdot 5$

27. $28 = 2 \cdot 2 \cdot 7$

29. $117 = 3 \cdot 3 \cdot 13$

31. $220 = 2 \cdot 2 \cdot 5 \cdot 11$

33. $\dfrac{1}{3} = \dfrac{1 \cdot 3}{3 \cdot 3}$

    $= \dfrac{3}{9}$

35. $\dfrac{4}{9} = \dfrac{4 \cdot 6}{9 \cdot 6}$

    $= \dfrac{24}{54}$

37. $\dfrac{7}{1} = \dfrac{7 \cdot 5}{1 \cdot 5}$

    $= \dfrac{35}{5}$

39. $\dfrac{6}{12} = \dfrac{\cancel{2} \cdot \cancel{3}}{\cancel{2} \cdot 2 \cdot \cancel{3}}$

    $= \dfrac{1}{2}$

41. $\dfrac{24}{18} = \dfrac{\cancel{2} \bullet 2 \bullet 2 \bullet \cancel{3}}{\cancel{2} \bullet \cancel{3} \bullet 3}$

$\quad = \dfrac{4}{3}$

43. $\dfrac{15}{20} = \dfrac{\cancel{5} \bullet 3}{2 \bullet 2 \bullet \cancel{5}}$

$\quad = \dfrac{3}{4}$

45. $\dfrac{72}{64} = \dfrac{\cancel{2} \bullet \cancel{2} \bullet \cancel{2} \bullet 3 \bullet 3}{\cancel{2} \bullet \cancel{2} \bullet \cancel{2} \bullet 2 \bullet 2 \bullet 2}$

$\quad = \dfrac{9}{8}$

47. $\dfrac{33}{56} = \dfrac{3 \bullet 11}{2 \bullet 2 \bullet 2 \bullet 7}$

$\quad = \dfrac{33}{56}$

It is in lowest terms.

49. $\dfrac{36}{225} = \dfrac{2 \bullet 2 \bullet \cancel{3} \bullet \cancel{3}}{\cancel{3} \bullet \cancel{3} \bullet 5 \bullet 5}$

$\quad = \dfrac{4}{25}$

51. $\dfrac{1}{2} \bullet \dfrac{3}{5} = \dfrac{1 \bullet 3}{2 \bullet 5}$

$\quad = \dfrac{3}{10}$

53. $\dfrac{4}{3}\left(\dfrac{6}{5}\right) = \dfrac{4(6)}{3(5)}$

$\quad = \dfrac{24}{15}$

$\quad = \dfrac{8}{5}$

55. $\dfrac{5}{12} \bullet \dfrac{18}{5} = \dfrac{5 \bullet 18}{12 \bullet 5}$

$\quad = \dfrac{90}{60}$

$\quad = \dfrac{3}{2}$

57. $\dfrac{21}{1}\left(\dfrac{10}{3}\right) = \dfrac{21 \bullet 10}{1 \bullet 3}$

$\quad = \dfrac{210}{3}$

$\quad = \dfrac{70}{1}$

$\quad = 70$

59. $7\dfrac{1}{2} \bullet 1\dfrac{2}{5} = \dfrac{15}{2} \bullet \dfrac{7}{5}$

$\quad = \dfrac{15 \bullet 7}{2 \bullet 5}$

$\quad = \dfrac{105}{10}$

$\quad = \dfrac{21}{2}$

$\quad = 10\dfrac{1}{2}$

61. $6 \bullet 2\dfrac{7}{24} = \dfrac{6}{1} \bullet \dfrac{55}{24}$

$\quad = \dfrac{6 \bullet 55}{1 \bullet 24}$

$\quad = \dfrac{330}{24}$

$\quad = \dfrac{55}{4}$

$\quad = 13\dfrac{3}{4}$

63. $\dfrac{3}{5} \div \dfrac{2}{3} = \dfrac{3}{5} \bullet \dfrac{3}{2}$

$\quad = \dfrac{9}{10}$

65. $\dfrac{3}{4} \div \dfrac{6}{5} = \dfrac{3}{4} \bullet \dfrac{5}{6}$

$\qquad = \dfrac{15}{24}$

$\qquad = \dfrac{5}{8}$

67. $\dfrac{21}{35} \div \dfrac{3}{14} = \dfrac{21}{35} \bullet \dfrac{14}{3}$

$\qquad = \dfrac{294}{105}$

$\qquad = \dfrac{14}{5}$

69. $6 \div \dfrac{3}{14} = \dfrac{6}{1} \bullet \dfrac{14}{3}$

$\qquad = \dfrac{84}{3}$

$\qquad = 28$

71. $3\dfrac{1}{3} \div 1\dfrac{5}{6} = \dfrac{10}{3} \div \dfrac{11}{6}$

$\qquad = \dfrac{10}{3} \bullet \dfrac{6}{11}$

$\qquad = \dfrac{60}{33}$

$\qquad = \dfrac{20}{11}$

$\qquad = 1\dfrac{9}{11}$

73. $8 \div 3\dfrac{1}{5} = \dfrac{8}{1} \div \dfrac{16}{5}$

$\qquad = \dfrac{8}{1} \bullet \dfrac{5}{16}$

$\qquad = \dfrac{40}{16}$

$\qquad = \dfrac{5}{2}$

$\qquad = 2\dfrac{1}{2}$

75. $\dfrac{3}{5} + \dfrac{3}{5} = \dfrac{3+3}{5}$

$\qquad = \dfrac{6}{5}$

77. $\dfrac{1}{6} + \dfrac{1}{24} = \dfrac{1}{6} \bullet \dfrac{4}{4} + \dfrac{1}{24}$

$\qquad = \dfrac{4}{24} + \dfrac{1}{24}$

$\qquad = \dfrac{4+1}{24}$

$\qquad = \dfrac{5}{24}$

79. $\dfrac{3}{5} + \dfrac{2}{3} = \dfrac{3}{5} \bullet \dfrac{3}{3} + \dfrac{2}{3} \bullet \dfrac{5}{5}$

$\qquad = \dfrac{9}{15} + \dfrac{10}{15}$

$\qquad = \dfrac{9+10}{15}$

$\qquad = \dfrac{19}{15}$

81. $\dfrac{5}{12} + \dfrac{1}{3} = \dfrac{5}{12} + \dfrac{1}{3} \bullet \dfrac{4}{4}$

$\qquad = \dfrac{5}{12} + \dfrac{4}{12}$

$\qquad = \dfrac{5+4}{12}$

$\qquad = \dfrac{9}{12}$

$\qquad = \dfrac{3}{4}$

83. $\dfrac{9}{4} - \dfrac{5}{6} = \dfrac{9}{4} \bullet \dfrac{3}{3} - \dfrac{5}{6} \bullet \dfrac{2}{2}$

$\qquad = \dfrac{27}{12} - \dfrac{10}{12}$

$\qquad = \dfrac{27-10}{12}$

$\qquad = \dfrac{17}{12}$

85. $\dfrac{7}{10} - \dfrac{1}{14} = \dfrac{7}{10} \bullet \dfrac{7}{7} - \dfrac{1}{14} \bullet \dfrac{5}{5}$

$= \dfrac{49}{70} - \dfrac{5}{70}$

$= \dfrac{49 - 5}{70}$

$= \dfrac{44}{70}$

$= \dfrac{22}{35}$

87. $\dfrac{5}{14} - \dfrac{4}{21} = \dfrac{5}{14} \bullet \dfrac{3}{3} - \dfrac{4}{21} \bullet \dfrac{2}{2}$

$= \dfrac{15}{42} - \dfrac{8}{42}$

$= \dfrac{15 - 8}{42}$

$= \dfrac{7}{42}$

$= \dfrac{1}{6}$

89. $3 - \dfrac{3}{4} = \dfrac{3}{1} \bullet \dfrac{4}{4} - \dfrac{3}{4}$

$= \dfrac{12}{4} - \dfrac{3}{4}$

$= \dfrac{9}{4}$

91. $3\dfrac{3}{4} - 2\dfrac{1}{2} = \dfrac{15}{4} - \dfrac{5}{2}$

$= \dfrac{15}{4} - \dfrac{5}{2} \bullet \dfrac{2}{2}$

$= \dfrac{15}{4} - \dfrac{10}{4}$

$= \dfrac{15 - 10}{4}$

$= \dfrac{5}{4} = 1\dfrac{1}{4}$

93. $8\dfrac{2}{9} - 7\dfrac{2}{3} = \dfrac{74}{9} - \dfrac{23}{3}$

$= \dfrac{74}{9} - \dfrac{23}{3} \bullet \dfrac{3}{3}$

$= \dfrac{74}{9} - \dfrac{69}{9}$

$= \dfrac{74 - 69}{9}$

$= \dfrac{5}{9}$

APPLICATIONS

95. BOTANY

a) Add the 2 widths of the growth rings.

$\dfrac{5}{32} + \dfrac{1}{16} = \dfrac{5}{32} + \dfrac{1}{16} \bullet \dfrac{2}{2}$

$= \dfrac{5}{32} + \dfrac{2}{32}$

$= \dfrac{7}{32}$

b) Subtract the 2 widths of the growth rings.

$\dfrac{5}{32} - \dfrac{1}{16} = \dfrac{5}{32} - \dfrac{1}{16} \bullet \dfrac{2}{2}$

$= \dfrac{5}{32} - \dfrac{2}{32}$

$= \dfrac{3}{32}$

97. FRAMES

Since there are 4 sides and each side is $10\frac{1}{8}$ inches long, multiply 4 times $10\frac{1}{8}$.

$$4\left(10\frac{1}{8}\right) = \frac{4}{1} \cdot \frac{81}{8}$$
$$= \frac{324}{8}$$
$$= \frac{81}{2}$$
$$= 40\frac{1}{2}$$

$40\frac{1}{2}$ in of moldding is needed.

WRITING

99. Answers will vary.

101. Answers will vary.

REVIEW

103. The difference of 7 and 5 is 2.

105. The quotient of 30 and 15 is 2.

107. $T = 15g$

Number of gears g	Number of teeth T
10	15(**10**) = **150**
12	15(**12**) = **180**

CHALLENGE PROBLEMS

109. Find a common denominator and compare numerators.

$$\frac{11}{12} \cdot \frac{3}{3} = \frac{33}{36}$$
$$\frac{8}{9} \cdot \frac{4}{4} = \frac{32}{36}$$
$$\frac{33}{36} > \frac{32}{36}$$
so $\frac{11}{12} > \frac{8}{9}$

Section 1.2

SECTION 1.3

VOCABULARY

1. The set of **whole** numbers is
   {0, 1, 2, 3, 4, 5, …}.

3. The set of **integers** is
   {…-2, -1, 0, 1, 2, …}.

5. Numbers less than zero are **negative**, and
   numbers greater than zero are **positive**.

7. A **rational** number is any number that can
   be expressed as a fraction with an integer
   numerator and a nonzero integer
   denominator.

9. An **irrational** number is a nonterminating,
   nonrepeating decimal.

11. Every point on the number line
    corresponds to exactly one **real** number.

CONCEPTS

13. opposites

15. $\dfrac{6}{1}, \dfrac{-9}{1}, \dfrac{-7}{8}, \dfrac{7}{2}, \dfrac{-3}{10}, \dfrac{283}{100}$

17. 13 and -3 are 8 units away from 5

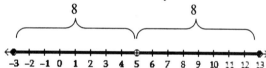

19.   a) $a < b$
      b) $b > a$
      c) $b > 0$ and $a < 0$
      d) $|a| > |b|$

21. π inches

NOTATION

23. $\sqrt{2}$ is read "the **square root** of 2."

25. The symbol $\neq$ means **is not equal to**.

27. The symbol π is a letter from the **Greek**
    alphabet.

29. $-\dfrac{4}{5} = \dfrac{-4}{5}$

        $= \dfrac{4}{-5}$

PRACTICE

31. $\dfrac{5}{8} = 8\overline{)5.000}\,^{0.625}$

     $= 0.625$

33. $\dfrac{1}{30} = 30\overline{)1.0000}\,^{0.0333}$

     $= 0.0\bar{3}$

35. $\dfrac{21}{50} = 50\overline{)21.00}\,^{0.42}$

     $= 0.42$

37. $\dfrac{5}{11} = 11\overline{)5.000000}\,^{0.454545}$

     $= 0.\overline{45}$

39. $5 > 4$

41. $-2 > -3$

43. $|3.4| > \sqrt{2}$

45. $|-1.1| < 1.2$

47. $-\dfrac{5}{8} < -\dfrac{3}{8}$

49. $\left|-\dfrac{15}{2}\right| = 7.5$

51. $\dfrac{99}{100} = 0.99$

53. $0.333\ldots > 0.3$

55. $1 > \left| -\dfrac{15}{16} \right|$

57. a) True
    b) False
    c) False
    d) True

59. a) $-5 > -6$
    b) $-25 < 16$

61.

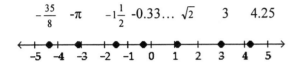

APPLICATIONS

63. DRAFTING

natural: 9
whole: 9
integers: 9
rational: $9, \dfrac{15}{16}, 3\dfrac{1}{8}, 1.765$
irrational: $2\pi, 3\pi, \sqrt{89}$
real: all

65. TARGET PRACTICE

Shell 1 landed farther from the target.

$|-6| > |5|$

67. TRADE

a) 2000 had a deficit of -$81 billion.
   2002 had a deficit of -$70 billion.
   1999 had a deficit of -$74 billion.

b) The smallest deficit was in 1990
   and was -$40 billion.

WRITING

69. Answers will vary.

71. Answers will vary.

REVIEW

73. $\dfrac{24}{54} = \dfrac{\cancel{2} \bullet 2 \bullet 2 \bullet \cancel{3}}{\cancel{2} \bullet \cancel{3} \bullet 3 \bullet 3}$

$= \dfrac{4}{9}$

75. $\dfrac{3}{4}\left(\dfrac{8}{5}\right) = \dfrac{3 \bullet 8}{4 \bullet 5}$

$= \dfrac{24}{20}$

$= \dfrac{6}{5}$

77. $\dfrac{3}{10} + \dfrac{2}{15} = \dfrac{3}{10} \bullet \dfrac{3}{3} + \dfrac{2}{15} \bullet \dfrac{2}{2}$

$= \dfrac{9}{30} + \dfrac{4}{30}$

$= \dfrac{13}{30}$

CHALLENGE PROBLEMS

79. $\{0, 1, 2, 3, 4, 5, \ldots\}$

Section 1.3

SECTION 1.4

VOCABULARY

1. Positive and negative numbers are called **signed** numbers.

3. Two numbers that are the same distance from 0 on a number line, but on opposite sides of it, are called **opposites**.

5. Since any number added to 0 remains the same (is identical), the number 0 is called the **identity** element for addition.

CONCEPTS

7. $-1 + (-3) = -4$

9. -5

11. -3

13. a) $a + (-a) = \boxed{0}$

   b) $a + 0 = \boxed{a}$

   c) $a + b = b + \boxed{a}$

   d) $(a + b) + c = a + \boxed{(b \ + \ c)}$

15. positive

17. a) $1 + (-5)$

   b) $-80.5 + 15$

   c) $(20 + 4)$

19. a) $5 + (-5) = 0$

   b) $-2.2 + 22 = 0$

   c) $0 + (-6) = -6$

   d) $-\dfrac{15}{16} + 0 = \dfrac{15}{16}$

   e) $-\dfrac{3}{4} + \dfrac{3}{4} = 0$

   f) $19 + (-19) = 0$

NOTATION

21. $x + y = y + x$

23. $8 + 9$

PRACTICE

25. $6 + (-8) = -2$

27. $-6 + 8 = 2$

29. $-4 + (-4) = -8$

31. $9 + (-1) = 8$

33. $-16 + 16 = 0$

35. $-65 + (-12) = -77$

37. $15 + (-11) = 4$

39. $300 + (-335) = -35$

41. $-10.5 + 2.3 = -8.2$

43. $-9.1 + (-11) = -20.1$

45. $0.7 + (-0.5) = 0.2$

47. $-\dfrac{9}{16} + \dfrac{7}{16} = -\dfrac{2}{16}$

$= -\dfrac{1}{8}$

49. $-\dfrac{1}{4} + \dfrac{2}{3} = -\dfrac{1}{4} \bullet \dfrac{3}{3} + \dfrac{2}{3} \bullet \dfrac{4}{4}$

$= -\dfrac{3}{12} + \dfrac{8}{12}$

$= \dfrac{5}{12}$

51. $-\dfrac{4}{5} + \left(-\dfrac{1}{10}\right) = -\dfrac{4}{5} \bullet \dfrac{2}{2} + \left(-\dfrac{1}{10}\right)$

$= -\dfrac{8}{10} + \left(-\dfrac{1}{10}\right)$

$= -\dfrac{9}{10}$

53. $8 + (-5) + 13 = 3 + 13$
$= 16$

55. $21 + (-27) + (-9) = -6 + (-9)$
$= -15$

57. $-27 + (-3) + (-13) + 22 = -30 + (-13) + 22$
$= -43 + 22$
$= -21$

59. $-20 + (-16 + 10) = -20 + (-6)$
$= -26$

61. $19 + (-20 + 1) = 19 + (-19)$
$= 0$

63. $(-7 + 8) + 2 + (-12 + 13) = 1 + 2 + 1$
$= 4$

65. $-7 + 5 + (-10) + 7 = -2 + (-10) + 7$
$= -12 + 7$
$= -5$

67. $-8 + 11 + (-11) + 8 + 1 = 3 + (-11) + 8 + 1$
$= -8 + 8 + 1$
$= 0 + 1$
$= 1$

69. $-2.1 + 6.5 + (-8.2) + 0.6 = 4.4 + (-8.2) + 0.6$
$= -3.8 + 0.6$
$= -3.2$

71. $-60 + 70 + (-10) + (-10) + 20$
$= 10 + (-10) + (-10) + 20$
$= 0 + (-10) + 20$
$= -10 + 20$
$= 10$

73. $-99 + (99 + 215) = (-99 + 99) + 215$
$= 0 + 215$
$= 215$

75. $(-112 + 56) + (-56) = -112 + [56 + (-56)]$
$= -112 + 0$
$= -112$

APPLICATIONS

77. MILITARY SCIENCE

$-1,500 + 2,400 + 1,250 = 900 + 1,250$
$= 2,150$
The net gain is 2, 150 m.

79. GOLF
Tiger Woods:
$-2 + (-6) + (-7) + (-3) = -8 + (-7) + (-3)$
$= -15 + (-3)$
$= -18$
Tom Kite:
$5 + (-3) + (-6) + (-2) = 2 + (-6) + (-2)$
$= -4 + (-2)$
$= -6$
Tommy Tolles:
$0 + 0 + 0 + (-5) = 0 + (-5)$
$= -5$
Tom Watson:
$3 + (-4) + (-3) + 0 = -1 + (-3) + 0$
$= -4 + 0$
$= -4$

81. CREDIT CARDS
a) Previous Balance: 3,660.66
New Purchases: 1,408.78
b) $-3,660.66 - 1,408.78 + 3,826.58$
$= -5,069.44 + 3,826.58$
$= -1,242.86$

83. MOVIE LOSSES

$-\$100 + \$11 = -\$89$ million

85. SAHARA DESERT

$-240 + 110 + 30 + (-55) + 100 + (-77)$
$= -130 + 30 + (-55) + 100 + (-77)$
$= -100 + (-55) + 100 + (-77)$
$= -155 + 100 + (-77)$
$= -55 + (-77)$
$= -132$

The net movement was 132 km southward.

87. PROFITS AND LOSSES

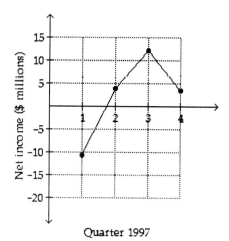

$$-10.7 + 4 + 12.2 + (-3.4)$$
$$= -6.7 + 12.2 + (-3.4)$$
$$= 5.5 + (-3.4)$$
$$= 2.1$$

The net income for 2001 was $2.1 million.

REVIEW

89. True

91. The numbers -9 and 3 are 6 units away from -3.

WRITING

93. Answers will vary.

CHALLENGE PROBLEMS

95. No. $1 + 1 = 2$, and 2 is not a member of the set.

SECTION 1.5

VOCABULARY

1. Two numbers that are the same distance from 0 on a number line, but on opposite sides of it, are called **opposites**, or additive **inverses**.

3. The difference between the maximum and the minimum value of a collection of measurements is called the **range** of the values.

CONCEPTS

5. a) -12

   b) $\dfrac{1}{5}$

   c) -2.71

   d) 0

7. a) The opposite of the opposite of a number is that **number**.
   b) To subtract two numbers, add the first number to the **opposite** of the number to be subtracted.

9. $-1 - 9 = -1 \boxed{+}\boxed{(-9)}$

11. $15 - (-8) = 15 + 8$
    $\qquad = 23$

    No. $15 - (-8) \neq 7$

13. $-10 + (-8) + (-23) + 5 + 34$

15. a) $-\left|-500\right| = -500$

    b. $-(-y) = y$

PRACTICE

17. $4 - 7 = 4 + (-7)$
    $\qquad = -3$

19. $2 - 15 = 2 + (-15)$
    $\qquad = -13$

21. $8 - (-3) = 8 + 3$
    $\qquad = 11$

23. $-12 - 9 = -12 + (-9)$
    $\qquad = -21$

25. $0 - 6 = 0 + (-6)$
    $\qquad = -6$

27. $0 - (-1) = 0 + 1$
    $\qquad = 1$

29. $10 - (-2) = 10 + 2$
    $\qquad = 12$

31. $-1 - (-3) = -1 + 3$
    $\qquad = 2$

33. $20 - (-20) = 20 + 20$
    $\qquad = 40$

35. $-3 - (-3) = -3 + 3$
    $\qquad = 0$

37. $-2 - (-7) = -2 + 7$
    $\qquad = 5$

39. $-4 - 5 = -4 + (-5)$
    $\qquad = -9$

41. $-44 - 44 = -44 + (-44)$
    $\qquad = -88$

43. $0 - (-12) = 0 + 12$
    $\qquad = 12$

45. $-25 - (-25) = -25 + 25$
    $\qquad = 0$

47. $0 - 4 = 0 + (-4)$
    $\qquad = -4$

49. $-19 - (-17) = -19 + 17$
    $\qquad = -2$

51. $-\dfrac{1}{8} - \dfrac{3}{8} = -\dfrac{1}{8} + \left(-\dfrac{3}{8}\right)$

    $\qquad = -\dfrac{4}{8}$

    $\qquad = -\dfrac{1}{2}$

53. $-\dfrac{9}{16}-\left(-\dfrac{1}{4}\right)=-\dfrac{9}{16}+\left(\dfrac{1}{4}\right)$

$$=-\dfrac{9}{16}+\left(\dfrac{1}{4}\bullet\dfrac{4}{4}\right)$$

$$=-\dfrac{9}{16}+\dfrac{4}{16}$$

$$=-\dfrac{5}{16}$$

55. $\dfrac{1}{3}-\dfrac{3}{4}=\dfrac{1}{3}+\left(-\dfrac{3}{4}\right)$

$$=\dfrac{1}{3}\bullet\dfrac{4}{4}+\left(-\dfrac{3}{4}\bullet\dfrac{3}{3}\right)$$

$$=\dfrac{4}{12}+\left(-\dfrac{9}{12}\right)$$

$$=-\dfrac{5}{12}$$

57. $-0.9 - 0.2 = -0.9 + (-0.2)$
$= -1.1$

59. $6.3 - 9.8 = 6.3 + (-9.8)$
$= -3.5$

61. $-1.5 - 0.8 = -1.5 + (-0.8)$
$= -2.3$

63. $2.8 - (-1.8) = 2.8 + 1.8$
$= 4.6$

65. $17 - (-5) = 17 + 5$
$= 22$

67. $-13 - 12 = -13 + (-12)$
$= -25$

69. $8 - 9 - 10 = 8 + (-9) + (-10)$
$= -1 + (-10)$
$= -11$

71. $-25 - (-50) - 75 = -25 + 50 + (-75)$
$= 25 + (-75)$
$= -50$

73. $-6 + 8 - (-1) - 10 = -6 + 8 + 1 + (-10)$
$= 2 + 1 + (-10)$
$= 3 + (-10)$
$= -7$

75. $61 - (-62) + (-64) - 60 = 61 + 62 + (-64) + (-60)$
$= 123 + (-64) + (-60)$
$= 59 + (-60)$
$= -1$

77. $-6 - 7 - (-3) + 9 = -6 + (-7) + 3 + 9$
$= -13 + 3 + 9$
$= -10 + 9$
$= -1$

79. $-20 - (-30) - 50 + 40 = -20 + 30 + (-50) + 40$
$= 10 + (-50) + 40$
$= -40 + 40$
$= 0$

APPLICATIONS

81. TEMPERATURE RECORDS

$108 - (-52) = 108 + 52$
$= 160°F$

83. LAND ELEVATIONS

$-282 - (-1,312) = -282 + 1,312$
$= 1,030 \text{ ft}$
$= -23$

85. RACING

$-2.25 - 3.5 = -5.75°$

87. HISTORY

$-347 - 81 = -428$
He was born in 428 B.C.

89. GAUGES

$5 - 7 - 6 = 5 + (-7) + (-6)$
$= -2 + (-6)$
$= -8$
The arrow would move left and the new reading would be -8.

WRITING

91. Answers will vary.

93. Answers will vary.

REVIEW

95. $30 = 2 \cdot 3 \cdot 5$

97. $\dfrac{3}{8} = \dfrac{3}{8} \bullet \dfrac{7}{7}$

$= \dfrac{21}{56}$

99. true

CHALLENGE PROBLEMS

101. a) true; positive + positive = positive

b) true; negative - positive = negative

c) true; the opposite of a positive is a
    negative

d) false; the opposite of a negative is
    a positive

SECTION 1.6

VOCABULARY

1. The answer to a multiplication problem is called a **product**. The answer to a division problem is called a **quotient**.

3. The **commutative** property of multiplication states that changing the order when multiplying does not affect the answer.

5. Division of a nonzero number by zero is **undefined**.

CONCEPTS

7. $4(\boxed{-5})$

9. The product of two negative numbers is **positive**.

11. The product of **1** and any number is that number.

13. a) One of the numbers is 0.
    b) They are reciprocals or multiplicative inverses.

15. a) negative
    b) not possible to tell
    c) positive
    d) negative

17. a) $a \bullet b = b \bullet \boxed{a}$
    b) $(ab)(c) = \boxed{a(bc)}$
    c) $0 \bullet a = \boxed{0}$
    d) $1 \bullet a = \boxed{a}$
    e) $a\left(\dfrac{1}{a}\right) = \boxed{1}$ $(a \neq 0)$

19. a) associative property of multiplication
    b) multiplicative inverse
    c) commutative property of multiplication
    d) multiplication property of 1

NOTATION

21. a) $-4(-5) = 20$

    b) $\dfrac{16}{-8} = -2$

PRACTICE

23. $-2 \cdot 8 = -16$

25. $(-6)(-9) = 54$

27. $12(-5) = -60$

29. $-6 \cdot 4 = -24$

31. $-20(40) = -800$

33. $(-6)(-6) = 36$

35. $-0.6(-4) = 2.4$

37. $1.2(-0.4) = -0.48$

39. $-1.1(-0.9) = 0.99$

41. $7.2(-2.1) = -15.12$

43. $\dfrac{1}{2}\left(-\dfrac{3}{4}\right) = -\dfrac{3}{8}$

45.
$$\left(-\dfrac{7}{8}\right)\left(-\dfrac{2}{21}\right) = \dfrac{14}{168}$$
$$= \dfrac{1}{12}$$

47.
$$-\dfrac{16}{25} \bullet \dfrac{15}{64} = -\dfrac{240}{1600}$$
$$= -\dfrac{3}{20}$$

49.
$$-1\dfrac{1}{4}\left(-\dfrac{3}{4}\right) = -\dfrac{5}{4}\left(-\dfrac{3}{4}\right)$$
$$= \dfrac{15}{16}$$

51. $-5.2 \cdot 100 = -520$

53. $0(-22) = 0$

55. $-3(-4)(0) = 0$

57. $3(-4)(-5) = -12(-5)$
$\qquad = 60$

59. $(-4)(3)(-7) = -12(-7)$
$\qquad = 84$

61. $(-2)(-3)(-4)(-5) = 6(-4)(-5)$
$\qquad = -24(-5)$
$\qquad = 120$

63.
$$\frac{1}{2}\left(-\frac{1}{3}\right)\left(-\frac{1}{4}\right) = -\frac{1}{6}\left(-\frac{1}{4}\right)$$
$$= \frac{1}{24}$$

65. $-2(-3)(-4)(-5)(-6) = 6(-4)(-5)(-6)$
$\qquad = -24(-5)(-6)$
$\qquad = 120(-6)$
$\qquad = -720$

67. $-30 \div (-3) = 10$

69. $\dfrac{-6}{-2} = 3$

71. $\dfrac{4}{-2} = -2$

73. $\dfrac{80}{-20} = -4$

75. $\dfrac{17}{-17} = -1$

77. $\dfrac{-110}{-110} = 1$

79. $\dfrac{-160}{-40} = -4$

81. $\dfrac{320}{-16} = -20$

83. $\dfrac{0.5}{-100} = -0.005$

85. $\dfrac{0}{150} = 0$

87. $\dfrac{-17}{0} =$ **undefined**

89.
$$-\frac{1}{3} \div \frac{4}{5} = -\frac{1}{3} \bullet \frac{5}{4}$$
$$= -\frac{5}{12}$$

91.
$$-\frac{9}{16} \div \left(-\frac{3}{20}\right) = -\frac{9}{16} \bullet \left(-\frac{20}{3}\right)$$
$$= \frac{180}{48}$$
$$= \frac{15}{4}$$

93.
$$-3\frac{3}{8} \div \left(-2\frac{1}{4}\right) = -\frac{27}{8} \div \left(-\frac{9}{4}\right)$$
$$= -\frac{27}{8}\left(-\frac{4}{9}\right)$$
$$= \frac{108}{72}$$
$$= \frac{3}{2}$$
$$= 1\frac{1}{2}$$

95. $\dfrac{-23.5}{5} = -4.7$

97. $\dfrac{-24.24}{-0.8} = 30.3$

99.
$$-\frac{1}{2}(2 \bullet 67) = \left(-\frac{1}{2} \bullet 2\right)(67)$$
$$= -1(67)$$
$$= -67$$

101.　　$-0.2(10 \cdot 3) = (-0.2 \cdot 10)3$
　　　　　　　　$= -2 \cdot 3$
　　　　　　　　$= -6$

APPLICATIONS

103.　TEMPERATURE CHANGE

　　　$12(-6) = -72°$

105.　GAMBLING

　　　First bet $40, loses $40
　　　Second bet double $40, loses $80
　　　Third bet doubles $80, loses $160

　　　$(-40) + (-80) + (-160) = -240 + (-40)$
　　　　　　　　　　　　　　　$= -280$

107.　PLANETS

　　　$\dfrac{-386}{2} = -193\,°\,F$

109.　ACCOUNTING

　　　$1.9(-22.8) = -\$43.32$ million

111.　THE QUEEN MARY
　　　$\dfrac{3,450,000 - 22,500,000}{31} = \dfrac{-19,050,000}{31}$
　　　　　　　　　　　　　　　$\approx -\$614,516$

113.　PHYSICS

　　a)　normal: 5, -10

　　b)　×0.5: $5(0.5) = 2.5$
　　　　　　$-10(0.5) = -5$

　　c)　×1.5: $5(1.5) = 7.5$
　　　　　　$-10(1.5) = -15$

　　d)　×2: $5(2) = 10$
　　　　　$-10(2) = -20$

WRITING

115.　Answers will vary.

117.　Answers will vary.

REVIEW

119.　$-3 + (-4) + (-5) + 4 + 3$
　　　　　　　$= -7 + (-5) + 4 + 3$
　　　　　　　$= -12 + 4 + 3$
　　　　　　　$= -8 + 3$
　　　　　　　$= -5$

121.　$\dfrac{1}{2} + \dfrac{1}{4} + \dfrac{1}{3} = \dfrac{1}{2}\left(\dfrac{6}{6}\right) + \dfrac{1}{4}\left(\dfrac{3}{3}\right) + \dfrac{1}{3}\left(\dfrac{4}{4}\right)$

　　　　　　　　$= \dfrac{6}{12} + \dfrac{3}{12} + \dfrac{4}{12}$

　　　　　　　　$= \dfrac{6+3+4}{12}$

　　　　　　　　$= \dfrac{13}{12}$

　　　　　　　　$= 12\overline{)13.0000}^{\,1.0833}$

　　　　　　　　$= 1.08\overline{3}$

123.　$\dfrac{2 \bullet 3 \bullet 5 \bullet 5}{2 \bullet 5 \bullet 5 \bullet 7} = \dfrac{3}{7}$

CHALLENGE PROBLEMS

125.　An odd number must be negative—
　　　one negative number, three negative
　　　numbers, or five negative numbers.

SECTION 1.7

VOCABULARY

1. In the exponential expression 3^2, 3 is the **base** and 2 is the **exponent**.

3. 7^5 is the fifth **power** of seven.

5. The rules for the **order** of operations guarantee that an evaluation of a numerical expression will result in a single answer.

CONCEPTS

7. a) $4 + 5 \cdot 6 = 9 \cdot 6$
$\qquad\qquad\quad = 54$
$\quad 4 + 5 \cdot 6 = 4 + 30$
$\qquad\qquad\quad = 34$
   b) The correct answer is 34. Multiplication is to be done before addition.

9. The innermost grouping symbols are the parentheses. The outermost grouping symbols are the brackets.

11. a) subtraction, power, addition, multiplication
    b) power, multiplication, subtraction, addition

13. a) subtraction
    b) division

15. a) subtraction
    b) power
    c) power
    d) power

NOTATION

17. a) $3^1 = \boxed{3}$
    b) $x^1 = \boxed{x}$
    c) $9 = 9^{\boxed{1}}$
    d) $y = y^{\boxed{1}}$

19. a) -5
    b) 5

21. $-19 - 2[(1 + 2) \cdot 3] = -19 - 2[\,\boxed{3}\, \cdot 3]$
$\qquad\qquad\qquad\qquad\quad = -19 - 2\,[\boxed{9}\,]$
$\qquad\qquad\qquad\qquad\quad = -19 - \boxed{18}$
$\qquad\qquad\qquad\qquad\quad = -37$

PRACTICE

23. 3^4

25. $10^2 k^3$

27. $8\pi r^3$

29. $6x^2 y^3$

31. $(-6)(-6) = 36$

33. $-(4)(4)(4)(4) = -256$

35. $(-5)(-5)(-5) = -125$

37. $-(-6)(-6)(-6)(-6) = -1{,}296$

39. $(-0.4)(-0.4) = 0.16$

41. $\left(-\dfrac{2}{5}\right)\left(-\dfrac{2}{5}\right)\left(-\dfrac{2}{5}\right) = -\dfrac{8}{125}$

43. $3 - 5 \cdot 4 = 3 - 20$
$\qquad\qquad\quad = -17$

45. $3 \cdot 8^2 = 3 \cdot 64$
$\qquad\quad = 192$

47. $8 \cdot 5^1 - 4 \div 2 = 8 \cdot 5 - 4 \div 2$
$\qquad\qquad\qquad = 40 - 2$
$\qquad\qquad\qquad = 38$

49. $100 - 8(10) + 60 = 100 - 80 + 60$
$\qquad\qquad\qquad\qquad = 20 + 60$
$\qquad\qquad\qquad\qquad = 80$

51. $-22 - (15 - 3) = -22 - 12$
$\qquad\qquad\qquad\quad = -34$

53. $-2(9) - 2(5) = -18 - 10$
$\qquad\qquad\qquad\quad = -28$

55. $5^2 + 13^2 = 25 + 169$
$\qquad\qquad\quad = 194$

57. $-4(6 + 5) = -4(11)$
$$= -44$$

59. $(9 - 3)(9 - 9) = (6)(0)$
$$= 0$$

61. $(-1 - 18)2 = (-19)2$
$$= -38$$

63. $-2(-1)^2 + 3(-1) - 3 = -2(1) + 3(-1) - 3$
$$= -2 - 3 - 3$$
$$= -5 - 3$$
$$= -8$$

65. $4^2 - (-2)^2 = 16 - 4$
$$= 12$$

67. $12 + 2\left(-\dfrac{9}{3}\right) - (-2) = 12 + 2(-3) + 2$
$$= 12 - 6 + 2$$
$$= 6 + 2$$
$$= 8$$

69. $\dfrac{-2 - 5}{-7 - (-7)} = \dfrac{-7}{-7 + 7}$
$$= \dfrac{-7}{0}$$
$$= \text{undefined}$$

71. $200 - (-6 + 5)^3 = 200 - (-1)^3$
$$= 200 - (-1)$$
$$= 200 + 1$$
$$= 201$$

73. $\left|5 \bullet 2^2 \bullet 4\right| - 30 = \left|5 \bullet 4 \bullet 4\right| - 30$
$$= \left|20 \bullet 4\right| - 30$$
$$= \left|80\right| - 30$$
$$= 80 - 30$$
$$= 50$$

75. $[6(5) - 5(5)]4 = [30 - 25]4$
$$= [5]4$$
$$= 20$$

77. $-6(130 - 4^3) = -6(130 - 64)$
$$= -6(66)$$
$$= -396$$

79. $(17 - 5 \cdot 2)^3 = (17 - 10)^3$
$$= (7)^3$$
$$= 343$$

81. $-5(-2)^3(3)^2 = -5(-8)(9)$
$$= 40(9)$$
$$= 360$$

83. $-2\left(\dfrac{15}{-5}\right) - \dfrac{6}{2} + 9 = -2(-3) - 3 + 9$
$$= 6 - 3 + 9$$
$$= 3 + 9$$
$$= 12$$

85. $\dfrac{5 \cdot 50 - 160}{-9} = \dfrac{250 - 160}{-9}$
$$= \dfrac{90}{-9}$$
$$= -10$$

87. $\dfrac{2(6 - 1)}{16 - (-4)^2} = \dfrac{2(5)}{16 - 16}$
$$= \dfrac{10}{0}$$
$$= \text{undefined}$$

89. $5(10 + 2) - 1 = 5(12) - 1$
$$= 60 - 1$$
$$= 59$$

91. $64 - 6[15 + (-3)3] = 64 - 6[15 - 9]$
$$= 64 - 6[6]$$
$$= 64 - 36$$
$$= 28$$

93. $(12 - 2)^3 = (10)^3$
$$= 1{,}000$$

95. $(-3)^3\left(\dfrac{-4}{2}\right)(-1) = -27(-2)(-1)$
$$= 54(-1)$$
$$= -54$$

97.
$$\frac{1}{2}\left(\frac{1}{8}\right)+\left(-\frac{1}{4}\right)^2=\frac{1}{16}+\frac{1}{16}$$
$$=\frac{2}{16}$$
$$=\frac{1}{8}$$

99.
$$-2\left|4-8\right|=-2\left|-4\right|$$
$$=-2(4)$$
$$=-8$$

101.
$$\left|7-8(4-7)\right|=\left|7-8(-3)\right|$$
$$=\left|7-(-24)\right|$$
$$=\left|7+24\right|$$
$$=\left|31\right|$$
$$=31$$

103.
$$3+2[-1-4(5)]=3+2[-1-20]$$
$$=3+2[-21]$$
$$=3+(-42)$$
$$=-39$$

105.
$$-3[5^2-(7-3)^2]=-3[5^2-(4)^2]$$
$$=-3[25-16]$$
$$=-3[9]$$
$$=-27$$

107.
$$-(2\cdot3-4)^3=-(6-4)^3$$
$$=-(2)^3$$
$$=-8$$

109.
$$\frac{(3+5)^2+\left|-2\right|}{-2(5-8)}=\frac{(8)^2+2}{-2(-3)}$$
$$=\frac{64+2}{6}$$
$$=\frac{66}{6}$$
$$=11$$

111.
$$\frac{2[-4-2(3-1)]}{3(-3)(-2)}=\frac{2\left[-4-2(2)\right]}{-9(-2)}$$
$$=\frac{2[-4-4]}{18}$$
$$=\frac{2[-8]}{18}$$
$$=\frac{-16}{18}$$
$$=-\frac{8}{9}$$

113.
$$\frac{\left|6-4\right|+2\left|-4\right|}{26-2^4}=\frac{2+2(4)}{26-16}$$
$$=\frac{2+8}{10}$$
$$=\frac{10}{10}$$
$$=1$$

115.
$$\frac{\left(4^3-10\right)+(-4)}{5^2-(-4)(-5)}=\frac{(64-10)+(-4)}{25-20}$$
$$=\frac{54+(-4)}{5}$$
$$=\frac{50}{5}$$
$$=10$$

117.
$$\frac{72-(2-2\bullet1)}{10^2-(90+2^2)}=\frac{72-(2-2)}{100-(90+4)}$$
$$=\frac{72-0}{100-94}$$
$$=\frac{72}{6}$$
$$=12$$

119. $$-\left(\frac{40-1^3-2^4}{3(2+5)+2}\right) = -\left(\frac{40-1-16}{3(7)+2}\right)$$

$$= -\left(\frac{39-16}{21+2}\right)$$

$$= -\left(\frac{23}{23}\right)$$

$$= -1$$

APPLICATIONS

121. LIGHT

2 yards = 2^2 square units
3 yards = 3^2 square units
4 yards = 4^2 square units

123. AUTO INSURANCE

$$2,672 + 1,680 + 2,485 = 6,837$$

$$1,370 + 2,737 + 1,692 = 5,799$$

$$\frac{6 \text{ premiums}}{6} = \frac{6,837 + 5,799}{6}$$

$$= \frac{12,636}{6}$$

$$= 2,106$$

The average is $2,106.

125. CASH AWARDS

a) $2,500 + 4(500) + 35(150) + 85(25)$

$$= 2,500 + 2,000 + 5,250 + 2,125$$

$$= 11,875$$

$11,875 will be awarded.

b) $\frac{11,875}{1+4+35+85} = \frac{11,875}{125}$

$$= 95$$

The average cash prize is $95.

127. WRAPPING GIFTS

$2(9) + 2(16) + 4(4) + 15$

$$= 18 + 32 + 16 + 15$$

$$= 81$$

81 inches of ribbon is needed.

WRITING

129. Answers will vary.

131. Answers will vary.

REVIEW

133. a) ii

b) iii

c) iv

d) i

CHALLENGE PROBLEMS

135. $\left(3^2\right)^4 = 9^4$

$$= 6,561$$

SECTION 1.8

VOCABULARY

1. Variables and/or numbers can be combined with the operations of arithmetic to create algebraic **expressions**.

3. A term, such as 27, that consists of a single number is called a **constant** term.

5. To **evaluate** an algebraic expression, we substitute the values for the variables and simplify.

7. $2x + 5$ is an example of an algebraic **expression**, whereas $2x + 5 = 7$ is an example of an **equation**.

CONCEPTS

9. $11x^2 - 6x - 9$
   a) 3 terms
   b) 11
   c) -6
   d) -9

11. a) term
    b) factor
    c) factor
    d) term
    e) factor
    f) term

13.
   a)

Number of weeks	Number of days
1	$1 \cdot 7 = 7$
2	$2 \cdot 7 = 14$
3	$3 \cdot 7 = 21$
w	$w \cdot 7 = 7w$

   b)

Number of seconds	Number of minutes
60	$60 \div 60 = 1$
120	$120 \div 60 = 2$
180	$180 \div 60 = 3$
s	$s \div 60 = \frac{s}{60}$

15. a) Let x = the weight of the car.
       $2x - 500$ = the weight of the van.
    b) Let $x = 2{,}000$.
       $$2(\mathbf{2{,}000}) - 500 = 4{,}000 - 500$$
       $$= 3{,}500$$
       The van weights 3,500 pounds.

17.

Type of coin	Number	·	Value (in cents)	=	Total value (in cents)
Nickel	6		5		30
Dime	d		10		$10d$
Half-dollar	$x + 5$		50		$50(x + 5)$

NOTATION

19. Let $a = 5$.
    $$9a - a^2 = 9(\boxed{5}) - (\,5\,)^2$$
    $$= 9(5) - 25\,\boxed{}$$
    $$= 45 \boxed{-}25$$
    $$= 20$$

21. a) $8y$
    b) $2cd$
    c) $15sx$
    b) $-9a^3b^2$

PRACTICE

23. $l + 15$

25. $50x$

27. $\dfrac{w}{l}$

29. $P + p$

31. $k^2 - 2{,}005$

33. $J - 500$

35. $\dfrac{1{,}000}{n}$

37. $p + 90$

39. $35 + h + 300$

41. $p - 680$

43. $4d - 15$

45. $2(200 + t)$

47. $|a - 2|$

49. a) $5(60) = 300$ minutes
    b) $h(60) = 60h$

51. a) $3y$

    b) $\dfrac{f}{3}$

53. $29x¢$

55. $\dfrac{c}{6}$

57. $5b$

59. $\$5(x + 2)$

61.

x	$x^3 - 1$
0	$(\mathbf{0})^3 - 1 = 0 - 1$ $= 0$
-1	$(\mathbf{-1})^3 - 1 = -1 - 1$ $= -2$
-3	$(\mathbf{-3})^3 - 1 = -27 - 1$ $= -28$

63.

S	$\dfrac{5s + 36}{s}$
1	$\dfrac{5(\mathbf{1}) + 36}{\mathbf{1}} = \dfrac{5 + 36}{1}$ $= \dfrac{41}{1}$ $= 41$
6	$\dfrac{5(\mathbf{6}) + 36}{\mathbf{6}} = \dfrac{30 + 36}{6}$ $= \dfrac{66}{6}$ $= 11$
-12	$\dfrac{5(\mathbf{-12}) + 36}{\mathbf{-12}} = \dfrac{-60 + 36}{-12}$ $= \dfrac{-24}{-12}$ $= 2$

65.

Input x	Output $2x - \dfrac{x}{2}$
100	$2(\mathbf{100}) - \dfrac{\mathbf{100}}{2} = 200 - 50$ $= 150$
-300	$2(\mathbf{-300}) - \dfrac{\mathbf{-300}}{2} = -600 - (-150)$ $= -600 + 150$ $= -450$

67.

x	$(x + 1)(x + 5)$
-1	$(\mathbf{-1} + 1)(\mathbf{-1} + 5) = (0)(4)$ $= 0$
-5	$(\mathbf{-5} + 1)(\mathbf{-5} + 5) = (-4)(0)$ $= 0$
-6	$(\mathbf{-6} + 1)(\mathbf{-6} + 5) = (-5)(-1)$ $= 5$

69. $\begin{aligned} 3y^2 - 6y - 4 &= 3(\mathbf{-2})^2 - 6(\mathbf{-2}) - 4 \\ &= 3(4) - 6(-2) - 4 \\ &= 12 + 12 - 4 \\ &= 24 - 4 \\ &= 20 \end{aligned}$

71. $\begin{aligned} (3 + x)y &= (3 + \mathbf{3}) \cdot \mathbf{-2} \\ &= 6 \cdot -2 \\ &= -12 \end{aligned}$

73. $\begin{aligned} (x + y)^2 - |z + y| &= (\mathbf{3} + \mathbf{-2})^2 - |\mathbf{-4} + \mathbf{-2}| \\ &= (1)^2 - |-6| \\ &= 1 - 6 \\ &= -5 \end{aligned}$

75. $\begin{aligned} (4x)^2 + 3y^2 &= (4 \cdot \mathbf{3})^2 + 3(\mathbf{-2})^2 \\ &= (12)^2 + 3(4) \\ &= 144 + 12 \\ &= 156 \end{aligned}$

77. $-\dfrac{2x+y^3}{y+2z} = -\dfrac{2(3)+(-2)^3}{-2+2(-4)}$

$= -\dfrac{6+(-8)}{-2+(-8)}$

$= -\dfrac{-2}{-10}$

$= -\dfrac{1}{5}$

79. $b^2 - 4ac = (5)^2 - 4(-1)(-2)$

$= 25 - (-4)(-2)$

$= 25 - 8$

$= 17$

81. $a^2 + 2ab + b^2 = (-5)^2 + 2(-5)(-1) + (-1)^2$

$= 25 + 2(-5)(-1) + 1$

$= 25 + 10 + 1$

$= 35 + 1$

$= 36$

83. $\dfrac{n}{2}\big[2a + (n-1)d\big]$

$= \dfrac{10}{2}\big[2(-4) + (10-1)6\big]$

$= 5\big[2(-4) + (9)6\big]$

$= 5\big[-8 + 54\big]$

$= 5\big[46\big]$

$= 230$

85. $\dfrac{a^2 + b^2}{2} = \dfrac{0^2 + (-10)^2}{2}$

$= \dfrac{0 + 100}{2}$

$= \dfrac{100}{2}$

$= 50$

APPLICATIONS

87. ROCKETRY

t	$h = 64t - 16t^2$
1	$64(1) - 16(1)^2 = 64(1) - 16(1)$ $= 64 - 16$ $= 48$
2	$64(2) - 16(2)^2 = 64(2) - 16(4)$ $= 128 - 64$ $= 64$
3	$64(3) - 16(3)^2 = 64(3) - 16(9)$ $= 192 - 144$ $= 48$
4	$64(4) - 16(4)^2 = 64(4) - 16(16)$ $= 256 - 256$ $= 0$

89. ANTIFREEZE

For -34° F:

$\dfrac{5(-34-32)}{9} = \dfrac{5(-66)}{9}$

$= \dfrac{-330}{9}$

$\approx -37°C$

For -84° F:

$\dfrac{5(-84-32)}{9} = \dfrac{5(-116)}{9}$

$= \dfrac{-580}{9}$

$\approx -64°C$

91. TOOLS

$$A = \frac{1}{2}h(b + d)$$

$$= \frac{1}{2}\left(\frac{3}{4}\right)\left(1\frac{1}{4} + 2\frac{3}{8}\right)$$

$$= \frac{1}{2}\left(\frac{3}{4}\right)\left(\frac{5}{4} + \frac{19}{8}\right)$$

$$= \frac{1}{2}\left(\frac{3}{4}\right)\left(\frac{5}{4} \cdot \frac{2}{2} + \frac{19}{8}\right)$$

$$= \frac{1}{2}\left(\frac{3}{4}\right)\left(\frac{10}{8} + \frac{19}{8}\right)$$

$$= \frac{1}{2}\left(\frac{3}{4}\right)\left(\frac{29}{8}\right)$$

$$= \frac{3}{8}\left(\frac{29}{8}\right)$$

$$= \frac{87}{64}$$

$$= 1\frac{23}{64} \text{ in.}^2$$

The area is $1\frac{23}{64}$ in.^{2}

WRITING

93. Answers will vary.

94. Answers will vary.

REVIEW

97. $12 = 2 \cdot 2 \cdot 3$
   $15 = 3 \cdot 5$
   common denominator $= 2 \cdot 2 \cdot 3 \cdot 5 = 60$

99. $(-2 \cdot 3)^3 - 10 + 1 = (-6)^3 - 10 + 1$
   $$= -216 - 10 + 1$$
   $$= -226 + 1$$
   $$= -225$$

CHALLENGE PROBLEMS

101. The answer would be 0, because
    $(8 - 8) = 0$ is a factor of the expression.

CHAPTER 1 KEY CONCEPTS

1. f

2. j

3. h

4. a

5. b

6. e

7. c

8. i

9. d

10. g

11. $C = p + t$ (Answers may vary depending on the variables chosen.)

12. $b = 2t$ (Answers may vary depending on the variables chosen.)

13. $x + 4$ = amount of business (\$ millions) in the year with the celebrity

14.

x	$3x^2 - 2x + 1$
0	$3(0)^2 - 2(0) + 1 = 3(0) - 2(0) + 1$ $\qquad\qquad\qquad = 0 - 0 + 1$ $\qquad\qquad\qquad = 1$
4	$3(4)^2 - 2(4) + 1 = 3(16) - 2(4) + 1$ $\qquad\qquad\qquad = 48 - 8 + 1$ $\qquad\qquad\qquad = 41$
6	$3(6)^2 - 2(6) + 1 = 3(36) - 2(6) + 1$ $\qquad\qquad\qquad = 108 - 12 + 1$ $\qquad\qquad\qquad = 97$

CHAPTER 1 REVIEW

SECTION 1.1
Introducing the Language of Algebra
REVIEW EXERCISES

1. horizontal axis is 1 hour
    vertical axis is 100 cars

2. 100

3. 7 P.M.

4. The difference of 15 and 3 is 12.

5. The sum of 15 and 3 is 18.

6. The quotient of 15 and 3 is 5.

7. The product of 15 and 3 is 45.

8. $4 \cdot 9$; $4(9)$

9. $\dfrac{9}{3}$

10. $8b$

11. Prt

12. equation

13. algebraic expression

14.

Number of Brackets (b)	Number of Nails (n)
5	$\mathbf{5} + 5 = 10$
10	$\mathbf{10} + 5 = 15$
20	$\mathbf{20} + 5 = 25$

SECTION 1.2
Fractions

15. a) 1(24), 2(12), 3(8), or 4(6)
    b) $2 \cdot 2 \cdot 6$ (answers will vary)
    c) 1, 2, 3, 4, 6, 8, 12, 24

16. $54 = 3^3 \cdot 2$

17. $147 = 7^2 \cdot 3$

18. $385 = 11 \cdot 7 \cdot 5$

19. $41 = $ prime

20. $\dfrac{12}{12} = 1$

21. $\dfrac{0}{10} = 0$

22. $\dfrac{20}{35} = \dfrac{2 \cdot 2 \cdot \cancel{5}}{\cancel{5} \cdot 7}$
    $= \dfrac{4}{7}$

23. $\dfrac{24}{18} = \dfrac{\cancel{2} \cdot 2 \cdot 2 \cdot \cancel{3}}{\cancel{2} \cdot \cancel{3} \cdot 3}$
    $= \dfrac{4}{3}$

24. $\dfrac{5}{8} \cdot \dfrac{8}{8} = \dfrac{40}{64}$

25. $\dfrac{12}{1} \cdot \dfrac{3}{3} = \dfrac{36}{3}$

26. $\dfrac{1}{8} \cdot \dfrac{7}{8} = \dfrac{1 \cdot 7}{8 \cdot 8}$
    $= \dfrac{7}{64}$

27. $\dfrac{16}{35} \cdot \dfrac{25}{48} = \dfrac{400}{1680}$
    $= \dfrac{5}{21}$

28. $\dfrac{1}{3} \div \dfrac{15}{16} = \dfrac{1}{3} \cdot \dfrac{16}{15}$
    $= \dfrac{16}{45}$

29.
$$16\frac{1}{4} \div 5 = \frac{65}{4} \cdot \frac{1}{5}$$
$$= \frac{65}{20}$$
$$= \frac{13}{4}$$
$$= 3\frac{1}{4}$$

30.
$$\frac{17}{25} - \frac{7}{25} = \frac{17-7}{25}$$
$$= \frac{10}{25}$$
$$= \frac{2}{5}$$

31.
$$\frac{8}{11} - \frac{1}{2} = \frac{8}{11} \cdot \frac{2}{2} - \frac{1}{2} \cdot \frac{11}{11}$$
$$= \frac{16}{22} - \frac{11}{22}$$
$$= \frac{16-11}{22}$$
$$= \frac{5}{22}$$

32.
$$\frac{1}{4} + \frac{2}{3} = \frac{1}{4} \cdot \frac{3}{3} + \frac{2}{3} \cdot \frac{4}{4}$$
$$= \frac{3}{12} + \frac{8}{12}$$
$$= \frac{3+8}{12}$$
$$= \frac{11}{12}$$

33.
$$4\frac{1}{9} - 3\frac{5}{6} = \frac{37}{9} - \frac{23}{6}$$
$$= \frac{37}{9} \cdot \frac{2}{2} - \frac{23}{6} \cdot \frac{3}{3}$$
$$= \frac{74}{18} - \frac{69}{18}$$
$$= \frac{74-69}{18}$$
$$= \frac{5}{18}$$

34. MACHINE SHOP
$$\frac{17}{24} - \frac{17}{32} = \frac{17}{24} \cdot \frac{4}{4} - \frac{17}{32} \cdot \frac{3}{3}$$
$$= \frac{68}{96} - \frac{51}{96}$$
$$= \frac{68-51}{96}$$
$$= \frac{17}{96} \text{ in.}$$

$\frac{17}{96}$ in. must be milled off.

SECTION 1.3
The Real Numbers

35. 0 is a whole number but is not a natural number.

36. -$65 billion

37. -206 feet

38. 0 $\boxed{<}$ 5

39. -12 $\boxed{>}$ -13

40. $\frac{7}{10}$

41. $\frac{14}{3}$

42. $250\overline{)1.000}^{\,0.004} = 0.004$

43. $22\overline{)17.00000}^{\,0.77272} = 0.7\overline{72}$

44. $-\frac{17}{4}$ -2 $0.\overline{3}$ $\frac{7}{8}$ $\sqrt{2}$ π 3.75

45. false

46. false

47. true

Chapter 1 Review

48. true

49. natural: 8
   whole: 0, 8
   integers: 0, -12, 8
   rational: $-\dfrac{4}{5}$, 99.99, 0, -12, $4\dfrac{1}{2}$, 0.666..., 8
   irrational: $\sqrt{2}$
   real: all

50. $\left|-6\right| > \left|5\right|$

51. $-9 < \left|-10\right|$

SECTION 1.4
Adding Real Numbers

52. -45 + (-37) = -82

53. 25 + (-13) = 12

54. 0 + (-7) = -7

55. -7 + 7 = 0

56. 12 + (-8) + (-15) = 4 + (-15)
                     = -11

57. -9.9 + (-2.4) = -12.3

58. $\dfrac{5}{16} + \left(-\dfrac{1}{2}\right) = \dfrac{5}{16} + \left(-\dfrac{1}{2} \bullet \dfrac{8}{8}\right)$
   $= \dfrac{5}{16} + \left(-\dfrac{8}{16}\right)$
   $= -\dfrac{3}{16}$

59. 35 + (-13) + (-17) + 6 = 22 + (-17) + 6
                             = 5 + 6
                             = 11

60. commutative property of addition

61. associative property of addition

62. addition property of opposites

63. addition property of 0

64. ATOMS

     3(1) + 4(-1) = 3 + (-4)
                  = -1

SECTION 1.5
Subtracting Real Numbers

65. -10

66. 3

67. $\dfrac{9}{16}$

68. -4

69. 45 − 64 = 45 + (-64)
               = -19

70. -17 − 32 = -17 + (-32)
                = -49

71. -27 − (-12) = -27 + 12
                  = -15

72. 3.6 − (-2.1) = 3.6 + 2.1
                = 5.7

73. 0 − 10 = 0 + (-10)
             = -10

74. -33 + 7 − 5 − (-2) = -33 + 7 + (-5) + 2
                         = -26 + (-5) + 2
                         = -31 + 2
                         = -29

75. GEOGRAPHY
      29,028 − (-36,205) = 29,028 + 36,205
                          = 65,233
   The difference is 65, 233 ft.

SECTION 1.6
Multiplying and Dividing Real Numbers

76. -8 · 7 = -56

77. (-9)(-6) = 54

78. 2(-3)(-2) = -6(-2)
                  = 12

79. $(-4)(-1)(-3)(-3) = 4(-3)(-3)$
$= -12(-3)$
$= 36$

80. $-1.2(-5.3) = 6.36$

81. $0.002(-1,000) = -2$

82. $-\dfrac{2}{3}\left(\dfrac{1}{5}\right) = -\dfrac{2 \bullet 1}{3 \bullet 5}$
$= -\dfrac{2}{15}$

83. $-6(-3)(0)(-1) = 18(0)(-1)$
$= 0(-1)$
$= 0$

84. associative property of multiplication

85. commutative property of multiplication

86. multiplication property of 1

87. inverse property of multiplication

88. 3

89. $-\dfrac{1}{3}$

90. $\dfrac{44}{-44} = -1$

91. $\dfrac{-100}{25} = -4$

92. $\dfrac{-81}{-27} = 3$

93. $-\dfrac{3}{5} \div \dfrac{1}{2} = -\dfrac{3}{5} \bullet \dfrac{2}{1}$
$= -\dfrac{6}{5}$

94. $\dfrac{-60}{0} =$ undefined

95. $\dfrac{-4.5}{1} = -4.5$

96. The high reading is 2;
the low reading is -3.

97. New high: $2(2) = 4$
New low: $2(-3) = -6$

SECTION 1.7
Exponents and Order of Operations

98. 8^5

99. a^4

100. $9\pi r^2$

101. $x^3 y^4$

102. $9 \cdot 9 = 81$

103. $\left(-\dfrac{2}{3}\right)\left(-\dfrac{2}{3}\right)\left(-\dfrac{2}{3}\right) = -\dfrac{8}{27}$

104. $2(2)(2)(2)(2) = 32$

105. 50

106. 4 operations; power, multiplication, subtraction, addition

107. $2 + 5 \cdot 3 = 2 + 15$
$= 17$

108. $24 - 3(6)(4) = 24 - 18(4)$
$= 24 - 72$
$= -48$

109. $-(6 - 3)^2 = -(3)^2$
$= -9$

110. $4^3 + 2(-6 - 2 \cdot 2) = 64 + 2(-6 - 4)$
$= 64 + 2(-10)$
$= 64 - 20$
$= 44$

111. $10 - 5[-3 - 2(5 - 7^2)] - 5$
$= 10 - 5[-3 - 2(5 - 49)] - 5$
$= 10 - 5[-3 - 2(-44)] - 5$
$= 10 - 5[-3 + 88] - 5$
$= 10 - 5[85] - 5$
$= 10 - 425 - 5$
$= -415 - 5$
$= -420$

112. $\dfrac{-4(4+2)-4}{\left|-18-4(5)\right|} = \dfrac{-4(6)-4}{\left|-18-20\right|}$
$= \dfrac{-24-4}{\left|-38\right|}$
$= \dfrac{-28}{38}$
$= -\dfrac{14}{19}$

113. $(-3)^3\left(\dfrac{-8}{2}\right) + 5 = -27(-4) + 5$
$= 108 + 5$
$= 113$

114. $-9^2 + (-9)^2 = -81 + 81$
$= 0$

115. $20(5) + 65(10) + 25(20) = 1,250$
$5(50) + 10(100) = 1,250$

$\dfrac{1,250+1,250}{20+65+25+5+10} = \dfrac{2,500}{125}$
$= 20$

The average donation is $20.

SECTION 1.8
Algebraic Expressions

116. 3

117. 1

118. 2, -5

119. 16, -5, 25

120. $\dfrac{1}{2}$, 1

121. 9.6, -1

122. $h + 25$

123. $x - 15$

124. $\dfrac{1}{2}t$

125. $(n + 4)$ in.

126. $(b - 4)$ in.

127. $10d$

128. $(x - 5)$ years

129.

Type of Coin	Number	Value (¢)	Total Value (¢)
Nickel	6	5	$6(5) = 30$
Dime	d	10	$10d$

130.

x	$20x - x^3$
0	$20(0) - (0)^3 = 0 - 0$ $= 0$
1	$20(1) - (1)^3 = 20 - 1$ $= 19$
-4	$20(-4) - (-4)^3 = -80 - (-64)$ $= -80 + 64$ $= -16$

131. $(-10)^2 - 4(3)(5) = 100 - 12(5)$
$= 100 - 60$
$= 40$

132. $\dfrac{19+17}{-(19)-(-18)} = \dfrac{36}{-19+18}$
$= \dfrac{36}{-1}$
$= -36$

CHAPTER 1 TEST

1. $24

2. 5 hours

3.

Area in square miles (a)	Number of fire stations (f) $f = \dfrac{a}{5}$
15	$\dfrac{15}{5} = 3$
100	$\dfrac{100}{5} = 20$
350	$\dfrac{350}{5} = 70$

4. $180 = 5 \cdot 3 \cdot 3 \cdot 2 \cdot 2$
   $= 2^2 \cdot 3^2 \cdot 5$

5. $\dfrac{42}{105} = \dfrac{2 \bullet \cancel{3} \bullet \cancel{7}}{\cancel{3} \bullet 5 \bullet \cancel{7}}$
   $= \dfrac{2}{5}$

6. $\dfrac{15}{16} \div \dfrac{5}{8} = \dfrac{15}{16} \bullet \dfrac{8}{5}$
   $= \dfrac{120}{80}$
   $= \dfrac{3}{2}$
   $= 1\dfrac{1}{2}$

7. $\dfrac{11}{12} - \dfrac{2}{9} = \dfrac{11}{12}\left(\dfrac{3}{3}\right) - \dfrac{2}{9}\left(\dfrac{4}{4}\right)$
   $= \dfrac{33}{36} - \dfrac{8}{36}$
   $= \dfrac{25}{36}$

8. $1\dfrac{2}{3} + 8\dfrac{2}{5} = \dfrac{5}{3} + \dfrac{42}{5}$
   $= \dfrac{5}{3}\left(\dfrac{5}{5}\right) + \dfrac{42}{5}\left(\dfrac{3}{3}\right)$
   $= \dfrac{25}{15} + \dfrac{126}{15}$
   $= \dfrac{151}{15}$
   $= 10\dfrac{1}{15}$

9. The oranges weigh 4.25 pounds.
   The cost would be $4.25(0.84) = \$3.57$

10. $\dfrac{5}{6} = 6\overline{)5.0000} = 0.8\overline{3}$ (quotient 0.8333)

11.
   -3.75   -3 $-1\dfrac{1}{4}$ 0.5 $\sqrt{2}$ $\dfrac{7}{2}$

   Number line: -5 -4 -3 -2 -1 0 1 2 3 4 5

12. a) true
    b) false
    c) true
    d) true

13. The set of real numbers corresponds to all points on a number line. A real number is any number that is either a rational number or an irrational number.

14. a) $-2 > -3$
    b) $-|\text{-}7| < 8$
    c) $|-4| < -(-5)$
    d) $\left|-\dfrac{7}{8}\right| > 0.5$

15. SWEEPS WEEK
    $$\dfrac{0.6 + (-0.3) + 1.7 + 1.5 + (-0.2) + 1.1 + (-0.2)}{7}$$
    $$= \dfrac{4.2}{7}$$
    $$= 0.6$$

    There was a gain of 0.6 of a rating point.

16. $(-6) + 8 + (-4) = 2 + (-4)$
$\qquad = -2$

17. $-\dfrac{1}{2} + \dfrac{7}{8} = -\dfrac{1}{2}\left(\dfrac{4}{4}\right) + \dfrac{7}{8}$
$\qquad = -\dfrac{4}{8} + \dfrac{7}{8}$
$\qquad = \dfrac{3}{8}$

18. $-10 - (-4) \quad = -10 + 4$
$\qquad\qquad\quad = -6$

19. $(-2)(-3)(-5) = 6(-5)$
$\qquad\qquad\quad = -30$

20. $\dfrac{-22}{-11} = 2$

21. $-6.1(0.4) = -2.44$

22. $\dfrac{0}{-3} = 0$

23. $0 - 3 = 0 + (-3)$
$\qquad\quad = -3$

24. $3 + (-3) = 0$

25. $-30 + 50 - 10 - (-40) = -30 + 50 + (-10) + 40$
$\qquad\qquad\qquad\qquad = 20 + (-10) + 40$
$\qquad\qquad\qquad\qquad = 10 + 40$
$\qquad\qquad\qquad\qquad = 50$

26. $\left(-\dfrac{3}{5}\right)^3 = \left(-\dfrac{3}{5}\right)\left(-\dfrac{3}{5}\right)\left(-\dfrac{3}{5}\right)$
$\qquad\quad = -\dfrac{3 \bullet 3 \bullet 3}{5 \bullet 5 \bullet 5}$
$\qquad\quad = -\dfrac{27}{125}$

27. $-12.5 - (-26.5) = -12.5 + 26.5$
$\qquad\qquad\qquad = 14$

28. associative property of addition

29. a) 9^5
    b) $3x^2z^3$

30. $8 + 2 \cdot 3^4 = 8 + 2 \cdot 81$
$\qquad\qquad = 8 + 162$
$\qquad\qquad = 170$

31. $9^2 - 3[45 - 3(6 + 4)] = 81 - 3[45 - 3(10)]$
$\qquad\qquad\qquad\qquad = 81 - 3[45 - 30]$
$\qquad\qquad\qquad\qquad = 81 - 3[15]$
$\qquad\qquad\qquad\qquad = 81 - 45$
$\qquad\qquad\qquad\qquad = 36$

32. $\dfrac{3(40 - 2^3)}{-2(6 - 4)^2} = \dfrac{3(40 - 8)}{-2(2)^2}$
$\qquad\qquad\quad = \dfrac{3(32)}{-2(4)}$
$\qquad\qquad\quad = \dfrac{96}{-8}$
$\qquad\qquad\quad = -12$

33. $-10^2 = -(10)(10)$
$\qquad\quad = -100$

34. $3(x{-}y) - 5(x{+}y) = 3(2 - (-5)) - 5(2 + (-5))$
$\qquad\qquad\qquad\quad = 3(2 + 5) - 5(2 + (-5))$
$\qquad\qquad\qquad\quad = 3(7) - 5(-3)$
$\qquad\qquad\qquad\quad = 21 - (-15)$
$\qquad\qquad\qquad\quad = 21 + 15$
$\qquad\qquad\qquad\quad = 36$

35.

x	$2x - \dfrac{30}{x}$
5	$2(5) - \dfrac{30}{5} = 10 - 6$ $= 4$
10	$2(10) - \dfrac{30}{10} = 20 - 3$ $= 17$
-30	$2(-30) - \dfrac{30}{-30} = -60 + 1$ $= -59$

36. $2w + 7$

37. $x - 2 =$ number of songs on the CD

38. $25q$¢

39. An equation is a mathematical sentence that contains an = sign. An expression does not contain an = sign.

40. There are 3 terms. The coefficient of the second term is 5.

SECTION 2.1

VOCABULARY

1. An **equation** is a statement indicating that two expressions are equal.

3. To **check** the solution of an equation, we substitute the value for the variable in the original equation and see whether the result is a true statement.

5. Equations with the same solutions are called **equivalent** equations.

7. To solve an equation, we **isolate** the variable on one side of the equal symbol.

CONCEPTS

9. a) $x + 5 = 7$
    b) Subtract 5 from both sides.

11. a) $x + 6$
    b) It is neither since we do not know the value of x.
    c) $x = 5$ is not a solution because when 5 is substituted for x, the left side is 11.
    d) $x = 6$ is a solution because when 6 is substituted for x, the left side is 12.

13. **24**

15. **n**

17. a) If $x = y$, then $x + \text{c} = y + \underline{\textbf{c}}$
    and $x - \text{c} = y - \underline{\textbf{c}}$.
    Adding (or **subtracting**) the same number to (or from) **both** sides of an equation does not change the solution.

    b) If $x = y$, then $cx = \underline{\textbf{c}}y$ and $\dfrac{x}{c} = \dfrac{y}{c}$.
    Multiplying (or **dividing**) both sides of an **equation** by the same nonzero number does not change the solution.

19. a) Simplify: $x + 7 - 7$, x
    b) Simplify: $y - 2 + 2$, y
    c) Simplify: $\dfrac{5t}{5}$, t
    d) Simplify: $6 \cdot \dfrac{h}{6}$, h

NOTATION

21. Solve:
$$x + 15 = 45$$
$$x + 15 - \mathbf{15} = 45 - \mathbf{15}$$
$$\boxed{x = 30}$$

    Check:
$$x + 15 = 45$$
$$30 + 15 \overset{?}{=} 45$$
$$45 = 45$$
    30 is a solution.

23. a) The $\overset{?}{=}$ symbol means "possibly equal to".
    b) twenty-seven degrees, **$27°$**

PRACTICE

25. $\underline{\mathbf{6}} + 12 \overset{?}{=} 18$
    $18 \overset{?}{=} 18$ true

27. $2(\mathbf{-8}) + 3 \overset{?}{=} -15$
    $-16 + 3 \overset{?}{=} -15$
    $-13 \overset{?}{=} -15$ false

28. $0.5(\mathbf{5}) \overset{?}{=} 2.9$
    $2.5 \overset{?}{=} 2.9$ false

31. $33 - \dfrac{-6}{2} \overset{?}{=} 30$
    $33 + 3 \overset{?}{=} 30$
    $36 \overset{?}{=} 30$ false

33. $|\,\mathbf{20} - 8\,| \overset{?}{=} 10$
    $|\,12\,| \overset{?}{=} 10$
    $12 \overset{?}{=} 10$ false

35. $3(\mathbf{12}) - 2 \overset{?}{=} 4(\mathbf{12}) - 5$
    $36 - 2 \overset{?}{=} 48 - 5$
    $34 \overset{?}{=} 43$ false

37. $(\mathbf{-3})^2 - (\mathbf{-3}) - 6 \overset{?}{=} 0$
    $9 + 3 - 6 \overset{?}{=} 0$
    $12 - 6 \overset{?}{=} 0$
    $6 \overset{?}{=} 0$ false

39. $\quad \dfrac{2}{1+1}+5 \stackrel{?}{=} \dfrac{12}{1+1}$

$\quad\quad \dfrac{2}{2}+5 \stackrel{?}{=} \dfrac{12}{2}$

$\quad\quad\quad 1+5 \stackrel{?}{=} 6$

$\quad\quad\quad\quad 6 \stackrel{?}{=} 6$ true

41. $\quad (-3-4)(-3+3) \stackrel{?}{=} 0$

$\quad\quad\quad (-7)(0) \stackrel{?}{=} 0$

$\quad\quad\quad\quad 0 \stackrel{?}{=} 0$ true

43. $\quad x+7=10$

$\quad x+7-7=10-7$

$\quad\quad \boxed{x=3}$

$\quad\quad 3+7 \stackrel{?}{=} 10$ true

45. $\quad a-5=66$

$\quad a-5+5=66+5$

$\quad\quad \boxed{a=71}$

$\quad\quad 71-5 \stackrel{?}{=} 66$ true

47. $\quad 0=n-9$

$\quad 0+9=n-9+9$

$\quad\quad \boxed{9=n}$

$\quad\quad 0 \stackrel{?}{=} 9-9$ true

49. $\quad 9+p=9$

$\quad 9+p-9=9-9$

$\quad\quad \boxed{p=0}$

$\quad\quad 9+0 \stackrel{?}{=} 9$ true

51. $\quad x-16=-25$

$\quad x-16+16=-25+16$

$\quad\quad \boxed{x=-9}$

$\quad\quad -9-16 \stackrel{?}{=} -25$ true

53. $\quad a+3=0$

$\quad a+3-3=0-3$

$\quad\quad \boxed{a=-3}$

$\quad\quad -3+3 \stackrel{?}{=} 0$ true

55. $\quad f+3.5=1.2$

$\quad f+3.5-3.5=1.2-3.5$

$\quad\quad \boxed{f=-2.3}$

$\quad\quad -2.3+3.5 \stackrel{?}{=} 1.2$ true

57. $\quad -8+p=-44$

$\quad -8+p+8=-44+8$

$\quad\quad \boxed{p=-36}$

$\quad\quad -8+(-36) \stackrel{?}{=} -44$ true

59. $\quad 8.9=-4.1+t$

$\quad 8.9+4.1=-4.1+4.1+t$

$\quad\quad \boxed{13=t}$

$\quad\quad 8.9 \stackrel{?}{=} -4.1+13$ true

61. $\quad d-\dfrac{1}{9}=\dfrac{7}{9}$

$\quad d-\dfrac{1}{9}+\dfrac{1}{9}=\dfrac{7}{9}+\dfrac{1}{9}$

$\quad\quad \boxed{d=\dfrac{8}{9}}$

$\quad\quad \dfrac{8}{9}-\dfrac{1}{9} \stackrel{?}{=} \dfrac{7}{9}$ true

63. $\quad s+\dfrac{4}{25}=\dfrac{11}{25}$

$\quad s+\dfrac{4}{25}-\dfrac{4}{25}=\dfrac{11}{25}-\dfrac{4}{25}$

$\quad\quad \boxed{s=\dfrac{7}{25}}$

$\quad\quad \dfrac{7}{25}+\dfrac{4}{25} \stackrel{?}{=} \dfrac{11}{25}$ true

65. $\quad 4x=16$

$\quad \dfrac{4x}{4}=\dfrac{16}{4}$

$\quad\quad \boxed{x=4}$

$\quad\quad 4(4) \stackrel{?}{=} 16$ true

67.

$$369 = 9c$$

$$\frac{369}{9} = \frac{9c}{9}$$

$$\boxed{41 = c}$$

$$369 \stackrel{?}{=} 9(41) \text{ true}$$

69.

$$4f = 0$$

$$\frac{4f}{4} = \frac{0}{4}$$

$$\boxed{f = 0}$$

$$4(0) \stackrel{?}{=} 0 \text{ true}$$

71.

$$23b = 23$$

$$\frac{23b}{23} = \frac{23}{23}$$

$$\boxed{b = 1}$$

$$23(1) \stackrel{?}{=} 23 \text{ true}$$

73.

$$-8h = 48$$

$$\frac{-8h}{-8} = \frac{48}{-8}$$

$$\boxed{h = -6}$$

$$-8(-6) \stackrel{?}{=} 48 \text{ true}$$

75.

$$100 = -5g$$

$$\frac{-100}{-5} = \frac{-5g}{-5}$$

$$\boxed{20 = g}$$

$$-100 \stackrel{?}{=} -5(20) \text{ true}$$

77.

$$-3.4y = -1.7$$

$$\frac{-3.4y}{-3.4} = \frac{-1.7}{-3.4}$$

$$\boxed{y = 0.5}$$

$$-3.4(0.5) \stackrel{?}{=} -1.7 \text{ true}$$

79.

$$\frac{x}{15} = 3$$

$$\frac{15x}{15} = 15 \cdot 3$$

$$\boxed{x = 45}$$

$$\frac{45}{15} \stackrel{?}{=} 3 \text{ true}$$

81.

$$0 = \frac{v}{11}$$

$$11 \cdot 0 = \frac{11v}{11}$$

$$\boxed{0 = v}$$

$$0 \stackrel{?}{=} \frac{0}{11} \text{ true}$$

83.

$$\frac{w}{-7} = 15$$

$$\frac{-7w}{-7} = -7 \cdot 15$$

$$\boxed{w = -105}$$

$$\frac{-105}{-7} \stackrel{?}{=} 15 \text{ true}$$

85.

$$\frac{d}{-7} = -3$$

$$\frac{-7d}{-7} = -7(-3)$$

$$\boxed{d = 21}$$

$$\frac{21}{-7} \stackrel{?}{=} -3 \text{ true}$$

87.

$$\frac{y}{0.6} = -4.4$$

$$\frac{0.6y}{0.6} = (0.6)(-4.4)$$

$$\boxed{y = -2.64}$$

$$\frac{-2.64}{0.6} \stackrel{?}{=} -4.4 \text{ true}$$

89.
$$a + 456{,}932 = 1{,}708{,}921$$
$$a + 456{,}932 - 456{,}932 = 1{,}708{,}921 - 456{,}932$$
$$\boxed{a = 1{,}251{,}989}$$

$$\mathbf{1{,}251{,}989} + 456{,}932 \overset{?}{=} 1{,}708{,}921 \text{ true}$$

91.
$$-1{,}563x = 43{,}764$$
$$\frac{-1{,}563x}{-1{,}563} = \frac{43{,}764}{-1{,}563}$$
$$\boxed{x = -28}$$

$$(-1{,}563)(-28) \overset{?}{=} 43{,}764 \text{ true}$$

APPLICATIONS

93. SYNTHESIZERS
The two angles are supplementary.
The two angles total 180°.

Let x = measure of unknown angle
$$x + 115 = 180$$
$$x + 115 - 115 = 180 - 115$$
$$\boxed{x = 65}$$

$$65 + 115 \overset{?}{=} 180 \text{ true}$$

95. AVIATION
The two angles are complementary.
The two angles total 90°.

Let x = measure of unknown angle
$$x + 52 = 90$$
$$x + 52 - 52 = 90 - 52$$
$$\boxed{x = 38}$$

$$38 + 52 \overset{?}{=} 90 \text{ true}$$

WRITING

97. Answers will vary.

99. Answers will vary.

REVIEW

101. Evaluate: $-9 - 3x$ for $x = -3$
$$-9 - 3(-3) = -9 + 9$$
$$= 0$$

103. Translate: Subtract x from 45.
$$45 - x$$

CHALLENGE PROBLEMS

105. If $a + 80 = 50$, what is $a - 80$?
First solve: $a + 80 = 50$
$$a + 80 - 80 = 50 - 80$$
$$a = -30$$
Now substitute -30 in $a - 80$
$$-30 - 80 = -110$$

SECTION 2.2

VOCABULARY

1. A letter that is used to represent a number is called a **variable**.

3. To solve an applied problem, we let a **variable** represent the unknown quantity. Then we write an **equation** that models the situation. Finally, we **solve** the equation for the variable to find the unknown.

5. In the statement "10 is 50% of 20," 10 is called the **amount**, 50% is the **percent**, and 20 is the **base**.

7. $x + 371 + 479 = 1,240$

9. $x + 11,000 = 13,500$

11. $x + 5 + 8 + 16 = 31$

13. $\boxed{}$ is $\boxed{}$ % of $\boxed{}$?

NOTATION

15. 12 is 40% of what number?
    $12 = 40\% \cdot x$

17. a) $35\% = 0.35$
    b) $3.5\% = 0.035$
    c) $350\% = 3.5$
    d) $\frac{1}{2}\% = 0.5\% = 0.005$

PRACTICE

19. What number is 48% of 650?
    $$x = 48\% \cdot 650$$
    $$x = 0.48 \cdot 650$$
    $$\boxed{x = 312}$$

21. What percent of 300 is 78?
    $$x \cdot 300 = 78$$
    $$300x = 78$$
    $$\frac{300x}{300} = \frac{78}{300}$$
    $$x = 0.26$$
    $$\boxed{x = 26\%}$$

23. 75 is 25% of what number?
    $$75 = 25\% \cdot x$$
    $$75 = 0.25x$$
    $$\frac{75}{0.25} = \frac{0.25x}{0.25}$$
    $$\boxed{300 = x}$$

25. What number is 92.4% of 50?
    $$x = 92.4\% \cdot 50$$
    $$x = 0.924 \cdot 50$$
    $$\boxed{x = 46.2}$$

27. What percent of 16.8 is 0.42?
    $$x \cdot 16.8 = 0.42$$
    $$16.8x = 0.42$$
    $$\frac{16.8x}{16.8} = \frac{0.42}{16.8}$$
    $$x = 0.025$$
    $$\boxed{x = 2.5\%}$$

29. 128.1 is 8.75% of what number?
    $$128.1 = 8.75\% \cdot x$$
    $$128.1 = 0.0875x$$
    $$\frac{128.1}{0.0875} = \frac{0.0875x}{0.0875}$$
    $$\boxed{1,464 = x}$$

APPLICATIONS

31. GRAVITY
    $6x = 330$

33. MONARCHY
    Given: George III reigned 59 years.
    Find: Queen Victoria ruled how long?
    Queen Victoria yrs − 4 = George III yrs
    Let x = of years ruled by QV
    $$x - 4 = 59$$
    $$x - 4 + 4 = 59 + 4$$
    $$\boxed{x = 63}$$

    Queen Victoria ruled 63 years.

35. ATM RECEIPT
Given: Bal = \$287.00, WD = \$35.00
Find: Previous balance
  PB − WD = Current Balance
    Let x = previous balance

$$x - 35 = 287$$
$$x - 35 + 35 = 287 + 35$$
$$\boxed{x = 322}$$

The previous balance was \$322.

37. TV NEWS
Given: aired 9-minute interview,
      over 3 days
Find: total length of interview
  T inter divided by 3 days = 9 min
  Let x = total minutes of interview

$$\frac{x}{3} = 9$$
$$\frac{3x}{3} = 9 \cdot 3$$
$$\boxed{x = 27}$$

The original interview was 27 minutes.

39. STATEHOOD
of states prior to 1800 = ?
# of states between 1800-1850 = 15
# of states between 1851-1900 = 14
# of states between 1901-1950 = 3
# of states between 1951-present = 2
    total states = 50
    Let x = # of states before 1800
$$x + 15 + 14 + 3 + 2 = 50$$
$$x + 34 = 50$$
$$x + 34 - 34 = 50 - 34$$
$$\boxed{x = 16}$$

16 states were part of the union prior to 1800.

41. THEATER
of scenes 1st act = ?
# of scenes 2nd act = 6
# of scenes 3rd act = 5
# of scenes 4th act = 5
# of scenes 5th act = 3
total scenes = 24
Let x = # of scenes for 1st act
$$x + 6 + 5 + 5 + 3 = 24$$
$$x + 19 = 24$$
$$x + 19 - 19 = 24 - 19$$
$$\boxed{x = 5}$$

There are 5 scenes in the first act.

43. ORCHESTRAS
of woodwinds = 19
# of brass   = 23
# of percussion = 2
# of strings   = ?
 total musicians = 98
  Let x = # of string musicians
$$19 + 23 + 2 + x = 98$$
$$x + 44 = 98$$
$$x + 44 - 44 = 98 - 44$$
$$\boxed{x = 54}$$

There are 54 musicians in the string section.

45. BERMUDA TRIANGLE
Given: triangle
perimeter = 3,075 miles
 1st side = 1,100 miles
 2nd side = 1,000 miles
 3rd side = ?

1st side + 2nd side + 3rd side = perimeter

Let x = length of 3rd side in miles
$$1{,}100 + 1{,}000 + x = 3{,}075$$
$$x + 2{,}100 = 3{,}075$$
$$x + 2{,}100 - 2{,}100 = 3{,}075 - 2{,}100$$
$$\boxed{x = 975}$$

It is 975 miles from Bermuda to Florida.

Section 2.2

47. SPACE TRAVEL

Given: 364 foot tall rocket
1st stage = 138 feet
2nd stage = 98feet
3rd stage = 46 feet
escape tower = 28 feet
lunar module = ?

$1^{st} + 2^{nd} + 3^{rd}$ + escape + module = total

Let x = height of module in feet

$$138 + 98 + 46 + 28 + x = 364$$
$$x + 310 = 364$$
$$x + 310 - 310 = 364 - 310$$
$$\boxed{x = 54}$$

The lunar module was 54 ft tall.

49. STOP SIGNS

Given: Octagonal STOP sign = 8 sides
Total of measures of angles = $1,080°$

Let x = measure of 1 of the 8 equal angles

$$8x = 1,080$$
$$\frac{8x}{8} = \frac{1,080}{8}$$
$$\boxed{x = 135}$$

Each angle is $135°$.

PERCENT PROBLEMS

51. ANTISEPTICS

Given: base = 16 fl. oz.
percent = 3%
amount = ?

amount = percent • base
$$x = 3\% \cdot 16$$
$$x = 0.03 \cdot 16$$
$$\boxed{x = 0.48}$$

53. FEDERAL OUTLAYS

Given: base = $1,900 billion
percent = 36%
amount = ?

amount = percent • base
$$x = 36\% \cdot 1,900$$
$$x = 0.36 \cdot 1,900$$
$$\boxed{x = 684}$$

The amount paid was $684 billion.

55. COLLEGE ENTRANCE EXAMS

Given: base = 1,600 point
math amount = 550 points
verbal amount = 700 points
percent = ?

amount = percent • base
$$550 + 700 = x \cdot 1,600$$
$$1,250 = 1,600x$$
$$\frac{1,250}{1,600} = \frac{1,600x}{1,600}$$
$$0.78125 = x$$
$$\boxed{78.125\% = x}$$

57. DENTAL RECORDS

Given: base = total # of teeth = 32
filling teeth amount = 6
percent = ? (round to nearest %)

amount = percent • base
$$6 = x \cdot 32$$
$$6 = 32x$$
$$\frac{6}{32} = \frac{32x}{32}$$
$$0.1875 = x$$
$$18.75\% = x$$
$$\boxed{19\% = x}$$

59. CHILD CARE

Given: base = ? = maximum # of children
amount = 84 registered children
percent = 70% (round to nearest %)

amount = percent • base

$$84 = 70\% \cdot x$$
$$84 = 0.7x$$
$$\frac{84}{0.7} = \frac{0.7x}{0.7}$$
$$\boxed{120 = x}$$

61. NUTRITION

Use the table
a) 5 g and 25%

b) Given: base = ? = total g of sat. fat
amount = 5 g
percent = 25% (round to nearest %)

amount = percent • base

$$5 = 25\% \cdot x$$
$$5 = 0.25x$$
$$\frac{5}{0.25} = \frac{0.25x}{0.25}$$
$$\boxed{20 = x}$$

63. EXPORTS

Looking at the bar graph, one sees that there are decreases between the years of 1994 and 1995 and between the years of 2000 and 2001. By just looking, one can't really tell which is the larger percent decrease, so we have to calculate both of them and then make a decision.

Find the difference between the amounts exported for years 1994 and 1995.
$$51 - 46 = 5$$

Now find the percent of decrease.
$$\frac{5}{51} \approx 0.098039$$
$$\approx 0.098$$
$$= 9.8\%$$

Find the difference between the amounts exported for the years 2000 and 2001.
$$111 - 101 = 10$$

63. (cont.)

Now find the percent of decrease.
$$\frac{10}{111} \approx 0.090090$$
$$\approx 0.090$$
$$= 9.0\%$$

The greater percent of decrease is between the years of 1994 and 1995, and it is 9.8%.

65. INSURANCE COSTS

Find the difference between $1,050 and $925.
$$\$1,050 - \$925 = \$125$$

Now find the percent of decrease.

$$\frac{125}{1,050} \approx 0.11904$$
$$\approx 0.12$$
$$= 12\%$$

WRITING

67. Answers will vary.

69. Answers will vary.

REVIEW

71.
$$-\frac{16}{25} \div \left(-\frac{4}{15}\right) = -\frac{16}{25} \cdot \left(-\frac{15}{4}\right)$$
$$= \frac{12}{5}$$
$$= 2\frac{2}{5}$$

73.
$$x + 15 = -49$$
$$-34 + 15 \overset{?}{=} -49$$
$$-19 = -49$$
$$\text{no}$$

CHALLENGE PROBLEMS

75.

$$100\% - 99\frac{44}{100}\% = \frac{56}{100}\%$$
$$= 0.56\%$$
$$= 0.0056$$

SECTION 2.3

VOCABULARY

1. We can use the associative property of multiplication to **simplify** the expression $5(6x)$.

3. We simplify **expressions**, and we solve **equations**.

5. We call $-(c + 9)$ the **opposite** of a sum.

7. The **coefficient** of the term $-23y$ is -23.

CONCEPTS

9. Fill in the blanks to simplify each product.

   a) $5 \cdot 6t = (\boxed{5} \cdot \boxed{6})t$
   $= \boxed{30}t$

   b) $-8(2x)4 = (\boxed{-8} \cdot \boxed{2} \cdot \boxed{4})x$
   $= \boxed{-64}x$

11. They are not like terms.

13. Fill in the blanks.

   $-(x - 10) = \boxed{-1}(x - 10)$
   Distributing the multiplication by -1 changes the **sign** of each term within the parentheses.

15. Identify any like terms.

   a) $3a , 2a$

   b) $10 , 12$

   c) none

   d) $9y^2 , -8y^2$

17. Fill in the blanks.
   a) $4m + 6m = (\boxed{4 + 6})m$
   $= \boxed{10m}$

   b) $30n - 50n = (\boxed{30 - 50})n$
   $= \boxed{-20n}$

   c) $12 - 32d + 15 = -32d + \boxed{27}$

NOTATION

19. Translate to symbols.
   a) Six times the quantity of h minus four.
   $6(h - 4)$

   b) The opposite of the sum of z and sixteen.
   $-(z + 16)$

21.

student's	book's	equivalent?
$10x$	$10 + x$	no
$3 + y$	$y + 3$	yes
$5 - 8a$	$8a - 5$	no
$3x + 4$	$3(x + 4)$	no
$3 - 2x$	$-2x + 3$	yes
$h^2 + (-16)$	$h^2 - 16$	yes

PRACTICE

23. $63m$

25. $-35q$

27. $300t$

29. $11.2x$

31. g

33. $5x$

35. $6y$

37. s

39. $-20r$

41. $60c$

43. $-96m$

45. $5x + 15$

47. $36c - 42$

49. $24t + 16$

51. $0.4x - 1.6$

53. $5t + 5$

55. $-12x - 20$

57. $-78c + 18$

59. $-2w + 4$

61. $9x + 10$

63. $9r - 16$

65. $-x + 7$

67. $5.6y - 7$

69. $40d + 50$

71. $-12r - 60$

73. $x + y - 5$

75. $6x - 21y - 16z$

77. $20x$

79. 0

81. 0

83. r

85. $37y$

87. $-s^3$

89. 5

91. $-10r$

93. $3a$

95. $-3x$

97. x

99. $\dfrac{4}{5}t$

101. $0.4r$

103. $2z + 5(z - 3) = 2z + 5z - 15$
$\qquad\qquad\qquad = 7z - 15$

105. $-2x + 5$

107. $10(2d - 7) + 4 = 20d - 70 + 4$
$\qquad\qquad\qquad\quad = 20d - 66$

109. $-(c + 7) - 2(c - 3) = -c - 7 - 2c + 6$
$\qquad\qquad\qquad\qquad\quad = -3c - 1$

111. $2(s - 7) - (s - 2) = 2s - 14 - s + 2$
$\qquad\qquad\qquad\qquad = s - 12$

113. $6 - 4(-3c - 7) = 6 + 12c + 28$
$\qquad\qquad\qquad\quad = 12c + 34$

115. $36\left(\dfrac{2}{9}x - \dfrac{3}{4}\right) + 36\left(\dfrac{1}{2}\right) = \dfrac{36}{1}\left(\dfrac{2}{9}x\right) - \dfrac{36}{1}\left(\dfrac{3}{4}\right) + \dfrac{36}{1}\left(\dfrac{1}{2}\right)$
$\qquad\qquad\qquad\qquad\qquad = 4(2x) - 9(3) + 18(1)$
$\qquad\qquad\qquad\qquad\qquad = 8x - 27 + 18$
$\qquad\qquad\qquad\qquad\qquad = 8x - 9$

APPLICATIONS

117. THE AMERICAN RED CROSS
The cross has 12 lengths of equal measurement, x. Its perimeter is $12x$.

119. PING-PONG
Width $= x$ ft
Length $= x + 4$ ft
$P = l + w + l + w$
$P = (x + 4) + x + (x + 4) + x$
$P = (4x + 8)$ft

WRITING

121. Answers will vary.

REVIEW

Evaluate each expression
for $x = -3$, $y = -5$, $z = 0$.

123. $x^2z(y^3 - z) = (-3)(-3)(0)[(-5)(-5)(-5) - 0]$
$\qquad\qquad\qquad = 0$

125.
$$\frac{x-y^2}{2y-1+x} = \frac{-3-(-5)^2}{2(-5)-1+(-3)}$$
$$= \frac{-3-(25)}{-10-1-3}$$
$$= \frac{-28}{-14}$$
$$= 2$$

CHALLENGE PROBLEMS

127.   2 inches + 2 feet + 2 yards
One can convert each of the different units into a common unit. Changing feet and yards into inches would be the easier conversion.

$$1 \text{ ft} = 12 \text{ inches}$$
$$2 \text{ ft} = 2(12 \text{ inches})$$
$$= 24 \text{ inches}$$

$$1 \text{ yd} = 36 \text{ inches}$$
$$2 \text{ yd} = 2(36 \text{ inches})$$
$$= 72 \text{ inches}$$

2 inches + 2 feet + 2 yards
$$2 \text{ in} + 24 \text{ in} + 72 \text{ in} = \boxed{98 \text{ inches}}$$

129.
$$-2[x+4(2x+1)] = -2[x+8x+4]$$
$$= -2[9x+4]$$
$$= -18x-8$$

SECTION 2.4

VOCABULARY

1. An equation is a statement indicating that two expressions are **equal**.

3. After solving an equation, we can check our result by substituting that value for the variable in the **original** equation.

5. An equation that is true for all values of its variable is called an **identity**.

CONCEPTS

7. To solve the equation $2x - 7 = 21$, we first undo the **subtraction** of 7 by adding 7 to both sides. Then we undo the **multiplication** by 2 by dividing both sides by 2.

9. To solve $\dfrac{s}{3} + \dfrac{1}{4} = -\dfrac{1}{2}$, we can clear the equation of the fractions by **multiplying** both sides by 12.

11. One method of solving $-\dfrac{4}{5}x = 8$ is to multiply both sides of the equations by the reciprocal of $-\dfrac{4}{5}$. What is the reciprocal of $-\dfrac{4}{5}$? , $-\dfrac{5}{4}$

13. LCD = 30

15. Multiply by LCD 6.

NOTATION

17.
$$2x - 7 = 21_1$$
$$2x - 7 + 7 = 21 + 7$$
$$2x = 28$$
$$\frac{2x}{2} = \frac{28}{2}$$
$$\boxed{x = 14}$$

19. a) $-x = \boxed{-1}x$

   b) $\dfrac{3x}{5} = \boxed{\dfrac{3}{5}}x$

PRACTICE

21.
$$2x + 5 = 17$$
$$2x + 5 - 5 = 17 - 5$$
$$2x = 12$$
$$\frac{2x}{2} = \frac{12}{2}$$
$$\boxed{x = 6}$$

23.
$$5q - 2 = 23$$
$$5q - 2 + 2 = 23 + 2$$
$$5q = 25$$
$$\frac{5q}{5} = \frac{25}{5}$$
$$\boxed{q = 5}$$

25.
$$-33 = 5t + 2$$
$$-33 - 2 = 5t + 2 - 2$$
$$-35 = 5t$$
$$\frac{-35}{5} = \frac{5t}{5}$$
$$\boxed{-7 = t}$$

27.
$$20 = -x$$
$$\frac{20}{-1} = \frac{-x}{-1}$$
$$\boxed{-20 = x}$$

29.
$$-g = -4$$
$$\frac{-g}{-1} = \frac{-4}{-1}$$
$$\boxed{g = 4}$$

31.
$$1.2 - x = -1.7$$
$$1.2 - x - 1.2 = -1.7 - 1.2$$
$$-x = -2.9$$
$$\frac{-x}{-1} = \frac{-2.9}{-1}$$
$$\boxed{x = 2.9}$$

33.
$$-3p + 7 = -3$$
$$-3p + 7 - 7 = -3 - 7$$
$$-3p = -10$$
$$\frac{-3p}{-3} = \frac{-10}{-3}$$
$$\boxed{p = \frac{10}{3}}$$

35.
$$0 - 2y = 8$$
$$-2y = 8$$
$$\frac{-2y}{-2} = \frac{8}{-2}$$
$$\boxed{y = -4}$$

37.
$$-8 - 3c = 0$$
$$-8 - 3c + 8 = 0 + 8$$
$$-3c = 8$$
$$\frac{-3c}{-3} = \frac{8}{-3}$$
$$\boxed{c = -\frac{8}{3}}$$

39.
$$\frac{5}{6}k = 10$$
$$\frac{6}{5}\left(\frac{5}{6}k\right) = \frac{6}{5}\left(\frac{10}{1}\right)$$
$$\boxed{k = 12}$$

41.
$$-\frac{7}{16}h = 21$$
$$-\frac{16}{7}\left(-\frac{7}{16}h\right) = -\frac{16}{7}\left(\frac{21}{1}\right)$$
$$\boxed{h = -48}$$

43.
$$-\frac{t}{3} + 2 = 6$$
$$-\frac{t}{3} + 2 - 2 = 6 - 2$$
$$-\frac{t}{3} = 4$$
$$\frac{-3}{1}\left(\frac{t}{3}\right) = -3(4)$$
$$\boxed{t = -12}$$

45.
$$2(-3) + 4y = 14$$
$$-6 + 4y = 14$$
$$-6 + 4y + 6 = 14 + 6$$
$$4y = 20$$
$$\frac{4y}{4} = \frac{20}{4}$$
$$\boxed{y = 5}$$

47.
$$7(0) - 4y = 17$$
$$0 - 4y = 17$$
$$-4y = 17$$
$$\frac{-4y}{-4} = \frac{17}{-4}$$
$$\boxed{y = -\frac{17}{4}}$$

49.
$$10.08 = 4(0.5x + 2.5)$$
$$10.08 = 2x + 10$$
$$10.08 - 10 = 2x + 10 - 10$$
$$0.08 = 2x$$
$$\frac{0.08}{2} = \frac{2x}{2}$$
$$\boxed{0.04 = x}$$

51.
$$-(4-m)=-10$$
$$-4+m=-10$$
$$-4+m+4=-10+4$$
$$\boxed{m=-6}$$

53.
$$15s+8-s=7+1$$
$$14s+8=8$$
$$14s+8-8=8-8$$
$$14s=0$$
$$\frac{14s}{14}=\frac{0}{14}$$
$$\boxed{s=0}$$

55.
$$-3(2y-2)-y=5$$
$$-6y+6-y=5$$
$$-7y+6=5$$
$$-7y+6-6=5-6$$
$$-7y=-1$$
$$\frac{-7y}{-7}=\frac{-1}{-7}$$
$$\boxed{y=\frac{1}{7}}$$

57.
$$3x-8-4x-7x=-2-8$$
$$-8x-8=-10$$
$$-8x-8+8=-10+8$$
$$-8x=-2$$
$$\frac{-8x}{-8}=\frac{-2}{-8}$$
$$\boxed{x=\frac{1}{4}}$$

59.
$$4(5b)+2(6b-1)=-34$$
$$20b+12b-2=-34$$
$$32b-2=-34$$
$$32b\;\;2+2=-34+2$$
$$32b=-32$$
$$\frac{32b}{32}=\frac{-32}{32}$$
$$\boxed{b=-1}$$

61.
$$9(x+11)+5(13-x)=0$$
$$9x+99+65-5x=0$$
$$4x+164=0$$
$$4x+164-164=0-164$$
$$4x=-164$$
$$\frac{4x}{4}=\frac{-164}{4}$$
$$\boxed{x=-41}$$

63.
$$60r-50=15r-5$$
$$60r-50-15r=15r-5-15r$$
$$45r-50=-5$$
$$45r-50+50=-5+50$$
$$45r=45$$
$$\frac{45r}{45}=\frac{45}{45}$$
$$\boxed{r=1}$$

65.
$$8y-3=4y+15$$
$$8y-3-4y=4y+15-4y$$
$$4y-3=15$$
$$4y-3+3=15+3$$
$$4y=18$$
$$\frac{4y}{4}=\frac{18}{4}$$
$$\boxed{y=\frac{9}{2}}$$

67.
$$5x + 7.2 = 4x$$
$$5x + 7.2 - 5x = 4x - 5x$$
$$7.2 = -x$$
$$\frac{7.2}{-1} = \frac{-x}{-1}$$
$$\boxed{-7.2 = x}$$

69.
$$8y + 328 = 4y$$
$$8y + 328 - 8y = 4y - 8y$$
$$328 = -4y$$
$$\frac{328}{-4} = \frac{-4y}{-4}$$
$$\boxed{-82 = y}$$

71.
$$15x = x$$
$$15x - x = x - x$$
$$14x = 0$$
$$\frac{14x}{14} = \frac{0}{14}$$
$$\boxed{x = 0}$$

73.
$$3(a + 2) = 2(a - 7)$$
$$3a + 6 = 2a - 14$$
$$3a + 6 - 2a = 2a - 14 - 2a$$
$$a + 6 = -14$$
$$a + 6 - 6 = -14 - 6$$
$$\boxed{a = -20}$$

75.
$$2 - 3(x - 5) = 4(x - 1)$$
$$2 - 3x + 15 = 4x - 4$$
$$-3x + 17 = 4x - 4$$
$$-3x + 17 + 3x = 4x - 4 + 3x$$
$$17 = 7x - 4$$
$$17 + 4 = 7x - 4 + 4$$
$$21 = 7x$$
$$\frac{21}{7} = \frac{7x}{7}$$
$$\boxed{3 = x}$$

77.
$$\frac{x + 5}{3} = 11$$
$$3\left(\frac{x + 5}{3}\right) = 3(11)$$
$$x + 5 = 33$$
$$x + 5 - 5 = 33 - 5$$
$$\boxed{x = 28}$$

79.
$$\frac{y}{6} + \frac{y}{4} = -1$$
$$12\left(\frac{y}{6}\right) + 12\left(\frac{y}{4}\right) = 12(-1)$$
$$2y + 3y = -12$$
$$5y = -12$$
$$\frac{5y}{5} = \frac{-12}{5}$$
$$\boxed{y = \frac{-12}{5}}$$

81.
$$-\frac{2}{9} = \frac{5x}{6} - \frac{1}{3}$$
$$18\left(-\frac{2}{9}\right) = 18\left(\frac{5x}{6}\right) - 18\left(\frac{1}{3}\right)$$
$$-4 = 15x - 6$$
$$-4 + 6 = 15x - 6 + 6$$
$$2 = 15x$$
$$\frac{2}{15} = \frac{15x}{15}$$
$$\boxed{\frac{2}{15} = x}$$

83.
$$\frac{2}{3}y + 2 = \frac{1}{5} + y$$
$$15\left(\frac{2}{3}y + 2\right) = 15\left(\frac{1}{5} + y\right)$$
$$10y + 30 = 3 + 15y$$
$$10y + 30 - 10y = 3 + 15y - 10y$$
$$30 = 3 + 5y$$
$$30 - 3 = 3 + 5y - 3$$
$$27 = 5y$$
$$\frac{27}{5} = \frac{5y}{5}$$
$$\boxed{\frac{27}{5} = y}$$

85.
$$-\frac{3}{4}n + 2n = \frac{1}{2}n + \frac{13}{3}$$
$$12\left(-\frac{3}{4}n\right) + 12(2n) = 12\left(\frac{1}{2}n\right) + 12\left(\frac{13}{3}\right)$$
$$-9n + 24n = 6n + 52$$
$$15n = 6n + 52$$
$$15n - 6n = 6n + 52 - 6n$$
$$9n = 52$$
$$\frac{9n}{9} = \frac{52}{9}$$
$$\boxed{n = \frac{52}{9}}$$

87.
$$\frac{10 - 5s}{3} = s$$
$$3\left(\frac{10 - 5s}{3}\right) = 3(s)$$
$$10 - 5s = 3s$$
$$10 - 5s + 5s = 3s + 5s$$
$$10 = 8s$$
$$\frac{10}{8} = \frac{8x}{8}$$
$$\boxed{\frac{5}{4} = x}$$

89.
$$\frac{5(1 - x)}{6} = -x$$
$$6\left[\frac{5(1 - x)}{6}\right] = 6(-x)$$
$$5(1 - x) = -6x$$
$$5 - 5x = -6x$$
$$5 - 5x + 5x = -6x + 5x$$
$$5 = -x$$
$$\frac{5}{-1} = \frac{-x}{-1}$$
$$\boxed{-5 = x}$$

91.
$$\frac{3(d - 8)}{4} = \frac{2(d + 1)}{3}$$
$$\frac{3d - 24}{4} = \frac{2d + 2}{3}$$
$$12\left[\frac{3d - 24}{4}\right] = 12\left[\frac{2d + 2}{3}\right]$$
$$3[3d - 24] = 4[2d + 2]$$
$$9d - 72 = 8d + 8$$
$$9d - 72 + 72 = 8d + 8 + 72$$
$$9d = 8d + 80$$
$$9d - 8d = 8d + 80 - 8d$$
$$\boxed{d = 80}$$

93.
$$\frac{1}{2}(x + 3) + \frac{3}{4}(x - 2) = x + 1$$
$$\frac{1}{2}x + \frac{3}{2} + \frac{3}{4}x - \frac{6}{4} = x + 1$$
$$\frac{4}{1}\left(\frac{1}{2}x\right) + \frac{4}{1}\left(\frac{3}{2}\right) + \frac{4}{1}\left(\frac{3}{4}x\right) - \frac{4}{1}\left(\frac{6}{4}\right) = 4(x + 1)$$
$$2x + 6 + 3x - 6 = 4x + 4$$
$$5x = 4x + 4$$
$$5x - 4x = 4x + 4 - 4x$$
$$\boxed{x = 4}$$

95.

$$8x + 3(2 - x) = 5(x + 2) - 4$$
$$8x + 6 - 3x = 5x + 10 - 4$$
$$5x + 6 = 5x + 6$$
$$5x + 6 - 5x = 5x + 6 - 5x$$
$$\boxed{6 = 6} \text{ true}$$

The terms involving x drop out and the result, $6 = 6$, is true. This means **all real numbers** are a solution, and this equation is an **identity**.

97.

$$-3(s + 2) = -2(s + 4) - s$$
$$-3s - 6 = -2s - 8 - s$$
$$-3s - 6 = -3s - 8$$
$$-3s - 6 + 3s = -3s - 8 + 3s$$
$$\boxed{-6 = -8} \text{ false}$$

no solution

99.

$$2(3z + 4) = 2(3z - 2) + 13$$
$$6z + 8 = 6z - 4 + 13$$
$$6z + 8 = 6z + 9$$
$$6z + 8 - 6z = 6z + 9 - 6z$$
$$\boxed{8 = 9} \text{ false}$$

no solution

101.

$$4(y - 3) - y = 3(y - 4)$$
$$4y - 12 - y = 3y - 12$$
$$3y - 12 = 3y - 12$$
$$3y - 12 - 3y = 3y - 12 - 3y$$
$$\boxed{-12 = -12} \text{ true}$$

all real numbers

103.

$$\frac{h}{709} - 23{,}898 = -19{,}678$$
$$709\left(\frac{h}{709}\right) - 709(23{,}898) = 709(-19{,}678)$$
$$h - 16{,}943{,}682 = -13{,}951{,}702$$

add $16{,}943{,}682$ to both sides
$$+16{,}943{,}682 = +16{,}943{,}682$$

$$\boxed{h = 2{,}991{,}980}$$

WRITING

105. Answers will vary.

107. Answers will vary.

REVIEW

109.
$$-8 - (-8) = -8 + 8$$
$$= 0$$

111. $\dfrac{1}{8} \cdot \dfrac{1}{8} = \dfrac{1}{64}$

113. $8x + 8 + 8x - 8 = 16x$

CHALLENGE PROBLEMS

115. $|x| = 2$ or $x^2 = 4$ both have solutions of 2 and -2

SECTION 2.5

VOCABULARY

1. A **formula** is an equation that is used to state a known relationship between two or more variables.

3. The distance around a geometric figure is called its **perimeter**.

5. A segment drawn from the center of a circle to a point on the circle is called a **radius**.

7. The perimeter of a circle is called its **circumference**.

CONCEPTS

9. Use variables to write the formula relating the following:
   a) $d = rt$
   b) $r = c + m$
   c) $p = r - c$
   d) $I = Prt$
   e) $C = 2\pi r$

11. Complete the table.

	Rate	• time	= distance
Light	186,282 mi/sec	60 sec	**11,176,920 mi**
Sound	1,088 ft/sec	60 sec	**65,280 ft**

13. Tell which concept, perimeter, area, circumference, or volume should be used to find the following.
   a) volume
   b) circumference
   c) area
   d) perimeter

15. Write an expression for perimeter.
   a) $P = l + w + l + w$
   $P = (x+3)\text{cm} + 2\text{cm} + (x+3)\text{ cm} + 2\text{cm}$
   $\boxed{P = (2x + 10) \text{ cm}}$

   Write an expression for area.
   b) $A = lw$
   $A = 2\text{cm} (x + 3)\text{cm}$
   $\boxed{A = (2x + 6) \text{ cm}^2}$

NOTATION

17. Solve $Ax + By = C$ for y.

$$Ax + By = C$$
$$Ax + By - \boxed{Ax} = C - \boxed{Ax}$$
$$\boxed{By} = C - Ax$$
$$\frac{By}{\boxed{B}} = \frac{C - Ax}{\boxed{B}}$$
$$y = \frac{C - Ax}{B}$$

19.
   a) $\pi \doteq 3.14$
   b) 98π means $98 \cdot \pi$
   c) $V = \pi r^2 h$

   r represents the radius of the cylinder

   h represents the height of the cylinder

PRACTICE

21. SWIMMING
   d is 1,826 miles
   t is 743 hours
   r is ? mph

$$rt = d$$
$$r \cdot 743 = 1,826$$
$$\frac{r \cdot 743}{743} = \frac{1,826}{743}$$
$$\boxed{r = 2.45}$$

   Rate is 2.5 mph.

23. HOLLYWOOD
   total receipts is $190 millions
   profits is $125 million
   cost is ? $

$$\text{cost} + \text{profit} = \text{total receipts}$$
$$\text{cost} + 125 = 190$$
$$\text{cost} + 125 - 125 = 190 - 125$$
$$\boxed{\text{cost} = \$65 \text{ million}}$$

25. ENTREPRENEURS

I is \$175
P is \$2,500
r is ?%
t is 2 years

$$P \cdot r \cdot t = I$$
$$2,500 \cdot r \cdot 2 = 175$$
$$\frac{2,500 \cdot r \cdot 2}{2,500 \cdot 2} = \frac{175}{2,500 \cdot 2}$$
$$r = 0.035$$
$$\boxed{r = 3.5\%}$$

Rate is 3.5%.

27. METALLURGY

C is $2,212°$

$$C = \frac{5}{9}(F - 32)$$
$$2,212 = \frac{5}{9}(F - 32)$$
$$\frac{9}{5}(2,212) = \frac{9}{5}\left[\frac{5}{9}(F - 32)\right]$$
$$\frac{19,908}{5} = F - 32$$
$$3981.6 + 32 = F - 32 + 32$$
$$4,013.6 = F$$
$$\boxed{4,014° = F}$$

The temperature is 4,014°F.

29. VALENTINE'S DAY

wholesale is \$12.95
retail is \$37.50
markup is \$?

$$\text{wholesale} + \text{markup} = \text{retail}$$
$$12.95 + \text{markup} = 37.50$$
$$12.95 + \text{markup} - 12.95 = 37.50 - 12.95$$
$$\boxed{\text{markup} = \$24.55}$$

31. YO-YOS

c is ? inches
π is 3.14
r is 21 inches
Answers will vary due
to approximation of π.

$$c = 2\pi r$$
$$c = 2 \cdot 3.14 \cdot 21$$
$$\boxed{c = 131.88}$$

The circumference is 132 in.

33. $E = IR$; for R

$$\frac{E}{I} = \frac{IR}{I}$$
$$\boxed{\frac{E}{I} = R}$$

35. $V = lwh$; for w

$$\frac{V}{lh} = \frac{lwh}{lh}$$
$$\boxed{\frac{V}{lh} = w}$$

37. $C = 2\pi r$; for r

$$\frac{C}{2\pi} = \frac{2\pi r}{2\pi}$$
$$\boxed{\frac{C}{2\pi} = r}$$

39. $A = \dfrac{Bh}{3}$; for h

$$3(A) = 3\left(\frac{Bh}{3}\right)$$
$$3A = Bh$$
$$\frac{3A}{B} = \frac{Bh}{B}$$
$$\boxed{\frac{3A}{B} = h}$$

41. $w = \dfrac{s}{f}$; for f

$$f(w) = f\left(\dfrac{s}{f}\right)$$

$$fw = s$$

$$\dfrac{fw}{w} = \dfrac{s}{w}$$

$$\boxed{f = \dfrac{s}{w}}$$

43. $P = a + b + c$, for b

$$P = a + b + c$$

$$P - a - c = a + b + c - a - c$$

$$\boxed{P - a - c = b}$$

45. $T = 2r + 2t$; for r

$$T = 2r + 2t$$

$$T - 2t = 2r + 2t - 2t$$

$$T - 2t = 2r$$

$$\dfrac{T - 2t}{2} = \dfrac{2r}{2}$$

$$\boxed{\dfrac{T - 2t}{2} = r}$$

47. $Ax + By = C$; for x

$$Ax + By = C$$

$$Ax + By - By = C - By$$

$$Ax = C - By$$

$$\dfrac{Ax}{A} = \dfrac{C - By}{A}$$

$$\boxed{x = \dfrac{C - By}{A}}$$

49. $K = \dfrac{1}{2}mv^2$; for m

$$K = \dfrac{1}{2}mv^2$$

$$2K = 2\left(\dfrac{1}{2}mv^2\right)$$

$$2K = mv^2$$

$$\dfrac{2K}{v^2} = \dfrac{mv^2}{v^2}$$

$$\boxed{\dfrac{2K}{v^2} = m}$$

51. $A = \dfrac{a + b + c}{3}$; for c

$$A = \dfrac{a + b + c}{3}$$

$$3A = 3\left(\dfrac{a + b + c}{3}\right)$$

$$3A = a + b + c$$

$$3A - a - b = a + b + c - a - b$$

$$\boxed{3A - a - b = c}$$

53. $2E = \dfrac{T - t}{9}$; for t

$$2E = \dfrac{T - t}{9}$$

$$9(2E) = 9\left(\dfrac{T - t}{9}\right)$$

$$18E = T - t$$

$$18E + t = T - t + t$$

$$18E + t = T$$

$$18E + t - 18E = T - 18E$$

$$\boxed{t = T - 18E}$$

55. $s = 4\pi r^2$; for r^2

$$s = 4\pi r^2$$

$$\frac{s}{4\pi} = \frac{4\pi r^2}{4\pi}$$

$$\boxed{\frac{s}{4\pi} = r^2}$$

57. $Kg = \dfrac{wv^2}{2}$; for v^2

$$Kg = \frac{wv^2}{2}$$

$$2(Kg) = 2\left(\frac{wv^2}{2}\right)$$

$$2Kg = wv^2$$

$$\frac{2Kg}{w} = \frac{wv^2}{w}$$

$$\boxed{\frac{2Kg}{w} = v^2}$$

59. $V = \dfrac{4}{3}\pi r^3$; for r^3

$$V = \frac{4}{3}\pi r^3$$

$$\frac{3}{4}(V) = \frac{3}{4}\left(\frac{4}{3}\pi r^3\right)$$

$$\frac{3V}{4} = \pi r^3$$

$$\frac{1}{\pi}\left(\frac{3V}{4}\right) = \frac{\pi r^3}{\pi}$$

$$\boxed{\frac{3V}{4\pi} = r^3}$$

61. $\dfrac{M}{2} - 9.9 = 2.1B$; for M

$$\frac{M}{2} - 9.9 = 2.1B$$

$$\frac{M}{2} - 9.9 + 9.9 = 2.1B + 9.9$$

$$\frac{M}{2} = 2.1B + 9.9$$

$$2\left(\frac{M}{2}\right) = 2(2.1B + 9.9)$$

$$\boxed{M = 4.2B + 19.8}$$

63. $S = 2\pi rh + 2\pi r^2$; for h

$$S = 2\pi rh + 2\pi r^2$$

$$S - 2\pi r^2 = 2\pi rh + 2\pi r^2 - 2\pi r^2$$

$$S - 2\pi r^2 = 2\pi rh$$

$$\frac{S - 2\pi r^2}{2\pi r} = \frac{2\pi rh}{2\pi r}$$

$$\boxed{\frac{S - 2\pi r^2}{2\pi r} = h}$$

Some students may attempt to cross out $2\pi r$. You can not divide out anything on the left side because the numerator contains two terms and the denominator contains a monomial factor.

65. $3x + y = 9$; for y

$$3x + y = 9$$

$$3x + y - 3x = 9 - 3x$$

$$\boxed{y = 9 - 3x}$$

67. $3y - 9 = x$; for y

$$3y - 9 = x$$

$$3y - 9 + 9 = x + 9$$

$$3y = x + 9$$

$$\frac{3y}{3} = \frac{x + 9}{3}$$

$$\boxed{y = \frac{1}{3}x + 3}$$

69. $4y + 16 = -3x$; for y

$$4y + 16 = -3x$$

$$4y + 16 - 16 = -3x - 16$$

$$4y = -3x - 16$$

$$\frac{4y}{4} = \frac{-3x - 16}{4}$$

$$\boxed{y = -\frac{3}{4}x - 4}$$

71. $A = \frac{1}{2}h(b + d)$; for b

$$A = \frac{1}{2}h(b + d)$$

$$2(A) = 2\left(\frac{1}{2}h(b + d)\right)$$

$$2A = h(b + d)$$

$$\frac{2A}{h} = \frac{h(b + d)}{h}$$

$$\frac{2A}{h} = b + d$$

$$\frac{2A}{h} - d = b + d - d$$

$$\boxed{\frac{2A}{h} - d = b}$$

or

$$\boxed{b = \frac{2A - hd}{h}}$$

73. $\frac{7}{8}c + w = 9$; for c

$$\frac{7}{8}c + w = 9$$

$$\frac{7}{8}c + w - w = 9 - w$$

$$\frac{7}{8}c = 9 - w$$

$$\frac{8}{7}\left(\frac{7}{8}c\right) = \frac{8}{7}(9 - w)$$

$$\boxed{c = \frac{72 - 8w}{7}}$$

APPLICATIONS

75. PROPERTIES OF WATER
In Fahrenheit, water freezes at 32° and boils at 212°.

In Celsius, water freezes at 0° and boils at 100°.

77. AVON PRODUCTS

Revenue	1,854.1	1,463.4
Cost of goods sold	679.5	506.5
Gross profit	**1,174.6**	**956.9**

79. CARPENTRY

$P = a + b + c$
$P = 10 + 10 + 16$
$\boxed{P = 36 \text{ ft}}$

$A = \frac{1}{2}bh$
$A = \frac{1}{2} \cdot 16 \cdot 6$
$\boxed{A = 48 \text{ ft}^2}$

81. ARCHERY

$C = 2\pi r$
$C = 2 \cdot 3.14 \cdot 8$
$C = 50.24$
$\boxed{C = 50.2 \text{ in.}}$

Answers will vary due to the approximation of π.

$A = \pi r^2$
$A = 3.14 \cdot 8 \cdot 8$
$A = 200.96$
$\boxed{A = 201.0 \text{ in.}^2}$

83. LANDSCAPE

$P = a + b + c + d$
$P = 10 + 12 + 10 + 24$
$\boxed{P = 56 \text{ in.}}$

$A = \frac{1}{2}h(B + b)$
$A = \frac{1}{2} 8(24 + 12)$
$A = 4(36)$
$\boxed{A = 144 \text{ in.}^2}$

85. MEMORIALS

Use the formula for the area of a triangle and multiply by 2.

$$A = 2\left(\frac{1}{2}bh\right)$$

$$A = bh$$

$$A = 245 \cdot 10$$

$$\boxed{A = 2,450 \text{ ft}^2}$$

87. RUBBER MEETS THE ROAD

$$P = 2l + 2w$$

$$P = 2 \cdot 6.375 + 2 \cdot 7.5$$

$$P = 12.75 + 15$$

$$\boxed{P = 27.75 \text{ in.}}$$

$$A = lw$$

$$A = 6.375 \cdot 7.5$$

$$\boxed{A = 47.8125 \text{ in.}^2}$$

89. FIREWOOD

$$A = lw$$

$$A = 8 \cdot 4$$

$$\boxed{A = 32 \text{ ft}^2}$$

$$V = lwh$$

$$V = 8 \cdot 4 \cdot 4$$

$$\boxed{V = 128 \text{ ft}^3}$$

91. IGLOOS

Use the formula for the volume of a sphere and multiply by ½.

$$V = \frac{1}{2}\left(\frac{4}{3}\pi r^3\right)$$

$$V = \frac{1}{2} \cdot \frac{4}{3} \cdot 3.14 \cdot 5.5 \cdot 5.5 \cdot 5.5$$

$$V = 348.28$$

$$\boxed{V = 348 \text{ ft}^3}$$

93. BARBECUING

$$A = \pi r^2$$

$$A = 3.14 \cdot 9 \cdot 9$$

$$A = 254.34$$

$$\boxed{A = 254 \text{ in.}^2}$$

Answers will vary due to the approximation of π.

95. PULLEYS

$$L = 2D + 3.25(r + R) \; ; \; \text{solve for } D$$

$$L = 2D + 3.25(r + R)$$

$$L = 2D + 3.25r + 3.25R$$

$$L - 3.25r - 3.25R = 2D - 3.25r - 3.25R$$

$$L - 3.25r - 3.25R = 2D$$

$$\frac{L - 3.25r - 3.25R}{2} = \frac{2D}{2}$$

$$\boxed{\frac{L - 3.25r - 3.25R}{2} = D}$$

WRITING

97. Answers will vary.

98. Answers will vary.

REVIEW

99. Find 82% of 168.

$$0.82 \cdot 168$$

$$\boxed{137.76}$$

100. What percent of 200 is 30?

$$x \cdot 200 = 30$$

$$\frac{x \cdot 200}{200} = \frac{30}{200}$$

$$x = 0.15$$

$$\boxed{x = 15\%}$$

CHALLENGE PROBLEMS

101. $-7(\alpha - \beta) - (4\alpha - \theta) = \dfrac{\alpha}{2}$

solve for α

$$-7(\alpha - \beta) - (4\alpha - \theta) = \dfrac{\alpha}{2}$$

$$-7\alpha + 7\beta - 4\alpha + \theta = \dfrac{\alpha}{2}$$

$$-11\alpha + 7\beta + \theta = \dfrac{\alpha}{2}$$

$$-11\alpha + 7\beta + \theta + 11\alpha = \dfrac{\alpha}{2} + 11\alpha$$

$$7\beta + \theta = \dfrac{\alpha}{2} + 11\alpha$$

$$7\beta + \theta = \dfrac{\alpha + 22\alpha}{2}$$

$$7\beta + \theta = \dfrac{23\alpha}{2}$$

$$2(7\beta + \theta) = 2\left(\dfrac{23\alpha}{2}\right)$$

$$14\beta + 2\theta = 23\alpha$$

$$\dfrac{14\beta + 2\theta}{23} = \dfrac{23\alpha}{23}$$

$$\boxed{\dfrac{14\beta + 2\theta}{23} = \alpha}$$

VOCABULARY

1. The **perimeter** of a triangle or a rectangle is the distance around it.

3. The equal sides of an isosceles triangle meet to form the **vertex** angle. The angles opposite the equal sides are called **base** angles, and they have equal measures.

CONCEPTS

5. Pipe
   a) 17 ft = total length
      $x + 2$ = length of middle sized section
      $3x$ = length of longest section
   b) $x = 3$, short section
      $x + 2 = 3 + 2 = 5$ft, middle section
      $3x = 3(3) = 9$ ft, long section
      $3 + 5 + 9 = 17$ ft, total length of pipe

7. The sum of the measures of the angles of any triangle is **180°**.

9. Principal = $30,000
   rate = 14%
   time = 1 year

11. Complete the table.

	r	$\cdot$	t	$=$	d
Husband	35		t		$35t$
Wife	45		t		$45t$

13. Complete the table.

a)

	Amount	$\cdot$ Strength	= Pure vinegar
Strong	x	0.06	**0.06x**
Weak	$10 - x$	0.03	**0.03(10 − x)**
Mixture	10	0.05	**0.05(10)**

b)

	Amount	$\cdot$ Strength	= Pure antifreeze
Strong	6	0.50	**0.50(6)**
Weak	x	0.25	**0.25(x)**
Mixture	$6 + x$	0.30	**0.30(6 + x)**

NOTATION

15. $100(0.08) = 8$

   To multiply a decimal by 100, move the decimal point two places to the right.

17. $2 \cdot 2w - 3 + 2w$ does not represent the perimeter of a rectangle because parentheses are needed $2 \cdot (2w - 3) + 2w$.

PRACTICE

19.
$$0.08x + 0.07(15,000 - x) = 1,110$$
$$0.08x + 1,050 - 0.07x = 1,110$$
$$0.01x + 1,050 = 1,110$$
subtract 1,050 from both sides
$$-1,050 = -1,050$$
$$0.01x = 60$$
$$\frac{0.01x}{0.01} = \frac{60}{0.01}$$
$$\boxed{x = 6,000}$$

APPLICATIONS

21. CARPENTRY

Short piece in feet	x
Long piece in feet	$2x$
Total feet	**12**

Let x = short piece in feet
$2x$ = long piece in feet
short + long = total
$$x + 2x = 12$$
$$3x = 12$$
$$\frac{3x}{3} = \frac{12}{3}$$
$$\boxed{x = 4}$$
$$2x = 2(4)$$
$$= 8$$
Short piece is 4 ft long.
Long piece is 8 ft long.

23. NATIONAL PARKS

Miles 1st day	x
Miles 2nd day	$x + 6$
Miles 3rd day	$x + 12$
Miles 4th day	$x + 18$
Total miles	444

Let $x = 1^{st}$ day's travel in miles
$x + 6 = 2^{nd}$ day's travel in miles
$x + 12 = 3^{rd}$ day's travel in miles
$x + 18 = 4^{th}$ day's travel in miles

$$1^{st} + 2^{nd} + 3^{rd} + 4^{th} = \text{total}$$
$$x + (x+6) + (x+12) + (x+18) = 444$$
$$4x + 36 = 444$$
$$4x + 36 - 36 = 444 - 36$$
$$4x = 408$$
$$\frac{4x}{4} = \frac{408}{4}$$
$$\boxed{x = 102}$$

1st day's travel is 102 miles
2nd day's travel is 108 miles
3rd day's travel is 114 miles
4th day's travel is 120 miles

25. TOURING

Weeks in Australia	x
Weeks in Japan	$x + 4$
Weeks in Sweden	$x - 2$
Total weeks	38

Let $x =$ weeks in Australia
$x + 4 =$ weeks in Japan
$x - 2 =$ weeks in Sweden

$$\text{Aus} + \text{Japan} + \text{Sweden} = 38$$
$$x + (x+4) + (x-2) = 38$$
$$3x + 2 = 38$$
$$3x + 2 - 2 = 38 - 2$$
$$3x = 36$$
$$\frac{3x}{3} = \frac{36}{3}$$
$$\boxed{x = 12}$$

12 weeks in Australia
16 weeks in Japan
10 weeks in Sweden

27. COUNTING CALORIES

Calories in ice cream	x
Calories in pie	$2x + 100$
Total calories	850

Let $x =$ calories in ice cream
$2x + 100 =$ calories in pie

$$\text{ice cream} + \text{pie} = \text{total}$$
$$x + (2x + 100) = 850$$
$$3x + 100 = 850$$
$$3x + 100 - 100 = 850 - 100$$
$$3x = 750$$
$$\frac{3x}{3} = \frac{750}{3}$$
$$\boxed{x = 250}$$

250 calories in ice cream.
600 calories in pie.

29. ACCOUNTING

1st quarter income	x
2nd quarter income	$x + 119$
3rd quarter income	$(x + 119) - 40$ $x + 79$
4th quarter income	$7.7x$
Total income	1,375

Let $x = 1^{st}$ qt income in millions
$x + 119 = 2^{nd}$ qt income in millions
$x + 79 = 3^{rd}$ qt income in millions
$7.7x = 4^{th}$ qt income in millions

$$1^{st} + 2^{nd} + 3^{rd} + 4^{th} = \text{total}$$
$$x + (x+119) + (x+79) + 7.7x = 1,375$$
$$10.7x + 198 = 1,375$$
$$10.7x + 198 - 198 = 1,375 - 198$$
$$10.7x = 1,177$$
$$\frac{10.7x}{10.7} = \frac{1,177}{10.7}$$
$$\boxed{x = 110}$$

1st qt income is $110 million
2nd qt income is $229 million
3rd qt income is $189 million
4th qt income is $847 million

31. ENGINEERING

3rd side in ft	x
1st equal side in ft	$x-4$
2nd equal side in ft	$x-4$
Perimeter in ft	25

Let $x = 3^{rd}$ side in ft
$x - 4 =$ one equal side in ft
$x - 4 =$ other equal side in ft

$$1^{st} + 2^{nd} + 3^{rd} = P$$
$$(x-4)+(x-4)+x = 25$$
$$3x - 8 = 25$$
$$3x - 8 + 8 = 25 + 8$$
$$3x = 33$$
$$\frac{3x}{3} = \frac{33}{3}$$
$$\boxed{x = 11}$$

3rd side is 11 ft.
One equal side is 7 ft.
Other equal side is 7 ft.

33. SWIMMING POOLS

width of pool in m	w
length of pool in m	$6w + 30$
Perimeter in m	1,110

Let $w =$ width in meters
$6w + 30 =$ length in meters

$$2l + 2w = P$$
$$2(6w+30)+2w = 1,110$$
$$12w+60+2w = 1,110$$
$$14w + 60 = 1,110$$
$$14w + 60 - 60 = 1,110 - 60$$
$$14w = 1,050$$
$$\frac{14w}{14} = \frac{1,050}{14}$$
$$\boxed{w = 75}$$

Width is 75 meters.
Length is 480 meters.

35. TV TOWERS

Vertex angle (3rd angle)	x
1st base angle	$4x$
2nd base angle	$4x$
Total degrees	180

Let $x =$ vertex angle
$4x =$ one base angle
$4x =$ other base angle

$$1^{st}\angle + 2^{nd}\angle + 3^{rd}\angle = \text{Total}$$
$$x + 4x + 4x = 180$$
$$9x = 180$$
$$\frac{9x}{9} = \frac{180}{9}$$
$$\boxed{x = 20}$$

The vertex angle is 20°.
One base angle is 80°.
Other base angle is 80°.

37. COMPLEMENTARY ANGLES

1st angle	$2x$
Complementary angle	$6x + 2$
Total degree	90

Let $x =$ the number
$2x = 1^{st}$ angle
$6x + 2 =$ complementary angle

$$1^{st}\angle + 2^{nd}\angle = \text{Total}$$
$$2x + (6x+2) = 90$$
$$8x + 2 = 90$$
$$8x + 2 - 2 = 90 - 2$$
$$8x = 88$$
$$\frac{8x}{8} = \frac{88}{8}$$
$$\boxed{x = 11}$$

The number is 11.
1st angle is 22°.
Complementary angle is 68°.

39. RENTALS

	#	• rent	= total
One-bedroom	x	$550	**550x**
Two bed-room	x	$700	**700x**
Three bedroom	x	$900	**900x**
All types			**$36,550**

Let x = number of rentals

1 bed + 2 bed + 3 bed = total

$$550x + 700x + 900x = 36,550$$

$$2,150x = 36,550$$

$$\frac{2,150x}{2,150} = \frac{36,550}{2,150}$$

$$\boxed{x = 17}$$

Number of rentals is 17.

41. SOFTWARE

	#	• cost	= total value
SS	x	$150	**150x**
DB	x	$195	**195x**
WP	$15 + 2x$	$210	**210(2x + 15)**
All types			**$72,000**

Let x = # of each sold for SS and DB
$2x + 15$ = # of WP sold

SS val + BD val + WP val = total val

$$150x + 195x + 210(2x + 15) = 72,000$$

$$150x + 195x + 420x + 3,150 = 72,000$$

$$765x + 3,150 = 72,000$$

$$765x + 3,150 - 3,150 = 72,000 - 3,150$$

$$765x = 68,850$$

$$\frac{765x}{765} = \frac{68,850}{765}$$

$$\boxed{x = 90}$$

90 SS were sold.

43. INVESTMENTS

	P	• r	• t	= I
7% int	x	0.07	1 yr	**0.07x**
8% int	x	0.08	1 yr	**0.08x**
10.5% int	x	0.105	1 yr	**0.105x**
T interest				**$1,249.50**

Let x = the principal of each investment

7% inv + 8% inv + 10.5% inv = Total inv

$$0.07x + 0.08x + 0.105x = 1,249.50$$

$$0.255x = 1,249.50$$

$$\frac{0.255x}{0.255} = \frac{1,249.50}{0.255}$$

$$\boxed{x = 4,900}$$

$4,900 was invested in each.

45. INVESTMENT PLANS

	P	• r	• t =	I
6.2% i	x	0.062	1 y	**0.062x**
12% i	$65,000 - x$	0.12	1 y	**0.12(65,000 − x)**
T i	65,000			**$6,477.60**

Let x = amt invested at 6.2%
$65,000 - x$ = amt invested at 12%

6.2% int + 12% int = Total int

$$0.062x + 0.12(65,000 - x) = 6,477.60$$

$$0.062x + 7,800 - 0.12x = 6,477.60$$

$$-0.058x + 7,800 = 6,477.60$$

$$-0.058x + 7,800 - 7,800 = 6,477.60 - 7,800$$

$$-0.058x = -1,322.40$$

$$\frac{-0.058x}{-0.058} = \frac{-1,322.40}{-0.058}$$

$$\boxed{x = 22,800}$$

$22,800 invested at 6.2%
$42,200 invested at 12%

47. FINANCIAL PLANNING

	P	$\bullet$	r	$\bullet$	t	$=$	I
11%	P		0.11		1		**0.11 P**
13%	P		0.13		1		**0.13P**
Total							**13% earns \$150 more**

Since the 13% investment earns \$150 more interest, add this amount to the 11% interest and then set them equal to each other.

Let P = amount invested in each

$$13\% \text{ int} = 11\% \text{ int} + 150$$
$$0.13P = 0.11P + 150$$
$$0.13P - 0.11P = 0.11P + 150 - 0.11P$$
$$0.02P = 150$$
$$\frac{0.02P}{0.02} = \frac{150}{0.02}$$
$$\boxed{P = 7,500}$$

\$7,500 invested in each

49. TORNADOS

	r	$\bullet$	t	$=$	d
East van	20 mph		t		**20t**
West van	25 mph		t		**25t**
Total					**90 miles**

Let t = the time in hours each travels

East dist + West dist = total dist
$$20t + 25t = 90$$
$$45t = 90$$
$$\frac{45t}{45} = \frac{90}{45}$$
$$\boxed{t = 2}$$

2 hours each travels.

51. AIR TRAFFIC CONTROL

	r	$\bullet$	t	$=$	d
1st plane	450 mph		t		**450t**
2nd plane	500 mph		t		**500t**
Total					**3,800 m**

Let t = the number of hours

1st dist + 2nd dist = total dist
$$450t + 500t = 3,800$$
$$950t = 3,800$$
$$\frac{950t}{950} = \frac{3,800}{950}$$
$$\boxed{t = 4}$$

4 hours each.

53. ROAD TRIPS

	r	$\bullet$	t	$=$	d
slow part	40		t		**40t**
fast part	50		$5 - t$		**50(5 − t)**
Total					**210 mi**

Let t = time of slow part in hours
slow dist + fast dist = total dist
$$40t + 50(5 - t) = 210$$
$$40t + 250 - 50t = 210$$
$$-10t + 250 = 210$$
$$-10t + 250 - 250 = 210 - 250$$
$$-10t = -40$$
$$\frac{-10t}{-10} = \frac{-40}{-10}$$
$$\boxed{t = 4}$$

It took 4 hours to travel the slow part.

55. PHOTOGRAPHIC CHEMICALS

	Amount	• Strength =	Pure acid
Weak	2	5%	**0.05(2)**
Strong	x	10%	**0.10(x)**
Mixture	$x + 2$	7%	**0.07(x + 2)**

Let x = number of liters of 10% acetic acid

weak acid + strong acid = mixture

$$0.05(2) + 0.10x = 0.07(x + 2)$$
$$0.1 + 0.10x = 0.07x + 0.14$$
$$0.1 + 0.10x - 0.1 = 0.07x + 0.14 - 0.1$$
$$0.10x = 0.07x + 0.04$$
$$0.10x - 0.07x = 0.07x + 0.04 - 0.07x$$
$$0.03x = 0.04$$
$$\frac{0.03x}{0.03} = \frac{0.04}{0.03}$$
$$\boxed{x = 1\frac{1}{3}}$$

$1\frac{1}{3}$ liters of 10% acetic acid

57. ANTISEPTIC SOLUTION

	Amount	• Strength =	Pure BC
BC solution	30	10%	**0.10(30)**
Water	x	0%	**0.0(x)**
Mixture	$x + 30$	8%	**0.08(x + 30)**

Let x = number of ounces of water

BC pure + water = Mixture pure

$$0.10(30) + 0.0x = 0.08(x + 30)$$
$$3 = 0.08x + 2.4$$
$$3 - 2.4 = 0.08x + 2.4 - 2.4$$
$$0.6 = 0.08x$$
$$\frac{0.6}{0.08} = \frac{0.08x}{0.08}$$
$$\boxed{7.5 = x}$$

7.5 ounces of water.

59. MIXING FUELS

	Amount	• Value =	Total value
$1.15 gas	x	1.15	**1.15(x)**
$0.85 gas	20	0.85	**20(0.85)**
Mixture	$x + 20$	1.00	**1.00(x + 20)**

Let x = number of gallon of $1.15 gas

$1.15 val + $0.85 val = mixture val

$$1.15x + 0.85(20) = 1.00(x + 20)$$
$$1.15x + 17 = x + 20$$
$$1.15x + 17 - 17 = x + 20 - 17$$
$$1.15x = x + 3$$
$$1.15x - x = x + 3 - x$$
$$0.15x = 3$$
$$\frac{0.15x}{0.15} = \frac{3}{0.15}$$
$$\boxed{x = 20}$$

20 gallon of $1.15 gas

61. BLENDING GRASS SEED

	Amount	• Value =	Total value
Blue grass	100	$6	**$6(100)**
Rye grass	x	$3	**$3x**
Mixture	$x + 100$	$5	**$5(x + 100)**

Let x = number of lbs of Ryegrass

Blue val + Rye val = Mixture val

$$6(100) + 3x = 5(x + 100)$$
$$600 + 3x = 5x + 500$$
$$600 + 3x - 500 = 5x + 500 - 500$$
$$3x + 100 = 5x$$
$$3x + 100 - 3x = 5x - 3x$$
$$100 = 2x$$
$$\frac{100}{2} = \frac{2x}{2}$$
$$\boxed{50 = x}$$

50 lbs of Ryegrass

63. MIXING CANDY

	Amount	• Value	=	Total value
Lemon Drops	x	$1.90		**$1.90x**
Jelly Beans	$100 - x$	$1.20		**$1.20(100–x)**
Mixture	100	$1.48		**$1.48(100)**

Let x = number of lbs. of lemon drops
$100 - x$ = number of lbs. of jelly beans

$$\text{LD val} + \text{JB val} = \text{Mixture val}$$
$$1.90x + 1.20(100 - x) = 1.48(100)$$
$$1.90x + 120 - 1.20x = 148$$
$$0.7x + 120 = 148$$
$$0.7x + 120 - 120 = 148 - 120$$
$$0.7x = 28$$
$$\frac{0.7x}{0.7} = \frac{28}{0.7}$$
$$\boxed{x = 40}$$

40 lbs. of lemon drops
60 lbs. of jelly beans

WRITING

65. Answers will vary.

67. Answers will vary.

REVIEW

69. $\quad -25(2x - 5) = -50x + 125$

71. $\quad -(-3x - 3) = 3x + 3$

73. $\quad (4y - 4)(4) = 16y - 16$

CHALLENGE PROBLEMS

75. EVAPORATION

	Amount	• Strength	=	Pure BC
Salt solution	300	2%		**0.02(300)**
Water	x	0%		**0.0(x)**
Mixture	300-x	3%		**0.03(300-x)**

Let x = ml of water to be boiled away

$$\text{Old pure} - \text{water} = \text{New pure}$$
$$300(0.02) + 0.0x = 0.03(300 - x)$$
$$6 = 9 - 0.03x$$
$$6 - 9 = 9 - 0.03x - 9$$
$$-3 = -0.03x$$
$$\frac{-3}{-0.03} = \frac{-0.03x}{-0.03}$$
$$\boxed{100 = x}$$

100 ml of water to be boiled away.

Section 2.6

VOCABULARY

1. An **inequality** is a statement that contains one of the following symbols: $>$, $\geq$, $<$, or $\leq$.

3. To **solve** an inequality means to find all the values of the variable that make the inequality true.

5. The solution set of $x > 2$ can be expressed in **interval** notation as $(2, \infty)$.

CONCEPTS

7.
   a) $35 \geq 34$, **true**
   b) $-16 \leq -17$, **false**
   c) $\dfrac{3}{4} \leq 0.75$, **true**
   d) $-0.6 \geq -0.5$, **false**

9.
   a) $17 \geq -2$, **$-2 \leq 17$**
   b) $32 < x$, **$x > 32$**

11. Fill in the blanks.
   a) Adding the same number to, or subtracting the $\boxed{\text{same}}$ number from, both sides of an inequality does not change the inequality.

   b) Multiplying or dividing both sides of an inequality by the same $\boxed{\text{positive}}$ number does not change the solutions.

13. Solve $x + 2 > 10$ and give the solutions set:
$$x + 2 > 10$$
$$x + 2 - 2 > 10 - 2$$
$$x > 8$$

   a) ───────(────►
                     8

   b) $(8, \infty)$

   c) All real numbers greater than 8.

15. To solve compound inequalities, the properties of inequalities are applied to all **three** parts of the inequality.

NOTATION

17.
   a) The symbol $<$ means "**is less than**,"
   $$x + 2 > 10$$
   $$x + 2 - 2 > 10 - 2$$
   $$x > 8$$
   and the symbol $>$ means "**is greater than**."

   b) The symbol $\geq$ means "**is greater than or equal to**," and symbol $\leq$ means "is less than **or equal to**."

19. Give an example of each symbol: bracket, parenthesis, infinity, negative infinity.

$$[\,,\,(\,,\,\infty\,,\,-\infty$$

21. Complete the solution to solve each inequality.

$$4x - 5 \geq 7$$
$$4x - 5 + \boxed{5} \geq 7 + \boxed{5}$$
$$4x \geq \boxed{12}$$
$$\frac{4x}{\boxed{4}} \geq \frac{12}{\boxed{4}}$$
$$x \geq 3$$

Solution set: $[\boxed{3}, \infty)$

PRACTICE

23. ◄──────)──► , $(-\infty, 5)$
               5

25. ◄─(────]──► , $(-3, 1]$
        -3   1

27. $x < -1$, $(-\infty, -1)$

29. $-7 < x \leq 2$, $(-7, 2]$

31.
$$x + 2 > 5$$
$$x + 2 - 2 > 5 - 2$$
$$x > 3$$
$$(3, \infty)$$

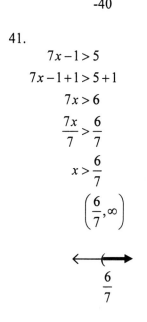

33.
$$3 + x < 2$$
$$3 + x - 3 < 2 - 3$$
$$x < -1$$
$$(-\infty, -1)$$

35.
$$g - 30 \geq -20$$
$$g - 30 + 30 \geq -20 + 30$$
$$x \geq 10$$
$$[10, \infty)$$

37.
$$\frac{2}{3}x \geq 2$$
$$\frac{3}{2}\left(\frac{2}{3}\right)x \geq \frac{3}{2}(2)$$
$$x \geq 3$$
$$[3, \infty)$$

39.
$$\frac{y}{4} + 1 \leq -9$$
$$\frac{y}{4} + 1 - 1 \leq -9 - 1$$
$$\frac{y}{4} \leq -10$$
$$\frac{4}{1}\left(\frac{y}{4}\right) \leq 4(-10)$$
$$y \leq -40$$
$$(-\infty, -40]$$

41.
$$7x - 1 > 5$$
$$7x - 1 + 1 > 5 + 1$$
$$7x > 6$$
$$\frac{7x}{7} > \frac{6}{7}$$
$$x > \frac{6}{7}$$
$$\left(\frac{6}{7}, \infty\right)$$

43.
$$0.5 \geq 2x - 0.3$$
$$0.5 + 0.3 \geq 2x - 0.3 + 0.3$$
$$0.8 \geq 2x$$
$$\frac{0.8}{2} \geq \frac{2x}{2}$$
$$0.4 \geq x$$
$$x \leq 0.4$$
$$(-\infty, 0.4]$$

45.

$$-30y \le -600$$

$$\frac{-30y}{-30} \ge \frac{-600}{-30}$$

$$y \ge 20$$

$$[20, \infty)$$

47.

$$-\frac{7}{8}x \le 21$$

$$-\frac{8}{7}\left(-\frac{7}{8}x\right) \ge -\frac{8}{7}(21)$$

$$x \ge -24$$

$$[-24, \infty)$$

49.

$$-1 \le -\frac{1}{2}n$$

$$-\frac{2}{1}((-1)) \ge -\frac{2}{1}\left(-\frac{1}{2}n\right)$$

$$2 \ge n$$

$$n \le 2$$

$$(-\infty, 2]$$

51.

$$\frac{m}{-42} - 1 > -1$$

$$\frac{m}{-42} - 1 + 1 > -1 + 1$$

$$\frac{m}{-42} > 0$$

$$\frac{-42}{1}\left(\frac{m}{-42}\right) < \frac{-42}{1}(0)$$

$$m < 0$$

$$(-\infty, 0)$$

53.

$$-x - 3 \le 7$$

$$-x - 3 + 3 \le 7 + 3$$

$$-x \le 10$$

$$\frac{-x}{-1} \ge \frac{10}{-1}$$

$$x \ge -10$$

$$[-10, \infty)$$

55.

$$-3x - 7 > -1$$

$$-3x - 7 + 7 > -1 + 7$$

$$-3x > 6$$

$$\frac{-3x}{-3} < \frac{6}{-3}$$

$$x < -2$$

$$(-\infty, -2)$$

57.

$$-4x + 6 > 17$$

$$-4x + 6 - 6 > 17 - 6$$

$$-4x > 11$$

$$\frac{-4x}{-4} < \frac{11}{-4}$$

$$x < -\frac{11}{4}$$

$$\left(-\infty, -\frac{11}{4}\right)$$

59.

$$2x + 9 \leq x + 8$$
$$2x + 9 - 9 \leq x + 8 - 9$$
$$2x \leq x - 1$$
$$2x - x \leq x \quad 1 \quad x$$
$$x \leq -1$$
$$(-\infty, -1]$$

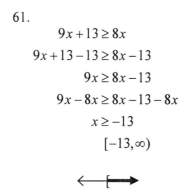

61.

$$9x + 13 \geq 8x$$
$$9x + 13 - 13 \geq 8x - 13$$
$$9x \geq 8x - 13$$
$$9x - 8x \geq 8x - 13 - 8x$$
$$x \geq -13$$
$$[-13, \infty)$$

-13

63.

$$8x + 4 > -(3x - 4)$$
$$8x + 4 > -3x + 4$$
$$8x + 4 - 4 > -3x + 4 - 4$$
$$8x > -3x$$
$$8x + 3x > -3x + 3x$$
$$11x > 0$$
$$\frac{11x}{11} > \frac{0}{11}$$
$$x > 0$$
$$(0, \infty)$$

0

65.

$$0.4x + 0.4 \leq 0.1x + 0.85$$
$$0.4x + 0.4 - 0.4 \leq 0.1x + 0.85 - 0.4$$
$$0.4x \leq 0.1x + 0.45$$
$$0.4x - 0.1 \leq 0.1x + 0.45 - 0.1x$$
$$0.3x \leq 0.45$$
$$\frac{0.3x}{0.3} \leq \frac{0.45}{0.3}$$
$$x \leq 1.5$$
$$(-\infty, 1.5]$$

1.5

67.

$$7 < \frac{5}{3}a - 3$$
$$7 + 3 < \frac{5}{3}a - 3 + 3$$
$$10 < \frac{5}{3}a$$
$$\frac{3}{5}(10) < \frac{3}{5}\left(\frac{5}{3}a\right)$$
$$6 < a$$
$$a > 6$$
$$(6, \infty)$$

6

69.

$$7 - x \leq 3x - 2$$
$$7 - x + 2 \leq 3x - 2 + 2$$
$$9 - x \leq 3x$$
$$9 - x + x \leq 3x + x$$
$$9 \leq 4x$$
$$\frac{9}{4} \leq \frac{4x}{4}$$
$$\frac{9}{4} \leq x$$
$$x \geq \frac{9}{4}$$
$$\left[\frac{9}{4}, \infty\right)$$

$$\frac{9}{4}$$

71.

$$8(5-x) \leq 10(8-x)$$
$$40-8x \leq 80-10x$$
$$40-8x-40 \leq 80-10x-40$$
$$-8x \leq 40-10x$$
$$-8x+10x \leq 40-10x+10x$$
$$2x \leq 40$$
$$\frac{2x}{2} \leq \frac{40}{2}$$
$$x \leq 20$$
$$(-\infty, 20]$$

$$20$$

73.

$$\frac{1}{2} + \frac{n}{5} > \frac{3}{4}$$
$$20\left(\frac{1}{2}\right) + 20\left(\frac{n}{5}\right) > 20\left(\frac{3}{4}\right)$$
$$10+4n > 15$$
$$10+4n-10 > 15-10$$
$$4n > 5$$
$$\frac{4n}{4} > \frac{5}{4}$$
$$n > \frac{5}{4}$$
$$\left(\frac{5}{4}, \infty\right)$$

$$\frac{5}{4}$$

75.

$$-\frac{2}{3} \geq \frac{2y}{3} - \frac{3}{4}$$
$$12\left(-\frac{2}{3}\right) \geq 12\left(\frac{2y}{3}\right) - 12\left(\frac{3}{4}\right)$$
$$-8 \geq 8y - 9$$
$$-8+9 \geq 8y-9+9$$
$$1 \geq 8y$$
$$\frac{1}{8} \geq \frac{8y}{8}$$
$$\frac{1}{8} \geq y$$
$$y \leq \frac{1}{8}$$
$$\left(-\infty, \frac{1}{8}\right]$$

$$\frac{1}{8}$$

77.

$$\frac{6x+1}{4} \leq x+1$$
$$4\left(\frac{6x+1}{4}\right) \leq 4(x+1)$$
$$6x+1 \leq 4x+4$$
$$6x+1-1 \leq 4x+4-1$$
$$6x \leq 4x+3$$
$$6x-4x \leq 4x+3-4x$$
$$2x \leq 3$$
$$\frac{2x}{2} \leq \frac{3}{2}$$
$$x \leq \frac{3}{2}$$
$$\left(-\infty, \frac{3}{2}\right]$$

$$\frac{3}{2}$$

79.

$$\frac{5}{2}(7x-15)+x \geq \frac{13}{2}x-\frac{3}{2}$$

$$2\left(\frac{5}{2}(7x-15)\right)+2(x) \geq 2\left(\frac{13}{2}x\right)-2\left(\frac{3}{2}\right)$$

$$5(7x-15)+2x \geq 13x-3$$

$$35x-75+2x \geq 13x-3$$

$$37x-75 \geq 13x-3$$

$$37x-75+75 \geq 13x-3+75$$

$$37x \geq 13x+72$$

$$37x-13x \geq 13x+72-13x$$

$$24x \geq 72$$

$$\frac{24x}{24} \geq \frac{72}{24}$$

$$x \geq 3$$

$$[3,\infty)$$

81.

$$2 < x-5 < 5$$

$$2+5 < x-5+5 < 5+5$$

$$7 < x < 10$$

$$(7,10)$$

83.

$$0 \leq x+10 \leq 10$$

$$0-10 \leq x+10-10 \leq 10-10$$

$$-10 \leq x \leq 0$$

$$[-10,0]$$

85.

$$-3 \leq \frac{c}{2} \leq 5$$

$$2(-3) \leq 2\left(\frac{c}{2}\right) \leq 2(5)$$

$$-6 \leq c \leq 10$$

$$[-6,10]$$

87.

$$3 \leq 2x-1 < 5$$

$$3+1 \leq 2x-1+1 < 5+1$$

$$4 \leq 2x < 6$$

$$\frac{4}{2} \leq \frac{2x}{2} < \frac{6}{2}$$

$$2 \leq x < 3$$

$$[2,3)$$

89.

$$4 < -2x < 10$$

$$-\frac{1}{2}(4) > -\frac{1}{2}(-2x) > -\frac{1}{2}(10)$$

$$-2 > x > -5$$

$$-5 < x < -2$$

$$(-5,-2)$$

91.

$$0 < 10-5x \leq 15$$

$$0-10 < 10-5x-10 \leq 15-10$$

$$-10 < -5x \leq 5$$

$$\frac{-10}{-5} > \frac{-5x}{-5} \geq \frac{5}{-5}$$

$$2 > x \geq -1$$

$$-1 \leq x < 2$$

$$[-1,2)$$

93.

$$9(0.05-0.3x)+0.162 \leq 0.081+15x$$

$$0.45-2.7x+0.162 \leq 0.081+15x$$

$$-2.7x+0.612-0.612 \leq 0.081+15x-0.612$$

$$-2.7x \leq 15x-0.531$$

$$-2.7x-15x \leq 15x-0.531-15x$$

$$-17.7x \leq -0.531$$

$$\frac{-17.7x}{-17.7} \geq \frac{-0.531}{-17.7}$$

$$x \geq 0.03$$

$$[0.03,\infty)$$

APPLICATIONS

95. **GRADES**
Let $x = \%$ score needed on last exam
$$\frac{68 + 75 + 79 + x}{4} \geq 80$$
$$\frac{222 + x}{4} \geq 80$$
$$4\left(\frac{222 + x}{4}\right) \geq 4(80)$$
$$222 + x \geq 320$$
$$222 + x - 222 \geq 320 - 222$$
$$x \geq 98$$

97. **FLEET AVERAGES**
Let $x =$ mpg for third model car
$$\frac{17 + 19 + x}{3} \geq 21$$
$$\frac{36 + x}{3} \geq 21$$
$$3\left(\frac{36 + x}{3}\right) \geq 3(21)$$
$$36 + x \geq 63$$
$$36 + x - 36 \geq 63 - 36$$
$$x \geq 27$$

99. **GEOMETRY**
Let $s =$ length of one equal side
$$0 < 3s \leq 57$$
$$\frac{0}{3} < \frac{3s}{3} \leq \frac{57}{3}$$
$$0 \text{ ft} < s \leq 19 \text{ ft}$$

101. **COUNTER SPACE**
Let $x =$ the acceptable value
$$2x + 2(x + 5) \geq 150$$
$$2x + 2x + 10 \geq 150$$
$$4x + 10 \geq 150$$
$$4x + 10 - 10 \geq 150 - 10$$
$$4x \geq 140$$
$$\frac{4x}{4} \geq \frac{140}{4}$$
$$x \geq 35 \text{ ft}$$

103. **SAFETY CODES**
a) Ramps or inclines
$$0° < a \leq 18°$$

b) Stairs
$$18° \leq a \leq 50°$$

c) Preferred range for stairs
$$30° \leq a \leq 37°$$

d) Ladders with cleats
$$75° \leq a < 90°$$

105. **WEIGHT CHART**
a) 10 months old, "heavy"
$$26 \text{ lb} \leq w \leq 31 \text{ lb}$$

b) 5 months old, "light"
$$12 \text{ lb} \leq w \leq 14 \text{ lb}$$

c) 8 months old, "average"
$$18.5 \text{ lb} \leq w \leq 20.5 \text{ lb}$$

d) 3 months old, moderately light"
$$11 \text{ lb} \leq w \leq 13 \text{ lb}$$

WRITING

107. Answers will vary.

REVIEW

109.
$$-5^3 = -(5)(5)(5)$$
$$= -125$$

111. Complete the table

x	$x^2 - 3$
2	$2^2 - 3$
	$4 - 3$
	1
0	$0^2 - 3$
	$0 - 3$
	-3
3	$3^2 - 3$
	$9 - 3$
	6

CHALLENGE PROBLEMS

113.

$$3 - x < 5 < 7 - x$$
$$3 - x + x < 5 + x < 7 - x + x$$
$$3 < 5 + x < 7$$
$$3 - 5 < 5 + x - 5 < 7 - 5$$
$$-2 < x < 2$$
$$(-2, 2)$$

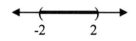

Section 2.7

CHAPTER 2: KEY CONCEPT
Simplify and Solve

1.
a) $-3x + 2 + 5x - 10 = 2x - 8$

b) $-3x + 2 + 5x - 10 = 4$
$$2x - 8 = 4$$
$$2x - 8 + 8 = 4 + 8$$
$$2x = 12$$
$$\frac{2x}{2} = \frac{12}{2}$$
$$\boxed{x = 6}$$

2.
a) $4(y + 2) - 3(y + 1) = 4y + 8 - 3y - 3$
$$= y + 5$$
b) $4(y + 2) = 3(y + 1)$
$$4y + 8 = 3y + 3$$
$$4y + 8 - 8 = 3y + 3 - 8$$
$$4y = 3y - 5$$
$$4y - 3y = 3y - 5 - 3y$$
$$\boxed{y = -5}$$

3.
a) $\frac{1}{3}a + \frac{1}{3}a = \frac{2}{3}a$

b) $\frac{1}{3}a + \frac{1}{3} = \frac{1}{2}$
$$6\left(\frac{1}{3}a + \frac{1}{3}\right) = 6\left(\frac{1}{2}\right)$$
$$2a + 2 = 3$$
$$2a + 2 - 2 = 3 - 2$$
$$2a = 1$$
$$\frac{2a}{2} = \frac{1}{2}$$
$$\boxed{a = \frac{1}{2}}$$

4.
a) $-(2x + 10) = -2x - 10$

b) $-2x \geq -10$
$$\frac{-2x}{-2} \leq \frac{-10}{-2}$$
$$\boxed{x \leq 5}$$

Reverse the symbol's direction.

5.
a) $\frac{2}{3}(x - 2) - \frac{1}{6}(4x - 8)$
$$= \frac{2}{3}x - \frac{4}{3} - \frac{4}{6}x + \frac{8}{6}$$
$$= \frac{4}{6}x - \frac{4}{6}x - \frac{8}{6} + \frac{8}{6}$$
$$= 0$$

b) $\frac{2}{3}(x - 2) - \frac{1}{6}(4x - 8) = 0$
$$\frac{2}{3}x - \frac{4}{3} - \frac{4}{6}x + \frac{8}{6} = 0$$
$$\frac{4}{6}x - \frac{4}{6}x - \frac{8}{6} + \frac{8}{6} = 0$$
$$\boxed{0 = 0}$$

all real numbers

6. The mistake is on the third line. The student made an equation out of the answer, which is $x - 6$, by writing "0 =" on the left. Then the student solved the equation.

CHAPTER 2: REVIEW

SECTION 2.1
Solving Equations
REVIEW EXERCISES

Tell whether the given number is a solution of the equation.

1.

$$x - 34 = 50, \quad 84$$

$$84 - 34 \overset{?}{=} 50$$

$$50 = 50$$

$$\text{yes}$$

2.

$$5y + 2 = 12, \quad 3$$

$$5(3) + 2 \overset{?}{=} 12$$

$$15 + 2 \overset{?}{=} 12$$

$$\text{no}$$

3.

$$\frac{x}{5} = 6, \quad -30$$

$$\frac{-30}{5} \overset{?}{=} 6$$

$$-6 = 6$$

$$\text{no}$$

4.

$$a^2 - a - 1 = 0, \quad 2$$

$$(2)^2 - 2 - 1 \overset{?}{=} 0$$

$$4 - 2 - 1 \overset{?}{=} 0$$

$$1 \neq 0$$

$$\text{no}$$

5.

$$5b - 2 = 3b - 8, \quad -3$$

$$5(-3) - 2 \overset{?}{=} 3(-3) - 8$$

$$-15 - 2 \overset{?}{=} -9 - 8$$

$$-17 = -17$$

$$\text{yes}$$

6.

$$\frac{2}{y+1} = \frac{12}{y+1} - 5, \quad 1$$

$$\frac{2}{1+1} \overset{?}{=} \frac{12}{1+1} - 5$$

$$\frac{2}{2} \overset{?}{=} \frac{12}{2} - 5$$

$$1 \overset{?}{=} 6 - 5$$

$$1 = 1$$

$$\text{yes}$$

7. To solve $x + 8 = 10$ means to find all the values of the **variable** that make the equation a **true** statement.

8.

$$x - 9 = 12$$

$$x - 9 + 9 = 12 + 9$$

$$\boxed{x = 21}$$

$$\text{check}$$

$$21 - 9 \overset{?}{=} 12$$

$$12 = 12$$

9.

$$y + 15 = -32$$

$$y + 15 - 15 = -32 - 15$$

$$\boxed{y = -47}$$

$$\text{check}$$

$$-47 + 15 \overset{?}{=} -32$$

$$-32 = -32$$

10.

$$a + 3.7 = 16.9$$

$$a + 3.7 - 3.7 = 16.9 - 3.7$$

$$\boxed{a = 13.2}$$

$$\text{check}$$

$$13.2 + 3.7 \overset{?}{=} 16.9$$

$$16.9 = 16.9$$

11.

$$100 = -7 + r$$
$$100 + 7 = -7 + r + 7$$
$$\boxed{107 = r}$$

check

$$100 \overset{?}{=} -7 + 107$$
$$100 = 100$$

12.

$$120 = 15c$$
$$\frac{120}{15} = \frac{15c}{15}$$
$$\boxed{8 = c}$$

check

$$120 \overset{?}{=} 15(8)$$
$$120 = 120$$

13.

$$t - \frac{1}{2} = \frac{1}{2}$$
$$t - \frac{1}{2} + \frac{1}{2} = \frac{1}{2} + \frac{1}{2}$$
$$\boxed{t = 1}$$

check

$$1 - \frac{1}{2} \overset{?}{=} \frac{1}{2}$$
$$\frac{1}{2} = \frac{1}{2}$$

14.

$$\frac{t}{8} = -12$$
$$8\left(\frac{t}{8}\right) = 8(-12)$$
$$\boxed{t = -96}$$

check

$$\frac{-96}{8} \overset{?}{=} -12$$
$$-12 = -12$$

15.

$$3 = \frac{q}{2.6}$$
$$2.6(3) = 2.6\left(\frac{q}{2.6}\right)$$
$$\boxed{7.8 = q}$$

check

$$3 \overset{?}{=} \frac{7.8}{2.6}$$
$$3 = 3$$

16.

$$6b = 0$$
$$\frac{6b}{6} = \frac{0}{6}$$
$$\boxed{b = 0}$$

check

$$6(0) \overset{?}{=} 0$$
$$0 = 0$$

17.

$$\frac{x}{14} = 0$$
$$14\left(\frac{x}{14}\right) = 14(0)$$
$$\boxed{x = 0}$$

check

$$\frac{0}{14} \overset{?}{=} 0$$
$$0 = 0$$

18. GEOMETRY

$$x + 20 = 180$$
$$x + 20 - 20 = 180 - 20$$
$$\boxed{x = 160°}$$

check

$$160 + 20 \overset{?}{=} 180$$
$$180 = 180$$

19.

goalkeepers	3
defenders	6
midfielders	6
forwards	x
Total	20

Let x = number of forwards

$3 + 6 + 6 + x = 20$

$x + 15 = 20$

$x + 15 - 15 = 20 - 15$

$\boxed{x = 5}$

check

$3 + 6 + 6 + 5 \overset{?}{=} 20$

$20 = 20$

20. HISORIC TOURS

To Philadelphia	296 mi
To Washington D.C.	133 mi
To Boston	x mi
Total	858 mi

Let x = miles back to Boston

$296 + 133 + x = 858$

$x + 429 = 858$

$x + 429 - 429 = 858 - 429$

$\boxed{x = 429}$

check

$296 + 133 + 429 \overset{?}{=} 858$

$858 = 858$

21. CHROME WHEELS

Let x = measure of one equal angle

$6x = 360$

$\dfrac{6x}{6} = \dfrac{360}{6}$

$\boxed{x = 60}$

check

$6(60) \overset{?}{=} 360$

$360 = 360$

22. ADVERTISING

Amount = Percent • Base

$A = 23.5\% \cdot \$231 \text{ billions}$

$A = 0.235 \cdot \$231 \text{ billion}$

$A = \$54.285 \text{ billion}$

$\boxed{A = \$54 \text{ billion}}$

23. COST OF LIVING

Amount = Percent • Base

$A = 3.5\% \cdot \$764$

$A = 0.035 \cdot \$764$

$\boxed{A = \$26.74}$

24. Amount = Percent • Base

$4.81 = 2.5\% \cdot b$

$4.81 = 0.025 \cdot b$

$\dfrac{4.81}{0.025} = \dfrac{0.025b}{0.025}$

$\boxed{192.4 = b}$

25. FAMILY BUDGET

Amount = Percent • Base

$A = 30\% \cdot \$1,890$

$A = 0.3 \cdot \$1,890$

$\boxed{A = \$567}$

No, their present monthly payment of $625 is more than 30% of their income.

26. COLLECTIBLES

Find the difference between $6 and $100.

$100 - 6 = 94$

Now find the percent of increase.

$\dfrac{94}{6} \approx 15.666$

≈ 15.67

$\approx 1,567\%$

The increase is 1,567%.

SECTION 2.3
Simplifying Algebraic Expressions

27. $-4(7w) = -28w$

28. $-3r(-5) = 15r$

29. $\begin{aligned} 3(-2x)(-4) &= -6x(-4) \\ &= 24x \end{aligned}$

30. $0.4(5.2f) = 2.08f$

31. $15(\frac{3}{5}a) = 9a$

32. $\frac{7}{2} \cdot \frac{2}{7}r = r$

33. $5(x+3) = 5x + 15$

34. $-2(2x + 3 - y) = -4x - 6 + 2y$

35. $-(a-4) = -a + 4$

36. $\frac{3}{4}(4c - 8) = 3c - 6$

37. $40\left(\frac{x}{2} + \frac{4}{5}\right) = 20x + 32$

38. $\begin{aligned} 2(-3c - 7)(2.1) &= (-6c - 14)(2.1) \\ &= -12.6c - 29.4 \end{aligned}$

39. $\begin{aligned} 8p + 5p - 4p &= 13p - 4p \\ &= 9p \end{aligned}$

40. $\begin{aligned} -5m + 2 - 2m - 2 &= -5m - 2m + 2 - 2 \\ &= -7m \end{aligned}$

41. $n + n + n + n = 4n$

42. $\begin{aligned} 5(p-2) - 2(3p + 4) &= 5p - 10 - 6p - 8 \\ &= 5p - 6p - 10 - 8 \\ &= -p - 18 \end{aligned}$

43. $55.7k - 55.6k = 0.1k$

44. $\begin{aligned} 8a^3 + 4a^3 - 20a^3 &= 12a^3 - 20a^3 \\ &= -8a^3 \end{aligned}$

45. $\begin{aligned} \frac{3}{5}w - \left(-\frac{2}{5}w\right) &= \frac{3}{5}w + \frac{2}{5}w \\ &= \frac{5}{5}w \\ &= w \end{aligned}$

46. $\begin{aligned} 36\left(\frac{1}{9}h - \frac{3}{4}\right) + 36\left(\frac{1}{3}\right) &= 4h - 27 + 12 \\ &= 4h - 15 \end{aligned}$

47. $\begin{aligned} (x+7) \text{ ft } + (2x - 3) \text{ ft} + x \text{ ft} \\ = [(x + 2x + x) + (7 - 3)] \text{ ft} \\ = (4x + 4) \text{ ft} \end{aligned}$

SECTION 2.4
More about Solving Equations

48. $$\begin{aligned} 5x + 4 &= 14 \\ 5x + 4 - 4 &= 14 - 4 \\ 5x &= 10 \\ \frac{5x}{5} &= \frac{10}{5} \\ \boxed{x = 2} \end{aligned}$$

check

$5(2) + 4 \overset{?}{=} 14$

$10 + 4 \overset{?}{=} 14$

$14 = 14$

49. $$\begin{aligned} 98.6 - t &= 129.2 \\ 98.6 - t + t &= 129.2 + t \\ 98.6 &= 129.2 + t \\ 98.6 - 129.2 &= 129.2 + t - 129.2 \\ \boxed{-30.6 = t} \end{aligned}$$

check

$98.6 - (-30.6) \overset{?}{=} 129.2$

$98.6 + 30.6 \overset{?}{=} 129.2$

$129.2 = 129.2$

50.

$$\frac{n}{5} - 2 = 4$$

$$\frac{n}{5} - 2 + 2 = 4 + 2$$

$$\frac{n}{5} = 6$$

$$5\left(\frac{n}{5}\right) = 5(6)$$

$$\boxed{n = 30}$$

check

$$\frac{30}{5} - 2 \overset{?}{=} 4$$

$$6 - 2 \overset{?}{=} 4$$

$$4 = 4$$

51.

$$\frac{b-5}{4} = -6$$

$$4\left(\frac{b-5}{4}\right) = 4(-6)$$

$$b - 5 = -24$$

$$b - 5 + 5 = -24 + 5$$

$$\boxed{b = -19}$$

check

$$\frac{-19 - 5}{4} \overset{?}{=} -6$$

$$\frac{-24}{4} \overset{?}{=} -6$$

$$-6 = -6$$

52.

$$5(2x - 4) - 5x = 0$$

$$10x - 20 - 5x = 0$$

$$5x - 20 = 0$$

$$5x - 20 + 20 = 0 + 20$$

$$5x = 20$$

$$\frac{5x}{5} = \frac{20}{5}$$

$$\boxed{x = 4}$$

check

$$5(2 \cdot 4 - 4) - 5 \cdot 4 \overset{?}{=} 0$$

$$5(8 - 4) - 20 \overset{?}{=} 0$$

$$5(4) - 20 \overset{?}{=} 0$$

$$20 - 20 \overset{?}{=} 0$$

$$0 = 0$$

53.

$$-2(x - 5) = 5(-3x + 4) + 3$$

$$-2x + 10 = -15x + 20 + 3$$

$$-2x + 10 = -15x + 23$$

$$-2x + 10 - 10 = -15x + 23 - 10$$

$$-2x = -15x + 13$$

$$-2x + 15x = -15x + 13 + 15x$$

$$13x = 13$$

$$\frac{13x}{13} = \frac{13}{13}$$

$$\boxed{x = 1}$$

check

$$-2(1 - 5) \overset{?}{=} 5(-3 \cdot 1 + 4) + 3$$

$$-2(-4) \overset{?}{=} 5(1) + 3$$

$$8 = 8$$

54.

$$\frac{3}{4} = \frac{1}{2} + \frac{d}{5}$$

$$20\left(\frac{3}{4}\right) = 20\left(\frac{1}{2}\right) + 20\left(\frac{d}{5}\right)$$

$$15 = 10 + 4d$$

$$15 - 10 = 10 + 4d - 10$$

$$5 = 4d$$

$$\frac{5}{4} = \frac{4d}{4}$$

$$\boxed{\frac{5}{4} = d}$$

check

$$\frac{3}{4} \overset{?}{=} \frac{1}{2} + \frac{\frac{5}{4}}{5}$$

$$\frac{3}{4} \overset{?}{=} \frac{1}{2} + \left(\frac{5}{4} \cdot \frac{1}{5}\right)$$

$$\frac{3}{4} \overset{?}{=} \frac{1}{2} + \frac{1}{4}$$

$$\frac{3}{4} \overset{?}{=} \frac{2}{4} + \frac{1}{4}$$

$$\frac{3}{4} = \frac{3}{4}$$

55.

$$-\frac{2}{3}f = 4$$

$$-\frac{3}{2}\left(-\frac{2}{3}f\right) = -\frac{3}{2}(4)$$

$$\boxed{f = -6}$$

check

$$-\frac{2}{3}(-6) \overset{?}{=} 4$$

$$4 = 4$$

56.

$$\frac{3(2-c)}{2} = \frac{-2(2c+3)}{5}$$

$$10\left(\frac{3(2-c)}{2}\right) = 10\left(\frac{-2(2c+3)}{5}\right)$$

$$5[3(2-c)] = 2[-2(2c+3)]$$

$$15(2-c) = -4(2c+3)$$

$$30 - 15c = -8c - 12$$

$$30 - 15c + 15c = -8c - 12 + 15c$$

$$30 = 7c - 12$$

$$30 + 12 = 7c - 12 + 12$$

$$42 = 7c$$

$$\frac{42}{7} = \frac{7c}{7}$$

$$\boxed{6 = c}$$

check

$$\frac{3(2-6)}{2} \overset{?}{=} \frac{-2(2 \cdot 6 + 3)}{5}$$

$$\frac{3(-4)}{2} \overset{?}{=} \frac{-2(15)}{5}$$

$$\frac{-12}{2} \overset{?}{=} \frac{-30}{5}$$

$$-6 = -6$$

57.

$$\frac{b}{3} + \frac{11}{9} + 3b = -\frac{5}{6}b$$

$$18\left(\frac{b}{3}\right) + 18\left(\frac{11}{9}\right) + 18(3b) = 18\left(-\frac{5}{6}b\right)$$

$$6b + 22 + 54b = -15b$$

$$60b + 22 = -15b$$

$$60b + 22 - 22 = -15b - 22$$

$$60b = -15b - 22$$

$$60b + 15b = -15b - 22 + 15b$$

$$75b = -22$$

$$\frac{75b}{75} = \frac{-22}{75}$$

$$\boxed{b = \frac{-22}{75}}$$

check

$$\frac{-\frac{22}{75}}{3} + \frac{11}{9} + 3\left(-\frac{22}{75}\right) \overset{?}{=} -\frac{5}{6}\left(-\frac{22}{75}\right)$$

$$-\frac{22}{75}\left(\frac{1}{3}\right) + \frac{11}{9} - \frac{22}{25} \overset{?}{=} \frac{11}{45}$$

$$-\frac{22}{225} + \frac{11}{9} - \frac{22}{25} \overset{?}{=} \frac{11}{45}$$

$$225\left(-\frac{22}{225} + \frac{11}{9} - \frac{22}{25}\right) \overset{?}{=} 225\left(\frac{11}{45}\right)$$

$$-22 + 275 - 198 \overset{?}{=} 55$$

$$55 = 55$$

58.

$$3(a+8) = 6(a+4) - 3a$$

$$3a + 24 = 6a + 24 - 3a$$

$$3a + 24 = 3a + 24$$

$$3a + 24a = 3a + 24 - 3a$$

$$\boxed{24 = 24}$$

identity; all real numbers

59.

$$2(y+10) + y = 3(y+8)$$

$$2y + 20 + y = 3y + 24$$

$$3y + 20 = 3y + 24$$

$$3y + 20 - 3y = 3y + 24 - 3y$$

$$\boxed{20 \neq 24}$$

contradiction; no solution

SECTION 2.5
Formulas

60.

Wholesale = $219

Markup = ?

Retail = $395

$$WS + MUp = Retail$$
$$219 + MUp = 395$$
$$219 + MUp - 219 = 395 - 219$$
$$\boxed{MUp = 176}$$

The markup is $176.

61.

Sales = $13,500

Profit = $1,700

Expenses = ?

$$Exp + Prof = Sales$$
$$Exp + 1,700 = 13,500$$
$$Exp + 1,700 - 1,700 = 13,500 - 1,700$$
$$\boxed{Exp = 11,800}$$

The expenses are $11,800.

62. INDY 500

d = 500 miles

r = 166.499 mph

t = ? hrs

$$rt = d$$
$$166.499t = 500$$
$$\frac{166.499t}{166.499} = \frac{500}{166.499}$$
$$t = 3.003$$
$$\boxed{t = 3.00}$$

The amount of time is about 3 hr.

63. JEWELRY MAKING

$$C^o = 1,056$$
$$\frac{5}{9}(F - 32) = C$$
$$\frac{5}{9}(F - 32) = 1,065$$
$$9\left(\frac{5}{9}(F - 32)\right) = 9(1,065)$$
$$5F - 160 = 9,585$$
$$5F - 160 + 160 = 9,585 + 160$$
$$5F = 9,745$$
$$\frac{5F}{5} = \frac{9,745}{5}$$
$$\boxed{F = 1,949}$$

The temperature is $1,949°F$.

64. CAMPING

l = 60 inches

w = 24 inches

P = ? inches

$$P = 2l + 2w$$
$$P = 2 \cdot 60 + 2 \cdot 24$$
$$P = 120 + 48$$
$$\boxed{P = 168}$$

Perimeter is 168 inches.

65. CAMPING

l = 60 inches

w = 24 inches

A = ? sq in

$$A = lw$$
$$A = 60 \cdot 24$$
$$\boxed{A = 1,440}$$

Area is 1,440 in.²

66.

b = 17 meters

h = 9 meters

A = ? sq m

$$A = \frac{1}{2}bh$$
$$A = 0.5 \cdot 17 \cdot 9$$
$$\boxed{A = 76.5}$$

Area is 76.5 m².

67. $b = 11$ inches

$d = 13$ inches

$h = 12$ inches

$A = ?$ sq in.

$A = \dfrac{1}{2}h(b+d)$

$A = 0.5 \cdot 12 \cdot (11+13)$

$A = 6(24)$

$\boxed{A = 144}$

Area is 144 in.^{2}

68. $r = 8$ cm

$C = ?$ cm

$C = 2\pi r$

$C = 2 \cdot 3.14 \cdot 8$

$\boxed{C = 50.24}$

Circumference is 50.24 cm.
Answers will vary due to
the approximation of π.

69. $r = 8$ cm

$A = ?$ sq cm

$A = \pi r^2$

$A = 3.14 \cdot 8 \cdot 8$

$\boxed{A = 200.96}$

Area is 200.96 cm^2.
Answers will vary due to
the approximation of π.

70. CAMPING

$l = 60$ inches

$w = 24$ inches

$h = 3$ inches

$V = ?$ cu in.

$V = lwh$

$V = (60)(24)(3)$

$\boxed{V = 4,320}$

Volume is 4,320 in.^{3}

71. $h = 12$ feet

$r = 0.5$ foot

$V = ?$ cu ft

$V = \pi r^2 h$

$V = 3.14 \cdot 0.5 \cdot 0.5 \cdot 12$

$V = 9.42$

$\boxed{V = 9.4}$

Volume is 9.4 ft^3.
Answers will vary due to
the approximation of π.

72. $b = 6$ feet

$h = 10$ feet

$V = ?$ cu ft

$V = \dfrac{1}{3}Bh$

$V = \dfrac{1}{3} \cdot 6 \cdot 6 \cdot 10$

$\boxed{V = 120}$

Volume is 120 ft^3.

73. HALLOWEEN

$d = 9$ inches

$r = 4.5$ inches

$V = ?$ cu in.

$V = \dfrac{4}{3}\pi r^3$

$V = \dfrac{4}{3} \cdot 3.14 \cdot 4.5 \cdot 4.5 \cdot 4.5$

$\boxed{V = 381.51}$

Volume is 381.51 in.^{3}
Answers will vary due to
the approximation of π.

74. $A = 2\pi rh$, for h

$A = 2\pi rh$

$\dfrac{A}{2\pi r} = \dfrac{2\pi rh}{2\pi r}$

$\boxed{\dfrac{A}{2\pi r} = h}$

75.

$$A - BC = \dfrac{G - K}{3}, \text{ for } G$$

$$A - BC = \dfrac{G - K}{3}$$

$$3(A - BC) = 3\left(\dfrac{G - K}{3}\right)$$

$$3(A - BC) = G - K$$

$$3A - 3BC + K = G - K + K$$

$$\boxed{3A - 3BC + K = G}$$

76.

$$a^2 + b^2 = c^2, \text{ for } b^2$$

$$a^2 + b^2 = c^2$$

$$a^2 + b^2 - a^2 = c^2 - a^2$$

$$\boxed{b^2 = c^2 - a^2}$$

77.

$$4y - 16 = 3x, \text{ for } y$$

$$4y - 16 = 3x$$

$$4y - 16 + 16 = 3x + 16$$

$$4y = 3x + 16$$

$$\dfrac{4y}{4} = \dfrac{3x + 16}{4}$$

$$\boxed{y = \dfrac{3}{4}x + 4}$$

SECTION 2.6
More about Problem Solving

78. SOUND SYSTEMS

one piece = 15 ft
other piece = x ft
last piece = $3x - 2$ ft
total = 45 feet

Let x = length of other piece in ft
$3x - 2$ = length of last piece in ft

$$x + (3x - 2) + 15 = 45$$

$$4x + 13 = 45$$

$$4x + 13 - 13 = 45 - 13$$

$$4x = 32$$

$$\dfrac{4x}{4} = \dfrac{32}{4}$$

$$\boxed{x = 8}$$

Shorter piece is 8 ft.

79. AUTOGRAPHS

	#	• cost	= value
Movie stars	x	$250	$250 x
TV celebrities	$x + 8$	$75	$75(x + 8)$
total			$1,900

Let x = # of movie star autographs
$x + 8$ = # of TV stars autographs

$$250x + 75(x + 8) = 1,900$$

$$250x + 75x + 600 = 1,900$$

$$325x + 600 = 1,900$$

$$325x + 600 - 600 = 1,900 - 600$$

$$325x = 1,300$$

$$\dfrac{325x}{325} = \dfrac{1,300}{325}$$

$$\boxed{x = 4}$$

$$\boxed{\begin{array}{l} x + 8 = 4 + 8 \\ = 12 \end{array}}$$

4 movie star autographs.
12 TV stars autographs.

80. ART HISTORY

l = 5 inches more than width
w = ? inches
$P = 109\dfrac{1}{2}$ = 109.5 inches

Let w = width of painting
$w + 5$ = length of painting

$$2x + 2(x + 5) = 109.5$$

$$2x + 2x + 10 = 109.5$$

$$4x + 10 = 109.5$$

$$4x + 10 - 10 = 109.5 - 10$$

$$4x = 99.5$$

$$\dfrac{4x}{4} = \dfrac{99.5}{4}$$

$$\boxed{x = 24.875}$$

$$\boxed{\begin{array}{l} x + 5 = 24.875 + 5 \\ = 29.875 \end{array}}$$

Width is 24.875 inches.
Length is 29.875 inches.

81.

base angle = ?°

vertex angle = 27°

total degrees = 180°

Let x = one base angle

$$x + x + 27 = 180$$
$$2x + 27 = 180$$
$$2x + 27 - 27 = 180 - 27$$
$$2x = 153$$
$$\frac{2x}{2} = \frac{153}{2}$$
$$\boxed{x = 76.5}$$

Both base angles are 76.5°.

82. $45x

83. INVESTMENT INCOME

	P	•	r	•	t =	I
7%	P		0.07		1	**0.07P**
9%	27,000 - P		0.09		1	**0.09(27,000 − P)**
	Total interest					**$2,110**

Let P = amount invested at 7%

27,000 - P = amount invested at 9%

7% int + 9% int = Total int

$$0.07P + 0.09(27,000 - P) = 2,110$$
$$0.07P + 2,430 - 0.09P = 2,110$$
$$-0.02P + 2,430 = 2,110$$
$$-0.02P + 2,430 - 2,430 = 2,110 - 2,430$$
$$-0.02P = -320$$
$$\frac{-0.02P}{-0.02} = \frac{-320}{-0.02}$$
$$\boxed{P = 16,000}$$
$$\boxed{\begin{aligned}27,000 - P &= 27,000 - 16,000 \\ &= 11,000\end{aligned}}$$

$16,000 earned at 7%.

$11,000 earned at 9%.

84. WALKING AND BICYCLING

	r	•	t =	d
Walker	3 mph		t	**3t**
Bike rider	12 mph		t	**12t**
Total				**5 miles**

Let t = the # of minutes they meet

1st dist + 2nd dist = total dist

$$3t + 12t = 5$$
$$15t = 5$$
$$\frac{15t}{15} = \frac{5}{15}$$
$$t = \frac{1}{3} \text{ , convert}$$
$$\boxed{\begin{aligned}t &= \frac{1}{3} \cdot 60 \text{ minutes} \\ &= 20\end{aligned}}$$

They will meet in 20 minutes.

85. MIXTURES

	Amount	•	Value	=	Total value
Candy	x		$0.90		**$0.90x**
Gum Drops	20 − x		$1.50		**$1.50(20−x)**
Mixture	20		$1.20		**$1.20(20)**

Let x = number of lbs. of candy

20 - x = number of lbs. of gumdrops

Candy val + GD val = Mixture val

$$0.90x + 1.50(20 - x) = 1.20(20)$$
$$0.90x + 30 - 1.50x = 24$$
$$-0.60x + 30 = 24$$
$$-0.60x + 30 - 30 = 24 - 30$$
$$-0.60x = -6$$
$$\frac{-0.60x}{-0.60} = \frac{-6}{-0.60}$$
$$\boxed{x = 10}$$
$$\boxed{\begin{aligned}20 - x &= 20 - 10 \\ &= 10\end{aligned}}$$

10 of lbs. of each needed.

86. SOLUTIONS

	Amount	• Strength =	Pure salt
Acid	x	12%	**0.12x**

$0.12x$ gallon is pure acetic acid.

SECTION 2.7
Inequalities

87.
$$3x + 2 < 5$$
$$3x + 2 - 2 < 5 - 2$$
$$3x < 3$$
$$\frac{3x}{3} < \frac{3}{3}$$
$$x < 1$$
$$(-\infty, 1)$$

88.
$$-\frac{3}{4}x \geq -9$$
$$-\frac{4}{3}\left(-\frac{3}{4}x\right) \leq -\frac{4}{3}(-9)$$
$$x \leq 12$$
$$(-\infty, 12]$$

89.
$$\frac{3}{4} < \frac{d}{5} + \frac{1}{2}$$
$$20\left(\frac{3}{4}\right) < 20\left(\frac{d}{5}\right) + 20\left(\frac{1}{2}\right)$$
$$15 < 4d + 10$$
$$15 - 10 < 4d + 10 - 10$$
$$5 < 4d$$
$$\frac{5}{4} < \frac{4d}{4}$$
$$\frac{5}{4} < d$$
$$d > \frac{5}{4}$$
$$\left(\frac{5}{4}, \infty\right)$$

90.
$$5(3 - x) \leq 3(x - 3)$$
$$15 - 5x \leq 3x - 9$$
$$15 - 5x - 15 \leq 3x - 9 - 15$$
$$-5x \leq 3x - 24$$
$$-5x - 3x \leq 3x - 24 - 3x$$
$$-8x \leq -24$$
$$\frac{-8x}{-8} \geq \frac{-24}{-8}$$
$$x \geq 3$$
$$[3, \infty)$$

91.
$$8 < x + 2 < 13$$
$$8 - 2 < x + 2 - 2 < 13 - 2$$
$$6 < x < 11$$
$$(6, 11)$$

92.
$$0 \le 3 - 2x < 10$$
$$0 - 3 \le 3 - 2x - 3 < 10 - 3$$
$$-3 \le -2x < 7$$
$$\frac{-3}{-2} \ge \frac{-2x}{-2} > \frac{7}{-2}$$
$$\frac{3}{2} \ge x > -\frac{7}{2}$$
$$-\frac{7}{2} < x \le \frac{3}{2}$$
$$\left(-\frac{7}{2}, \frac{3}{2}\right]$$

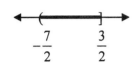

93. $2.40 \text{ g} \le w \le 2.53 \text{ g}$

94.   Let l = the maximum length

   w = 18 inches

   P = not greater than 132 inches

$$2l + 2w \le P$$
$$2l + 2(18) \le 132$$
$$2l + 36 \le 132$$
$$2l + 36 - 36 \le 132 - 36$$
$$2l \le 96$$
$$\frac{2l}{2} \le \frac{96}{2}$$
$$\boxed{l \le 48}$$

   Length is ≤ 48 inches.

CHAPTER 2 TEST

1. Is 3 a solution of $x + 3 = 4x - 6$?

$$x + 3 = 4x - 6$$
$$3 + 3 \overset{?}{=} 4(3) - 6$$
$$9 \overset{?}{=} 12 - 6$$
$$9 \neq 6, \text{ no}$$

2. Let x = the angle

Angle + 60 = 180
$$x + 60 = 180$$
$$x + 60 - 60 = 180 - 60$$
$$\boxed{x = 120}$$

The angle is 120°.

3. Let x = # of multiple birth in 1997
$7x$ = # of multiple birth in 2000
7,322 = total # of birth of in 2000

$$7x = 7,322$$
$$\frac{7x}{7} = \frac{7,322}{7}$$
$$\boxed{x = 1,046}$$

1,046 births in 1971.

4. Let x = selling price of house
$15\%x$ = down payment
$11,400 = down payment

$$0.15x = 11,400$$
$$\frac{0.15x}{0.15} = \frac{11,400}{0.15}$$
$$\boxed{x = 76,000}$$

Selling price of house is $76,000.

5. Bulls win-lost record 72-10.
Find total games played. 82
Find games won. 72
Find winning percentage. $\frac{72}{82}$

$$\frac{72}{82} = 0.8780$$
$$= 0.878$$

Dolphins win-lost record 14-0.
Find total games played. 14
Find games won. 14
Find winning percentage. $\frac{14}{14}$

$$\frac{14}{14} = 1.000$$

6. Starting temperature. 98.6°F
Ending temperature. 101.6°F
Find the difference. 3°F
Find percent increase.

$$\frac{3}{98.6} = 0.0304$$
$$= 0.03$$
$$= 3\%$$

7. $P = 2l + 2w$
$l = (x + 3)$ feet
$w = x$ feet

$$P = 2(x + 3) + 2x$$
$$P = 2x + 6 + 2x$$
$$\boxed{P = (4x + 6) \text{ feet}}$$

8. $2(x + 7) = 2x + 2(7)$
The distributive property.

Simplify each expression.

9. $5(-4x) = -20x$

10. $-8(-7t)(4) = -224t$

11. $\dfrac{4}{5}(15a+5)-16a = 12a+4-16a$

$\qquad\qquad\qquad\qquad = -4a+4$

12. $-1.1d^2 - 3.8d^2 - d^2 = -5.9d^2$

Solve each equation.

13. $5h+8-3h+h = 8$

$\qquad\qquad 3h+8 = 8$

$\qquad 3h+8-8 = 8-8$

$\qquad\qquad\quad 3h = 0$

$\qquad\qquad \dfrac{3h}{3} = \dfrac{0}{3}$

$\boxed{h = 0}$

14.

$\dfrac{4}{5}t = -4$

$\dfrac{5}{4}\left(\dfrac{4}{5}t\right) = \dfrac{5}{4}(-4)$

$\boxed{t = -5}$

15.

$\dfrac{11(b-1)}{5} = 3b-2$

$\dfrac{11b-11}{5} = 3b-2$

$5\left(\dfrac{11b-11}{5}\right) = 5(3b-2)$

$11b-11 = 15b-10$

$11b-11+11 = 15b-10+11$

$11b = 15b+1$

$11b-15b = 15b+1-15b$

$-4b = 1$

$\dfrac{-4b}{-4} = \dfrac{1}{-4}$

$\boxed{b = -\dfrac{1}{4}}$

16.

$0.8x+1.4 = 2.9+0.2x$

$0.8x+1.4-1.4 = 2.9+0.2x-1.4$

$0.8x = 1.5+0.2x$

$0.8x-0.2x = 1.5+0.2x-0.2x$

$0.6x = 1.5$

$\dfrac{0.6x}{0.6} = \dfrac{1.5}{0.6}$

$\boxed{x = 2.5}$

17.

$\dfrac{m}{2}-\dfrac{1}{3} = \dfrac{1}{4}$

$12\left(\dfrac{m}{2}-\dfrac{1}{3}\right) = 12\left(\dfrac{1}{4}\right)$

$6m-4 = 3$

$6m-4+4 = 3+4$

$6m = 7$

$\dfrac{6m}{6} = \dfrac{7}{6}$

$\boxed{m = \dfrac{7}{6}}$

18.

$23-5(x+10) = -12$

$23-5x-50 = -12$

$-5x-27 = -12$

$-5x-27+27 = -12+27$

$-5x = 15$

$\dfrac{-5x}{-5} = \dfrac{15}{-5}$

$\boxed{x = -3}$

19. $A = P+Prt$, for r

$A-P = P+Prt-P$

$A-P = Prt$

$\dfrac{A-P}{Pt} = \dfrac{Prt}{Pt}$

$\boxed{\dfrac{A-P}{Pt} = r}$

20.
Revenue = $445.

Cost = $295.

Profit = ?

Profit = Revenue - Cost

$$P = 445 - 295$$

$$\boxed{P = 150}$$

Profit is $150.

21.
$$F = 14^o$$

$$C = \frac{5}{9}(F - 32)$$

$$C = \frac{5}{9}(14 - 32)$$

$$C = \frac{5}{9}(-18)$$

$$\boxed{C = -10}$$

22. PETS

$$d = 10 \text{ in.}$$

$$r = 5 \text{ in.}$$

$$\text{tank} = \frac{3}{4} \text{ full}$$

$$V = \frac{4}{3}\pi r^3$$

$$V = \frac{3}{4}\left[\frac{4}{3}(3.14)(5 \text{ in.})^3\right]$$

$$V = (3.14)(125) \text{ in.}^3$$

$$V = 392.5$$

$$\boxed{V = 393}$$

Volume is 393 in³.

Answers may vary due to the approximation of π.

23. TRAVEL TIMES

	r	$\cdot\ t\ =$	d
slow car	55	t	**55t**
fast car	65	t	**65t**
Total miles			**72**

Let t = time in hr for each car

slow dis + fast dis = total dis

$$55t + 65t = 72$$

$$120t = 72$$

$$\frac{120t}{120} = \frac{72}{120}$$

$$\boxed{t = \frac{3}{5} \text{ hr}}$$

24. SALT SOLUTION

	Amount	$\cdot$ Strength =	Pure salt
Weak	x	2%	**0.02x**
Strong	30	10%	**0.10(30)**
Mixture	$x + 30$	8%	**0.08(x + 30)**

Let x = # of liters of 2% brine

weak salt + strong salt = mixture

$$0.02x + 0.10(30) = 0.08(x + 30)$$

$$0.02x + 3 = 0.08x + 2.4$$

$$0.02x + 3 - 2.4 = 0.08x + 2.4 - 2.4$$

$$0.02x + 0.6 = 0.08x$$

$$-0.02x + 0.6 - 0.02x = 0.08x - 0.02x$$

$$0.6 = 0.06x$$

$$\frac{0.6}{0.06} = \frac{0.06x}{0.06}$$

$$\boxed{10 = x}$$

10 liters of 2% brine needed.

25. GEOMETRY

Vertex angle (3rd angle)	44
1st base angle	x
2nd base angle	x
Total degrees	180

Let x = measure of both base angles

vertex $\angle$ + base $\angle$ + base $\angle$ = total

$$44 + x + x = 180$$
$$2x + 44 = 180$$
$$2x + 44 - 44 = 180 - 44$$
$$2x = 136$$
$$\frac{2x}{2} = \frac{136}{2}$$
$$\boxed{x = 68}$$

Each base angle is $68°$.

26. INVESTMENT PROBLEM

	P •	r •	$t =$	I
6.2%	P	0.08	1	0.08P
12%	$13,750 - P$	0.09	1	0.09(13,750 − P)
Total				$1,185

Let P = amount invested at 8%

8% int + 9% int = Total int

$$0.08P + 0.09(13,750 - P) = 1,185$$
$$0.08P + 1,237.50 - 0.09P = 1,185$$
$$-0.01P + 1,237.50 = 1,185$$
$$-0.01P + 1,237.50 - 1,237.50 = 1,185 - 1,237.50$$
$$-0.01P = -52.50$$
$$\frac{-0.01P}{-0.01} = \frac{-52.5}{-0.01}$$
$$\boxed{P = 5,250}$$

$5,250 invested at 8%

27.

$$-8x - 20 \le 4$$
$$-8x - 20 + 20 \le 4 + 20$$
$$-8x \le 24$$
$$\frac{-8x}{-8} \ge \frac{24}{-8}$$
$$x \ge -3$$
$$[-3, \infty)$$

28.

$$-4 \le 2(x+1) < 10$$
$$\frac{-4}{2} \le \frac{2(x+1)}{2} < \frac{10}{2}$$
$$-2 \le x + 1 < 5$$
$$-2 - 1 \le x + 1 - 1 < 5 - 1$$
$$-3 \le x < 4$$
$$[-3, 4)$$

29.

$$1.497 - 0.001 \le w \le 1.497 + 0.001$$
$$\boxed{1.496 \text{ in.} \le w \le 1.498 \text{ in.}}$$

30.

Solve:
$$2(y - 7) - 3y = -(y - 3) - 17$$
$$2y - 14 - 3y = -y + 3 - 17$$
$$-y - 14 = -y - 14$$
$$-y - 14 + y = -y - 14 + y$$
$$\boxed{-14 = -14}, \quad \text{true}$$

When the left side is equal to the right side, this means that any real number is a solution.

CHAPTERS 1-2
CUMULATIVE REVIEW EXERCISES

1. a) $4m - 3 + 2m$, expression

    b) $4m = 3 + 2m$, equation

2.

Weight (lb)	Cooking times (hr)
15	$\dfrac{15}{5} = 3$
20	$\dfrac{20}{5} = 4$
25	$\dfrac{25}{5} = 5$

3. Prime Factorization of 200.

    $2 \cdot 2 \cdot 2 \cdot 5 \cdot 5 = 2^3 \cdot 5^2$

4. $\dfrac{24}{36} = \dfrac{2}{3}$

5. $\dfrac{11}{21}\left(-\dfrac{14}{33}\right) = -\dfrac{2}{9}$

6. $\dfrac{3}{4} \div \dfrac{1}{8} = \dfrac{3}{4} \cdot \dfrac{8}{1} = 6$

    $6 - \dfrac{1}{8}$ cups of flour.

7. $\dfrac{4}{5} + \dfrac{2}{3} = \dfrac{12}{15} + \dfrac{10}{15}$

    $\phantom{\dfrac{4}{5} + \dfrac{2}{3}} = \dfrac{22}{15}$

    $\phantom{\dfrac{4}{5} + \dfrac{2}{3}} = 1\dfrac{7}{15}$

8. $42\dfrac{1}{8} - 29\dfrac{2}{3} = 42\dfrac{3}{24} - 29\dfrac{16}{24}$

    $\phantom{42\dfrac{1}{8} - 29\dfrac{2}{3}} = 41\dfrac{27}{24} - 29\dfrac{16}{24}$

    $\phantom{42\dfrac{1}{8} - 29\dfrac{2}{3}} = 12\dfrac{11}{24}$

9. Write as a decimal.

    $\dfrac{15}{16} = 0.9375$

10. $0.45(100) = 45$

11. a) $|-65| = 65$

     b) $-|-12| = -12$

12. $x \cdot 5 = 5x$

     The commutative property
     of multiplication.

13. 3

     natural number, whole number,
     integer, rational number, real number

14. -1.95

     rational number, real number

15. $\dfrac{17}{20}$

     rational number, real number

16. π

     irrational number, real number

17. a) $4 \cdot 4 \cdot 4 = 4^3$

     b) $\pi \cdot r \cdot r \cdot h = \pi r^2 h$

18. a) $-6 + (-12) + 8 = -18 + 8$
$$= -10$$
   b) $-15 - (-1) = -15 + 1$
$$= -14$$
   c) $2(-32) = -64$
   d) $\dfrac{0}{35} = 0$
   e) $\dfrac{-11}{11} = -1$

19. a) The sum of the width w and 12.
$$w + 12$$
   b) Four less than a number n.
$$n - 4$$

20. total days $= 4 + 8 + 3 + 2 + 0 + 5 + 4 + 6$
$$= 32$$
   number of people $= 8$
   average number of sick days $= \dfrac{32}{8}$
$$= 4$$

21.

x	$x^2 - 3$
-2	$(-2)^2 - 3$ $4 - 3$ 1
0	$(0)^2 - 3$ $0 - 3$ -3
3	$(3)^2 - 3$ $9 - 3$ 6

22. $l = \dfrac{2{,}000}{d^2}$

Answers may vary depending
on the variable chosen.

23. a) $A = lw$
$$A = (2 \text{ ft})(3 \text{ ft})$$
$$A = 32 \text{ ft}^2$$
   b) $A = \pi r^2$
$$A = (3.14)(0.625 \text{ ft})(0.625 \text{ ft})$$
$$A = 1.22$$
$$A = 1.2 \text{ ft}^2$$
   Answers will vary due to
   the approximation of π.
   c) Percent of area of red dot
   to area of the whole flag.
$$\dfrac{1.2}{6} = 0.2$$
$$= 20\%$$

24. 45 is 15% of what number?
$$45 = 0.15 \cdot x$$
$$\dfrac{45}{0.15} = \dfrac{0.15x}{0.15}$$
$$\boxed{300 = x}$$

25. Let $x = -5$, $y = 3$, and $z = 0$.
$$(3x - 2y)z = [(3 \cdot -5) - (2 \cdot 3)]0$$
$$= 0$$

26. Let $x = -5$, $y = 3$, and $z = 0$.
$$\dfrac{x - 3y + |z|}{2 - x} = \dfrac{-5 - 3(3) + |0|}{2 - (-5)}$$
$$= \dfrac{-5 - 9 + 0}{2 + 5}$$
$$= \dfrac{-14}{7}$$
$$= -2$$

27. Let $x = -5$, $y = 3$, and $z = 0$.
$$x^2 - y^2 + z^2 = (-5)^2 - (3)^2 + 0^2$$
$$= 25 - 9$$
$$= 16$$

28.   Let $x = -5$, $y = 3$, and $z = 0$.

$$\frac{x}{y} + \frac{y+2}{3-z} = \frac{-5}{3} + \frac{3+2}{3-0}$$
$$= \frac{-5}{3} + \frac{5}{3}$$
$$= 0$$

29. $-8(4d) = -32d$

30. $5(2x - 3y + 1) = 10x - 15y + 5$

31. $2x + 3x - x = 4x$

32. $3a^2 + 6a^2 - 17a^2 = -8a^2$

33. $\dfrac{2}{3}(15t - 30) + t - 30 = 10t - 20 + t - 30$
$$= 11t - 50$$

34. $5(t - 4) + 3t = 5t - 20 + 3t$
$$= 8t - 20$$

35. $(x + 3)\text{ft}$

36. $(x + 3)\text{ft} + (x - 3)\text{ft} + x\ \text{ft} = 3x\ \text{ft}$

37. $3x - 4 = 23$
$$3x - 4 + 4 = 23 + 4$$
$$3x = 27$$
$$\frac{3x}{3} = \frac{27}{3}$$
$$\boxed{x = 9}$$

38. $\dfrac{x}{5} + 3 = 7$
$$\frac{x}{5} + 3 - 3 = 7 - 3$$
$$\frac{x}{5} = 4$$
$$5\left(\frac{x}{5}\right) = 5(4)$$
$$\boxed{x = 20}$$

39. $-5p + 0.7 = 3.7$
$$-5p + 0.7 - 0.7 = 3.7 - 0.7$$
$$-5p = 3$$
$$\frac{-5p}{-5} = \frac{3}{-5}$$
$$p = -\frac{3}{5}$$
$$p = -0.6$$

40. $\dfrac{y - 4}{5} = 3 - y$
$$5\left(\frac{y - 4}{5}\right) = 5(3 - y)$$
$$y - 4 = 15 - 5y$$
$$y - 4 + 4 = 15 - 5y + 4$$
$$y = -5y + 19$$
$$y + 5y = -5y + 19 + 5y$$
$$6y = 19$$
$$\frac{6y}{6} = \frac{19}{6}$$
$$\boxed{y = \frac{19}{6}}$$

41. $-\dfrac{4}{5}x = 16$
$$5\left(-\frac{4}{5}x\right) = 5(16)$$
$$-4x = 80$$
$$\frac{-4x}{-4} = \frac{80}{-4}$$
$$\boxed{x = -20}$$

42.

$$\frac{1}{2}+\frac{x}{5}=\frac{3}{4}$$

$$20\left(\frac{1}{2}\right)+20\left(\frac{x}{5}\right)=20\left(\frac{3}{4}\right)$$

$$10+4x=15$$

$$10+4x-10=15-10$$

$$4x=5$$

$$\frac{4x}{4}=\frac{5}{4}$$

$$\boxed{x=\frac{5}{4}}$$

43.

$$-9(n+2)-2(n-3)=10$$

$$-9n-18-2n+6=10$$

$$-11n-12=10$$

$$-11n-12+12=10+12$$

$$-11n=22$$

$$\frac{-11n}{-11}=\frac{22}{-11}$$

$$\boxed{n=-2}$$

44.

$$\frac{2}{3}(r-2)=\frac{1}{6}(4r-1)+1$$

$$6\left(\frac{2}{3}(r-2)\right)=6\left(\frac{1}{6}(4r-1)\right)+6(1)$$

$$4(r-2)=(4r-1)+6$$

$$4r-8=4r-1+6$$

$$4r-8=4r+5$$

$$4r-8-4r=4r+5-4r$$

$$\boxed{-8=5}$$

no solution, contradiction

45.

width = 5 meters

length = 13 meters

$$A = lw$$

$$A = 5(13)$$

$$\boxed{A=65 \text{ m}^2}$$

46.

diameter = 12 cm

radius = 6 cm

height = 10 cm

$$V = \frac{1}{3}\pi r^2 h$$

$$V = \frac{1}{3}(3.14)(6 \text{ cm})(6 \text{ cm})(10 \text{ cm})$$

$$\boxed{V = 376.80 \text{ cm}^3}$$

Answers will vary due to the approximation of π.

47.

$$V = \frac{1}{3}\pi r^2 h$$

$$3(V) = 3\left(\frac{1}{3}\pi r^2 h\right)$$

$$3V = \pi r^2 h$$

$$\frac{3V}{\pi h}=\frac{\pi r^2 h}{\pi h}$$

$$\boxed{\frac{3V}{\pi h}=r^2}$$

48. WORK

distance = 3 ft.

force = 12.5 lb

work = ?

$$W = Fd$$

$$W = (12.5)(3)$$

$$\boxed{W = 37.5 \text{ft-lb}}$$

49. WORK

distance = 3 ft.

weight (Force) = ?

work = 28.35 ft-lb

$$W = Fd$$
$$28.35 = 3F$$
$$\frac{28.35}{3} = \frac{3F}{3}$$
$$\boxed{9.45 = F}$$

1 – gallon of paint weighs 9.45 lb.

50.

both base angles = x

vertex angle = 70°

base $\angle$ + base $\angle$ + vertex $\angle$ = total

$$x + x + 70 = 180$$
$$2x + 70 = 180$$
$$2x + 70 - 70 = 180 - 70$$
$$2x = 110$$
$$\frac{2x}{2} = \frac{110}{2}$$
$$\boxed{x = 55}$$

The base angles are 55°.

51. INVESTING

	P	•	r	•	t =	I
8%	P		0.08		1	**0.08P**
9%	10,000 - P		0.09		1	**0.09(10,000 – P)**
T						**$860**

Let P = amount invested at 8%

10,000 - P = amount invested at 9%

8% int + 9% int = Total int

$$0.08P + 0.09(10,000 - P) = 860$$
$$0.08P + 900 - 0.09P = 860$$
$$-0.01P + 900 = 860$$
$$-0.01P + 900 - 900 = 860 - 900$$
$$-0.01P = -40$$
$$\frac{-0.01P}{-0.01} = \frac{-40}{-0.01}$$
$$\boxed{P = 4,000}$$

$4,000 invested at 8%.

52. GOLDSMITH

	Amount	• Strength =	Pure gold
40%	x	40%	**0.40x**
10%	10	10%	**0.10(10)**
25%	$x + 10$	25%	**0.25(x + 10)**

Let x = oz of 40% gold

40% gold + 10% gold = 25% gold

$$0.40x + 0.10(10) = 0.25(x + 10)$$
$$0.40x + 1 = 0.25x + 2.5$$
$$0.40x + 1 - 1 = 0.25x + 2.5 - 1$$
$$0.40x = 0.25x + 1.5$$
$$0.40x - 0.25x = 0.25x + 1.5 - 0.25x$$
$$0.15x = 1.5$$
$$\frac{0.15x}{0.15} = \frac{1.5}{0.15}$$
$$\boxed{x = 10}$$

10 oz of 40% gold needed.

53.

$$x - 4 > -6$$
$$x - 4 + 4 > -6 + 4$$
$$x > -2$$
$$(-2, \infty)$$

-2

54.

$$-6x \geq -12$$
$$\frac{-6x}{-6} \leq \frac{-12}{-6}$$
$$x \leq 2$$
$$(-\infty, 2]$$

2

55.
$$8x + 4 \geq 5x + 1$$
$$8x + 4 - 4 \geq 5x + 1 - 4$$
$$8x \geq 5x - 3$$
$$8x - 5x \geq 5x - 3 - 5x$$
$$3x \geq -3$$
$$\frac{3x}{3} \geq \frac{-3}{3}$$
$$x \geq -1$$
$$[-1, \infty)$$

-1

56.
$$-1 \leq 2x + 1 < 5$$
$$-1 - 1 \leq 2x + 1 - 1 < 5 - 1$$
$$-2 \leq 2x < 4$$
$$\frac{-2}{2} \leq \frac{2x}{2} < \frac{4}{2}$$
$$-1 \leq x < 2$$
$$[-1, 2)$$

-1   2

SECTION 3.1

VOCABULARY

1. (-1, -5) is called an **ordered** pair.

3. A rectangular coordinate system is formed by two perpendicular number lines called the **x-axis** and the **y-axis**. The point where the axes cross is called the **origin**.

5. The point with coordinates (4, 2) can be graphed on a **rectangular** coordinate system.

CONCEPTS

7. a) To plot the point with coordinates (-5, 4), we start at the **origin** and move 5 units to the **left** and then move 4 units **up**.

   b) To plot the point with coordinates $\left(6, -\frac{3}{2}\right)$, we start at the **origin** and move 6 units to the **right** and then move $\frac{3}{2}$ units **down**.

9. a) Quadrants **I and II**.
   b) Quadrants **II and III**.
   c) Quadrant **II**.
   d) Quadrant **IV**.

11. 60 beats/min

13. 140 beats/min

15. 5 min and 50 min after starting

17. No difference

19. Europe sold about 17 million and the United States sold about 11 million so the difference is **about 6 million.**

NOTATION

21. (3, 5) is an ordered pair
    3(5) indicates multiplication
    5(3 +5) an expression containing grouping symbols

23. Yes, $\left(2.5, -\frac{7}{2}\right), \left(2\frac{1}{2}, -3.5\right), \left(2.5, -3\frac{1}{2}\right)$ all name the same point because all of the x-values are equal to each other and all the y-values are equal to each other.

25. The 4 of the ordered pair (4, 5) is associated with the **horizontal** axis.

PRACTICE

27.

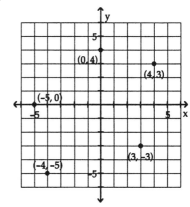

29.

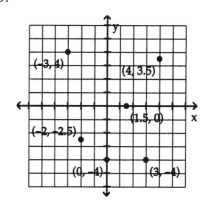

APPLICATIONS

31. CONSTRUCTION
    rivets: (-6, 0), (-2, 0), (2, 0), (6, 0)
    welds: (-4, 3), (0, 3), (4, 3)
    anchors: (-6, -3), (6, -3)

33. BATTLESHIP
    (E, 4), (F, 3), (G, 2)

35. MAPS

Rockford (5, B), Mount Carroll (1, C),
Harvard (7, A), intersection (5, E)

37. GEOMETRY

Fourth vertex is (2, 4)
Area = lw
Area = (3)(4)
Area = 12 sq. units

39. THE MILITARY

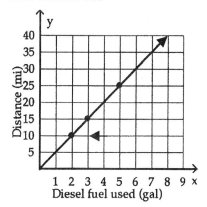

a) 35 mi
b) 4 gal
c) 32.5 mi

41. DEPRECIATION

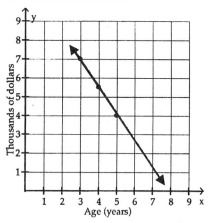

a) A 3-year old car is worth **$7,000**
b) Value of 7-year old car is **$1,000**
c) $2,500 worth after **6 yr**.

43. CENSUS

(1950, 152 million)
(1960, 179 million)
(1970, 202 million)
(1980, 227 million)
(1990, 252 million)
(2000, 277 million)

WRITING

45. Answers will vary.

47. Answers will vary.

REVIEW

49. Solve $AC = \dfrac{2}{3}h - T$ for h.

$$AC = \frac{2}{3}h - T$$

$$AC + T = \frac{2}{3}h - T + T$$

$$AC + T = \frac{2}{3}h$$

$$\left(\frac{3}{2}\right)(AC + T) = \left(\frac{3}{2}\right)\left(\frac{2}{3}h\right)$$

$$\boxed{\frac{3(AC + T)}{2} = h}$$

51. Evaluate: $\dfrac{-4(4+2) - 2^3}{|-12-20|} = \dfrac{-4(6) - 8}{|-32|}$

$$= \frac{-24 - 8}{|-32|}$$

$$= \frac{-32}{32}$$

$$\boxed{= -1}$$

CHALLENGE PROBLEMS

53. Sum of its coordinates is negative and product of its coordinates is positive. Such coordinates lie in **Quadrant III**.
Example: (-2, -5)

$$-2 + -5 = -10$$
$$(-2)(-5) = 10$$

SECTION 3.2

VOCABULARY

1. We say $y = 2x + 5$ is an equation in **two** variables, x and y.

3. Solutions of equations in two variables are often listed in a **table** of solutions.

5. The equation $y = 3x + 8$ is said to be **linear** because its graph is a line..

7. A linear equation in two variables has **infinitely** many solutions.

CONCEPTS

9. Consider the equation $y = -2x + 6$
    a) There **two** variables, x and y.
    b) Yes, (4, -2) satisfies the equation.
    c) No, (-3, 0) is not a solution.
    d) Infinitely many.

11. Every point on the graph represents an or-dered-pair **solution** of $y = -2x - 3$ and every ordered-pair solution is a **point** on the graph.

13.
    a) At least one point is in error. The 3 points should lie on a straight line. One of the points is not correct. Check the computations.
    b) The line is too short. Arrowheads are not drawn.

15. Solve each equation for y.
    a) $5y = 10x + 5$
        $$\frac{5y}{5} = \frac{10x}{5} + \frac{5}{5}$$

    $$\boxed{y = 2x + 1}$$

    b) $3y = -5x - 6$
        $$\frac{3y}{3} = \frac{-5x}{3} - \frac{6}{3}$$

    $$\boxed{y = -\frac{5}{3}x - 2}$$

    c) $-7y = -x + 21$
        $$\frac{-7y}{-7} = \frac{-x}{-7} + \frac{21}{-7}$$

    $$\boxed{y = \frac{1}{7}x - 3}$$

NOTATION

17. Verify (-2,6) is a solution of $y = -x + 4$

    $$y = -x + 4$$
    $$6 \overset{?}{=} -(\textbf{-2}) + 4$$
    $$6 \overset{?}{=} 2 + 4$$
    $$\textbf{6 = 6}$$
    (-2, 6) is a solution

19. Complete the labeling of the table of solutions and graph of $c = -a + 4$.

a	c	(a, c)
-1	5	(-1, 5)
0	4	(0, 4)
2	2	(2, 2)

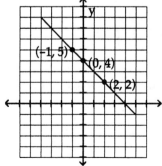

21. $y = 5x - 4$; (1,1)
    $1 \overset{?}{=} 5(1) - 4$
    $1 \overset{?}{=} 5 - 4$
    $1 = 1$
    yes

23. $y = \left(-\dfrac{3}{4}\right)x + 8$; (-8, 12)

    $12 \overset{?}{=} \left(-\dfrac{3}{4}\right)(-8) + 8$

    $12 \overset{?}{=} 6 + 8$
    $12 \overset{?}{=} 14$
      no

25. $7x - 2y = 3$; (2, 6)
    $7(2) - 2(6) \overset{?}{=} 3$
    $14 - 12 \overset{?}{=} 3$
    $2 \overset{?}{=} 3$
      no

27. $x + 12y = -12$; (0, -1)
    $0 + 12(-1) \overset{?}{=} -12$
    $0 - 12 \overset{?}{=} -12$
    $-12 \overset{?}{=} -12$
      yes

29. $y = -5x - 4$; (-3, ?)
    $y = -5(-3) - 4$
    $y = 15 - 4$
    $y = 11$
    (-3, **11**)

31. $4x - 5y = -4$; (?, 4)
    $4x - 5(4) = -4$
    $4x - 20 = -4$
    $4x - 20 + 20 = -4 + 20$
    $4x = 16$
    $x = 4$
    (**4**, 4)

33. $y = 2x - 4$

x	y	(x, y)
8	$y = 2x - 4$ $y = 2(8) - 4$ $y = 16 - 4$ **$y = 12$**	(8, **12**)
$y = 2x - 4$ $8 = 2x - 4$ $8+4 = 2x - 4+4$ **$6 = x$**	8	(**6**, 8)

35. $3x - y = -2$

x	y	(x, y)
-5	**-13**	(-5, **-13**)
-1	-1	(**-1**, -1)

37. $y = 2x - 3$

x	y	(x, y)
-2	$y = 2x - 3$ $y = 2(-2) - 3$ $y = -4 - 3$ $y = -7$	(-2, -7)
0	$y = 2x - 3$ $y = 2(0) - 3$ $y = 0 - 3$ $y = -3$	(0, -3)
2	$y = 2x - 3$ $y = 2(2) - 3$ $y = 4 - 3$ $y = 1$	(2, 1)

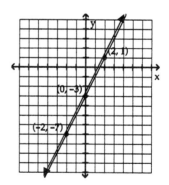

39. $y = 5x - 4$

x	y	(x, y)
-1	-9	(-1, -9)
0	-4	(0, -4)
1	1	(1, 1)

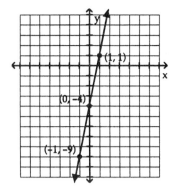

41. $y = x$

x	y	(x, y)
-3	-3	(-3, -3)
0	0	(0, 0)
4	4	(4, 4)

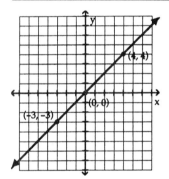

43. $y = -3x + 2$

x	y	(x, y)
-1	5	(-1, 5)
0	2	(0, 2)
1	-1	(1, -1)

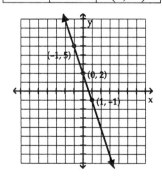

45. $y = -x - 1$

x	y	(x, y)
-2	1	(-2, 1)
0	-1	(0, -1)
2	-3	(2, -3)

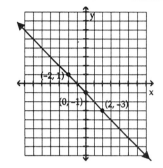

47. $y = \dfrac{x}{3}$

x	y	(x, y)
-3	-1	(-3, -1)
0	0	(0, 0)
3	1	(3, 1)

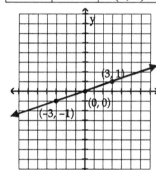

49. $y = -\dfrac{1}{2}x$

x	y	(x, y)
-2	1	(-2, 1)
0	0	(0, 0)
2	-1	(2, -1)

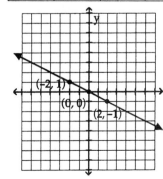

Section 3.2

51. $y = \dfrac{3}{8}x - 6$

x	y	(x, y)
-8	-9	(-8, -9)
0	-6	(0, -6)
8	-3	(8, -3)

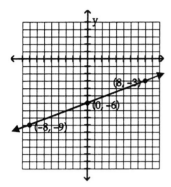

53. $y = \dfrac{2}{3}x - 2$

x	y	(x, y)
-3	-4	(-3, -4)
0	-2	(0, -2)
3	0	(3, 0)

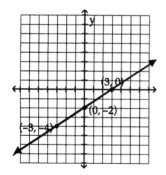

55. $7y = -2x$; divide both terms by 7

$y = -\dfrac{2}{7}x$

x	y	(x, y)
x	2	(-7, 2)
x	0	(0, 0)
x	-2	(7, -2)

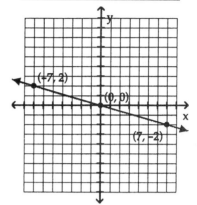

57. $3y = 12x + 15$; divide all 3 terms by 3

$y = 4x + 5$

x	y	(x, y)
-1	1	(-1, 1)
0	5	(0, 5)
1	9	(1, 9)

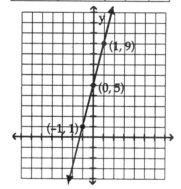

59. $5y = x + 20$; divide all 3 terms by 5

$$y = \frac{x}{5} + 4$$

x	y	(x, y)
-5	3	(-5, 3)
0	4	(0, 4)
5	5	(5, 5)

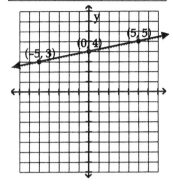

61. $y + 1 = 7x$; subtract 1 from both sides
$$y = 7x - 1$$

x	y	(x, y)
-1	-8	(-1, -8)
0	-1	(0, -1)
1	6	(1, 6)

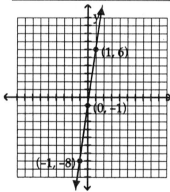

APPLICATIONS

63. BILLARDS

$y = 2x - 4$				$y = -2x + 12$		
x	y	(x, y)		x	y	(x, y)
1	-2	**(1, -2)**		4	4	**(4, 4)**
2	0	**(2, 0)**		6	0	**(6, 0)**
4	4	**(4, 4)**		8	-4	**(8, -4)**

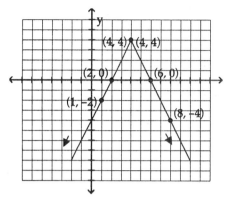

Did you notice that the angle the ball makes as it approaches the rail is the same angle when the ball rebounds off the rail and heads towards the hole?

65. HOUSEKEEPING

The problem wants us to estimate (not calculate) the amount of polish left in the bottle after **650** sprays. The numbers selected for "**n**" were used to calculate the values for "**A**" but you will notice that 650 was not a selected value to be used in the calculations.

Two different scales are used to keep the graph within reason. See the graph for the answer to the number of ounces of spray left after 650 sprays.

n = number of sprays (x-axis)
scale of 100
A = ounces remaining (y-axis)
scale of 1

$$A = -0.02n + 16$$

n	A	(n, A)
100	-0.02(100) + 16 -2 + 16 14	(100, 14)
200	12	(200, 12)
300	-0.02(300) + 16 -6 + 16 10	(300, 10)
500	6	(500, 6)

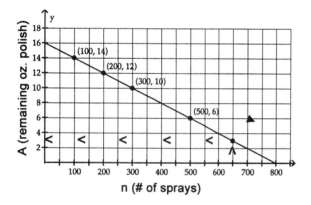

67. NFL TICKETS

This problem wants us to use a value for "***t***" that requires a little thought. The problem states the years (1990 – 2002) are to be considered for use with the formula, but it doesn't want us to use those large numbers. So we assign the following numbers to the stated years: 0 = 1990, 1 = 1991, 2 = 1992, and so on. Since we are asked to predict the price for the year 2010, we have to find the appropriate number for the year 2010. Based upon the previous pattern, we find **20 = 2010**.

t = number of years (x-axis)
p = price of ticket (y-axis)

$$p = \frac{9}{4}t + 23$$

t	p	(t, p)
0	\$23	(0, \$23)
4	$\left(\frac{9}{4}\right)(4) + 23$ $9 + 23$ \$32	(4, \$32)
8	$\left(\frac{9}{4}\right)(8) + 23$ $18 + 23$ \$41	(8, \$41)
12	$\left(\frac{9}{4}\right)(12) + 23$ $27 + 23$ \$50	(12, \$50)
16	\$59	(16, \$59)

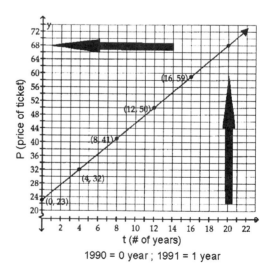

1990 = 0 year ; 1991 = 1 year

69. RAFFLES

You are to predict the number of raffle tickets to be sold for $6.

p = price of raffle ticket (x-axis)
n = # of tickets sold (y-axis)

$$n = -20p + 300$$

p	n	(p, n)
$1	-20(1) + 300 -20 + 300 280	($1, 280)
$3	-20(3) + 300 -60 + 300 240	($3, 240)
$5	-20(5) + 300 -100 + 300 200	($5, 200)

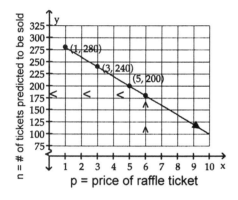

The predicted number of tickets is about 180.

WRITING

71. Answers will vary.

73. Answers will vary.

75. Answers will vary.

REVIEW

77. Simplify:
$-(-5 - 4c) = 5 + 4c$

79. $-2^2 + 2^2 = -(2)(2) + (2)(2)$
$\qquad = -4 + 4$
$\qquad = 0$

81. $1 + 2[-3 - 4(2 - 8^2)] = 1 + 2[-3 - 4(2 - 64)]$
$\qquad = 1 + 2[-3 - 4(-62)]$
$\qquad = 1 + 2[-3 + 248]$
$\qquad = 1 + 2[245]$
$\qquad = 1 + 490$
$\qquad = 491$

CHALLENGE PROBLEMS

83.

x	$y = x^2 + 1$	(x, y)
-3	$(-3)^2 + 1 = 9 + 1$ $= 10$	(-3, 10)
-2	$(-2)^2 + 1 = 4 + 1$ $= 5$	(-2, 5)
-1	$(-1)^2 + 1 = 1 + 1$ $= 2$	(-1, 2)
0	$0^2 + 1 = 0 + 1$ $= 1$	(0, 1)
1	$1^2 + 1 = 1 + 1$ $= 2$	(1, 2)
2	$2^2 + 1 = 4 + 1$ $= 5$	(2, 5)
3	$3^2 + 1 = 9 + 1$ $= 10$	(3, 10)

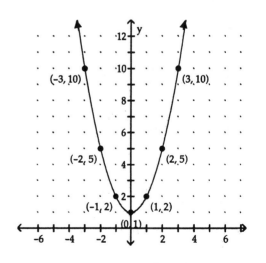

What did you notice about these values? This type of graph is called a **Parabola** and it has lines of symmetry.

85.

x	$y = \lvert x \rvert - 2$	(x, y)
-3	$\lvert -3 \rvert - 2$ $3 - 2$ 1	(-3, 1)
-2	$\lvert -2 \rvert - 2$ $2 - 2$ 0	(-2, 0)
-1	$\lvert -1 \rvert - 2$ $1 - 2$ -1	(-1, -1)
0	$\lvert 0 \rvert - 2$ $0 - 2$ -2	(0, -2)
1	$\lvert 1 \rvert - 2$ $1 - 2$ -1	(1, -1)
2	$\lvert 2 \rvert - 2$ $2 - 2$ 0	(2, 0)
3	$\lvert 3 \rvert - 2$ $3 - 2$ 1	(3, 1)

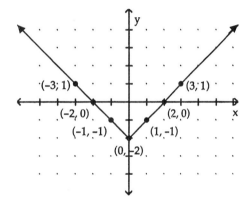

Do you notice anything interesting about the ordered pairs? This kind of graph is called an **Absolute Value** graph, and it has line of symmetry.

SECTION 3.3

VOCABULARY

1. We say $5x + 3y = 10$ is an equation in **two** variables, x and y.

3. The equation $2x - 3y = 7$ is written in **general or standard** form.

5. The y-intercept of a line is the point where the line **intersects or crosses** the y-axis.

CONCEPTS

7. The x-intercept is $(4, 0)$ and the y-intercept is $(0, 3)$.

9. The x-intercept is $(-5, 0)$ and the y-intercept is $(0, -4)$.

11. The y-intercept is $(0, 2)$. There is no x-intercept because the line does not cross the x-axis.

13. The x-intercept is $\left(-2\frac{1}{2}, 0\right)$ and the y-intercept is $\left(0, \frac{2}{3}\right)$.

15. a) The term $3x$ would be equal to 0.
    b) Let $x = 0$ for $3x + 2y = 6$:
$$3(0) + 2y = 6$$
$$0 + 2y = 6$$
$$2y = 6$$
$$y = 3$$
So the y-intercept is $(0, 3)$

17. A line may have at most **two** intercepts. A line must have at least **one** intercept.

19. a) ii
    b) iv
    c) vi
    d) i
    e) iii
    f) v

NOTATION

21. a) The point $(0, 6)$ lies on the y-axis.
    b) Yes, the point $(0, 0)$ lies on both the x-axis and y-axis.

23. The equation of the x-axis is $y = 0$. The equation of the y-axis is $x = 0$.

PRACTICE

25. $4x + 5y = 20$

y-intercept:	x-intercept:
If $x = 0$,	If $y = 0$
$4(0) + 5y = 20$	$4x + 5(0) = 20$
$5y = 20$	$4x = 20$
$y = 4$	$x = 5$

The y-intercept is $(0, 4)$, and the x-intercept is $(5, 0)$.

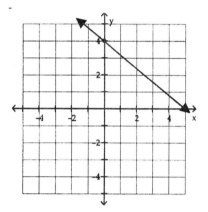

27. $x - y = -3$

y-intercept:	x-intercept:
If $x = 0$,	If $y = 0$
$(0) - y = -3$	$x - (0) = -3$
$-y = -3$	$x = -3$
$y = 3$	

The y-intercept is $(0, 3)$, and the x-intercept is $(-3, 0)$.

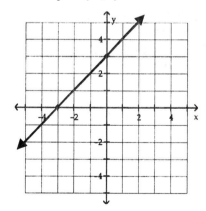

29. $5x + 15y = -15$

y-intercept: x-intercept:
If $x = 0$,   If $y = 0$
$5(0) + 15y = -15$ $5x + 15(0) = -15$
$15y = -15$ $5x = -15$
$y - -1$ $x = -3$

The y-intercept is (0, -1), and the x-intercept is (-3, 0).

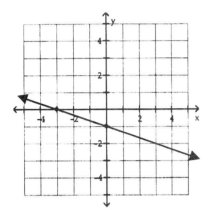

31. $x + 2y = -2$

y-intercept: x-intercept:
If $x = 0$,   If $y = 0$
$(0) + 2y = -2$ $x + 2(0) = -2$
$2y = -2$ $x = -2$
$y = -1$

The y-intercept is (0, -1), and the x-intercept is (-2, 0).

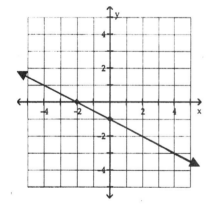

33. $4x - 3y = 12$

y-intercept: x-intercept:
If $x = 0$,   If $y = 0$
$4(0) - 3y = 12$ $4x - 3(0) = 12$
$-3y = 12$ $4x = 12$
$y - -4$ $x - 3$

The y-intercept is (0, -4), and the x-intercept is (3, 0).

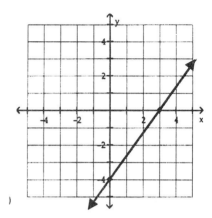

35. $3x + y = -3$

y-intercept: x-intercept:
If $x = 0$,   If $y = 0$
$3(0) + y = -3$ $3x + (0) = -3$
$y = -3$ $3x = -3$
             $x = -1$

The y-intercept is (0, -3), and the x-intercept is (-1, 0).

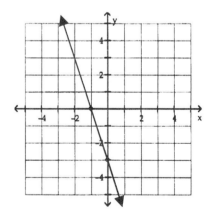

37. $9x - 4y = -9$

      y-intercept: *x*-intercept:

      If $x = 0$,   If $y = 0$

      $9(0) - 4y = -9$ $9x - 4(0) = -9$

      $-4y = -9$ $9x = -9$

      $y = \dfrac{9}{4}$ $x = -1$

The *y*-intercept is $(0, \dfrac{9}{4})$, and the *x*-intercept is $(-1, 0)$.

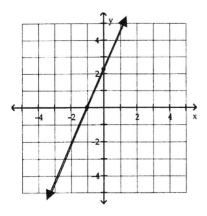

39. $8 = 3x + 4y$

      y-intercept: *x*-intercept:

      If $x = 0$,   If $y = 0$

      $8 = 3(0) + 4y$ $8 = 3x + 4(0)$

      $8 = 4y$ $8 = 3x$

      $2 = y$ $\dfrac{8}{3} = x$

The *y*-intercept is $(0, 2)$, and the *x*-intercept is $(\dfrac{8}{3}, 0)$.

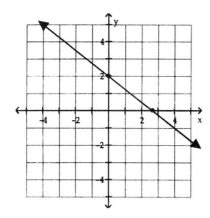

41. $4x - 2y = 6$

      y-intercept: *x*-intercept:

      If $x = 0$,   If $y = 0$

      $4(0) - 2y = 6$ $4x - 2(0) = 6$

      $-2y = 6$ $4x = 6$

      $y = -3$ $x = \dfrac{6}{4} = \dfrac{3}{2}$

The *y*-intercept is $(0, -3)$, and the *x*-intercept is $(\dfrac{3}{2}, 0)$.

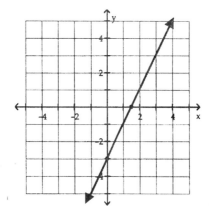

43. $3x - 4y = 11$

      y-intercept: *x*-intercept:

      If $x = 0$,   If $y = 0$

      $3(0) - 4y = 11$ $3x - 4(0) = 11$

      $-4y = 11$ $3x = 11$

      $y = -\dfrac{11}{4}$ $x = \dfrac{11}{3}$

The *y*-intercept is $(0, -\dfrac{11}{4})$, and the *x*-intercept is $(\dfrac{11}{3}, 0)$.

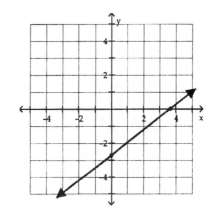

45. $9x + 3y = 10$

y-intercept: x-intercept:
If $x = 0$,   If $y = 0$
$9(0) + 3y = 10$ $9x + 3(0) = 10$
$3y = 10$ $9x = 10$
$y = \dfrac{10}{3}$ $x = \dfrac{10}{9}$

The y-intercept is $(0, \dfrac{10}{3})$, and the

x-intercept is $(\dfrac{10}{9}, 0)$.

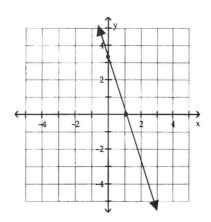

47. $3x = -15 - 5y$

y-intercept: x-intercept:
If $x = 0$,   If $y = 0$
$3(0) = -15 - 5y$ $3x = -15 - 5(0)$
$0 = -15 - 5y$ $3x = -15$
$5y = -15$ $x = -5$
$y = -3$

The y-intercept is $(0, -3)$, and the
x-intercept is $(-5, 0)$.

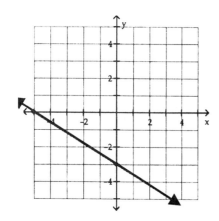

49. $-4x = 8 - 2y$

y-intercept: x-intercept:
If $x = 0$,   If $y = 0$
$-4(0) = 8 - 2y$ $-4x = 8 - 2(0)$
$0 = 8 - 2y$ $-4x = 8$
$2y = 8$ $x = -2$
$y = 4$

The y-intercept is $(0, 4)$, and the
x-intercept is $(-2, 0)$.

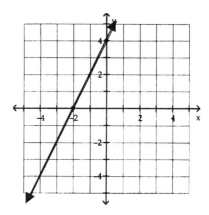

51. $7x = 4y - 12$

y-intercept: x-intercept:
If $x = 0$,   If $y = 0$
$7(0) = 4y - 12$ $7x = 4(0) - 12$
$0 = 4y - 12$ $7x = -12$
$-4y = -12$ $x = -\dfrac{12}{7}$
$y = 3$

The y-intercept is $(0, 3)$, and the

x-intercept is $(-\dfrac{12}{7}, 0)$.

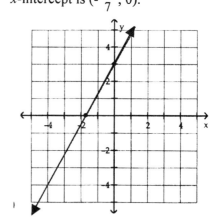

53. $y - 3x = -\frac{4}{3}$

y-intercept: x-intercept:
If $x = 0$,   If $y = 0$

$y - 3(0) = -\frac{4}{3}$ $0 - 3x = -\frac{4}{3}$

$y = -\frac{4}{3}$ $-3x = -\frac{4}{3}$

$x = -\frac{4}{3} \cdot -\frac{1}{3} = \frac{4}{9}$

The y-intercept is $(0, -\frac{4}{3})$, and the

x-intercept is $(\frac{4}{9}, 0)$.

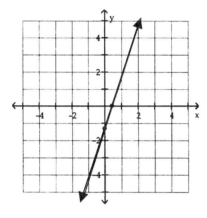

55. $y = 4$ is a horizontal line through $(0, 4)$.

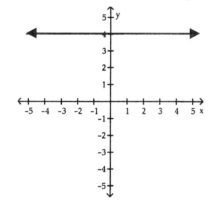

57. $x = -2$ is a vertical line through $(-2, 0)$

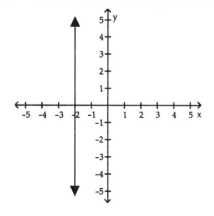

59. $y = -\frac{1}{2}$ is a horizontal line through $(0, -\frac{1}{2})$.

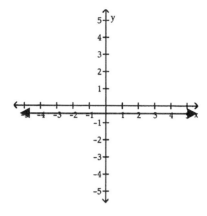

61. $x = \frac{4}{3}$ is a vertical line through $(\frac{4}{3}, 0)$

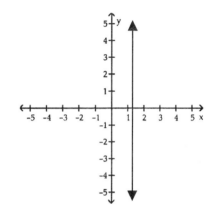

63. $y - 2 = 0$
$y - 2 + 2 = 0 + 2$
$y = 2$ is a horizontal line through $(0, 2)$.

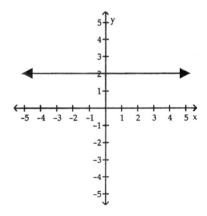

65. $-2x + 3 = 11$
   $-2x + 3 - 3 = 11 - 3$
   $-2x = 8$
   $\dfrac{-2x}{-2} = \dfrac{8}{-2}$
   $x = -4$ is a vertical line through (-4, 0)

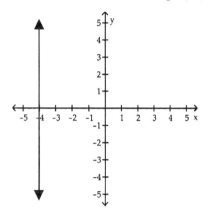

75. $\dfrac{\cancel{3} \cdot \cancel{5} \cdot \cancel{5}}{\cancel{3} \cdot \cancel{5} \cdot \cancel{5} \cdot 5} = \dfrac{1}{5}$

77. $2x - 6$

CHALLENGE PROBLEMS

79. Line $y = b$ will intersect line $x = a$ at the ordered pair (a, b).

APPLICATIONS

67. CHEMISTRY

   a) Absolute zero is approximately **-270° C**.
   b) When the temperature is absolute zero, the volume of the gas is **zero milliliters**.

69. LANDSCAPING

   a) The y-intercept is (0, 200) which indicates that if only shrubs are purchased, he can buy 200.
   b) The x-intercept is (100, 0) which indicates that if only trees are purchased, he can buy 100.

WRITING

71. Answers will vary.

73. Answers will vary.

SECTION 3.4

VOCABULARY

1. A **ratio** is the quotient of two numbers.

3. The **slope** of a line is defined to the ratio of the change in y to the change in x.

5. The rate of **change** of a linear relationship can be found by finding the slope of the graph of the line and attaching the proper units.

CONCEPTS

7. a) A line with *positive* slope **rises** from left to right.

   b). A line with *negative* slope **falls** from left to right.

9. A rise of 2 with a run of 15 would be written as a ratio of $\dfrac{2}{15}$.

11. a) The rise is –3.
    b) The run is 8.
    c) The slope would be the ratio of the rise over the run or $-\dfrac{3}{8}$.

13. a) Count the lines starting at point **A** and move up that vertical line. Stop when you reach the horizontal line that contains point **B**. Now count the lines until you reach point B. The rise is positive 3 and the run is positive 6. The slope is $\dfrac{3}{6}$, reduced to $\dfrac{1}{2}$.

    b) Count the lines starting at point **B** and move up that vertical line. Stop when you reach the horizontal line that contains point **C**. Now count the lines until you reach point C. The rise is positive 1 and the run is positive 2. The slope is $\dfrac{1}{2}$.

13.c) Count the lines starting at point **A** and move up that vertical line. Stop when you reach the horizontal line that contains point **C**. Now count the lines until you reach point C. The rise is positive 4 and the run is positive 8. The slope is $\dfrac{4}{8}$, reduced to $\dfrac{1}{2}$.

   d) Using any two points on the line will determine the same slope because all the points belong to the same line.

15. a) The fraction can be written simply as 0.
    b) When the slope contains a 0 in the denominator, we write it as *undefined*.

17. To convert a ratio to a percent, change the fraction to a decimal and then change the decimal to a percent.

    a) $\dfrac{2}{5} = 0.4 = 40\%$

    b) $\dfrac{3}{20} = 0.15 = 15\%$

NOTATION

19. The formula for finding the slope is given by:
$$m = \frac{y_2 - y_1}{x_2 - x_1}$$

21. y^2 means to raise y to the second power. y_2 means y sub two.

PRACTICE

23. $m = \dfrac{2}{3}$

25. $m = \dfrac{4}{3}$

27. $m = -2$

29. $m = 0$

31. $m = -\dfrac{1}{5}$

33. $(2,4)$ and $(1,3)$

(x_1, y_1) and (x_2, y_2)

$$m = \frac{y_2 - y_1}{x_2 - x_1}$$

$$= \frac{3-4}{1-2}$$

$$= \frac{-1}{-1}$$

$$= 1$$

35. $m = -3$

37. $m = \frac{5}{4}$

39. $(-3,5)$ and $(-5,6)$

(x_1, y_1) and (x_2, y_2)

$$m = \frac{y_2 - y_1}{x_2 - x_1}$$

$$= \frac{6-5}{-5-(-3)}$$

$$= \frac{1}{-5+3}$$

$$= \frac{1}{-2}$$

41. $m = \frac{3}{5}$

43. $(5,7)$ and $(-4,7)$

(x_1, y_1) and (x_2, y_2)

$$m = \frac{y_2 - y_1}{x_2 - x_1}$$

$$= \frac{7-7}{-4-5}$$

$$= \frac{0}{-9}$$

$$= 0$$

45. $(8,-4)$ and $(8,-3)$

(x_1, y_1) and (x_2, y_2)

$$m = \frac{y_2 - y_1}{x_2 - x_1}$$

$$= \frac{-3-(-4)}{8-8}$$

$$= \frac{1}{0}$$

$m = $ undefined

47. $m = -\frac{2}{3}$

49. $(-2.5, 1.75)$ and $(-0.5, -7.75)$

(x_1, y_1) and (x_2, y_2)

$$m = \frac{y_2 - y_1}{x_2 - x_1}$$

$$= \frac{-7.75 - 1.75}{-0.5 - (-2.5)}$$

$$= \frac{-9.5}{2}$$

$$= -4.75$$

51. $m = \frac{3}{4}$

53. $m = 0$

55. $y = -2$, This means that all the points of the line have y-coordinates of -2 and many different x-coordinates. Using the slope formula, means that the y values will always produce a **0** for an answer. And this means that one is dividing into **0**, which is a valid operation. **Slope is 0.**

57. $m = $ undefined

59. $m = 0$

61. $m = $ undefined

APPLICATIONS

63. DRAINAGE

First determine the rise of the slope by looking at the ruler. **3 inches**
The run is determined from the length of the 2-by-4. **10 feet**
Since the two unit measurements are different, one must convert one to the other. Since converting **feet into inch** stays a whole number, this may keep the math simpler. 10 feet times 12 inches = **120 inches**.
Compute the ratio.

$$m = \frac{\text{rise}}{\text{run}}$$

$$= \frac{3}{120}$$

$$= \frac{1}{40}$$

Determine that the slope rises from left to right thus indicating a **positive** slope.

$$\boxed{m = \frac{1}{40}}$$

65. GRADE OF A ROAD

$$m = \frac{\text{change in } y}{\text{change in } x}$$

$$= \frac{264}{5280}$$

$$= \frac{1}{20}$$

Change the fraction to a percent by dividing: $\dfrac{1}{20} = 0.05 = 5\%$

67. ENGINEERING

a) The change in y is 2 and the change in x is 16.

$$m = \frac{\text{change in } y}{\text{change in } x}$$

$$= \frac{2}{16}$$

$$= \frac{1}{8}$$

b) For each ramp, the change in y is 1 and the change in x is 12.

$$m = \frac{\text{change in } y}{\text{change in } x}$$

$$= \frac{1}{12}$$

c) Design #1 would be less expensive, but it would also be steeper. Design #2 is not as steep but would be more expensive.

69. IRRIGATION

Find an order pair for each of the end-points: (0, 8000) and (8, 1000). Then apply the following formula:

$$m = \frac{y_2 - y_1}{x_2 - x_1}$$

$$= \frac{1000 - 8000}{8 - 0}$$

$$= \frac{-7000}{8}$$

$$= -875$$

The rate of change would be -875 gallons per hour.

71. MILK PRODUCTION

Find an order pair for each of the end-points: (1993, 15,700) and (2002, 18,400). Then apply the following formula:

$$m = \frac{y_2 - y_1}{x_2 - x_1}$$
$$= \frac{18,400 - 15,700}{2002 - 1993}$$
$$= \frac{2,700}{9}$$
$$= 300$$

So, the rate of change would be 300 pounds per year.

WRITING

73. Answers will vary.

75. Answers will vary.

REVIEW

77. HALLOWEEN CANDY

Let x = the pounds of licorice.
Then $60 - x$ = the pounds of gumdrops.
The value of the licorice would be $1.90x and the value of the gumdrops would be $2.20(60 - x). The value of the mixture would be $2(60).

$$1.90x + 2.20(60 - x) = 2(60)$$
$$1.90x + 2.20(60) + 2.20(-x) = 2(60)$$
$$1.90x + 132 - 2.20x = 120$$
$$1.90x - 2.20x + 132 = 120$$
$$-0.3x + 132 = 120$$
$$-0.3x + 132 - 132 = 120 - 132$$
$$-0.3x = -12$$
$$\boxed{x = 40}$$

Thus, he would need 40 lbs of licorice and $60 - (40) = 20$ lbs of gumdrops.

CHALLENGE PROBLEMS

79. Find the slope from A to B:

$$m = \frac{y_2 - y_1}{x_2 - x_1}$$
$$= \frac{0 - (-10)}{20 - (-50)}$$
$$= \frac{10}{70}$$
$$= \frac{1}{7}$$

Find the slope from A to C:

$$m = \frac{y_2 - y_1}{x_2 - x_1}$$
$$= \frac{2 - (-10)}{34 - (-50)}$$
$$= \frac{12}{84}$$
$$= \frac{1}{7}$$

Find the slope from C to B:

$$m = \frac{y_2 - y_1}{x_2 - x_1}$$
$$= \frac{0 - 2}{20 - 34}$$
$$= \frac{-2}{-14}$$
$$= \frac{1}{7}$$

Since the slope is the same for each, all three points would lie on the same line.

SECTION 3.5

VOCABULARY

1. The equation $y = mx + b$ is called the **slope-intercept** form of the equation of a line.

3. **Parallel** lines do not intersect.

5. The numbers $\dfrac{5}{6}$ and $-\dfrac{6}{5}$ are called negative **reciprocals**. Their product is –1.

CONCEPTS

7.
   a) No. The equation needs to be solved y for the equation to be in slope-intercept form.
   b) No. The equation needs to be solved y for the equation to be in slope-intercept form.
   c) Yes
   d) No. The equation needs to be solved y for the equation to be in slope-intercept form.
   e) Yes
   f) Yes, the right side of the equation just doesn't have a visible y-intercept. The equation could be written as $y = 2x + 0$.

9. a)
$$5y = 10x + 20$$
$$\frac{5y}{5} = \frac{10x}{5} + \frac{20}{5}$$
$$\boxed{y} = \boxed{2x} + \boxed{4}$$
   b)
$$-2y = 6x - 12$$
$$\frac{-2y}{-2} = \frac{6x}{-2} - \frac{12}{-2}$$
$$\boxed{y} = \boxed{-3x}\ \boxed{+}\ 6$$

11. Select the point that the line goes through and also lies on the y-axis. The y-intercept is "**0**". Start from this point (0, 0), count **down** (negative) **5** places and then count **right** (positive) **4** places stopping at the second point on the line. Slope is $-\dfrac{5}{4}$.

Now put these values into the standard equation of $y = mx + b$.
$$y = -\frac{5}{4}x + 0$$
$$y = -\frac{5}{4}x$$

13.
   a) Two different lines with the same slope are **parallel**.
   b) If the slopes of two lines are negative reciprocals, the lines are **perpendicular**.
   c) The product of the slopes of perpendicular lines is **–1**.

15.
   a) Line 1's slope is 2. Line 2 is perpendicular to Line 1; therefore the slope of Line 2 is $-\dfrac{1}{2}$.
   b) Line 2's slope is $-\dfrac{1}{2}$. Line 3 is perpendicular to Line 1; therefore the slope of Line 3 is 2. Line 3 is also parallel to Line 1 thus making the two slopes the same.
   c) Line 3's slope is 2. Line 4 is perpendicular to Line 3; therefore the slope of Line 4 is $-\dfrac{1}{2}$. Line 4 is also parallel to Line 2 thus making the two slopes the same.
   d) Lines 1 and 2 have the same y-intercept because they both intersect the y-axis at the same place.

17.
$$2x + 5y = 15$$
$$2x + 5y - \boxed{2x} = \boxed{-2x} + 15$$
$$\boxed{5y} = -2x + 15$$
$$\frac{5y}{\boxed{5}} = \frac{-2x}{\boxed{5}} + \frac{15}{\boxed{5}}$$
$$y = -\frac{2}{5}x + \boxed{3}$$

The slope is $\boxed{-\dfrac{2}{5}}$ and the y-intercept is $\boxed{(0,3)}$.

19. a) $\dfrac{8x}{2} = 4x$

   b) $\dfrac{8x}{6} = \dfrac{4x}{3}$

   c) $\dfrac{-8x}{-8} = -x$

   d) $\dfrac{-16}{8} = -2$

21. **This is the right angle symbol**. It states that two lines are perpendicular to each other.

PRACTICE

23. $y = 4x + 2$
   slope = 4 ; y-intercept = (0, 2)

25. $y = -5x - 8$
   slope = -5 ; y-intercept = (0, -8)

27. $4x - 2 = y$
   slope = 4 ; y-intercept = (0, -2)
   It doesn't matter if the "y" is on the other side of the equal marks, the coefficient (4) of the "x" is still the slope and the constant (-2) is still the y-intercept.

29. $y = \dfrac{x}{4} - \dfrac{1}{2}$
   slope = $\dfrac{1}{4}$; y-intercept = $\left(0, -\dfrac{1}{2}\right)$
   Same reason as for #27.

31. $y = \dfrac{1}{2}x + 6$
   slope = $\dfrac{1}{2}$; y-intercept = (0, 6)
   Same reason as for #27.

33. $y = 6 - x$
   slope = -1 ; y-intercept = (0, 6)
   It doesn't matter if the "x" comes after the constant, the coefficient of the "x" (-1) is still the slope and the constant (6) is still the y-intercept.

35. $x + y = 8$
   Isolate the "y" and then find the slope and y-intercept.
   $$y = -x + 8$$
   slope = -1 ; y-intercept = (0, 8)

37. $6y = x - 6$
   Isolate the "y" and then find the slope and y-intercept.
   $$y = \frac{1}{6}x - 1$$
   slope = $\dfrac{1}{6}$; y-intercept = (0, -1)

39. $7y = -14x + 49$
   Solve for "y" and then find the slope and y-intercept.
   $$y = -2x + 7$$
   slope = -2 ; y-intercept = (0, 7)

41. $-4y = 6x - 4$
   Solve for "y" and then find the slope and y-intercept.
   $$y = -\frac{3}{2}x + 1$$
   slope = $-\dfrac{3}{2}$; y-intercept = (0, 1)

43. $2x + 3y = 6$
   $$3y = -2x + 6$$
   $$y = -\frac{2}{3}x + 2$$
   slope = $-\dfrac{2}{3}$; y-intercept = (0, 2)

45. $3x - 5y = 15$

$$-5y = -3x + 15$$

$$y = \frac{3}{5}x - 3$$

slope = $\frac{3}{5}$; y-intercept = (0, -3)

47. $-6x + 6y = -11$

$$6y = 6x - 11$$

$$y = x - \frac{11}{6}$$

slope = 1 ; y-intercept = $\left(0, -\frac{11}{6}\right)$

49. $y = x$

You can put a constant of "0" after the x and make the equation look like the previous equations.

$y = x + 0$

slope = 1 ; y-intercept = (0, 0)

51. $y = -5x$

$y = -5x + 0$

slope = -5 ; y-intercept = (0,0)

53. $y = -2$

You may put "$0x$" in the equation so it will look like the previous one.

$y = 0x - 2$

slope = 0 ; y-intercept = (0, -2)

This is a special line, which is horizontal and passes through the y-axis at -2.

55.

$$-5y - 2 = 0$$

$$-5y = 2$$

$$y = -\frac{2}{5}$$

$$y = 0x - \frac{2}{5}$$

slope = 0 ; y-intercept = $\left(0, -\frac{2}{5}\right)$

57. $y = 5x - 3$

Plot the y-intercept (0, -3) first. Use this as the starting point for placing the other two points by using the slope ($m = 5$).

Slope is rise over run. So $m = 5 = \frac{5}{1}$.

Start at (0, -3), go **up** the vertical axis 5 places and then go **right** 1 place, **stop**.

$m = \frac{5}{1} = \frac{-5}{-1}$ (Change both values to their opposites.) Start at (0, -3), go **down** the vertical axis 5 places and then go **left** 1 place, **stop**. This gives you three points in a straight line.

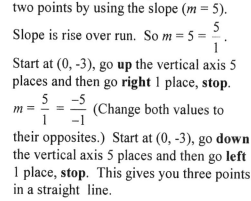

59. $y = \frac{1}{4}x - 2$

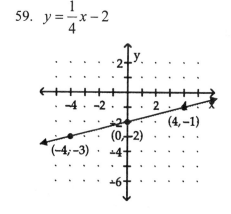

61. $y = -3x + 6$

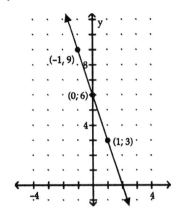

63. $y = -\dfrac{8}{3}x + 5$

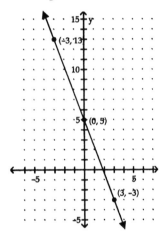

65. $m = 3$, y-intercept $= (0,3)$

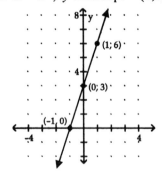

67. $m = -\dfrac{1}{2}$, y-intercept $= (0,2)$

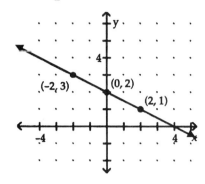

69. $m = -3$, y-intercept $= (0,0)$

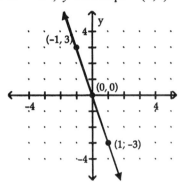

71. $m = -4$, y-intercept $= (0,-4)$
Remember to solve for "y" first.

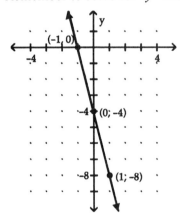

73. $m = -\dfrac{3}{4}$, y-intercept = $(0,4)$

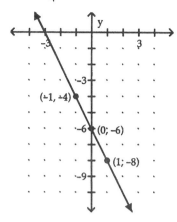

75. $m = 2$, y-intercept = $(0,-1)$

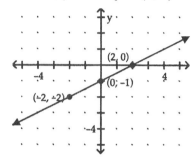

77. To determine if a pair of equations is parallel, perpendicular, or neither, just find each slope and then compare. If the slopes are equal, then the lines are parallel. If the slopes are negative reciprocals of each other or if their product is -1, then the lines are perpendicular.
$y = 6x + 8$, $m = 6$
$y = 6x$, $m = 6$

Slopes are equal; lines are parallel.

79. $y = x$, $m = 1$
$y = -x$, $m = -1$

Product is -1; lines are perpendicular.

81. $y = -2x - 9$, $m = -2$
$y = 2x - 9$, $m = 2$

Slopes are not equal nor is their product -1; therefore they are neither.

83.
$x - y = 12$, $m = 1$
$-2x + 2y = -23$, $m = 1$
Parallel.

85. $x = 9$, $m =$ undefined, vertical line
$y = 8$, $m = 0$, horizontal line

Perpendicular.

APPLICATIONS

87. PRODUCTION COSTS:
   a) basic fee = $5,000 (constant)
      extra cost = $2,000/hr (not constant)
      Total Prod Cost = extra cost + basic cost
      $c = \$2,000h + \$5,000$

   b) The cost for 8 hours of filming is found by replacing "h" with 8 and then calculating the results.
      $c = \$2,000h + \$5,000$
      $c = \$2,000(8) + \$5,000$
      $c = \$16,000 + \$5,000$
      $c = \$21,000$

89. CHEMISTRY
   Step 1: $-10°\,F$ (constant)
   Step 2: raise temperature $5°\,F$ every minute (not constant)
   Find final temperature F for t minutes

   $F = (5°\,F)(t \text{ minutes}) + (-10)$
   $F = 5t - 10$

91. EMPLOYMENT SERVICE
   Step 1: $500 fee (constant)
   Step 2: $20/month (not constant)
   Find actual cost "c" after m months

   $c = (\$20)(m \text{ months}) + \$500$
   $c = 20m + 500$

93. SEWING COSTS
   a) Step 1: basic fee = $20 (constant)
      Step 2: $5/letter (not constant)
      Find total "c" with x letters

   $$c = (\$5/\text{letter})(x \text{ letters}) + \$20$$
   $$\mathbf{c = 5x + 20}$$

   b) Solid line is the graph for part **"b"**.

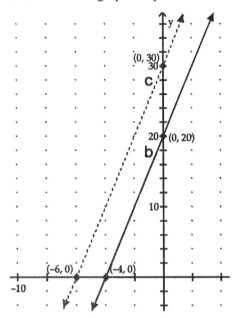

   c) Dotted line of the graph is part **"c"**.

95. PROFESSIONAL HOCKEY
   Step 1: average price = $33.50 (constant)
   Step 2: increase of $2.00/year
                        (not constant)
   Find average cost "c" for t years
   from 1995-2002

   $$c = (\$2/\text{year})(t \text{ years}) + \$33.50$$
   $$\mathbf{c = 2t + 33.50}$$

WRITING

97. Answers will vary.

REVIEW

99. CABLE TV
   Step 1: 186 foot cable cut into 4 pieces
   Step 2: Each piece is 3 ft longer than the
              previous one.

   Let x = 1st piece in feet
   $1^{ST} + 3$ = 2nd piece in feet
   $2^{nd} + 3$ = 3rd piece in feet
   $3^{rd} + 3$ = 4th piece in feet

   $$x + (1^{st} + 3) + (2^{nd} + 3) + (3^{rd} + 3) = 186$$

   $$x + (x + 3) + [(x + 3) + 3] + \{[(x +3) + 3] + 3\} = 186$$

   $$x + (x + 3) + (x + 6) + (x + 9) = 186$$

   $$4x + 18 = 186$$

   $$4x = 168$$

$x = 42$ ft
$x + 3 = 42 + 3 = 45$ ft
$x + 6 = 42 + 6 = 48$ ft
$x + 9 = 42 + 9 = 51$ ft

   1st piece is 42 feet.
   2nd piece is 45 feet.
   3rd piece is 48 feet.
   4th piece is 51 feet.

CHALLENGE PROBLEMS

101. If one would draw several lines such that
     they went through quadrants I, II, and IV,
     one would see that the **slope** of all the
     lines would **negative** and that all the
     y-intercepts would be **positive**.
     $m < 0; b > 0$

VOCABULARY

1. $y - y_1 = m(x - x_1)$ is called the **point-slope** form of the equation of a line.

CONCEPTS

3. a) $y - 4 = 2(x - 5)$; **point–slope form**
   b) $y = 2x + 15$; **slope-intercept form**

5. a) $x - (-6) = x + 6$
   b) $y - (-9) = y + 9$

7. Find the slope given (2, -3) and (-4, 12)

$$m = \frac{y_2 - y_1}{x_2 - x_1}$$
$$= \frac{12 - (-3)}{-4 - 2}$$
$$= \frac{15}{-6}$$
$$= -\frac{5}{2}$$

9. a) Does $y = 4x + 7$ have the required slope?
   Yes, it does. According to $y = mx + b$, $m = 4$.
   b) Does $y = 4x + 7$ pass through the point (-1, 3)? You can verify this by substituting the given points into the given equation. $3 \overset{?}{=} 4(-1) + 7$
   $$3 \overset{?}{=} -4 + 7$$
   $$3 \overset{?}{=} 3 \text{ , yes}$$

11. a) Bottom left point (-2, -3)
    b) $m = \dfrac{5}{6}$
    c) Second point (4, 2)

13. a) No, because it is only one point.
    b) No, because along with the slope one point needs to be known.
    c) Yes, because one point and one slope is all that is needed to write the equation of a line.
    d) Yes, because more than one point is given and the slope can be determined from the two points.

15. Choose the two points that lie on the line. (67, 170) and (79, 220)

NOTATION

17. Point-slope form: $y - y_1 = m(x - x_1)$

19. The original equation was in **point-slope** form. After solving for y, we obtain an equation in **slope-intercept** form.

PRACTICE

21. $y - 1 = 3(x - 2)$

23. $y + 1 = -\dfrac{4}{5}(x + 5)$

25.
$$y - 1 = \frac{1}{5}(x - 10)$$
$$y - 1 = \frac{1}{5}x - 2$$
$$y - 1 + 1 = \frac{1}{5}x - 2 + 1$$
$$y = \frac{1}{5}x - 1$$

27.
$$y - 8 = -5(x - (-9))$$
$$y - 8 = -5(x + 9)$$
$$y - 8 = -5x - 45$$
$$y - 8 + 8 = -5x - 45 + 8$$
$$y = -5x - 37$$

29.
$$y - (-4) = -\frac{4}{3}(x - 6)$$
$$y + 4 = -\frac{4}{3}x + 8$$
$$y + 4 - 4 = -\frac{4}{3}x + 8 - 4$$
$$y = -\frac{4}{3}x + 4$$

31.

$$y - (-6) = -\frac{11}{6}(x - 2)$$

$$y + 6 = -\frac{11}{6}x + \frac{11}{3}$$

$$y + 6 - 6 = -\frac{11}{6}x + \frac{11}{3} - 6$$

$$y = -\frac{11}{6}x + \frac{11}{3} - \frac{18}{3}$$

$$y = -\frac{11}{6}x - \frac{7}{3}$$

33.

$$y - 0 = -\frac{2}{3}(x - 3)$$

$$y = -\frac{2}{3}x + 2$$

35.

$$y - 4 = 8(x - 0)$$

$$y - 4 = 8x$$

$$y - 4 + 4 = 8x + 4$$

$$y = 8x + 4$$

37.

$$y - 0 = -3(x - 0)$$

$$y = -3x$$

39. Use the slope formula to find the slope.

$$m = \frac{y_2 - y_1}{x_2 - x_1}$$

$$= \frac{7 - 1}{1 - (-2)}$$

$$= \frac{6}{3}$$

$$= 2$$

Use the point $(1, 7)$ with the slope-point formula:

$$y - y_1 = m(x - x_1)$$

$$y - 7 = 2(x - 1)$$

$$y - 7 = 2x - 2$$

$$y - 7 + 7 = 2x - 2 + 7$$

$$y = 2x + 5$$

41. Use the slope formula to find the slope.

$$m = \frac{y_2 - y_1}{x_2 - x_1}$$

$$= \frac{3 - 0}{-4 - 2}$$

$$= \frac{3}{-6}$$

$$= -\frac{1}{2}$$

Use the point $(2, 0)$ with the slope-point formula:

$$y - y_1 = m(x - x_1)$$

$$y - 0 = -\frac{1}{2}(x - 2)$$

$$y = -\frac{1}{2}x + 1$$

43. Use the slope formula to find the slope.

$$m = \frac{y_2 - y_1}{x_2 - x_1}$$

$$= \frac{5 - 5}{7 - 5}$$

$$= \frac{0}{2}$$

$$= 0$$

Use the point $(5, 5)$ with the slope-point formula:

$$y - y_1 = m(x - x_1)$$

$$y - 5 = 0(x - 5)$$

$$y - 5 = 0$$

$$y - 5 + 5 = 0 + 5$$

$$y = 5$$

45. Use the slope formula to find the slope.

$$m = \frac{y_2 - y_1}{x_2 - x_1}$$

$$= \frac{1 - 0}{5 - (-5)}$$

$$= \frac{1}{10}$$

Use the point $(5, 1)$ with the slope-point formula:

$$y - y_1 = m(x - x_1)$$

$$y - 1 = \frac{1}{10}(x - 5)$$

$$y - 1 = \frac{1}{10}x - \frac{1}{2}$$

$$y - 1 + 1 = \frac{1}{10}x - \frac{1}{2} + 1$$

$$y = \frac{1}{10}x - \frac{1}{2} + \frac{2}{2}$$

$$y = \frac{1}{10}x + \frac{1}{2}$$

47. Use the slope formula to find the slope.

$$m = \frac{y_2 - y_1}{x_2 - x_1}$$

$$= \frac{17 - 2}{-8 - (-8)}$$

$$= \frac{15}{0}$$

$$= \text{undefined}$$

A horizontal line has an undefined slope.

The equation is $x = -8$

49. Use the slope formula to find the slope.

$$m = \frac{y_2 - y_1}{x_2 - x_1}$$

$$= \frac{-4 - 0}{-11 - 5}$$

$$= \frac{-4}{-16}$$

$$= \frac{1}{4}$$

Use the point $(5, 0)$ with the slope-point formula:

$$y - y_1 = m(x - x_1)$$

$$y - 0 = \frac{1}{4}(x - 5)$$

$$y = \frac{1}{4}x - \frac{5}{4}$$

51. $x = 4$

The equation of a **vertical line** has the form $\underline{x} = a$.

53. $y = 5$

The equation of a **horizontal line** has the form $y = b$.

55.

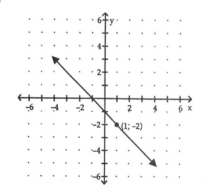

57.

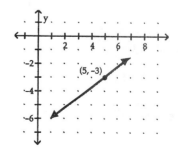

59.

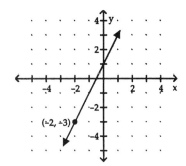

61.

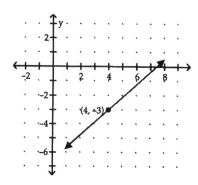

APPLICATIONS

63. ANATOMY

In this problem, we are given two different letters r and h to use other than x and y. One has to determine which letter correlates with which letter.

$r = x$ and $\qquad h = y$.

Find the slope which is given as 3.9 inches for each 1-inch for the radius.

$$m = \frac{\text{rise}}{\text{run}}$$
$$= \frac{3.9}{1}$$
$$= 3.9$$

Given: 64-inch tall woman is h_1.
Given: 9-inch-long radius bone is r_1.
Now use the point-slope formula.

$$h - h_1 = m(r - r_1)$$
$$h - 64 = 3.9(r - 9)$$
$$h - 64 = 3.9r - 35.1$$

$$h - 64 + 64 = 3.9r - 35.1 + 64$$

$$\boxed{h = 3.9r + 28.9}$$

65. POLE VAULTING

Part 1: Given $(5, 2)$ and $(10, 0)$
Find the slope:
$$m = \frac{y_2 - y_1}{x_2 - x_1}$$
$$= \frac{2 - 0}{5 - 10}$$
$$= \frac{2}{-5}$$
$$= -\frac{2}{5}$$

Now pick one point $(10, 0)$ and use
$$y - y_1 = m(x - x_1)$$
$$y - 0 = -\frac{2}{5}(x - 10)$$
$$\boxed{y = -\frac{2}{5}x + 4}$$

Part 2: Given $(9, 7)$ and $(10, 0)$
Find the slope:
$$m = \frac{y_2 - y_1}{x_2 - x_1}$$
$$= \frac{7 - 0}{9 - 10}$$
$$= \frac{7}{-1}$$
$$= -7$$

Now pick one point $(10, 0)$ and use
$$y - y_1 = m(x - x_1)$$
$$y - 0 = -7(x - 10)$$
$$\boxed{y = -7x + 70}$$

Part 3: Given $(10, 7)$ and $(10, 0)$
Find the slope:
$$m = \frac{y_2 - y_1}{x_2 - x_1}$$
$$= \frac{7 - 0}{10 - 10}$$
$$= \frac{7}{0}$$

Slope is undefined.

Undefined slope means the line is vertical and $x = a$ is the equation. $\boxed{x = 10}$

67. TOXIC CLEANUP

a) In this problem, we are given two different letters m and y (yard) to use for x and y. One has to determine which letter correlates with which letter, $m = x$ and $y = y$.

Find the slope which is given as two ordered pairs (3, 800) and (5, 720). The reason for the numbers 3 and 5 is because after "3 months" and then "2 months later".

$$m = \frac{y_2 - y_1}{m_2 - m_1}$$
$$= \frac{800 - 720}{3 - 5}$$
$$= \frac{80}{-2}$$
$$= -40$$

Now pick one point (3, 800) and use

$$y - y_1 = m(m - m_1)$$
$$y - 800 = -40(m - 3)$$
$$y - 800 = -40m + 120$$
$$y - 800 + 800 = -40m + 120 + 800$$
$$\boxed{y = -40m + 920}$$

b) To predict the number of cubic yards of waste on the site one year after the cleanup project began, let
$$m = 1 \text{ year} = 12 \text{ months.}$$

$$y = -40(12) + 920$$
$$= -480 + 920$$
$$= 440 \text{ yd}^3$$

69. TRAMPOLINES

In this problem, we are given two different letters r and l to use for x and y. One has to determine which letter correlates with which letter, $r = x$ and $l = y$.

Find the slope which is given as two ordered pairs (3, 19) and (7, 44).

$$m = \frac{l_2 - l_1}{r_2 - r_1}$$
$$= \frac{44 - 19}{7 - 3}$$
$$= \frac{25}{4}$$

Now pick one point (3, 19) and use
$$l - l_1 = m(r - r_1)$$

$$l - 19 = \frac{25}{4}(r - 3)$$
$$l - 19 = \frac{25}{4}r - \frac{75}{4}$$
$$l - 19 + 19 = \frac{25}{4}r - \frac{75}{4} + 19$$
$$l = \frac{25}{4}r - \frac{75}{4} + \frac{76}{4}$$
$$\boxed{l = \frac{25}{4}r - \frac{1}{4}}$$

Section 3.6

71. GOT MILK

a) Two points must be selected (yr, gal). (11, 26.5) and (19, 24.5) were selected because they both lie on the line. Now you must find the slope of the two selected ordered pairs.

$$m = \frac{y_2 - y_1}{x_2 - x_1}$$

$$= \frac{26.5 - 24.5}{11 - 19}$$

$$= \frac{2}{-8}$$

$$= -\frac{1}{4}$$

Now pick one point (11, 26.5) and use

$$y - y_1 = m(x - x_1)$$

$$y - 26.5 = -\frac{1}{4}(x - 11)$$

$$y - 26.5 = -\frac{1}{4}x + \frac{11}{4}$$

$$y - 26.5 + 26.5 = -\frac{1}{4}x + 2.75 + 26.5$$

$$y = -\frac{1}{4}x + 29.25$$

$$\boxed{y = -0.25 + 29.25}$$

or

$$\boxed{y = -\frac{1}{4}x + \frac{117}{4}}$$

b) Use $x = 40$ (the year 2015 correlates with 40) to calculate the number of gallons of milk to be consumed.

$$y = -0.25x + 29.25$$

$$y = -0.25 \cdot 40 + 29.25$$

$$y = -10 + 29.25$$

$$\boxed{y = 19.25 \text{ gal}}$$

WRITING

73. Answers will vary.

75. Answers will vary.

REVIEW

77. FRAMING PICTURES

Given: length is 5 inches more than 2(width)
perimeter is 112 inches

Implies: width = w

Formula: $l + w + l + w = P$

Let w = width in inches
$2w + 5$ = length in inches

$$l + w + l + w = P$$
$$(2w + 5) + w + (2w + 5) + w = 112$$
$$6w + 10 = 112$$
$$6w + 10 - 10 = 112 - 10$$
$$6w = 102$$
$$\frac{6}{6}w = \frac{102}{6}$$
$$\boxed{w = 17}$$

$$l = 2w + 5$$
$$l = 2 \cdot 17 + 5$$
$$l = 34 + 5$$
$$\boxed{l = 39}$$

Dimensions: 17 in. by 39 in.

CHALLENGE PROBLEMS

79. Given (2, 5)
Find equation of line parallel to $y = 4x - 7$.

First, find the slope from $y = 4x - 7$, $m = 4$.

The two lines are parallel and have the same slope, so use $m = 4$ and (2, 5) to write the equation of the line.

$$y - y_1 = m(x - x_1)$$
$$y - 5 = 4(x - 2)$$
$$y - 5 = 4x - 8$$
$$y - 5 + 5 = 4x - 8 + 5$$
$$\boxed{y = 4x - 3}$$

VOCABULARY

1. An **inequality** is a statement that contains one of the symbols $<$, $\leq$, $>$, or $\geq$

3. A **solution** of a linear inequality is an ordered pair of numbers that makes the inequality true.

5. In the graph, the line $2x - y = 4$ is the **boundary**.

7. When graphing a linear inequality, we determine which half-plane to shade by substituting the coordinates of a test **point** into the inequality.

CONCEPTS

9. a) False
   b) False
   c) True
   d) False

11. a)
$$x + 4y < -1$$
$$(3) + 4(1) \overset{?}{<} -1$$
$$3 + 4 \overset{?}{<} -1$$
$$7 \overset{?}{<} -1$$
No, (3, 1) is not a solution

   b)
$$x + 4y < -1$$
$$(-2) + 4(0) \overset{?}{<} -1$$
$$-2 + 0 \overset{?}{<} -1$$
$$-2 < -1$$
Yes, (-2, 0) is a solution.

   c)
$$x + 4y < -1$$
$$(-0.5) + 4(0.2) \overset{?}{<} -1$$
$$-0.5 + 0.8 \overset{?}{<} -1$$
$$0.3 \overset{?}{<} -1$$
No, (-0.5, 0.2) is not a solution.

11. d)
$$x + 4y < -1$$
$$(-2) + 4\left(\frac{1}{4}\right) \overset{?}{<} -1$$
$$-2 + 1 \overset{?}{<} -1$$
$$-1 \overset{?}{<} -1$$
No, $\left(-2, \frac{1}{4}\right)$ is not a solution.

13. a) **No**, $y > -x$ would not include the boundary line.
    b) **Yes**, $5x - 3y \leq -2$ would include the boundary line.

15. If the test point is false, you shade the half-plane opposite that in which the test point lies.

17. a) **Yes**, the graph is shaded over (2, 1).
    b) **No**, the graph is not shaded over (-2, -4).
    c) **No**, the graph is not shaded over (4, -2).
    d) **Yes**, the graph is shaded over (-3, 4).

19. a) The graph of $y = 5$ is a horizontal line.
    b) The graph of $x = -6$ is a vertical line.

NOTATION

21. a) $<$ is the symbol for less than
    b) $>$ is the symbol for greater than
    c) $\leq$ is the symbol for less than or equal to
    d) $\geq$ is the symbol for greater than or equal to

23. The symbols $\leq$ and $\geq$ are associated with a solid boundary line.

25. $x - y \geq -2$

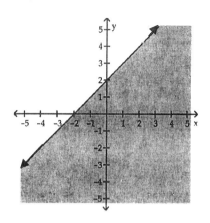

27. $y > 2x - 4$

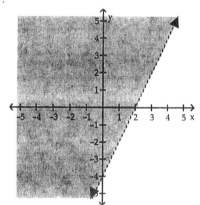

29. $x - 2y \geq 4$

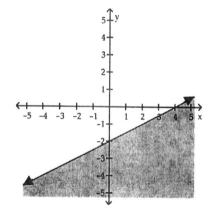

31. $y \leq 4x$

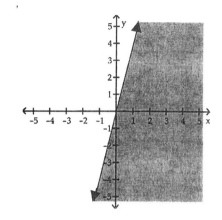

33. $x + y \geq 3$

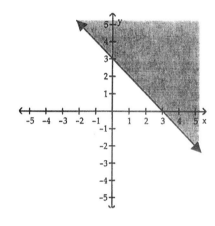

35. $3x - 4y > 12$

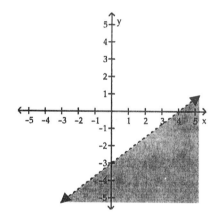

37. $2x + 3y \leq -12$

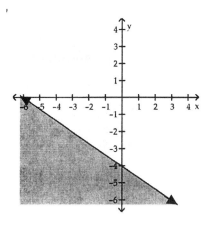

39. $y < 2x - 1$

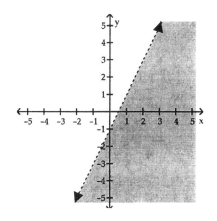

41. $y < -3x + 2$

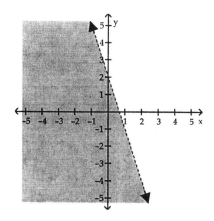

43. $y \geq -\frac{3}{2}x + 1$

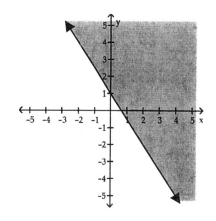

45. $x - 2y \geq 4$

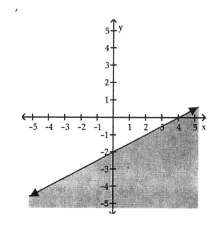

47. $2y - x < 8$

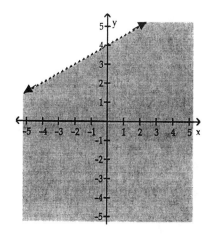

Section 3.7

49. $y + x < 0$

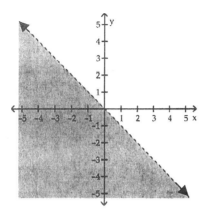

51. $y \geq 2x$

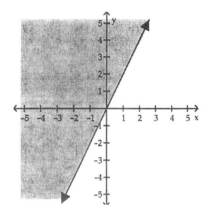

53. $y < -\dfrac{1}{2}x$

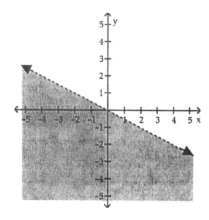

55. $x < 2$

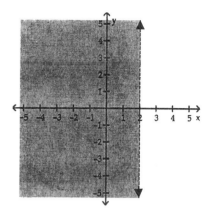

57. $y \leq 1$

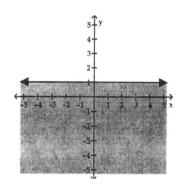

59. $x \leq 0$

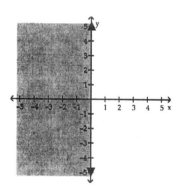

61. $7x - 2y < 21$

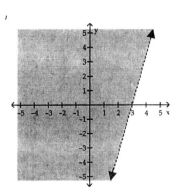

63. $2x - 3y \geq 4$

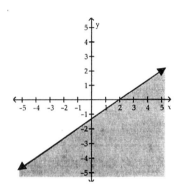

65. $5x + 3y < 0$

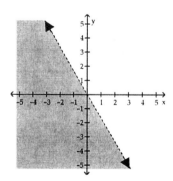

APPLICATIONS

67. ii—On all of the shaded areas, the sum exceeds 6.

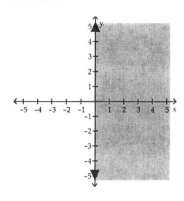

69. $3x + 4y \leq 120$

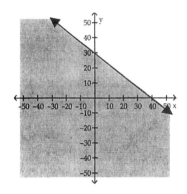

71. $100x + 88y \geq 4,400$

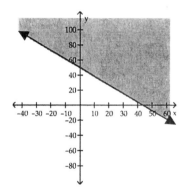

WRITING

73. Answers will vary.

75. Answers will vary.

REVIEW

77.
$$A = P + Prt$$
$$A - P = Prt$$
$$\frac{A - P}{Pr} = \frac{Prt}{Pr}$$
$$\frac{A - P}{Pr} = t$$

79.
$$40\left(\frac{3}{8}x - \frac{1}{4}\right) + 40\left(\frac{4}{5}\right) = 40\left(\frac{3}{8}x\right) + 40\left(-\frac{1}{4}\right) + 40\left(\frac{4}{5}\right)$$
$$= 15x - 10 + 32$$
$$= 15x + 22$$

CHALLENGE PROBLEMS

81. $3x - 2y \geq 6$

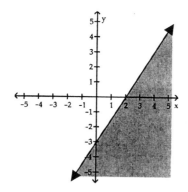

CHAPTER 3: KEY CONCEPTS

1. Use point-slope form of a line:
$$y - y_1 = m(x - x_1)$$
$$y - (-4) = -3(x - 0)$$
$$y + 4 = -3x + 0$$
$$y + 4 - 4 = -3x + 0 - 4$$
$$y = -3x - 4$$

2. Find the slope of the line containing the two points:
$$m = \frac{y_2 - y_1}{x_2 - x_1} = \frac{0 - 2}{-5 - 5} = \frac{-2}{-10} = \frac{1}{5}$$

Use point-slope form of a line:
$$y - y_1 = m(x - x_1)$$

$$y - 2 = \frac{1}{5}(x - 5)$$

$$y - 2 = \frac{1}{5}x - 1$$

$$y - 2 + 2 = \frac{1}{5}x - 1 + 2$$

$$y = \frac{1}{5}x + 1$$

3. Let $c = 160$.
$$T = \frac{1}{4}c + 40$$

$$T = \frac{1}{4}(160) + 40$$

$$T = 40 + 40$$

$$T = 80$$

So T = 80° F

4. $y = 209x + 2,660$

To find the expenditure in 2020,
let $x = 2020 - 1990 = 30$.
$$y = 209(30) + 2,660$$
$$y = 6,270 + 2,660$$
$$y = 8,930$$
The expenditure in 2020 would be $8,930.

5. $2x - 4y = 8$
Let $x = 0$ and solve for y.
$$2(0) - 4y = 8$$
$$0 - 4y = 8$$
$$-4y = 8$$
$$y = -2$$

Let $y = 0$ and solve for x.
$$2x - 4(0) = 8$$
$$2x - 0 = 8$$
$$2x = 8$$
$$x = 4$$
Let $x = -2$ and solve for y.
$$2(-2) - 4y = 8$$
$$-4 - 4y = 8$$
$$-4 - 4y + 4 = 8 + 4$$
$$-4y = 12$$
$$y = -3$$

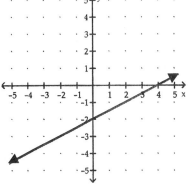

6. a) The y-intercept indicates that the cost of the press is $40,000 when new.
   b) The slope of the line is $-5,000$. This indicates that the value of the press decreased $5,000 each year.

7. a) Answers may vary. Possible answers are (-4, 2), (-3, 0), (-2, -2), (-1, -4) or (0, -6)
   b) Pick two points and use the formula for slope:
$$m = \frac{y_2 - y_1}{x_2 - x_1} = \frac{0 - 2}{-3 - (-4)} = \frac{-2}{1} = -2$$
   c) Use the point-slope form of a line with $m = -2$ and any point on the line such as (-3, 0).
$$y - y_1 = m(x - x_1)$$
$$y - 0 = -2(x - (-3))$$
$$y = -2(x + 3)$$
$$y = -2x - 6$$

8. Parallel lines have the same slope. A line parallel to the graphed line would be a vertical line going through the point (1, -1). The equation of a vertical line through that point would be $x = 1$.

CHAPTER 3 REVIEW

SECTION 3.1
Graphing Using the Rectangular Coordinate System

1.

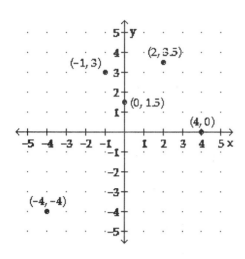

2. HAWAIIAN ISLANDS

   The coordinates of Oahu are approximately (158, 21.5).

3. Since both coordinates are negative, the point (-3, -4) lies in Quadrant III.

4. The coordinates of the origin are (0, 0).

5. GEOMETRY

   The coordinates of the fourth vertex would be (1, 4). Each side of the square is 6 units long.
   A = (length)(width)
   A = (6)(6)
   A = 36 square units

6. COLLEGE ENROLLMENTS

   a) The maximum enrollment was 2,500 students and occurred in Week 2.
   b) Two weeks before the semester began, the number of students enrolled was 1,000.
   c) The enrollment was 2,250 during week 1 and week 5.

SECTION 3.2
Graphing Linear Equations

7. Check to see if (-3, -2) is a solution of $y = 2x + 4$ by substituting –3 for x and –2 for y.

   $$-2 \stackrel{?}{=} 2(-3) + 4$$
   $$-2 \stackrel{?}{=} -6 + 4$$
   $$-2 \stackrel{?}{=} -2$$
   $$-2 = -2$$

   So, (-3, -2) is a solution of $y = 2x + 4$.

8. For $x = -2$, substitute –2 in for x and solve for y.
   $$3x + 2y = -18$$
   $$3(-2) + 2y = -18$$
   $$-6 + 2y = -18$$
   $$-6 + 6 + 2y = -18 + 6$$
   $$2y = -12$$
   $$y = -6$$

   The ordered pair would be (-2, -6).

   If $y = 3$, then substitute in 3 for y and solve the equation for x.
   $$3x + 2(3) = -18$$
   $$3x + 6 = -18$$
   $$3x + 6 - 6 = -18 - 6$$
   $$3x = -24$$
   $$x = -8$$

   The ordered pair would be (-8, 3).

9. The equations ii and v are not linear equations because the x term in both equations have term that contains an exponent higher than 1.

10. $y = 4x - 2$

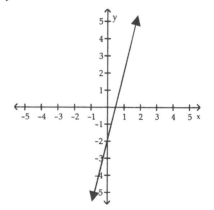

11. Solve the equation for y.
$$5y = -5x + 15$$
$$\frac{5y}{5} = \frac{-5x}{5} + \frac{15}{5}$$
$$y = -x + 3$$

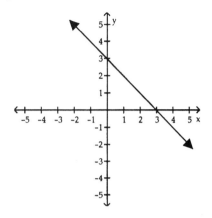

12. a) Since point A lies on the line, substituting the coordinates into the equation would result in a **true** statement.
    b) Since point B does not lie on the line, substituting the coordinates into the equation would result in a **false** statement.

13. BIRTHDAYS PARTIES

$$c = 8n + 50$$

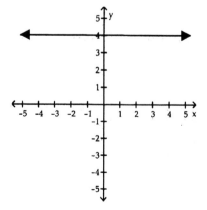

According to the graph, the cost for a party with 18 children attending would be approximately $195.

SECTION 3.3
More about Graphing Linear Equations

14. The x-intercept is the point the line crosses the x-axis, so the x-intercept is (-3, 0). The y-intercept is the point the line crosses the y-axis, so the x-intercept is (0, 2.5).

15. To find the x-intercept, let $y = 0$.
$$-4x + 2y = 8$$
$$-4x + 2(0) = 8$$
$$-4x + 0 = 8$$
$$-4x = 8$$
$$x = -2$$
So, the x-intercept is (-2, 0).

To find the y-intercept, let $x = 0$.
$$-4x + 2y = 8$$
$$-4(0) + 2y = 8$$
$$0 + 2y = 8$$
$$2y = 8$$
$$y = 4$$
So, the y-intercept is (0, 4).

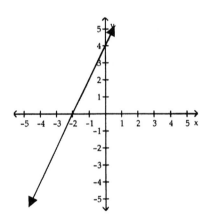

16. The graph $y = 4$ is a horizontal line through (0, 4).

17. The graph $x = -1$ is a vertical line through $(-1, 0)$.

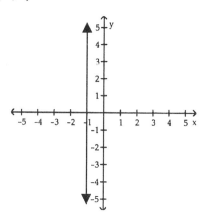

18. DEPRECIATION

The y-intercept is $(0, 25{,}000)$. This indicates that when the equipment was new, it was originally valued at \$25,000. The x-intercept is $(10, 0)$. This indicates that in 10 years, the equipment will have no value.

SECTION 3.4
The Slope of a Line

19. Use $(-4, -4)$ and $(4, -2)$ for the points (x_1, y_1) and (x_2, y_2) and substitute the values into the equation:

$$m = \frac{y_2 - y_1}{x_2 - x_1}$$
$$= \frac{-2 - (-4)}{4 - (-4)}$$
$$= \frac{-2 + 4}{4 + 4}$$
$$= \frac{2}{8}$$
$$= \frac{1}{4}$$

20. The slope of a horizontal line is always **zero**. You can pick any two points such as $(1, -3)$ and $(4, -3)$ to substitute into the equation:

$$m = \frac{y_2 - y_1}{x_2 - x_1}$$
$$= \frac{-3 - (-3)}{1 - 4}$$
$$= \frac{-3 + 3}{1 - 4}$$
$$= \frac{0}{-3}$$
$$= 0$$

21. Use $(2, -3)$ and $(4, -17)$ for the points (x_1, y_1) and (x_2, y_2) and substitute the values into the equation:

$$m = \frac{y_2 - y_1}{x_2 - x_1}$$
$$= \frac{-17 - (-3)}{4 - 2}$$
$$= \frac{-17 + 3}{4 - 2}$$
$$= \frac{-14}{2}$$
$$= -7$$

22. Use $(1, -4)$ and $(3, -7)$ for the points (x_1, y_1) and (x_2, y_2) and substitute the values into the equation:

$$m = \frac{y_2 - y_1}{x_2 - x_1}$$
$$= \frac{-7 - (-4)}{3 - 1}$$
$$= \frac{-7 + 4}{3 - 1}$$
$$= \frac{-3}{2}$$
$$= -\frac{3}{2}$$

23. CARPENTRY

$$m = \frac{\text{change in } y}{\text{change in } x}$$
$$= \frac{6}{16}$$
$$= \frac{3}{8}$$

24. HANDICAP ACCESSIBILITY

$$m = \frac{\text{change in } y}{\text{change in } x}$$

$$= \frac{2}{24}$$

$$= \frac{1}{12}$$

$$\approx 0.083$$

$$\approx 8.3\%$$

25. TOURISM

a) The y values would be the number of visitors and the x values would be the years.

$$m = \frac{\text{change in } y}{\text{change in } x}$$

$$= \frac{44.8 - 47.3}{1994 - 1992}$$

$$= \frac{-2.5}{2}$$

$$= -1.25$$

The rate of change was −1.25 million people per year.

b) The y values would be the number of visitors and the x values would be the years.

$$m = \frac{\text{change in } y}{\text{change in } x}$$

$$= \frac{34.1 - 26.0}{1988 - 1986}$$

$$= \frac{8.1}{2}$$

$$= 4.05$$

The rate of change was 4.05 million people per year.

SECTION 3.5
Slope-Intercept Form

26. Use the formula $y = mx + b$.
In $y = \frac{3}{4}x - 2$, $m = \frac{3}{4}$ and $b = -2$. So, the slope is $\frac{3}{4}$ and the y-intercept is (0, -2).

27. Use the formula $y = mx + b$.
In $y = -4x$, $m = -4$ and $b = 0$. So, the slope is -4 and the y-intercept is (0, 0).

28. Use the formula $y = mx + b$. Rewrite $\frac{x}{8}$ as $\frac{1}{8}x$. Then the equation is $y = \frac{1}{8}x + 10$, so $m = \frac{1}{8}$ and $b = 10$. So, the slope is $\frac{1}{8}$ and the y-intercept is (0, 10).

29. Solve the equation for y.
$$7x + 5y = -21$$
$$7x - 7x + 5y = -7x - 21$$
$$5y = -7x - 21$$
$$\frac{5y}{5} = \frac{-7x}{5} - \frac{21}{5}$$
$$y = -\frac{7}{5}x - \frac{21}{5}$$
$$m = -\frac{7}{5} \text{ and } b = -\frac{21}{5}$$

The slope is $-\frac{7}{5}$, and the y-intercept is $(0, -\frac{21}{5})$.

30. Let $m = -6$ and $b = 4$ and substitute the values into the equation:
$$y = mx + b$$
$$y = -6x + 4$$

31. Solve the equation for y.
$$9x - 3y = 15$$
$$9x - 9x - 3y = -9x + 15$$
$$-3y = -9x + 15$$
$$\frac{-3y}{-3} = \frac{-9x}{-3} + \frac{15}{-3}$$
$$y = 3x - 5$$

Since $m = 3$ and $b = -5$, the slope is 3 and the y-intercept is (0, -5).

32. COPIERS

a) 300 is the slope (rate of change) and 75,000 is the y-intercept. The equation would be
$$c = 300w + 75,000.$$

b) Let $w = 52$ and substitute it into the equation $c = 300w + 75,000$.
$$c = 300(\mathbf{52}) + 75,000$$
$$c = 15,600 + 75,000$$
$$c = 90,600$$
In 1 year (52 weeks), the number of copies made would be 90,600.

33. a) Both lines have a slope of $-\dfrac{2}{3}$. Since the slope is identical for both lines, the lines must be **parallel**.

   b) Solve the first equation for y to find the slope.

$$x + 5y = -10$$
$$x + 5y - x = -x - 10$$
$$5y = -x - 10$$
$$\frac{5y}{5} = \frac{-x}{5} - \frac{10}{5}$$
$$y = -\frac{1}{5}x - 2$$

The slope of the first equation is $-\dfrac{1}{5}$ and the slope of the second equation is 5. Since $-\dfrac{1}{5}$ and 5 are negative reciprocals, the lines must be **perpendicular**.

SECTION 3.6
Point-Slope Form

34. $m = 3$ and $(x_1, y_1) = (1, 5)$
   Use point-slope form of the equation:
$$y - y_1 = m(x - x_1)$$
$$y - \mathbf{5} = \mathbf{3}(x - \mathbf{1})$$
$$y - 5 = 3(x) + 3(-1)$$
$$y - 5 = 3x - 3$$
$$y - 5 + 5 = 3x - 3 + 5$$
$$y = 3x + 2$$

35. $m = -\dfrac{1}{2}$ and $(x_1, y_1) = (-4, -1)$
   Use point-slope form of the equation:
$$y - y_1 = m(x - x_1)$$
$$y - (\mathbf{-1}) = -\frac{1}{2}(x - (\mathbf{-4}))$$
$$y + 1 = -\frac{1}{2}(x + 4)$$
$$y + 1 = -\frac{1}{2}(x) - \frac{1}{2}(4)$$
$$y + 1 = -\frac{1}{2}x - 2$$
$$y + 1 - 1 = -\frac{1}{2}x - 2 - 1$$
$$y = -\frac{1}{2}x - 3$$

36. Find the slope of the line containing the two points $(3, 7)$ and $(-6, 1)$:
$$m = \frac{y_2 - y_1}{x_2 - x_1}$$
$$= \frac{1 - 7}{-6 - 3}$$
$$= \frac{-6}{-9}$$
$$= \frac{2}{3}$$

Use the slope of ⅔ and one of the points to substitute into the point-slope form of a line.
Let $m = ⅔$ and $(x_1, y_1) = (3, 7)$.
$$y - y_1 = m(x - x_1)$$
$$y - 7 = \frac{2}{3}(x - 3)$$
$$y - 7 = \frac{2}{3}(x) + \frac{2}{3}(-3)$$
$$y - 7 = \frac{2}{3}x - 2$$
$$y - 7 + 7 = \frac{2}{3}x - 2 + 7$$
$$y = \frac{2}{3}x + 5$$

37. The slope of a horizontal line is always zero. Use the point-slope formula with $m = 0$ and $(x_1, y_1) = (6, -8)$.
$$y - y_1 = m(x - x_1)$$
$$y - (-8) = 0(x - 6)$$
$$y + 8 = 0(x) + 0(-6)$$
$$y + 8 = 0x - 0$$
$$y + 8 - 8 = 0x - 0 - 8$$
$$y = 0x - 8$$
$$y = -8$$

38. CAR REGISTRATION

Let the x-values be the number of years old and the y-values be the amount of the registration fee. As an ordered pair, it would be (2, 380) and (4, 310). First find the rate of change, or slope.

$$m = \frac{y_2 - y_1}{x_2 - x_1}$$

$$= \frac{310 - 380}{4 - 2}$$

$$= \frac{-70}{2}$$

$$= -35$$

Use $m = -35$ and the point (2, 380) to substitute the values into the point-slope formula.

$$y - y_1 = m(x - x_1)$$

$$y - \mathbf{380} = -35(x - \mathbf{2})$$

$$y - 380 = -35(x) + -35(-2)$$

$$y - 380 = -35x + 70$$

$$y - 380 + 380 = -35x + 70 + 380$$

$$y = -35x + 450$$

SECTION 3.9
Graphing Linear Inequalities

39. a) $2x - y \le -4$

$2(0) - \mathbf{5} \stackrel{?}{=} -4$

$0 - 5 \stackrel{?}{=} -4$

$-5 \le -4$

So yes, (0, 5) is a solution.

b) $2x - y \le -4$

$2(\mathbf{2}) - \mathbf{8} \stackrel{?}{=} -4$

$4 - 8 \stackrel{?}{=} -4$

$-4 \le -4$

So yes, (2, 8) is a solution.

c) $2x - y \le -4$

$2(\mathbf{-3}) - (\mathbf{-2}) \stackrel{?}{=} -4$

$-6 + 2 \stackrel{?}{=} -4$

$-4 \le -4$

So yes, (-3, -2) is a solution.

d) $2x - y \le -4$

$2(-½) - (\mathbf{-5}) \stackrel{?}{=} -4$

$-1 + 5 \stackrel{?}{=} -4$

$4 \ge -4$

No, (-½, -5) is not a solution.

40. Solve the equation for y:

$x - y < 5$

$-y < -x + 5$

$\underline{-y} < \underline{-x} + \underline{5}$

$-1 \quad -1 \quad -1$

$y > x - 5$

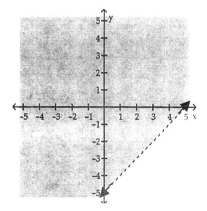

41. Solve the equation for y.

$2x - 3y \ge 6$

$-3y \ge -2x + 6$

$\frac{-3y}{-3} \ge \frac{-2x}{-3} + \frac{6}{-3}$

$y \le \frac{2}{3}x - 2$

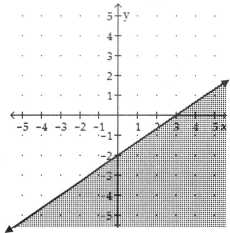

42. $y \leq -2x$

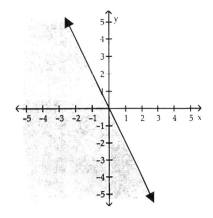

43. $y < -4$

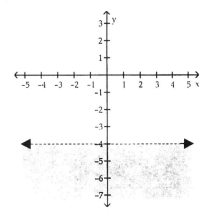

44. a) true
    b) false
    c) false

45. WORK SCHEDULES
    Answers may vary.
    Possible answers are (2, 4), (5, 3), and
    (6, 2).

46. An equation contains an equal sign (=).
    An inequality contains an inequality
    symbol such as $<, >, \leq,$ or $\geq$.

CHAPTER 3 TEST

1. 10

2. 60

3. There were 30 dogs in the kennel **1 day before** the holiday began and on the **3rd day** of the holiday.

4. The *y*-intercept shows there were **50** dogs in the kennel when the holiday began.

5.

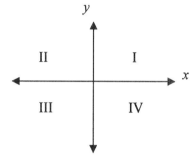

6. Substitute in 2 for *x* and solve for *y*:
   $x + 4y = 6$
   $\mathbf{2} + 4y = 6$
   $2 - 2 + 4y = 6 - 2$
   $4y = 4$
   $y = 1$
   So the ordered pair would be (2, 1).

   Substitute in 3 for *y* and solve for *x*:
   $x + 4y = 6$
   $x + 4(\mathbf{3}) = 6$
   $x + 12 = 6$
   $x + 12 - 12 = 6 - 12$
   $x = -6$
   So the ordered pair would be (-6, 3).

7. Substitute $x = -3$ and $y = -4$ into the equation:
   $3x - 4y = 7$
   $3(\mathbf{-3}) - 4(\mathbf{-4}) \overset{?}{=} 7$
   $-9 + 16 \overset{?}{=} 7$
   $7 = 7$
   Yes, (-3, -4) is a solution.

8. a) The result would be **false** because the point does not lie on the line.
   b) The result would be **true** because the point does lie on the line.

9. $y = \dfrac{1}{3}x$

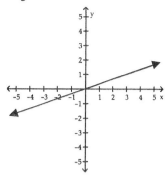

10. Solve the equation for *y*.
    $8x + 4y = -24$
    $8x + 4y - 8x = -24 - 8x$
    $4y = -8x - 24$
    $\dfrac{4y}{4} = \dfrac{-8x}{4} - \dfrac{24}{4}$
    $y = -2x - 6$

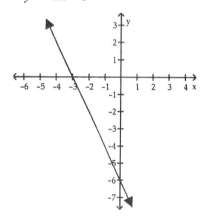

11. To find the *x*-intercept, let $y = 0$.
    $2x - 3y = 6$
    $2x - 3(\mathbf{0}) = 6$
    $2x - 0 = 6$
    $2x = 6$
    $\dfrac{2x}{2} = \dfrac{6}{2}$
    $x = 3$
    The *x*-intercept would be (3, 0).

    To find the *y*-intercept, let $x = 0$.
    $2x - 3y = 6$
    $2(\mathbf{0}) - 3y = 6$
    $0 - 3y = 6$
    $-3y = 6$
    $\dfrac{-3y}{-3} = \dfrac{6}{-3}$
    $y = -2$
    The *y*-intercept would be (0, -2).

12. Pick two points such as (4, 4) and (-3, -4) and use the formula for slope.

$$m = \frac{y_2 - y_1}{x_2 - x_1}$$

$$= \frac{-4 - 4}{-3 - 4}$$

$$= \frac{-8}{-7}$$

$$= \frac{8}{7}$$

13. For (-1, 3) and (3, -1):

$$m = \frac{y_2 - y_1}{x_2 - x_1}$$

$$= \frac{-1 - 3}{3 - (-1)}$$

$$= \frac{-4}{4}$$

$$= -1$$

14. The slope of a vertical line is **undefined**.

15. Perpendicular lines have slopes that are negative reciprocals of each other.

Therefore, if the slope of one line is $-\frac{7}{8}$, the line perpendicular would have a slope of $-\frac{8}{7}$.

16. For both lines, the slope is 2. Therefore, the lines would be **parallel**.

17.

$$m = \frac{\text{change in } y}{\text{change in } x}$$

$$= \frac{y_2 - y_1}{x_2 - x_1}$$

$$= \frac{250 - 100}{2 - 12}$$

$$= \frac{150}{-10}$$

$$= -15$$

So the rate of change would be −15 feet per mile.

18.

$$m = \frac{\text{change in } y}{\text{change in } x}$$

$$= \frac{y_2 - y_1}{x_2 - x_1}$$

$$= \frac{100 - 300}{12 - 20}$$

$$= \frac{-200}{-8}$$

$$= 25$$

So the rate of change would be 25 feet per mile.

19. $x = -4$

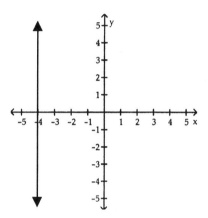

20.

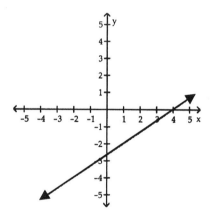

21. Solve the equation for y and use slope-intercept form: $y = mx + b$.

$$x + 2y = 8$$
$$x - x + 2y = -x + 8$$
$$2y = -x + 8$$
$$\frac{2y}{2} = \frac{-x}{2} + \frac{8}{2}$$
$$y = -\tfrac{1}{2}x + 4$$

So, $m = -\tfrac{1}{2}$ and the y-intercept is (0, 4).

22. $m = 7$ and $(x_1, y_1) = (-2, 5)$

Use point-slope form of the equation:
$$y - y_1 = m(x - x_1)$$
$$y - \mathbf{5} = 7(x - (\mathbf{-2}))$$
$$y - 5 = 7(x + 2)$$
$$y - 5 = 7(x) + 7(2)$$
$$y - 5 = 7x + 14$$
$$y - 5 + 5 = 7x + 14 + 5$$
$$y = 7x + 19$$

23. DEPRECIATION

$$y = -1500x + 15{,}000$$

24. Substitute (6, 1) into the inequality.
$$2x - 4y \geq 8$$
$$2(\mathbf{6}) - 4(\mathbf{1}) \overset{?}{\geq} 8$$
$$12 - 4 \overset{?}{\geq} 8$$
$$8 \geq 8$$
Yes, (6, 1) is a solution.

25. WATER HEATERS

Find the slope of the line using the two points (140, 13) and (180, 5):

$$m = \frac{y_2 - y_1}{x_2 - x_1}$$
$$= \frac{5 - 13}{180 - 140}$$
$$= \frac{-8}{40}$$
$$= -\frac{1}{5}$$

Use the slope of $-\frac{1}{5}$ and one of the points to substitute into the point-slope form of a line.

Let $m = -\frac{1}{5}$ and $(x_1, y_1) = (180, 5)$.

$$y - y_1 = m(x - x_1)$$
$$y - \mathbf{5} = -\frac{1}{5}(x - \mathbf{180})$$
$$y - 5 = -\frac{1}{5}(x) + -\frac{1}{5}(-180)$$
$$y - 5 = -\frac{1}{5}x + 36$$
$$y - 5 + 5 = -\frac{1}{5}x + 36 + 5$$
$$y = -\frac{1}{5}x + 41$$

26. Substitute (6, 1) into the inequality.
$$2x - 4y \geq 8$$
$$2(\mathbf{6}) - 4(\mathbf{1}) \overset{?}{\geq} 8$$
$$12 - 4 \overset{?}{\geq} 8$$
$$8 \geq 8$$
Yes, (6, 1) is a solution

27. Solve the inequality for y.
$$x - y > -2$$
$$-y > -x - 2$$
$$\frac{-y}{-1} > \frac{-x}{-1} - \frac{2}{-1}$$
$$y < x + 2$$

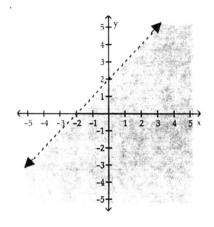

CHAPTERS 1—3

CUMULATIVE REVIEW EXERCISES

1. United Airlines
   a) The greatest quarter loss was **Q3 in 2001**.
   b) In 2002, the corporation's losses were approximately **$3.1 billion**.

2. $108 = 2 \cdot 2 \cdot 3 \cdot 3 \cdot 3 = 2^2 \cdot 3^3$

3. $\dfrac{1}{250} = 250\overline{)1.000}^{\,0.004} = 0.0004$

4. a) **True**, every whole number is an integer.
   b) **True**, every integer is a real number.
   c) **True**, 0 is a whole number, an integer, and a rational number.

5. PENNIES

   781 is 71% of what?
   $781 = 0.71x$
   $\dfrac{781}{0.71} = \dfrac{0.71x}{0.71}$
   $1100 = x$
   So, **1,100** people participated in the survey.

6.
$$\left|\frac{(6-5)^4 - (-21)}{-27 + 4^2}\right| = \left|\frac{(1)^4 - (-21)}{-27 + 16}\right|$$
$$= \left|\frac{1 + 21}{-27 + 16}\right|$$
$$= \left|\frac{22}{-11}\right|$$
$$= |-2|$$
$$= 2$$

7. $b^2 - 4ac$ for $a = 2$, $b = -8$, and $c = 4$
   $(-8)^2 - 4(2)(4) = 64 - 4(2)(4)$
   $\qquad\qquad\qquad = 64 - 32$
   $\qquad\qquad\qquad = 32$

8. $500 - x$

9. The algebraic expression $3x^2 - 2x + 1$ has **3 terms**. The coefficient of the second term is **-2**.

10. a) $2(x + 4) = 2(x) + 2(4)$
       $\qquad\qquad = 2x + 8$

    b) $2(x - 4) = 2(x) + 2(-4)$
       $\qquad\qquad = 2x - 8$

    c) $-2(x + 4) = -2(x) - 2(4)$
       $\qquad\qquad\ = -2x - 8$

    d) $-2(x - 4) = -2(x) - 2(-4)$
       $\qquad\qquad\ = -2x + 8$

11. $5a + 10 - a = (5a - a) + 10$
    $\qquad\qquad\quad = 6a + 10$

12. $-7(9t) = -63t$

13. $-2b^2 + 6b^2 = 4b^2$

14. $5(-17)(0)(2) = 0$

15. $(a + 2) - (a - 2) = a + 2 - a + 2$
    $\qquad\qquad\qquad\ = (a - a) + (2 + 2)$
    $\qquad\qquad\qquad\ = 0a + 4$
    $\qquad\qquad\qquad\ = 4$

16. $-4(-5)(-8a) = -160a$

17. $-y - y - y = -3y$

18. $\dfrac{3}{2}(4x - 8) + x = \dfrac{3}{2}(4x) + \dfrac{3}{2}(-8) + x$
    $\qquad\qquad\qquad = 6x - 12 + x$
    $\qquad\qquad\qquad = (6x + x) - 12$
    $\qquad\qquad\qquad = 7x - 12$

19. $3x - 5 = 13$
    $3x - 5 + 5 = 13 + 5$
    $3x = 18$
    $\dfrac{3x}{3} = \dfrac{18}{3}$
    $\boxed{x = 6}$

20. $1.2 - x = -1.7$

$1.2 - x - 1.2 = -1.7 - 1.2$

$-x = -2.9$

$\dfrac{-x}{-1} = \dfrac{-2.9}{-1}$

$\boxed{x = 2.9}$

21. $\dfrac{2}{3}x - 2 = 4$

$3\left(\dfrac{2}{3}x - 2\right) = 3(4)$

$3\left(\dfrac{2}{3}x\right) + 3(-2) = 3(4)$

$2x - 6 = 12$

$2x - 6 + 6 = 12 + 6$

$2x = 18$

$\boxed{x = 9}$

22. $\dfrac{y-2}{7} = -3$

$7\left(\dfrac{y-2}{7}\right) = 7(-3)$

$y - 2 = -21$

$y - 2 + 2 = -21 + 2$

$\boxed{y = -19}$

23. $-3(2y - 2) - y = 5$

$-3(2y) - 3(-2) - y = 5$

$-6y + 6 - y = 5$

$(-6y - y) + 6 = 5$

$-7y + 6 = 5$

$-7y + 6 - 6 = 5 - 6$

$-7y = -1$

$\boxed{y = \dfrac{1}{7}}$

24. $9y - 3 = 6y$

$9y - 3 - 9y = 6y - 9y$

$-3 = -3y$

$\boxed{1 = y}$

25. $\dfrac{1}{3} + \dfrac{c}{5} = -\dfrac{3}{2}$

$30\left(\dfrac{1}{3} + \dfrac{c}{5}\right) = 30\left(-\dfrac{3}{2}\right)$

$30\left(\dfrac{1}{3}\right) + 30\left(\dfrac{c}{5}\right) = 30\left(-\dfrac{3}{2}\right)$

$10 + 6c = -45$

$10 + 6c - 10 = -45 - 10$

$6c = -55$

$\boxed{c = -\dfrac{55}{6}}$

26. $5(x + 2) = 5x - 2$

$5(x) + 5(2) = 5x - 2$

$5x + 10 = 5x - 2$

$5x + 10 - 5x = 5x - 2 - 5x$

$10 \neq -2$

The equation is a **contradiction** and has $\boxed{\textbf{no solution.}}$

27. $-x = -99$

$\dfrac{-x}{-1} = \dfrac{-99}{-1}$

$\boxed{x = 99}$

28. $3c - 2 = \dfrac{11(c-1)}{5}$

$5(3c - 2) = 5\left(\dfrac{11(c-1)}{5}\right)$

$15c - 10 = 11(c - 1)$

$15c - 10 = 11c - 11$

$15c - 10 - 11c = 11c - 11 - 11c$

$4c - 10 = -11$

$4c - 10 + 10 = -11 + 10$

$4c = -1$

$\boxed{c = -\dfrac{1}{4}}$

Chapter 3 Cumulative Review

29. HIGN HEELS

The sum of the angles in a triangle is always 180°.

$$x + x + 90 = 180$$
$$2x + 90 = 180$$
$$2x + 90 - 90 = 180 - 90$$
$$2x = 90$$
$$\boxed{x = 45}$$

30.

$$S = 2\pi rh + 2\pi r^2$$
$$S - 2\pi r^2 = 2\pi rh + 2\pi r^2 - 2\pi r^2$$
$$S - 2\pi r^2 = 2\pi rh$$
$$\frac{S - 2\pi r^2}{2\pi r} = \frac{2\pi rh}{2\pi r}$$
$$\boxed{\frac{S - 2\pi r^2}{2\pi r} = h}$$

31. Perimeter

$$P = 2l + 2w$$
$$P = 2\left(\frac{13}{16}\right) + 2\left(\frac{3}{4}\right)$$
$$P = \frac{13}{8} + \frac{3}{2}$$
$$P = \frac{13}{8} + \frac{12}{8} = \frac{25}{8}$$
$$\boxed{P = 3\tfrac{1}{8} \text{ in.}}$$

Area:

$$A = L(W)$$
$$A = \frac{13}{16}\left(\frac{3}{4}\right)$$
$$\boxed{A = \frac{39}{64} \text{ in.}^2}$$

32.
If the equal sides are 3 feet, then
$$3 + 3 + 4 = \textbf{10 feet.}$$
If the equal sides are 4 feet, then
$$3 + 4 + 4 = \textbf{11 feet.}$$

33.
50% solution: **0.50x**
25% solution: **0.25(13 − x)**
30% mixture: **0.30(13)**

34. ROAD TRIPS

d = rt
The rate for the bus is 60 and let the time traveled by *t*, so the distance for the truck would 60*t*.
The rate for the truck is 50 and let the time traveled by *t*, so the distance for the truck would 50*t*.
Truck's distance + 75 = Bus' distance
$$50t + 75 = 60t$$
$$50t + 75 = 60t - 50t$$
$$75 = 10t$$
$$\boxed{7.5 = t}$$

35. MIXING CANDY

Let *x* = the pounds of candy corn used.
The value of the candy corn would be $1.90*x*.
(200 − *x*) = the pounds of gumdrops used.
The value of the gumdrops would be $1.20(200 − *x*). The value of the mixture would be $1.48(200).

$$1.90x + 1.20(200 - x) = 1.48(200)$$
$$1.90x + 240 - 1.20x = 296$$
$$(1.90x - 1.20x) + 240 = 296$$
$$0.70x + 240 = 296$$
$$0.70x + 240 - 240 = 296 - 240$$
$$0.70x = 56$$
$$x = 80$$
$$200 - 80 = 120$$

80 lbs. of candy corn and 120 lbs. of gumdrops should be used.

36.

$$-\frac{3}{16}x \geq -9$$
$$16\left(-\frac{3}{16}x\right) \geq 16(-9)$$
$$-3x \geq -144$$
$$\frac{-3x}{-3} \geq \frac{-144}{-3}$$
$$\boxed{x \leq 48}$$
$$(-\infty, 48]$$

48

37. $8x + 4 > 3x + 4$
  $8x + 4 - 3x > 3x + 4 - 3x$
    $5x + 4 > 4$
  $5x + 4 - 4 > 4 - 4$
      $5x > 0$
      $x > 0$
      $(0, \infty)$

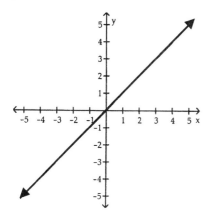

38. Substitute (-2, 4) into the equation.
  $y = 2x - 8$
  $4 \overset{?}{=} 2(\mathbf{-2}) - 8$
  $4 \overset{?}{=} -4 - 8$
  $4 \overset{?}{=} -12$
  Since $4 \neq -12$, (-2, 4) is NOT a solution.

39. $y = x$

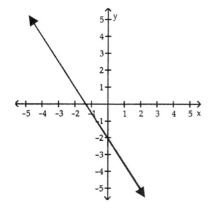

40. Solve the equation for y.
  $4y + 2x = -8$
  $4y = -2x - 8$
  $y = -\dfrac{1}{2}x - 2$

41. $y = 5$ is a horizontal line through the point (0, 5) and the slope of all horizontal lines is **zero**.

42. Use the formula for slope.
  $$m = \frac{y_2 - y_1}{x_2 - x_1}$$
  $$= \frac{-6 - 4}{5 - (-2)}$$
  $$= \frac{-10}{7}$$
  $$= -\frac{10}{7}$$

43. $m = \dfrac{\text{change in } y}{\text{change in } x}$
  $$= \frac{7}{12}$$

44. Solve the equation for y:
  $4x - 6y = -12$
  $-6y = -4x - 12$
  $y = \tfrac{2}{3}x + 2$
  So, $m = \tfrac{2}{3}$ and $b = (0, 2)$

45. Let $m = -2$ and $b = 1$. Then,
  $y = mx + b$
  $y = \mathbf{-2}x + \mathbf{1}$

46. $y - y_1 = m(x - x_1)$
  $$y - (-9) = -\frac{7}{8}(x - 2)$$
  $$y + 9 = -\frac{7}{8}(x - 2)$$

47. Substitute (-6, 0) into the inequality.
  $y \geq -x - 6$
  $$0 \overset{?}{\geq} -(\mathbf{-6}) - 6$$
  $$0 \overset{?}{\geq} 6 - 6$$
  $$0 \geq 0$$
  Yes, (-6, 0) is a solution.

48. $x < 4$

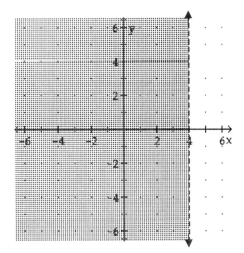

SECTION 4.1

VOCABULARY

1. Expressions such as x^4, 10^3, and $(5t)^2$, are called **exponential** expressions.

3. The expression x^4 represents a repeated multiplication where x is to be written as a **factor** four times.

5. $(h^3)^7$ is a **power** of an exponential expression.

CONCEPTS

7. a) $(3x)^4$ means $\boxed{3x} \cdot \boxed{3x} \cdot \boxed{3x} \cdot \boxed{3x}$.
    b) $(-5y)(-5y)(-5y)$ can be written
    as $\boxed{(-5y)^3}$.

9. a) Answers may vary. An example is $(3x^2)^6$.

    b) Answers may vary. An example is $\left(\dfrac{3a^3}{b}\right)^2$.

11. a) $2x^2$
    b) 0
    c) x^4
    d) a^7

13. a) doesn't simplify
    b) doesn't simplify
    c) x^5
    d) x

15. a) $(-4)^2 = -4 \cdot -4 = 16$
    b) $-4^2 = -(4 \cdot 4) = -16$

NOTATION

17. $\left(x^4 x^2\right)^3 = \left(\boxed{x^6}\right)^3$
    $= x^{\boxed{18}}$

19. base 4, exponent 3

21. base x, exponent 5

23. base $-3x$, exponent 2

25. base y, exponent 6

27. base $y + 9$, exponent 4

29. base $-3ab$, exponent 7

31. $x \cdot x \cdot x \cdot x \cdot x$

33. $\left(\dfrac{t}{2}\right)\left(\dfrac{t}{2}\right)\left(\dfrac{t}{2}\right)$

35. $(x - 5)(x - 5)$

37. $(4t)^4$

39. $-4t^3$

41. $(x - y)^3$

PRACTICE

43. $12^3 \cdot 12^4 = 12^{3+4}$
    $= 12^7$

45. $2(2^3)(2^2) = 2^{1+3+2}$
    $= 2^6$

47. $a^3 \cdot a^3 = a^{3+3}$
    $= a^6$

49. $x^4 x^3 = x^{4+3}$
    $= x^7$

51. $a^3 a a^5 = a^{3+1+5}$
    $= a^9$

53. $(-7)^2(-7)^3 = (-7)^{2+3}$
    $= (-7)^5$

55. $(8t)^{20}(8t)^{40} = (8t)^{20+40}$
    $= (8t)^{60}$

57. $(n-1)^2(n-1) = (n-1)^{2+1}$
    $= (n-1)^3$

59. $y^3(y^2 y^4)$ $\quad = y^3(y^{2+4})$
$\qquad\qquad = y^3(y^6)$
$\qquad\qquad = y^{3+6}$
$\qquad\qquad = y^9$

61. $\dfrac{8^{12}}{8^4} = 8^{12-4}$
$\qquad\quad = 8^8$

63. $\dfrac{x^{15}}{x^3} = x^{15-3}$
$\qquad\quad = x^{12}$

65. $\dfrac{c^{10}}{c^9} = c^{10-9}$
$\qquad\quad = c^1$
$\qquad\quad = c$

67. $\dfrac{(k-2)^{15}}{(k-2)} = (k-2)^{15-1}$
$\qquad\qquad\quad = (k-2)^{14}$

69. $(3^2)^4 = 3^{2 \bullet 4}$
$\qquad\quad = 3^8$

71. $(y^5)^3 = y^{5 \bullet 3}$
$\qquad\quad = y^{15}$

73. $(m^{50})^{10} = m^{50 \bullet 10}$
$\qquad\qquad = m^{500}$

75. $(a^2 b^3)(a^3 b^3) = a^{2+3} b^{3+3}$
$\qquad\qquad\qquad = a^5 b^6$

77. $(cd^4)(cd) = c^{1+1} d^{4+1}$
$\qquad\qquad\quad = c^2 d^5$

79. $xy^2 \bullet x^2 y = x^{1+2} y^{2+1}$
$\qquad\qquad = x^3 y^3$

81. $\dfrac{y^3 y^4}{y y^2} = \dfrac{y^{3+4}}{y^{1+2}}$
$\qquad\quad = \dfrac{y^7}{y^3}$
$\qquad\quad = y^{7-3}$
$\qquad\quad = y^4$

83. $\dfrac{c^3 d^7}{cd} = c^{3-1} d^{7-1}$
$\qquad\quad = c^2 d^6$

85. $(x^2 x^3)^5 = (x^{2+3})^5$
$\qquad\qquad = (x^5)^5$
$\qquad\qquad = x^{5 \bullet 5}$
$\qquad\qquad = x^{25}$

87. $(3zz^2 z^3)^5 = (3z^6)^5$
$\qquad\qquad = 3^5 z^{6 \bullet 5}$
$\qquad\qquad = 243 z^{30}$

89. $(3n^8)^2 = 3^2 n^{8 \bullet 2}$
$\qquad\quad = 9 n^{16}$

91. $(uv)^4 = u^{1 \bullet 4} v^{1 \bullet 4}$
$\qquad\quad = u^4 v^4$

93. $(a^3 b^2)^3 = a^{3 \bullet 3} b^{2 \bullet 3}$
$\qquad\qquad = a^9 b^6$

95. $(-2r^2 s^3)^3 = (-2)^3 r^{2 \bullet 3} s^{3 \bullet 3}$
$\qquad\qquad\quad = -8 r^6 s^9$

97. $\left(\dfrac{a}{b}\right)^3 = \dfrac{a^3}{b^3}$

99. $\left(\dfrac{x^2}{y^3}\right)^5 = \dfrac{x^{2\bullet5}}{y^{3\bullet5}}$

$= \dfrac{x^{10}}{y^{15}}$

101. $\left(\dfrac{-2a}{b}\right)^5 = \dfrac{(-2)^5 a^{1\bullet5}}{b^{1\bullet5}}$

$= \dfrac{-32a^5}{b^5}$

103. $\dfrac{(6k)^7}{(6k)^4} = (6k)^{7-4}$

$= (6k)^3$

$= 6^3 k^3$

$= 216k^3$

105. $\dfrac{(a^2b)^{15}}{(a^2b)^9} = (a^2b)^{15-9}$

$= (a^2b)^6$

$= a^{2\bullet6} b^{1\bullet6}$

$= a^{12}b^6$

107. $\dfrac{a^2 a^3 a^4}{(a^4)^2} = \dfrac{a^{2+3+4}}{a^{4\bullet2}}$

$= \dfrac{a^9}{a^8}$

$= a^{9-8}$

$= a$

109. $\dfrac{(ab^2)^3}{(ab)^2} = \dfrac{a^{1\bullet3} b^{2\bullet3}}{a^{1\bullet2} b^{1\bullet2}}$

$= \dfrac{a^3 b^6}{a^2 b^2}$

$= a^{3-2} b^{6-2}$

$= ab^4$

111. $\dfrac{(r^4 s^3)^4}{(rs^3)^3} = \dfrac{r^{4\bullet4} s^{3\bullet4}}{r^{1\bullet3} s^{3\bullet3}}$

$= \dfrac{r^{16} s^{12}}{r^3 s^9}$

$= r^{16-3} s^{12-9}$

$= r^{13} s^3$

113. $\left(\dfrac{y^3 y}{2yy^2}\right)^3 = \left(\dfrac{y^{3+1}}{2y^{1+2}}\right)^3$

$= \left(\dfrac{y^4}{2y^3}\right)^3$

$= \left(\dfrac{y^{4-3}}{2}\right)^3$

$= \left(\dfrac{y}{2}\right)^3$

$= \dfrac{y^{1\bullet3}}{2^3}$

$= \dfrac{y^3}{8}$

115.
$$\left(\frac{3t^3t^4t^5}{4t^2t^6}\right)^3 = \left(\frac{3t^{12}}{4t^8}\right)^3$$

$$= \left(\frac{3t^{12-8}}{4}\right)^3$$

$$= \left(\frac{3t^4}{4}\right)^3$$

$$= \frac{3^3 t^{4\cdot 3}}{4^3}$$

$$= \frac{27t^{12}}{64}$$

APPLICATIONS

117.   Area of a square:
$$A = s^2$$
$$= (a^5)^2$$
$$= a^{5\cdot 2}$$
$$= a^{10} \text{ mi}^2$$

119.   Volume of a prism:
$$V = blh$$
$$= x^4 \cdot x^3 \cdot x^2$$
$$= x^{4+3+2}$$
$$= x^9 \text{ ft}^3$$

121.   ART HISTORY

a) $(5x)(5x) = 25x^2 \text{ ft}^2$
b) $\pi(3a)^2 = 9a^2\pi \text{ ft}^2$

123.   BOUNCING BALLS

$$32\left(\frac{1}{2}\right) = 16 \text{ ft}$$

$$32\left(\frac{1}{2}\right)^2 = 32\left(\frac{1}{4}\right)$$
$$= 8 \text{ ft}$$

$$32\left(\frac{1}{2}\right)^3 = 32\left(\frac{1}{8}\right)$$
$$= 4 \text{ ft}$$

$$32\left(\frac{1}{2}\right)^4 = 32\left(\frac{1}{16}\right)$$
$$= 2 \text{ ft}$$

WRITING

125.   Answers will vary.

127.   Answers will vary.

REVIEW

129.   c

131.   d

CHALLENGE PROBLEMS

133.   a) $x^{2m}x^{3m} = x^{2m+3m}$
$$= x^{5m}$$

b) $\left(y^{5c}\right)^4 = y^{5c\bullet 4}$
$$= y^{20c}$$

c) $\dfrac{m^{8x}}{m^{4x}} = m^{8x-4x}$
$$= m^{4x}$$

d) $\left(2a^{6y}\right)^4 = 2^4 a^{6y\bullet 4}$
$$= 16a^{24y}$$

SECTION 4.2

VOCABULARY

1. In the expression 8^{-3}, 8 is the **base**, and -3 is the **exponent**.

3. Another way to write 2^{-3} is to write its **reciprocal** and to change the sign of the exponent:

$$2^{-3} = \frac{1}{2^{\boxed{3}}}$$

CONCEPTS

5. a) $\dfrac{6^4}{6^4} = 6^{\boxed{4-4}} \qquad \dfrac{6^4}{6^4} = \dfrac{\boxed{6} \bullet \boxed{6} \bullet \boxed{6} \bullet \boxed{6}}{6 \bullet 6 \bullet 6 \bullet 6}$

$\qquad\qquad = 6^{\boxed{0}} \qquad\qquad\quad = \boxed{1}$

b) So we define 6^0 to be $\boxed{1}$, and in general, if x is any nonzero real number, then $x^0 = \boxed{1}$.

7.

x	3^x
2	$3^2 = 9$
1	$3^1 = 3$
0	$3^0 = 1$
-1	$3^{-1} = \dfrac{1}{3}$
-2	$3^{-2} = \dfrac{1}{3^2}$ $= \dfrac{1}{9}$

9.

a) $x^m \bullet x^n = \boxed{x^{m+n}}$

b) $\dfrac{x^m}{x^n} = \boxed{x^{m-n}}$

c) $\left(x^m\right)^n = \boxed{x^{mn}}$

d) $\left(xy\right)^n = \boxed{x^n y^n}$

e) $\left(\dfrac{x}{y}\right)^n = \boxed{\dfrac{x^n}{y^n}}$

f) $x^{-n} = \boxed{\dfrac{1}{x^n}}$

g) $\dfrac{1}{x^{-n}} = \boxed{x^n}$

h) $\dfrac{x^{-m}}{y^{-n}} = \boxed{\dfrac{y^n}{x^m}}$

i) $x^0 = \boxed{1}$

11. Each base occurs only **once**.
There are **no** parentheses
There are no negative or zero **exponents**

NOTATION

13. $\left(y^5 y^3\right)^{-5} = \left(\boxed{y^8}\right)^{-5}$

$\qquad\qquad = y^{\boxed{-40}}$

$\qquad\qquad = \dfrac{1}{y^{\boxed{40}}}$

15.

Expression	Base	Exponent
4^{-2}	**4**	**-2**
$6x^{-5}$	x	**-5**
$\left(\dfrac{3}{y}\right)^{-8}$	$\dfrac{3}{y}$	**-8**
-7^{-1}	**7**	**-1**
$(-2)^{-3}$	**-2**	**-3**
$10a^0 b$	a	**0**

PRACTICE

17. $7^0 = 1$

19. $\left(\dfrac{1}{4}\right)^0 = 1$

Section 4.1

21. $2x^0 = 2 \cdot 1$
$\quad\quad = 2$

23. $(-x)^0 = 1$

25. $\left(\dfrac{a^2 b^3}{ab^4}\right)^0 = 1$

27. $\dfrac{5}{2x^0} = \dfrac{5}{2(1)}$
$\quad\quad\quad = \dfrac{5}{2}$

29. $-15x^0 y = -15(1)(y)$
$\quad\quad\quad\quad = -15y$

31. $12^{-2} = \dfrac{1}{12^2}$
$\quad\quad\quad = \dfrac{1}{144}$

33. $(-4)^{-1} = \dfrac{1}{(-4)^1}$
$\quad\quad\quad\quad = -\dfrac{1}{4}$

35. $44g^{-6} = 44 \bullet \dfrac{1}{g^6}$
$\quad\quad\quad\quad = \dfrac{44}{g^6}$

37. $(-10)^{-3} = \dfrac{1}{(-10)^3}$
$\quad\quad\quad\quad = -\dfrac{1}{1,000}$

39. $-4^{-3} = -\dfrac{1}{4^3}$
$\quad\quad\quad = -\dfrac{1}{64}$

41. $-(-4)^{-3} = -\dfrac{1}{(-4)^3}$
$\quad\quad\quad\quad = -\dfrac{1}{-64}$
$\quad\quad\quad\quad = \dfrac{1}{64}$

43. $x^{-2} = \dfrac{1}{x^2}$

45. $-b^{-5} = -\dfrac{1}{b^5}$

47. $\left(\dfrac{1}{6}\right)^{-2} = \left(\dfrac{6}{1}\right)^2$
$\quad\quad\quad\quad = \dfrac{36}{1}$
$\quad\quad\quad\quad = 36$

49. $\left(-\dfrac{1}{2}\right)^{-3} = \left(-\dfrac{2}{1}\right)^3$
$\quad\quad\quad\quad = -\dfrac{8}{1}$
$\quad\quad\quad\quad = -8$

51. $\left(\dfrac{7}{8}\right)^{-1} = \left(\dfrac{8}{7}\right)^1$
$\quad\quad\quad\quad = \dfrac{8}{7}$

53. $\dfrac{1}{5^{-3}} = 5^3$
$\quad\quad\quad = 125$

55. $\dfrac{2^{-4}}{3^{-1}} = \dfrac{3^1}{2^4}$
$\quad\quad\quad = \dfrac{3}{16}$

57. $\dfrac{a^{-5}}{b^{-2}} = \dfrac{b^2}{a^5}$

59. $\dfrac{r^{-50}}{r^{-70}} = r^{-50-(-70)}$

$= r^{-50+70}$

$= r^{20}$

61. $-\dfrac{1}{p^{-10}} = -p^{10}$

63. $\dfrac{h^{-5}}{h^2} = h^{-5-2}$

$= h^{-7}$

$= \dfrac{1}{h^7}$

65. $\dfrac{8}{s^{-1}} = 8s$

67. $\dfrac{h}{h^{-6}} = h^{1-(-6)}$

$= h^{1+6}$

$= h^7$

69. $(2y)^{-4} = \dfrac{1}{(2y)^4}$

$= \dfrac{1}{16y^4}$

71. $(ab^2)^{-3} = \dfrac{1}{(ab^2)^3}$

$= \dfrac{1}{a^{1\bullet 3}b^{2\bullet 3}}$

$= \dfrac{1}{a^3 b^6}$

73. $2^5 \bullet 2^{-2} = 2^{5+(-2)}$

$= 2^3$

$= 8$

75. $\left(\dfrac{y^4}{3}\right)^{-2} = \left(\dfrac{3}{y^4}\right)^2$

$= \dfrac{3^2}{y^{4\bullet 2}}$

$= \dfrac{9}{y^8}$

77. $\dfrac{y^4}{y^5} = y^{4-5}$

$= y^{-1}$

$= \dfrac{1}{y}$

79. $\dfrac{(r^2)^3}{(r^3)^4} = \dfrac{r^6}{r^{12}}$

$= r^{6-12}$

$= r^{-6}$

$= \dfrac{1}{r^6}$

81. $\dfrac{4s^{-5}}{t^{-2}} = \dfrac{4t^2}{s^5}$

83. $(5d^{-2})^3 = 5^3 d^{-2\bullet 3}$

$= 125d^{-6}$

$= \dfrac{125}{d^6}$

85. $\dfrac{-2a^{-4}}{a^{-8}} = -2a^{-4-(-8)}$

$= -2a^{-4+8}$

$= -2a^4$

Section 4.1

87.
$$x^{-3}x^{-3}x^{-3} = x^{-3+(-3)+(-3)}$$
$$= x^{-9}$$
$$= \frac{1}{x^9}$$

89.
$$\frac{t\left(t^{-2}\right)^{-2}}{t^{-5}} = \frac{t(t^4)}{t^{-5}}$$
$$= \frac{t^{1+4}}{t^{-5}}$$
$$= \frac{t^5}{t^{-5}}$$
$$= t^{5-(-5)}$$
$$= t^{10}$$

91.
$$\frac{y^4 y^3}{y^4 y^{-2}} = \frac{y^{4+3}}{y^{4+(-2)}}$$
$$= \frac{y^7}{y^2}$$
$$= y^{7-2}$$
$$= y^5$$

93.
$$\frac{2a^4 a^{-2}}{a^2 a^0} = \frac{2a^{4+(-2)}}{a^{2+0}}$$
$$= \frac{2a^2}{a^2}$$
$$= 2a^0$$
$$= 2(1)$$
$$= 2$$

95.
$$\left(ab^2\right)^{-2} = \frac{1}{\left(ab^2\right)^2}$$
$$= \frac{1}{a^{1\cdot2}b^{2\cdot2}}$$
$$= \frac{1}{a^2 b^4}$$

97.
$$\left(x^2 y\right)^{-3} = \frac{1}{\left(x^2 y\right)^3}$$
$$= \frac{1}{x^{2\cdot3}y^{1\cdot3}}$$
$$= \frac{1}{x^6 y^3}$$

99.
$$\left(x^{-4}x^3\right)^3 = \left(x^{-4+3}\right)^3$$
$$= \left(x^{-1}\right)^3$$
$$= x^{-3}$$
$$= \frac{1}{x^3}$$

101.
$$\left(-2x^3 y^{-2}\right)^{-5} = \frac{1}{\left(-2x^3 y^{-2}\right)^5}$$
$$= \frac{1}{-32x^{15}y^{-10}}$$
$$= -\frac{y^{10}}{32x^{15}}$$

103.
$$\left(\frac{a^3}{a^{-4}}\right)^2 = \left(a^{3-(-4)}\right)^2$$
$$= \left(a^7\right)^2$$
$$= a^{14}$$

105.
$$\left(\frac{4x^2}{3x^{-5}}\right)^4 = \left(\frac{4x^{2-(-5)}}{3}\right)^4$$
$$= \left(\frac{4x^7}{3}\right)^4$$
$$= \frac{4^4 x^{7\cdot4}}{3^4}$$
$$= \frac{256x^{28}}{81}$$

107. $\left(\dfrac{y^3 z^{-2}}{3 y^{-4} z^3}\right)^2 = \left(\dfrac{y^{3-(-4)} z^{-2-3}}{3}\right)^2$

$= \left(\dfrac{y^7 z^{-5}}{3}\right)^2$

$= \left(\dfrac{y^7}{3z^5}\right)^2$

$= \dfrac{y^{7 \cdot 2}}{3^2 z^{5 \cdot 2}}$

$= \dfrac{y^{14}}{9 z^{10}}$

109. $\dfrac{2^{-2} g^{-2} h^{-3}}{9^{-1} h^{-3}} = \dfrac{9^1 h^3}{2^2 g^2 h^3}$

$= \dfrac{9 h^{3-3}}{4 g^2}$

$= \dfrac{9}{4 g^2}$

111. $5^0 + (-7)^0 = 1 + 1$

$\qquad\qquad\quad = 2$

113. $2^{-2} + 4^{-1} = \dfrac{1}{2^2} + \dfrac{1}{4^1}$

$= \dfrac{1}{4} + \dfrac{1}{4}$

$= \dfrac{2}{4}$

$= \dfrac{1}{2}$

115. $9^0 - 9^{-1} = 1 - \dfrac{1}{9}$

$= \dfrac{9}{9} - \dfrac{1}{9}$

$= \dfrac{8}{9}$

APPLICATIONS

117. THE DECIMAL NUMERATION SYSTEM

$10^2, 10^1, 10^0, 10^{-1}, 10^{-2}, 10^{-3}, 10^{-4}$

119. RETIREMENT

$P = A(1+i)^{-n}$

$P = \mathbf{100,000}(1 + \mathbf{0.08})^{-40}$

$P = 100,000(1.08)^{-40}$

$P = 100,000 \left(\dfrac{1}{1.08}\right)^{40}$

$P \approx \$4,603.09$

121. BIOLOGY

If $n = 0$, it gives the initial number of bacteria b.

WRITING

123. Answers will vary.

REVIEW

125. IQ TESTS

$\mathbf{135} = \dfrac{x}{\mathbf{10}} \bullet 100$

$135 = 10x$

$13.5 = x$

The mental age is 13.5 years

127. $y = mx + b$

$y = \dfrac{3}{4} x - 5$

CHALLENGE PROBLEMS

129. a) $r^{5m}r^{-6m} = r^{5m+(-6m)}$

$$= r^{-m}$$

$$= \frac{1}{r^m}$$

b) $\dfrac{x^{3n}}{x^{6n}} = \dfrac{x^{3n-(6n)}}{1}$

$$= \frac{x^{-3n}}{1}$$

$$= \frac{1}{x^{3n}}$$

VOCABULARY

1. 4.84×10^9 and 1.05×10^{-2} are written in **scientific** notation.

3. Scientific **notation** provides a compact way of writing very large or very small numbers.

5. 10^{34}, 10^{50}, and 10^{-14} are **powers** of 10.

NOTATION

7. When we multiply a decimal by 10^5, the decimal point moves 5 places to the **right**. When we multiply a decimal by 10^{-7}, the decimal point moves 7 places to the **left**.

9. a) 10^{-7}
   b) 10^9

11. a) When a real number great than 1 is written in scientific notation, the exponent on 10 is a **positive** number.

    b) When a real number between 0 and 1 is written in scientific notation, the exponent on 10 is a **negative** number.

13. a) $7,700 = \boxed{7.7} \times 10^3$
    b) $500,000 = \boxed{5.0} \times 10^5$
    c) $114,000,000 = 1.14 \times 10^{\boxed{8}}$

15. a) $(5.1 \times 1.5)(10^9 \times 10^{22})$

    b) $\dfrac{8.8}{2.2} \times \dfrac{10^{30}}{10^{19}}$

NOTATION

17. A positive number is written in scientific notation when it is written in the form $N \times 10^n$, where $\underline{1} \le N < \underline{10}$ and n is an **integer**.

19. 230; Since the exponent is 2, the decimal moves 2 places to the right.

21. 812,000; Since the exponent is 5, the decimal moves 5 places to the right.

23. 0.00115; Since the exponent is -3, the decimal moves 3 places to the left.

25. 0.000976; Since the exponent is -4, the decimal moves 4 places to the left.

27. 6,001,000; Since the exponent is 6, the decimal moves 6 places to the right.

29. 2.718; Since the exponent is 0, the decimal does not move.

31. 0.06789; Since the exponent is -2, the decimal moves 2 places to the left.

33. 0.00002; Since the exponent is -5, the decimal moves 5 places to the left.

35. 9,000,000,000; Since the exponent is 9, the decimal moves 9 places to the right.

37. 2.3×10^4 ; Move the decimal 4 places to the left.

39. 1.7×10^6 ; Move the decimal 6 places to the left.

41. 6.2×10^{-2} ; Move the decimal 2 places to the right.

43. 5.1×10^{-6} ; Move the decimal 6 places to the right.

45. 5.0×10^9 ; Move the decimal 9 places to the left.

47. 3.0×10^{-7} ; Move the decimal 7 places to the right.

49. 9.09×10^8 ; Move the decimal 8 places to the left.

51. 3.45×10^{-2} ; Move the decimal 2 places to the right.

53. 9.0×10^0; The decimal does not move.

55. 1.718×10^{18}; Move the decimal 18 places to the left.

57. 1.23×10^{-14}; Move the decimal 14 places to the right.

59. $(3.4 \times 10^2)(2.1 \times 10^3)$
$= (3.4 \times 2.1)(10^2 \times 10^3)$
$= 7.14 \times 10^5$
$= 714,000$

61. $(8.4 \times 10^{-13})(4.8 \times 10^9)$
$= (8.4 \times 4.8)(10^{-13} \times 10^9)$
$= 40.32 \times 10^{-4}$
$= 0.004032$

63. $\dfrac{9.3 \times 10^2}{3.1 \times 10^{-2}} = \dfrac{9.3}{3.1} \times \dfrac{10^2}{10^{-2}}$
$= 3.0 \times 10^{2-(-2)}$
$= 3.0 \times 10^4$
$= 30,000$

65. $\dfrac{1.29 \times 10^{-6}}{3 \times 10^{-4}} = \dfrac{1.29}{3} \times \dfrac{10^{-6}}{10^{-4}}$
$= 0.43 \times 10^{-6-(-4)}$
$= 0.43 \times 10^{-2}$
$= 0.0043$

67. $(6 \times 10^{-9})(5.5 \times 10^6)$
$= (5.6 \times 5.5)(10^{-9} \times 10^6)$
$= 30.8 \times 10^{-3}$
$= 0.0308$

69. $\dfrac{9.6 \times 10^4}{(1.2 \times 10^4)(4 \times 10^{-5})} = \dfrac{9.6 \times 10^4}{(1.2 \times 4)(10^4 \times 10^{-5})}$
$= \dfrac{9.6 \times 10^4}{4.8 \times 10^{4+(-5)}}$
$= \dfrac{9.6 \times 10^4}{4.8 \times 10^{-1}}$
$= \dfrac{9.6}{4.8} \times \dfrac{10^4}{10^{-1}}$
$= 2.0 \times 10^{4-(-1)}$
$= 2.0 \times 10^5$
$= 200,000$

71. $(4.564 \times 10^2)^6 = (4.564)^6 \times (10^2)^6$
$= 9038.030748 \times 10^{12}$
$= 9.038030748 \times 10^{15}$

73. $(2.25 \times 10^2)^{-5} = (2.25)^{-5} \times (10^2)^{-5}$
$= 0.0173415299 \times 10^{-10}$
$= 1.73415299 \times 10^{-12}$

APPLICATIONS

75. ASTRONOMY

2.57×10^{13} miles; move the decimal to the left 13 places

77. GEOGRAPHY

63,800,000 mi^2; move the decimal to the right 7 places

79. LENGTH OF A METER

6.22×10^{-3} miles; move the decimal to the right 3 places

81. WAVELENGTHS

gamma ray, x-ray, ultraviolet, visible light, infrared, microwave, radio wave

83. PROTONS

$(1.7 \times 10^{-24})(1 \times 10^6)$
$= (1.7 \times 1)(10^{-24} \times 10^6)$
$= 1.7 \times 10^{-24+6}$
$= 1.7 \times 10^{-18}$ g

85. LIGHT YEARS

$(5.87 \times 10^{12}$ miles$)(5.28 \times 10^3$ feet$)$
$= (5.87 \times 5.28)(10^{12} \times 10^3)$
$= 30.9936 \times 10^{12+3}$
$= 30.9936 \times 10^{15}$
$= 3.09936 \times 10^{16}$ feet

87. INTEREST

$I = Prt$
$= (5.2 \times 10^{12})(0.04)(1)$
$= (5.2 \times 10^{12})(4 \times 10^{-2})(1)$
$= (5.2 \times 4 \times 1)(10^{12} \times 10^{-2})$
$= 20.88 \times 10^{12 + (-2)}$
$= 20.88 \times 10^{10}$
$= 2.088 \times 10^{11}$ dollars

89. THE MILITARY

a) 1.7×10^6 ; 1,700,000
b) 1986 ; 2.05×10^6 ; 2,050,000

WRITING

91. Answers will vary.

93. Answers will vary.

REVIEW

95. $-5y^{55} = -5(-1)^{55}$
$= -5(-1)$
$= 5$

97. $c = 30t + 45$

CHALLENGE PROBLEMS

99. a) -2.5×10^{-4}

b) $\dfrac{1}{2.5 \times 10^{-4}} = \dfrac{1 \times 10^0}{2.5 \times 10^{-4}}$

$\qquad = \dfrac{1}{2.5} \times \dfrac{10^0}{10^{-4}}$

$\qquad = 0.4 \times 10^{0-(-4)}$

$\qquad = 0.4 \times 10^4$

$\qquad = 4.0 \times 10^3$

Section 4.3

SECTION 4.4

VOCABULARY

1. A **polynomial** is a term or a sum of terms in which all variable have whole number exponents.

3. The degree of a polynomial is the same as the degree of its **term** with the highest degree.

5. The **degree** of the monomial $3x^7$ is 7.

7. $x^3 - 6x^2 + 9x - 2$ is a polynomial in $\underline{x}$ and is written in **decreasing or descending** powers of x.

9. Because the graph of $y = x^3$ is not a straight line, we call $y = x^3$ a **nonlinear** equation.

CONCEPTS

11. a) Yes
    b) No
    c) No
    d) Yes
    e) Yes
    f) Yes

13. $y = x^2 + 3$

$2 \overset{?}{=} (-1)^2 + 3$

$2 \overset{?}{=} 1 + 3$

$2 \overset{?}{=} 4$

No, $(-1, 2)$ is not a solution.

15. Not enough ordered pairs were found—the correct graph is a parabola.

NOTATION

17. $-2x^2 + 3x - 1 = -2(\boxed{-2})^2 + 3(\boxed{-2}) - 1$
$= -2(\boxed{4}) + 3(\boxed{-2}) - 1$
$= \boxed{-8} + (-6) - 1$
$= \boxed{-14} - 1$
$= \boxed{-15}$

19. $y = x^2 - 1$ and $y = x^3 - 1$

PRACTICE

21. two terms - binomial

23. three terms - trinomial

25. one term - monomial

27. two terms - binomial

29. three terms - trinomial

31. four terms - none of these

33. three terms - trinomial

35. 4th

37. 2nd

39. 1st

41. 4th (2+2 and 3+1)

43. 12th

45. 0th

47. $-9x + 1$
    a) $-9(\mathbf{5}) + 1 = -45 + 1$
    $= -44$

    b) $-9(\mathbf{-4}) + 1 = 36 + 1$
    $= 37$

49. $3x^2 - 2x + 8$
    a) $3(\mathbf{1})^2 - 2(\mathbf{1}) + 8 = 3(1) - 2(1) + 8$
    $= 3 - 2 + 8$
    $= 1 + 8$
    $= 9$

    b) $3(\mathbf{0})^2 - 2(\mathbf{0}) + 8 = 3(0) - 2(0) + 8$
    $= 0 - 0 + 8$
    $= 8$

51. $-x^2 - 6$
    a) $-(\mathbf{-4})^2 - 6 = -(16) - 6$
    $= -22$

    b) $-(\mathbf{20})^2 - 6 = -(400) - 6$
    $= -406$

53. $x^3 + 3x^2 + 2x + 4$

    a) $(2)^3 + 3(2)^2 + 2(2) + 4$
$$= 8 + 3(4) + 2(2) + 4$$
$$= 8 + 12 + 4 + 4$$
$$= 28$$

    b) $(-2)^3 + 3(-2)^2 + 2(-2) + 4$
$$= -8 + 3(4) + 2(-2) + 4$$
$$= -8 + 12 - 4 + 4$$
$$= 4$$

55. $x^4 - x^3 + x^2 + 2x - 1$

    a) $(1)^4 - (1)^3 + (1)^2 + 2(1) - 1$
$$= 1 - 1 + 1 + 2(1) - 1$$
$$= 1 - 1 + 1 + 2 - 1$$
$$= 2$$

    b) $(-1)^4 - (-1)^3 + (-1)^2 + 2(-1) - 1$
$$= 1 - (-1) + 1 + 2(-1) - 1$$
$$= 1 + 1 + 1 - 2 - 1$$
$$= 0$$

57.

x	$y = x^2$
-3	$(-3)^2 = 9$
-2	$(-2)^2 = 4$
-1	$(-1)^2 = 1$
0	$(0)^2 = 0$
1	$(1)^2 = 1$
2	$(4)^2 = 4$
3	$(3)^2 = 9$

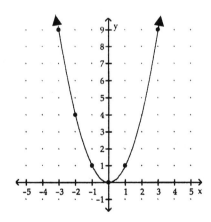

59.

x	$y = x^2 + 1$
-2	$(-2)^2 + 1 = 4 + 1$ $= 5$
-1	$(-1)^2 + 1 = 1 + 1$ $= 2$
0	$(0)^2 + 1 = 0 + 1$ $= 1$
1	$(1)^2 + 1 = 1 + 1$ $= 2$
2	$(2)^2 + 1 = 4 + 1$ $= 5$

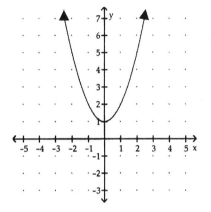

61.

x	$y = -x^2 - 2$
-2	$-(-2)^2 - 2 = -4 - 2$ $= -6$
-1	$-(-1)^2 - 2 = -1 - 2$ $= -3$
0	$-(0)^2 - 2 = -0 - 2$ $= -2$
1	$-(1)^2 - 2 = -1 - 2$ $= -3$
2	$-(2)^2 - 2 = -4 - 2$ $= -6$

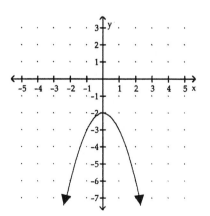

Section 4.4

63.

x	$y = 2x^2 - 3$
-2	$2(\mathbf{-2})^2 - 3 = 2(4) - 3$ $= 8 - 3$ $= 5$
-1	$2(\mathbf{-1})^2 - 3 = 2(1) - 3$ $= 2 - 3$ $= -1$
0	$2(\mathbf{0})^2 - 3 = 2(0) - 3$ $= 0 - 3$ $= -3$
1	$2(\mathbf{1})^2 - 3 = 2(1) - 3$ $= 2 - 3$ $= -1$
2	$2(\mathbf{2})^2 - 3 = 2(4) - 3$ $= 8 - 3$ $= 5$

65.

x	$y = x^3 + 2$
-2	$(\mathbf{-2})^3 + 2 = -8 + 2$ $= -6$
-1	$(\mathbf{-1})^3 + 2 = -1 + 2$ $- 1$
0	$(\mathbf{0})^3 + 2 = 0 + 2$ $= 2$
1	$(\mathbf{1})^3 + 2 = 1 + 2$ $= 3$
2	$(\mathbf{2})^3 + 2 = 8 + 2$ $= 10$

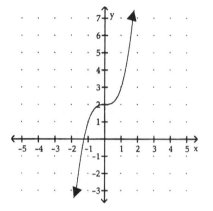

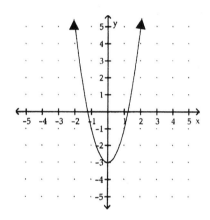

67.

x	$y = -x^3 - 1$
-2	$-(-2)^3 - 1 = -(-8) - 1$ $= 7$
-1	$-(-1)^3 - 1 = -(-1) - 1$ $= 0$
0	$-(0)^3 - 1 = -0 - 1$ $= -1$
1	$-(1)^3 - 1 = -1 - 1$ $= -2$
2	$-(2)^3 - 1 = -8 - 1$ $= -9$

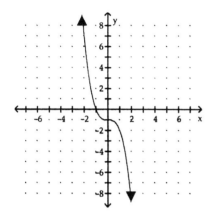

APPLICATIONS

69. SUPERMARKETS

Use the formula from Example 3 and let $c = 6$.

$$\frac{1}{3}c^3 + \frac{1}{2}c^2 + \frac{1}{6}c = \frac{1}{3}(6)^3 + \frac{1}{2}(6)^2 + \frac{1}{6}(6)$$
$$= \frac{1}{3}(216) + \frac{1}{2}(36) + \frac{1}{6}(6)$$
$$= 72 + 18 + 1$$
$$= 91$$

91 cantaloupes are used.

71. STOPPING DISTANCE

$$0.04(30)^2 + 0.9(30) = 0.04(900) + 27$$
$$= 36 + 27$$
$$= 63$$

The distance is 63 feet.

73. SCIENCE HISTORY

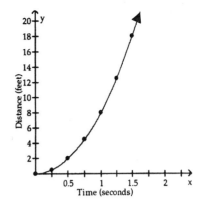

75. MANUFACTURING

a) It costs 8¢ to make a 2-inch bolt.
b) 12¢
c) a 4-inch bolt

WRITING

77. Answers will vary.

REVIEW

79. $$-4(3y + 2) \le 28$$
$$-4(3y) - 4(2) \le 28$$
$$-12y - 8 \le 28$$
$$-12y - 8 + 8 \le 28 + 8$$
$$-12y \le 36$$
$$y \ge -3$$
$$[-3, \infty)$$

81. $$\left(x^2 x^4\right)^3 = \left(x^{2+4}\right)^3$$
$$= \left(x^6\right)^3$$
$$= x^{6 \cdot 3}$$
$$= x^{18}$$

83. $\left(\dfrac{y^2 y^5}{y^4}\right)^3 = \left(\dfrac{y^{2+5}}{y^4}\right)^3$

$\qquad\qquad = \left(\dfrac{y^7}{y^4}\right)^3$

$\qquad\qquad = \left(y^{7-4}\right)^3$

$\qquad\qquad = \left(y^3\right)^3$

$\qquad\qquad = y^{3\cdot3}$

$\qquad\qquad = y^9$

CHALLENGE PROBLEMS

85. Answers will vary.
Example: $x^2 + x - 1$

VOCABULARY

1. The expressions $(b^3 - b^2 - 9b + 1) + (b^3 - b^2 - 9b + 1)$ is the sum of two **polynomials**.

3. **Like** terms have the same variables with the same exponents.

5. The polynomial $2t^4 + 3t^3 - 4t^2 + 5t - 6$ is written in **descending** powers of t.

CONCEPTS

7. To add polynomials, **combine** like terms.

9.
$$\begin{array}{r} 8x^2 - 7x - 1 \\ + \boxed{-4x^2} - 6x \boxed{+} 9 \\ \hline \end{array}$$

11. a) $2x^2 + 3x^2 = (2 + 3)x^2 = 5x^2$
    b) $15m^3 - m^3 = (15 - 1)m^3 = 14m^3$
    c) $8a^3b - ab$; unlike terms, so you cannot combine them
    d) $6cd + 4c^2d$; unlike terms, so you cannot combine them

13. a) $-(5x^2 - 8x + 23) = -5x^2 + 8x - 23$
    b) $-(-5y^4 + 3y^2 - 7) = 5y^4 - 3y^2 + 7$

NOTATION

15. $(6x^2 + 2x + 3) - (4x^2 - 7x + 1)$
$$= 6x^2 + 2x + 3 \boxed{-} 4x^2 \boxed{+} 7x \boxed{-} 1$$
$$= \boxed{2x^2} + 9x + \boxed{2}$$

17. True

PRACTICE

19. $8t^2 + 4t^2 = (8 + 4)t^2$
$$= 12t^2$$

21. $-32u^3 - 16u^3 = (-32 - 16)u^3$
$$= -48u^3$$

23. $1.8x - 1.9x = (1.8 - 1.9)x$
$$= -0.1x$$

25. $\dfrac{1}{2}st + \dfrac{3}{2}st = \left(\dfrac{1}{2} + \dfrac{3}{2}\right)st$
$$= \dfrac{4}{2}st$$
$$= 2st$$

27. $3r - 4r + 7r = (3 - 4 + 7)r$
$$= 6r$$

29. $-4ab + 4ab - ab = (-4 + 4 - 1)ab$
$$= -ab$$

31. $10x^2 - 8x + 9x - 9x^2$
$$= (10 - 9)x^2 + (-8 + 9)x$$
$$= 1x^2 + 1x$$
$$= x^2 + x$$

33. $6x^3 + 8x^4 + 7x^3 + (-8x^4)$
$$= (8 - 8)x^4 + (6 + 7)x^3$$
$$= 0x^4 + 13x^3$$
$$= 13x^3$$

35. $4x^2y + 5 - 6x^3y - 3x^2y + 2x^3y$
$$= (-6 + 2)x^3y + (4 - 3)x^2y + 5$$
$$= -4x^3y + x^2y + 5$$

37. $\dfrac{2}{3}d^2 - \dfrac{1}{4}c^2 + \dfrac{5}{6}c^2 - \dfrac{1}{2}cd + \dfrac{1}{3}d^2$
$$= \left(-\dfrac{1}{4} + \dfrac{5}{6}\right)c^2 - \dfrac{1}{2}cd + \left(\dfrac{2}{3} + \dfrac{1}{3}\right)d^2$$
$$= \dfrac{14}{24}c^2 - \dfrac{1}{2}cd + \dfrac{3}{3}d^2$$
$$= \dfrac{7}{12}c^2 - \dfrac{1}{2}cd + d^2$$

39. $(3x + 7) + (4x - 3) = (3x + 4x) + (7 - 3)$
$$= 7x + 4$$

41. $(9a^2 + 3a) - (2a - 4a^2)$
$$= 9a^2 + 3a - 2a + 4a^2$$
$$= (9a^2 + 4a^2) + (3a - 2a)$$
$$= 13a^2 + a$$

43. $(2x + 3y) + (5x - 10y)$
$$= (2x + 5x) + (3y - 10y)$$
$$= 7x - 7y$$

45. $(-8x - 3y) - (-11x + y)$
$$= -8x - 3y + 11x - y$$
$$= (-8x + 11x) + (-3y - y)$$
$$= 3x - 4y$$

47. $(3x^2 - 3x - 2) + (3x^2 + 4x - 3)$
$$= (3x^2 + 3x^2) + (-3x + 4x) + (-2 - 3)$$
$$= 6x^2 + x - 5$$

49. $(2b^2 + 3b - 5) - (2b^2 - 4b - 9)$
$$= 2b^2 + 3b - 5 - 2b^2 + 4b + 9$$
$$= (2b^2 - 2b^2) + (3b + 4b) + (-5 + 9)$$
$$= 0b^2 + 7b + 4$$
$$= 7b + 4$$

51. $(2x^2 - 3x + 1) - (4x^2 - 3x + 2) + (2x^2 + 3x + 2)$
$$= 2x^2 - 3x + 1 - 4x^2 + 3x - 2 + 2x^2 + 3x + 2$$
$$= (2x^2 - 4x^2 + 2x^2) + (-3x + 3x + 3x) + (1 - 2 + 2)$$
$$= 3x + 1$$

53. $(-4h^3 + 5h^2 + 15) - (h^3 - 15)$
$$= -4h^3 + 5h^2 + 15) - h^3 + 15$$
$$= (-4h^3 - h^3) + 5h^2 + (15 + 15)$$
$$= -5h^3 + 5h^2 + 30$$

55. $(1.04x^2 + 2.07x - 5.01) + (1.33x - 2.98x^2 + 5.02)$
$$= (1.04 - 2.98)x^2 + (2.07 + 1.33)x + (-5.01 + 5.02)$$
$$= 1.94x^2 + 3.4x + 0.01$$

57. $\left(\dfrac{7}{8}r^4 + \dfrac{5}{3}r^2 - \dfrac{9}{4}\right) - \left(-\dfrac{3}{8}r^4 - \dfrac{2}{3}r^2 - \dfrac{1}{4}\right)$

$$= \dfrac{7}{8}r^4 + \dfrac{5}{3}r^2 - \dfrac{9}{4} + \dfrac{3}{8}r^4 + \dfrac{2}{3}r^2 + \dfrac{1}{4}$$

$$= \left(\dfrac{7}{8}r^4 + \dfrac{3}{8}r^4\right) + \left(\dfrac{5}{3}r^2 + \dfrac{2}{3}r^2\right) + \left(-\dfrac{9}{4} + \dfrac{1}{4}\right)$$

$$= \dfrac{10}{8}r^4 + \dfrac{7}{3}r^2 - \dfrac{8}{4}$$

$$= \dfrac{5}{4}r^4 + \dfrac{7}{3}r^2 - 2$$

59.
$$\begin{array}{r} 3x^2 + 4x + 5 \\ +\ \underline{2x^2 - 3x + 6} \\ 5x^2 +\ x + 11 \end{array}$$

61.
$$\begin{array}{r} 3x^2 + 4x - 5 \\ -\ \underline{-2x^2 - 2x + 3} \end{array}$$

$$\begin{array}{r} 3x^2 + 4x - 5 \\ +\ \underline{2x^2 + 2x - 3} \\ 5x^2 + 6x - 8 \end{array}$$

63.
$$\begin{array}{r} 4x^3 + 4x^2 - 3x + 10 \\ -\ \underline{5x^3 - 2x^2 - 4x -\ 4} \end{array}$$

$$\begin{array}{r} 4x^3 + 4x^2 - 3x + 10 \\ +\ \underline{-5x^3 + 2x^2 + 4x +\ 4} \\ -x^3 + 6x^2 +\ x\ + 14 \end{array}$$

65.
$$\begin{array}{r} -3x^3 + 4x^2 - 4x + 9 \\ +\ \underline{2x^3 \qquad + 9x - 3} \\ -x^3 + 4x^2 + 5x + 6 \end{array}$$

67.
$$\begin{array}{r} 3x^3 + 4x^2 + 7x + 12 \\ -\ \underline{4x^3 + 6x^2 + 9x -\ 3} \end{array}$$

$$\begin{array}{r} 3x^3 + 4x^2 + 7x + 12 \\ +\ \underline{4x^3 - 6x^2 - 9x +\ 3} \\ 7x^3 - 2x^2 - 2x + 15 \end{array}$$

69. $2(x + 3) + 4(x - 2)$
$$= 2x + 6 + 4x - 8$$
$$= (2x + 4x) + (6 - 8)$$
$$= 6x - 2$$

71. $-2(x^2 + 7x - 1) - 3(x^2 - 2x + 2)$
$$= -2x^2 - 14x + 2 - 3x^2 + 6x - 6$$
$$= (-2x^2 - 3x^2) + (-14x + 6x) + (2 - 6)$$
$$= -5x^2 - 8x - 4$$

73. $2(2y^2 - 2y + 2) - 4(3y^2 - 4y - 1) + 4(y^3 - y^2 - y)$
$$= 4y^2 - 4y + 4 - 12y^2 + 16y + 4 + 4y^3 - 4y^2 - 4y$$
$$= 4y^3 + (4 - 12 - 4)y^2 + (-4 + 16 - 4)y + (4 + 4)$$
$$= 4y^3 - 12y^2 + 8y + 8$$

75. $(5s^2 - s + 9) - (s^2 + 4s + 2)$
$$= 5s^2 - s + 9 - s^2 - 4s - 2$$
$$= (5s^2 - s^2) + (-s - 4s) + (9 - 2)$$
$$= 4s^2 - 5s + 7$$

77. $(2y^5 - y^4) - (-y^5 + 5y^4 - 1.2)$
$$= 2y^5 - y^4 + y^5 - 5y^4 + 1.2$$
$$= (2y^5 + y^5) + (-y^4 - 5y^4) + 1.2$$
$$= 3y^5 - 6y^4 + 1.2$$

79. $(3t^3 + t^2) + (-t^3 + 6t - 3) - (t^3 - 2t^2 + 2)$
$$= 3t^3 + t^2 - t^3 + 6t - 3 - t^3 + 2t^2 - 2$$
$$= (3t^3 - t^3 - t^3) + (t^2 + 2t^2) + 6t + (-3 - 2)$$
$$= t^3 + 3t^2 + 6t - 5$$

81. $(-2x^2-7x+1)+(-4x^2+8x-1)+(3x^2+4x-7)$
$= -2x^2 - 7x +1 - 4x^2 + 8x - 1 + 3x^2 + 4x - 7$
$= (-2x^2 - 4x^2 + 3x^2) + (-7x + 8x + 4x) + (1-1-7)$
$= -3x^2 + 5x - 7$

APPLICATIONS

83. GREEK ARCHITECTURE

a) $(x^2 - 3x + 2) - (5x - 10)$
$= x^2 - 3x + 2 - 5x + 10$
$= x^2 + (-3x - 5x) + (2 + 10)$
$= (x^2 - 8x + 12)$ feet

b) $(x^2 - 3x + 2) + (5x - 10)$
$= x^2 + (-3x + 5x) + (2 - 10)$
$= (x^2 + 2x - 8)$ feet

85. JETS

$(9x - 15) + (2x + 3)$
$= (9x + 2x) + (-15 + 3)$
$= (11x - 12)$ ft

87. READING BLUEPRINTS

a) length:
$(x^2 - x + 6) + (4x + 3) = x^2 + 3x + 9$
width:
$(x^2 - 6x + 3) + (3x + 1) = x^2 - 3x + 4$

length − width =
$(x^2 + 3x + 9) - (x^2 - 3x + 4)$
$= x^2 + 3x + 9 - x^2 + 3x + 4$
$= (x^2 - x^2) + (3x + 3x) + (9 + 4)$
$= (6x + 13)$ ft

b) Use the length and widths from part a:
$2(x^2 + 3x + 9) + 2(x^2 - 3x + 4)$
$= 2x^2 + 6x + 18 + 2x^2 - 6x + 8$
$= (2x^2 + 2x^2) + (6x - 6x) + (18 + 8)$
$= (4x^2 + 26)$ ft

89. NAVAL OPERATIONS

a) $(-16t^2 + 150t + 40) - (-16t^2 + 128t + 20)$
$= -16t^2 + 150t + 40 + 16t^2 - 128t - 20$
$= (-16t^2 + 16t^2) + (150t - 128t) + (40 - 20)$
$= (22t + 20)$ ft

b) Let $t = 4$ and substitute it into the
answer from part a.
$22(4) + 20 = 88 + 20$
$= 108$ ft

WRITING

91. Answers will vary.

93. Answers will vary.

95. Answers will vary.

REVIEW

97. The sum of the measure of the angles of a
triangle is 180°.

99. $2x + 3y = 9$
$2x + 3y - 2x = 9 - 2x$
$3y = -2x + 9$
$y = -\dfrac{2}{3}x + 3$

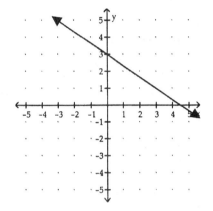

CHALLENGE PROBLEMS

101. $(6x^2 - 7x - 8) - (2x^2 - x + 3)$
$= 6x^2 - 7x - 8 - 2x^2 + x - 3$
$= (6x^2 - 2x^2) + (-7x + x) + (-8 - 3)$
$= 4x^2 - 6x - 11$

VOCABULARY

1. The expression $(2x^3)(3x^4)$ is the product of two **monomials**.

3. The expression $(2a - 4)(3a + 5)$ is the product of two **binomials**.

5. To simplify $y^2 + 3y + 2y + 6$, we combine the **like** terms $3y$ and $2y$ to get $y^2 + 5y + 6$.

CONCEPTS

7. a) To multiply two monomials, multiply the numerical factors (the **coefficients**) and then multiply the **variable** factors.
    b) To multiply two polynomials, multiply **each** term of one polynomial by **each** term of the other polynomial, and then combine like terms.
    c) When multiplying three polynomials, we begin by multiplying **any** two of them, and then we multiply that result by the **third** polynomial.
    d) To find the area of a rectangle, multiply the **length** by the width.

9. $(2x + 5)(3x + 4) = \boxed{6x^2} + \boxed{8x} + \boxed{15x} + \boxed{20}$

11. a) $(x - 4) + (x + 8) = 2x + 4$

    b) $(x - 4) - (x + 8) = x - 4 - x - 8$
    $$= -12$$

    c) $(x - 4)(x + 8) = x^2 + 8x - 4x - 32$
    $$= x^2 + 4x - 32$$

13. $7x(3x^2 - 2x + 5) = \boxed{7x}(3x^2) - \boxed{7x}(2x) + \boxed{7x}(5)$
$$= \boxed{21x^3} - 14x^2 + 35x$$

PRACTICE

15. $15m(m) = 15m^2$

17. $3x^2(4x^3) = (3 \cdot 4)(x^2 \cdot x^3)$
$$= 12x^5$$

19. $(3b^2)(-2b)(4b^3) = (3 \cdot -2 \cdot 4)(b^2 \cdot b \cdot b^3)$
$$= -24b^6$$

21. $(2x^2y^3)(3x^3y^2) = (2 \cdot 3)(x^2y^3 \cdot x^3y^2)$
$$= 6x^5y^5$$

23. $(8a^5)\left(-\frac{1}{4}a^6\right) = \left(8 \bullet -\frac{1}{4}\right)(a^5 a^6)$
$$= -2a^{11}$$

25. $(1.2c^3)(5c^3) - (1.2 \cdot 5)(c^3 c^3)$
$$= 6c^6$$

27. $\left(\frac{1}{2}a\right)\left(\frac{1}{4}a^4\right)(a^5) = \left(\frac{1}{2} \bullet \frac{1}{8}\right)(a a^4 a^5)$
$$= \frac{1}{16}a^{10}$$

29. $(x^2y^5)(x^2z^5)(-3z^3) = -3(x^2x^2y^5z^5z^3)$
$$= -3x^4y^5z^8$$

31. $3(x + 4) = 3(x) + 3(4)$
$$= 3x + 12$$

33. $-4(t^2 + 7) = -4(t^2) - 4(7)$
$$= -4t^2 - 28$$

35. $3x(x - 2) = 3x(x) + 3x(-2)$
$$= 3x^2 - 6x$$

37. $-2x^2(3x^2 - x) = -2x^2(3x^2) - 2x^2(-x)$
$$= -6x^4 + 2x^3$$

39. $(x^2 - 12x)(6x^{12}) = x^2(6x^{12}) - 12x(6x^{12})$
$$= 6x^{14} - 72x^{13}$$

41. $\frac{5}{8}t^2(t^6 + 8t^2) = \frac{5}{8}t^2(t^6) + \frac{5}{8}t^2(8t^2)$
$$= \frac{5}{8}t^8 + 5t^4$$

43. $0.3p^5(0.4p^4 - 6p^2) = 0.3p^5(0.4p^4) + 0.3p^5(-6p^2)$
$$= 0.12p^9 - 1.8p^7$$

45. $-4x^2z(3x^2 - z) = -4x^2z(3x^2) - 4x^2z(-z)$
$$= -12x^4z + 4x^2z^2$$

47. $2x^2(3x^2 + 4x - 7)$
$$= 2x^2(3x^2) + 2x^2(4x) + 2x^2(-7)$$
$$= 6x^4 + 8x^3 - 14x^2$$

49. $3a(4a^2 + 3a - 4)$
$$= 3a(4a^2) + 3a(3a) + 3a(-4)$$
$$= 12a^3 + 9a^2 - 12a$$

51. $(-2a^2)(-3a^3)(3a-2) = (6a^5)(3a-2)$
$= 6a^5(3a) + 6a^5(-2)$
$= 18a^6 - 12a^5$

53. $(y-3)(y+5) = y(y) + y(5) - 3(y) - 3(5)$
$= y^2 + 5y - 3y - 15$
$= y^2 + 2y - 15$

55. $(t+4)(2t-3) = t(2t) + t(-3) + 4(2t) + 4(-3)$
$= 2t^2 - 3t + 8t - 12$
$= 2t^2 + 5t - 12$

57. $(2y-5)(3y+7)$
$= 2y(3y) + 2y(7) - 5(3y) - 5(7)$
$= 6y^2 + 14y - 15y - 35$
$= 6y^2 - y - 35$

59. $(2x+3)(2x-5)$
$= 2x(2x) + 2x(-5) + 3(2x) + 3(-5)$
$= 4x^2 - 10x + 6x - 15$
$= 4x^2 - 4x - 15$

61. $(a+b)(a+b)$
$= a(a) + a(b) + b(a) + b(b)$
$= a^2 + ab + ab + b^2$
$= a^2 + 2ab + b^2$

63. $(3a-2b)(4a+b)$
$= 3a(4a) + 3a(b) - 2b(4a) - 2b(b)$
$= 12a^2 + 3ab - 8ab - 2b^2$
$= 12a^2 - 5ab - 2b^2$

65. $(t^2-3)(t^2+4)$
$= t^2(t^2) + t^2(4) - 3(t^2) - 3(4)$
$= t^4 + 4t^2 - 3t^2 - 12$
$= t^4 + t^2 - 12$

67. $(-3t+2s)(2t-3s)$
$= -3t(2t) - 3t(-3s) + 2s(2t) + 2s(-3s)$
$= -6t^2 + 9st + 4st - 6s^2$
$= -6t^2 + 13st - 6s^2$

69. $\left(4a - \dfrac{5}{9}r\right)\left(2a + \dfrac{3}{4}r\right)$

$= 4a(2a) + 4a\left(\dfrac{3}{4}r\right) - \dfrac{5}{9}r(2a) - \dfrac{5}{9}r\left(\dfrac{3}{4}r\right)$

$= 8a^2 + 3ar - \dfrac{10}{9}ar - \dfrac{5}{12}r^2$

$= 8a^2 + \dfrac{17}{9}ar - \dfrac{5}{12}r^2$

71. $4(2x+1)(x-2)$
$= 4(2x \cdot x + 2x \cdot -2 + 1 \cdot x + 1 \cdot -2)$
$= 4(2x^2 - 4x + x - 2)$
$= 4(2x^2 - 3x - 2)$
$= 4(2x^2) + 4(-3x) + 4(-2)$
$= 8x^2 - 12x - 8$

73. $3a(a+b)(a-b)$
$= 3a(a \cdot a + a \cdot -b + b \cdot a + b \cdot -b)$
$= 3a(a^2 - ab + ab - b^2)$
$= 3a(a^2 - b^2)$
$= 3a^3 - 3ab^2$

75. $(x+2)(x^2-2x+3)$
$= x(x^2-2x+3) + 2(x^2-2x+3)$
$= x \cdot x^2 + x \cdot -2x + x \cdot 3 + 2 \cdot x^2 + 2 \cdot -2x + 2 \cdot 3$
$= x^3 - 2x^2 + 3x + 2x^2 - 4x + 6$
$= x^3 + (-2x^2 + 2x^2) + (3x - 4x) + 6$
$= x^3 - x + 6$

77. $(4t+3)(t^2+2t+3)$
$= 4t(t^2+2t+3) + 3(t^2+2t+3)$
$= 4t^3 + 8t^2 + 12t + 3t^2 + 6t + 9$
$= 4t^3 + (8t^2 + 3t^2) + (12t + 6t) + 9$
$= 4^3 + 11t^2 + 18t + 9$

79. $(x^2+6x+7)(2x-5)$
$= x^2(2x-5) + 6x(2x-5) + 7(2x-5)$
$= 2x^3 - 5x^2 + 12x^2 - 30x + 14x - 35$
$= 2x^3 + (-5x^2 + 12x^2) + (-30x + 14x) - 35$
$= 2x^3 + 7x^2 - 16x - 35$

81. $(-3x+y)(x^2-8xy+16y^2)$
$= -3x(x^2-8xy+16y^2) + y(x^2-8xy+16y^2)$
$= -3x^3 + 24x^2y - 48xy^2 + x^2y - 8xy^2 + 16y^3$
$= -3x^3 + (24x^2y + x^2y) + (-48xy^2 - 8xy^2) + 16y^3$
$= -3x^3 + 25x^2y - 56xy^2 + 16y^3$

83. $(r^2-r+3)(r^2-4r-5)$
$= r^2(r^2-4r-5) - r(r^2-4r-5) + 3(r^2-4r-5)$
$= r^4 - 4r^3 - 5r^2 - r^3 + 4r^2 + 5r + 3r^2 - 12r - 15$
$= r^4 - 4r^3 - r^3 - 5r^2 + 4r^2 + 3r^2 + 5r - 12r - 15$
$= r^4 - 5r^3 + 2r^2 - 7r - 15$

85.
$$\begin{array}{r} x^2 - 2x + 1 \\ x + 2 \\ \hline 2x^2 - 4x + 2 \\ x^3 - 2x^2 + x \\ \hline x^3 + 0x^2 - 3x + 2 \\ x^3 - 3x + 2 \end{array}$$

87.
$$\begin{array}{r} 4x^2 + 3x - 4 \\ 3x + 2 \\ \hline 8x^2 + 6x - 8 \\ 12x^3 + 9x^2 - 12x \\ \hline 12x^3 + 17x^2 - 6x - 8 \end{array}$$

89.
$$\begin{array}{r} 2a^2 + 3a + 1 \\ 3a^2 - 2a + 4 \\ \hline 8a^2 + 12a + 4 \\ -4a^3 - 6a^2 - 2a \\ 6a^4 + 9a^3 + 3a^2 \\ \hline 6a^4 + 5a^3 + 5a^2 + 10a + 4 \end{array}$$

APPLICATIONS

91. STAMPS

$(2x + 1)(3x - 1)$
$= 2x \cdot 3x + 2x \cdot -1 + 1 \cdot 3x + 1 \cdot -1$
$= 6x^2 - 2x + 3x - 1$
$= (6x^2 + x - 1) \text{ cm}^2$

93. $A = s^2$

$(3x + 1)(3x + 1)$
$= 3x \cdot 3x + 3x \cdot 1 + 1 \cdot 3x + 1 \cdot 1$
$= 9x^2 + 3x + 3x + 1$
$= (9x^2 + 6x + 1) \text{ ft}^2$

95. $A = \frac{1}{2} bh$

$\frac{1}{2}(4x - 2)(2x - 2) = \frac{1}{2}(8x^2 - 8x - 4x + 4)$
$\qquad\qquad = \frac{1}{2}(8x^2 - 12x + 4)$
$\qquad\qquad = (4x^2 - 6x + 2) \text{cm}^2$

97. SUNGLASSES

$2 \cdot 3.14(x + 1)(x - 1)$
$= 2 \cdot 3.14(x^2 - x + x - 1)$
$= 2 \cdot 3.14(x^2 - 1)$
$= 6.28(x^2 - 1)$
$= (6.28x^2 - 6.28) \text{ in}^2$

99. TOYS

$(5x + 4)(7x + 3) = 35x^2 + 15x + 28x + 12$
$\qquad\qquad = (35x^2 + 43x + 12) \text{ cm}^2$

101. LUGGAGE

$x(x - 3)(2x + 2) = x(2x^2 + 2x - 6x - 6)$
$\qquad\qquad = x(2x^2 - 4x - 6)$
$\qquad\qquad = (2x^3 - 4x^2 - 6x) \text{ in}^3$

WRITING

103. Answers will vary.

105. Answers will vary.

REVIEW

107. Use any two points on the line such as $(-2, 0)$ and $(0, 2)$:

$$m = \frac{y_2 - y_1}{x_2 - x_1}$$
$$= \frac{2 - 0}{0 - (-2)}$$
$$= \frac{2}{2}$$
$$= 1$$

109. Use any two points on the line such as $(-1, 0)$ and $(2, -2)$:

$$m = \frac{y_2 - y_1}{x_2 - x_1}$$
$$= \frac{-2 - 0}{2 - (-1)}$$
$$= -\frac{2}{3}$$

111. Line 1 intersects the y-axis at $(0, 2)$.

CHALLENGE PROBLEMS

113.   a) $(x - 1)(x + 1) = x^2 - 1$

b) $(x - 1)(x^2 + x + 1)$
$\qquad = x(x^2 + x + 1) - 1(x^2 + x + 1)$
$\qquad = x^3 + x^2 + x - x^2 - x - 1$
$\qquad = x^3 - 1$

c) $(x - 1)(x^3 + x^2 + x + 1)$
$\qquad = x(x^3 + x^2 + x + 1) - 1(x^3 + x^2 + x + 1)$
$\qquad = x^4 + x^3 + x^2 + x - x^3 - x^2 - x - 1$
$\qquad = x^4 - 1$

SECTION 4.7

VOCABULARY

1. The first **term** of $3x + 6$ is $3x$ and the second **term** is 6.

3. $(b + 1)(b - 1)$ is the product of the **sum** and difference of two terms.

5. An expression of the form $x^2 - y^2$ is called a **difference** of two squares.

CONCEPTS

7. a) $(x^m)^n = \boxed{x^{m \cdot n}}$
   b) $(xy)^n = \boxed{x^n y^n}$

9. a) $(a + 5)(a + 5) = (a + 5)^{\boxed{2}}$
   b) $(n - 12)(n - 12) = (n - 12)^{\boxed{2}}$

11. $(x + y)(x - y) = x^2 - y^2$

The square of the
second term
The **square** of the first term

NOTATION

13. $(x + 4)^2 = \boxed{x}^2 + 2(x)(\boxed{4}) + \boxed{4}^2$
    $= x^2 + \boxed{8x} + 16$

15. $(s + 5)(s - 5) = \boxed{s}^2 \boxed{-} \boxed{5}^2$
    $= s^2 - 25$

PRACTICE

17. $(x + 1)^2 = x^2 + 2(x)(1) + 1^2$
    $= x^2 + 2x + 1$

19. $(r + 2)^2 = r^2 + 2(r)(2) + 2^2$
    $= r^2 + 4r + 4$

21. $(m - 6)^2 = m^2 - 2(m)(6) + (-6)^2$
    $= m^2 - 12m + 36$

23. $(f - 8)^2 = f^2 - 2(f)(8) + (-8)^2$
    $= f^2 - 16f + 64$

25. $(d + 7)(d - 7) = d^2 - 7^2$
    $= d^2 - 49$

27. $(n + 6)(n - 6) = n^2 - 6^2$
    $= n^2 - 36$

29. $(4x + 5)^2 = (4x)^2 + 2(4x)(5) + (5)^2$
    $= 16x^2 + 40x + 25$

31. $(7m - 2)^2 = (7m)^2 - 2(7m)(2) + (-2)^2$
    $= 49m^2 - 28m + 4$

33. $(y^2 + 9)^2 = (y^2)^2 + 2(y^2)(9) + (9)^2$
    $= y^4 + 18y^2 + 81$

35. $(2v^3 - 8)^2 = (2v^3)^2 - 2(2v^3)(8) + (-8)^2$
    $= 4v^6 - 32v^3 + 64$

37. $(4f + 0.4)(4f - 0.4) = (4f)^2 - (0.4)^2$
    $= 16f^2 - 0.16$

39. $(3n + 1)(3n - 1) = (3n)^2 - (1)^2$
    $= 9n^2 - 1$

41. $(1 - 3y)^2 = (1)^2 - 2(1)(3y) + (-3y)^2$
    $= 1 - 6y + 9y^2$

43. $(x - 2y)^2 = (x)^2 - 2(x)(2y) + (-2y)^2$
    $= x^2 - 4xy + 4y^2$

45. $(2a - 3b)^2 = (2a)^2 - 2(2a)(3b) + (-3b)^2$
    $= 4a^2 - 12ab + 9b^2$

47. $\left(s + \dfrac{3}{4}\right)^2 = s^2 + 2(s)\left(\dfrac{3}{4}\right) + \left(\dfrac{3}{4}\right)^2$

$= s^2 + \dfrac{6}{4}s + \dfrac{9}{16}$

$= s^2 + \dfrac{3}{2}s + \dfrac{9}{16}$

49. $(a + b)^2 = a^2 + 2(a)(b) + b^2$
    $= a^2 + 2ab + b^2$

51. $(r - s)^2 = r^2 - 2(r)(s) + (-s)^2$
    $= r^2 - 2rs + s^2$

53. $\left(6b + \dfrac{1}{2}\right)\left(6b - \dfrac{1}{2}\right) = (6b)^2 - \left(\dfrac{1}{2}\right)^2$

$= 36b^2 - \dfrac{1}{4}$

55. $(r + 10s)^2 = (r)^2 + 2(r)(10s) + (10s)^2$
    $= r^2 + 20rs + 100s^2$

57. $(6 - 2d^3)^2 = (6)^2 - 2(6)(2d^3) + (-2d^3)^2$
$= 36 - 24d^3 + 4d^6$

59. $-(8x + 3)^2 = -[(8x)^2 + 2(8x)(3) + (3)^2]$
$= -[64x^2 + 48x + 9]$
$= -64x^2 - 48x - 9$

61. $-(5 - 6g)(5 + 6g) = -[(5)^2 - (6g)^2]$
$= -[25 - 36g^2]$
$= -25 + 36g^2$

63. $3x(2x + 3)^2 = 3x[(2x)^2 + 2(2x)(3) + (3)^2]$
$= 3x[4x^2 + 12x + 9]$
$= 12x^3 + 36x^2 + 27x$

65. $-5d(4d - 1)^2 = -5d[(4d)^2 - 2(4d)(1) + (-1)^2]$
$= -5d[16d^2 - 8d + 1]$
$= -80d^3 + 40d^2 - 5d$

67. $4d(d^2 + g^3)(d^2 - g^3) = 4d[(d^2)^2 - (g^3)^2]$
$= 4d[d^4 - g^6]$
$= 4d^5 - 4dg^6$

69. $(x + 4)^3 = (x + 4)(x + 4)^2$
$= (x + 4)[x^2 + 2(x)(4) + 4^2]$
$= (x + 4)[x^2 + 8x + 16]$
$= x(x^2 + 8x + 16) + 4(x^2 + 8x + 16)$
$= x^3 + 8x^2 + 16x + 4x^2 + 32x + 64$
$= x^3 + 12x^2 + 48x + 64$

71. $(n - 6)^3 = (n - 6)(n - 6)^2$
$= (n - 6)[n^2 - 2(n)(6) + (-6)^2]$
$= (n - 6)[n^2 - 12n + 36]$
$= n(n^2 - 12n + 36) - 6(n^2 - 12n + 36)$
$= n^3 - 12n^2 + 36n - 6n^2 + 72n - 216$
$= n^3 - 18n^2 + 108n - 216$

73. $(2g - 3)^3 = (2g - 3)(2g - 3)^2$
$= (2g - 3)[(2g)^2 - 2(2g)(3) + (-3)^2]$
$= (2g - 3)[4g^2 - 12g + 9]$
$= 2g(4g^2 - 12g + 9) - 3(4g^2 - 12g + 9)$
$= 8g^3 - 24g^2 + 18g - 12g^2 + 36g - 27$
$= 8g^3 - 36g^2 + 54g - 27$

75. $(n - 2)^4$
$= (n-2)^2 (n-2)^2$
$= [(n)^2 - 2(n)(2) + (-2)^2][(n)^2 - 2(n)(2) + (-2)^2]$
$= [n^2 - 4n + 4][n^2 - 4n + 4]$
$= n^2(n^2 - 4n + 4) - 4n(n^2 - 4n + 4) + 4(n^2 - 4n + 4)$
$= n^4 - 4n^3 + 4n^2 - 4n^3 + 16n^2 - 16n + 4n^2 - 16n + 16$
$= n^4 - 8n^3 + 24n^2 - 32n + 16$

77. $2t(t + 2) + (t - 1)(t + 9)$
$= 2t^2 + 4t + t^2 + 9t - 1t - 9$
$= 3t^2 + 12t - 9$

79. $(x + y)(x - y) + x(x + y)$
$= (x^2 - y^2) + x^2 + xy$
$= 2x^2 + xy - y^2$

81. $(3x - 2)^2 + (2x + 1)^2$
$= [(3x)^2 - 2(3x)(2) + (2)^2] + [(2x)^2 + 2(2x)(1) + 1^2]$
$= 9x^2 - 12x + 4 + 4x^2 + 4x + 1$
$= 13x^2 - 8x + 5$

83. $(m + 10)^2 - (m - 8)^2$
$= [(m)^2 + 2(m)(10) + (10)^2] - [(m)^2 - 2(m)(8) + 8^2]$
$= [m^2 + 20m + 100] - [m^2 - 16m + 64]$
$= m^2 + 20m + 100 - m^2 + 16m - 64$
$= 36m + 36$

APPLICATIONS

85. DINNER PLATES

$$\frac{\pi}{4}(D - d)(D + d) = \frac{\pi}{4}(D^2 - d^2)$$
$$= \frac{\pi}{4}D^2 - \frac{\pi}{4}d^2$$

87. PLAYPENS

$$(x + 6)^2 = (x^2 + 2(x)(6) + 6^2)$$
$$= (x^2 + 12x + 36) \text{ in.}^2$$

89. PAINTING

Find the length: $(4x + 2) + (8x + 4) = 12x + 6$

Multiply the length times the width:

$(4x + 1)(12x + 6)$
$= 48x^2 + 24x + 12x + 6$
$= 48x^2 + 36x + 6$

Find the area of one window: $x \cdot x = x^2$

Area of 12 windows = $12x^2$

Subtract the area of the 12 windows from the total area:

$48x^2 + 36x + 6 - 12x^2$
$= (36x^2 + 36x + 6) \text{ ft}^2$

WRITING

91. Answers will vary.

93. Answers will vary.

REVIEW

95. $189 = 3 \cdot 3 \cdot 3 \cdot 7$
$\quad\quad = 3^3 \cdot 7$

97. $\dfrac{30}{36} = \dfrac{\cancel{2} \bullet \cancel{3} \bullet 5}{\cancel{2} \bullet 2 \bullet \cancel{3} \bullet 3} = \dfrac{5}{6}$

99. $\dfrac{7}{8} \bullet \dfrac{3}{5} = \dfrac{7 \bullet 3}{8 \bullet 5}$
$\quad\quad\quad = \dfrac{21}{40}$

CHALLENGE PROBLEMS

100. a) $(x + 1)(x - 1) = x^2 - 1$
      b) $(x + 1)(x + 1) = x^2 + 2x + 1$
      c) $(x + a)(x + b) = x^2 + ax + bx + ab$

SECTION 4.8

VOCABULARY

1. The **numerator** of $\frac{15x^2-25x}{5x}$ is $15x^2 - 25x$
    and the **denominator** is $5x$.

3. The expression $\frac{6x^3y-4x^2y^2+8xy^3-2y^4}{2x^4}$ is a
    polynomial divided by a monomial.

5. The powers of x in $2x^4 + 3x^3 + 4x^2 - 7x - 8$
    are written in **descending** order.

7. The expression $5x^2 + 6$ is missing an
    x-term. We can insert a **placeholder** $0x$
    term and write it as $5x^2 + 0x + 6$.

CONCEPTS

9. a) $\dfrac{x^m}{x^n} = x^{m-n}$

    b) $x^{-n} = \dfrac{1}{x^n}$

11. a) $7x^3 + 5x^2 - 3x - 9$

    b) $6x^4 - x^3 + 2x^2 + 9x$

13. $-9x - (-7x) = -9x + 7x$
    $= -2x$

15. It is correct.
    $(3x + 2)(x + 2) = 3x^2 + 6x + 2x + 4$
    $= 3x^2 + 8x + 4$

NOTATION

17. $\dfrac{28x^5 - x^3 + 7x^2}{7x^2} = \dfrac{28x^5}{\boxed{7x^2}} - \dfrac{\boxed{x^3}}{7x^2} + \dfrac{7x^2}{\boxed{7x^2}}$

    $= 4x^{5-2} - \dfrac{x^{3-2}}{\boxed{7}} + x^{\boxed{2-2}}$

    $= \boxed{4x^3} - \dfrac{x}{7} + \boxed{1}$

19. a) $5x^4 + 0x^3 + 2x^2 + 0x - 1$

    b) $-3x^5 + 0x^4 - 2x^3 + 0x^2 + 4x - 6$

21. $\dfrac{8}{6} = \dfrac{4}{3}$

23. $\dfrac{x^5}{x^2} = x^{5-2}$

    $= x^3$

25. $\dfrac{45m^{10}}{9m^5} = 5m^{10-5}$

    $= 5m^5$

27. $\dfrac{12h^8}{9h^6} = \dfrac{4h^{8-6}}{3}$

    $= \dfrac{4h^2}{3}$

29. $\dfrac{-3d^4}{15d^8} = \dfrac{-1}{5}d^{4-8}$

    $= -\dfrac{1}{5}d^{-4}$

    $= -\dfrac{1}{5d^4}$

31. $\dfrac{r^3s^2}{rs^3} = r^{3-1}s^{2-3}$

    $= r^2s^{-1}$

    $= \dfrac{r^2}{s}$

33. $\dfrac{8x^3y^2}{4xy^3} = 2x^{3-1}y^{2-3}$

    $= 2x^2y^{-1}$

    $= \dfrac{2x^2}{y}$

35. $\dfrac{-16r^3y^2}{-4r^2y^4} = 4r^{3-2}y^{2-4}$

$\qquad = 4r^1y^{-2}$

$\qquad = \dfrac{4r}{y^2}$

37. $\dfrac{-65rs^2t}{15r^2s^3t} = -\dfrac{13}{3}r^{1-2}s^{2-3}t^{1-1}$

$\qquad = -\dfrac{13}{3}r^{-1}s^{-1}t^0$

$\qquad = -\dfrac{13}{3rs}$

39. $\dfrac{6x+9}{3} = \dfrac{6x}{3} + \dfrac{9}{3}$

$\qquad = 2x+3$

41. $\dfrac{8x^9 - 32x^6}{4x^4} = \dfrac{8x^9}{4x^4} - \dfrac{32x^6}{4x^4}$

$\qquad = 2x^5 - 8x^2$

43. $\dfrac{6h^{12} + 48h^9}{24h^{10}} = \dfrac{6h^{12}}{24h^{10}} + \dfrac{48h^9}{24h^{10}}$

$\qquad = \dfrac{1}{4}h^{12-10} + 2h^{9-10}$

$\qquad = \dfrac{1}{4}h^2 + 2h^{-1}$

$\qquad = \dfrac{h^2}{4} + \dfrac{2}{h}$

45. $\dfrac{-18w^6 - 9}{9w^4} = \dfrac{-18w^6}{9w^4} - \dfrac{9}{9w^4}$

$\qquad = -2w^{6-4} - 1w^{-4}$

$\qquad = -2w^2 - \dfrac{1}{w^4}$

47. $\dfrac{9s^8 - 18s^5 + 12s^4}{3s^3} = \dfrac{9s^8}{3s^3} - \dfrac{18s^5}{3s^3} + \dfrac{12s^4}{3s^3}$

$\qquad = 3s^{8-3} - 6s^{5-3} + 4s^{4-3}$

$\qquad = 3s^5 - 6s^2 + 4s$

49. $\dfrac{7c^5 + 21c^4 - 14c^3 - 35c}{7c^2}$

$\qquad = \dfrac{7c^5}{7c^2} + \dfrac{21c^4}{7c^2} - \dfrac{14c^3}{7c^2} - \dfrac{35c}{7c^2}$

$\qquad = c^{5-2} + 3c^{4-2} - 2c^{3-2} - 5c^{1-2}$

$\qquad = c^3 + 3c^2 - 2c^1 - 5c^{-1}$

$\qquad = c^3 + 3c^2 - 2c - \dfrac{5}{c}$

51. $\dfrac{5x - 10y}{25xy} = \dfrac{5x}{25xy} - \dfrac{10y}{25xy}$

$\qquad = \dfrac{1}{5y} - \dfrac{2}{5x}$

53. $\dfrac{15a^3b^2 - 10a^2b^3}{5a^2b^2} = \dfrac{15a^3b^2}{5a^2b^2} - \dfrac{10a^2b^3}{5a^2b^2}$

$\qquad = 3a^{3-2}b^{2-2} - 2a^{2-2}b^{3-2}$

$\qquad = 3a^1b^0 - 2a^0b^1$

$\qquad = 3a - 2b$

55. $\dfrac{12x^3y^2 - 8x^2y - 4x}{4xy}$

$\qquad = \dfrac{12x^3y^2}{4xy} - \dfrac{8x^2y}{4xy} - \dfrac{4x}{4xy}$

$\qquad = 3x^{3-1}y^{2-1} - 2x^{2-1}y^{1-1} - 4x^{1-1}y^{-1}$

$\qquad = 3x^2y - 2x - \dfrac{1}{y}$

57. $\dfrac{-25x^2y + 30xy^2 - 5xy}{-5xy}$

$\qquad = \dfrac{-25x^2y}{-5xy} + \dfrac{30xy^2}{-5xy} + \dfrac{-5xy}{-5xy}$

$\qquad = 5x^{2-1}y^{1-1} - 6x^{1-1}y^{2-1} + 1x^{1-1}y^{1-1}$

$\qquad = 5x^1y^0 - 6x^0y^1 + 1x^0y^0$

$\qquad = 5x - 6y + 1$

59. $x + 6$

$$
\begin{array}{r}
x + 6 \\
x + 2 \overline{)\, x^2 + 8x + 12} \\
\underline{x^2 + 2x} \\
6x + 12 \\
\underline{6x + 12} \\
0
\end{array}
$$

61. $y + 12$

$$
\begin{array}{r}
y + 12 \\
y + 1 \overline{)\, y^2 + 13y + 12} \\
\underline{y^2 + 1y} \\
12y + 12 \\
\underline{12y + 12} \\
0
\end{array}
$$

63. $3a - 2$

$$
\begin{array}{r}
3a - 2 \\
2a + 3 \overline{)\, 6a^2 + 5a - 6} \\
\underline{6a^2 + 9a} \\
-4a - 6 \\
\underline{-4a - 6} \\
0
\end{array}
$$

65. $b + 3$

$$
\begin{array}{r}
b + 3 \\
3b + 2 \overline{)\, 3b^2 + 11b + 6} \\
\underline{3b^2 + 2b} \\
9b + 6 \\
\underline{9b + 6} \\
0
\end{array}
$$

67. $2x + 1$

Write the dividend in descending order.

$$
\begin{array}{r}
2x + 1 \\
5x + 3 \overline{)\, 10x^2 + 11x + 3} \\
\underline{10x^2 + 6x} \\
5x + 3 \\
\underline{5x + 3} \\
0
\end{array}
$$

69. $x - 7$

Write the dividend and the divisor in descending order.

$$
\begin{array}{r}
x - 7 \\
2x + 4 \overline{)\, 2x^2 - 10x - 28} \\
\underline{2x^2 + 4x} \\
-14x - 28 \\
\underline{-14x - 28} \\
0
\end{array}
$$

71. $3x + 2$

Write the dividend in descending order.

$$
\begin{array}{r}
3x + 2 \\
2x - 1 \overline{)\, 6x^2 + x - 2} \\
\underline{6x^2 - 3x} \\
4x - 2 \\
\underline{4x - 2} \\
0
\end{array}
$$

73. $x^2 + 2x - 1$

$$
\begin{array}{r}
x^2 + 2x - 1 \\
2x+3\overline{\smash{\big)}\,2x^3 + 7x^2 + 4x - 3} \\
\underline{2x^3 + 3x^2} \\
4x^2 + 4x \\
\underline{4x^2 + 6x} \\
-2x - 3 \\
\underline{-2x - 3} \\
0
\end{array}
$$

75. $2x^2 + 2x + 1$

$$
\begin{array}{r}
2x^2 + 2x + 1 \\
3x+2\overline{\smash{\big)}\,6x^3 + 10x^2 + 7x + 2} \\
\underline{6x^3 + 4x^2} \\
6x^2 + 7x \\
\underline{6x^2 + 4x} \\
3x + 2 \\
\underline{3x + 2} \\
0
\end{array}
$$

77. $x^2 + x + 1$

$$
\begin{array}{r}
x^2 + x + 1 \\
2x+1\overline{\smash{\big)}\,2x^3 + 3x^2 + 3x + 1} \\
\underline{2x^3 + x^2} \\
2x^2 + 3x \\
\underline{2x^2 + x} \\
2x + 1 \\
\underline{2x + 1} \\
0
\end{array}
$$

79. $x + 1$

Insert a $0x$ term as a placeholder.

$$
\begin{array}{r}
x + 1 \\
x-1\overline{\smash{\big)}\,x^2 + 0x - 1} \\
\underline{x^2 - x} \\
x - 1 \\
\underline{x - 1} \\
0
\end{array}
$$

81. $2x - 3$

Insert a $0x$ term as a placeholder.

$$
\begin{array}{r}
2x - 3 \\
2x+3\overline{\smash{\big)}\,4x^2 + 0x - 9} \\
\underline{4x^2 + 6x} \\
-6x - 9 \\
\underline{-6x - 9} \\
0
\end{array}
$$

83. $x^2 - x + 1$

Insert a $0x^2$ term and a $0x$ term as placeholders.

$$
\begin{array}{r}
x^2 - x + 1 \\
x+1\overline{\smash{\big)}\,x^3 + 0x^2 + 0x + 1} \\
\underline{x^3 + x^2} \\
-x^2 + 0x \\
\underline{-x^2 - x} \\
x + 1 \\
\underline{x + 1} \\
0
\end{array}
$$

85. $a^2 - 3a + 10 + \frac{-30}{a+3}$

Insert a $0a^2$ term and a 0 constant term as placeholders.

$$
\begin{array}{r}
a^2 - 3a + 10 + \frac{-30}{a+3} \\
a+3\overline{\smash{\big)}\,a^3 + 0a^2 + a + 0} \\
\underline{a^3 + 3a^2} \\
-3a^2 + a \\
\underline{-3a^2 - 9a} \\
10a + 0 \\
\underline{10a + 30} \\
-30
\end{array}
$$

87. $x + 1 + \frac{-1}{2x+3}$

$$\begin{array}{r} x + 1 + \frac{-1}{2x+3} \\ 2x+3 \overline{\smash{\big)}\ 2x^2 + 5x + 2} \\ \underline{2x^2 + 3x} \\ 2x + 2 \\ \underline{2x + 3} \\ -1 \end{array}$$

89. $2x + 2 + \frac{-3}{2x+1}$

$$\begin{array}{r} 2x + 2 + \frac{-3}{2x+1} \\ 2x+1 \overline{\smash{\big)}\ 4x^2 + 6x - 1} \\ \underline{4x^2 + 2x} \\ 4x - 1 \\ \underline{4x + 2} \\ -3 \end{array}$$

91. $x^2 + 2x - 1 + \frac{6}{2x+3}$

$$\begin{array}{r} x^2 + 2x - 1 + \frac{6}{2x+3} \\ 2x+3 \overline{\smash{\big)}\ 2x^3 + 7x^2 + 4x + 3} \\ \underline{2x^3 + 3x^2} \\ 4x^2 + 4x \\ \underline{4x^2 + 6x} \\ -2x + 3 \\ \underline{-2x - 3} \\ 6 \end{array}$$

93. $2x^2 + x + 1 + \frac{2}{3x-1}$

$$\begin{array}{r} 2x^2 + x + 1 + \frac{2}{3x-1} \\ 3x-1 \overline{\smash{\big)}\ 6x^3 + x^2 + 2x + 1} \\ \underline{6x^3 - 2x^2} \\ 3x^2 + 2x \\ \underline{3x^2 - x} \\ 3x + 1 \\ \underline{3x - 1} \\ 2 \end{array}$$

APPLICATIONS

95. POOL

$$\frac{6x^2 - 3x + 9}{3} = \frac{6x^2}{3} - \frac{3x}{3} + \frac{9}{3}$$
$$= \left(2x^2 - x + 3\right) \text{in.}$$

97. a) $d = rt$

$$\frac{d}{r} = \frac{rt}{r}$$
$$\frac{d}{r} = t$$
$$t = \frac{d}{r}$$

b) $t = \frac{d}{r}$

$$= \frac{6x^3}{2x}$$
$$= 3x^{3-1}$$
$$= 3x^2$$

c) $t = \frac{d}{r}$

$$= \frac{x^2 + x - 12}{x - 3}$$

$$= x-3 \overline{\smash{\big)}\ x^2 + x - 12} \begin{array}{r} x + 4 \end{array}$$
$$\underline{x^2 - 3x}$$
$$4x - 12$$
$$\underline{4x - 12}$$
$$0$$

99. AIR CONDITIONING

Multiply $3x \cdot 4x = 12x^2$. Then divide $12x^2$ into the volume of $(36x^3 - 24x^2)$ ft^3.

$$\frac{36x^3 - 24x^2}{12x^2} = \frac{36x^3}{12x^2} - \frac{24x^2}{12x^2}$$
$$= \left(3x - 2\right) \text{ft}$$

101. COMMUNICATION

$$4x^2 + 3x + 7$$

$$
\begin{array}{r}
4x^2 + 3x + 7 \\
2x - 3\overline{\smash{\big)}\,8x^3 - 6x^2 + 5x - 21} \\
\underline{8x^3 - 12x^2} \\
6x^2 + 5x \\
\underline{6x^2 - 9x} \\
14x - 21 \\
\underline{14x - 21} \\
0
\end{array}
$$

WRITING

103. Answers will vary.

105. Answers will vary.

REVIEW

107.
$$y - y_1 = m(x - x_1)$$
$$y - (-6) = -\frac{11}{6}(x - 2)$$
$$y + 6 = -\frac{11}{6}x - \frac{11}{6}(-2)$$
$$y + 6 = -\frac{11}{6}x + \frac{11}{3}$$
$$y + 6 - 6 = -\frac{11}{6}x + \frac{11}{3} - 6$$
$$y = -\frac{11}{6}x + \frac{11}{3} - \frac{18}{3}$$
$$y = -\frac{11}{6}x - \frac{7}{3}$$

109.
$$-10\left(18 - 4^2\right)^3 = -10(18 - 16)^3$$
$$= -10(2)^3$$
$$= -10(8)$$
$$= -80$$

CHALLENGE PROBLEMS

111.
$$\frac{6x^{6m}y^{6n} + 15x^{4m}y^{7n} - 24x^{2m}y^{8n}}{3x^{2m}y^n}$$
$$= \frac{6x^{6m}y^{6n}}{3x^{2m}y^n} + \frac{15x^{4m}y^{7n}}{3x^{2m}y^n} - \frac{24x^{2m}y^{8n}}{3x^{2m}y^n}$$
$$= 2x^{6m-2m}y^{6n-n} + 5x^{4m-2m}y^{7n-n}$$
$$\qquad - 8x^{2m-2m}y^{8n-n}$$
$$= 2x^{4m}y^{5n} + 5x^{2m}y^{6n} - 8y^{7n}$$

CHAPTER 4: KEY CONCEPTS

1. a) This is a polynomial in **x**. It is written in **descending** powers of x.
   b) It has 4 terms.
   c) The degrees of each term are 3, 2, 1, and 0.
   d) The degree of the polynomial is 3.
   e) The coefficients are 1, -2, 6, and -8.

2. a) binomial
   b) none of these
   c) trinomial
   d) monomial

3. $4x^3 + 3x^3 = (4+3)x^3$
   $= 7x^3$

4. $7m^{10} + (-6m^{10}) = (7-6)m^{10}$
   $= 1m^{10}$
   $= m^{10}$

5. $7a^2b - 9a^2b = (7-9)a^2b$
   $= -2a^2b$

6. $(6y^5)(-7y^8) = (6 \cdot -7)y^{5+8}$
   $= -42y^{13}$

7. $\dfrac{16c^4d^5}{8c^2d^6} = 2c^{4-2}d^{5-6}$
   $= 2c^2d^{-1}$
   $= \dfrac{2c^2}{d}$

8. $\left(5f^3\right)^2 = 5^2 f^{3 \cdot 2}$
   $= 25f^6$

9. To add polynomials, **combine** like terms.

10. To subtract two polynomials, change the **signs** of the terms of the polynomial being subtracted, and combine like terms.

11. To multiply two polynomials, multiply **each** term of one polynomial by **each** term of the other polynomial and combine like terms.

12. To divide a polynomial by a monomial, divide each **term** of the polynomial by the monomial.

13. To divide two polynomials, use the **long** division method.

14. $(8x^3 + 4x^2 - 8x + 1) + (6x^3 - 5x^2 - 2x + 3)$
    $= (8x^3 + 6x^3) + (4x^2 - 5x^2) + (-8x - 2x) + (1+3)$
    $= 14x^3 - x^2 - 10x + 4$

15. $(20s^3t + s^2t^2 - 6st^3) - (8s^3t - 9s^2t^2 + 12st^3)$
    $= 20s^3t + s^2t^2 - 6st^3 - 8s^3t + 9s^2t^2 - 12st^3$
    $= (20s^3t - 8s^3t) + (s^2t^2 + 9s^2t^2) + (-6st^3 - 12st^3)$
    $= 12s^3t + 10s^2t^2 - 18st^3$

16. $(2x+3)(x-8) = 2x(x) + 2x(-8) + 3(x) + 3(-8)$
    $= 2x^2 - 16x + 3x - 24$
    $= 2x^2 - 13x - 24$

17. $(2x^2 + 3)^2 = (2x^2)^2 + 2(2x^2)(3) + 3^2$
    $= 4x^4 + 12x^2 + 9$

18. $(4h^5 + 8t)(4h^5 + 8t) = (4h^5)^2 - (8t)^2$
    $= 16h^{10} - 64t^2$

19. $(y^2+y-6)(y+3) = y^2(y+3) + y(y+3) - 6(y+3)$
    $= y^3 + 3y^2 + y^2 + 3y - 6y - 18$
    $= y^3 + (3y^2+y^2) + (3y-6y) - 18$
    $= y^3 + 4y^2 - 3y - 18$

20. $\dfrac{9x^6 + 27x^7 - 18x^5}{3x^2} = \dfrac{9x^6}{3x^2} + \dfrac{27x^7}{3x^2} - \dfrac{18x^5}{3x^2}$
    $= 3x^{6-2} + 9x^{7-2} - 6x^{5-2}$
    $= 3x^4 + 9x^5 - 6x^3$

21. $x^2 + 2x + 3$

$$
\begin{array}{r}
x^2 + 2x + 3 \\
x+1\overline{)x^3 + 3x^2 + 5x + 3} \\
\underline{x^3 + x^2} \\
2x^2 + 5x \\
\underline{2x^2 + 2x} \\
3x + 3 \\
\underline{3x + 3} \\
0
\end{array}
$$

CHAPTER 4 REVIEW

SECTION 4.1
Rules for Exponents

1. a) $-3x^4 = -3 \cdot x \cdot x \cdot x \cdot x$

    b) $\left(\dfrac{1}{2}pq\right)^3 = \left(\dfrac{1}{2}pq\right)\left(\dfrac{1}{2}pq\right)\left(\dfrac{1}{2}pq\right)$

2. a) base x, exponent 6
    b) base $2x$, exponent 6

3. $5^3 = 5 \cdot 5 \cdot 5$
       $= 125$

4. $(-8)^2 = (-8)(-8)$
        $= 64$

5. $-8^2 = -(8)(8)$
        $= -64$

6. $(5-3)^2 = (2)^2$
           $= 2(2)$
           $= 4$

7. $7^4 \cdot 7^8 = 7^{4+8}$
           $= 7^{12}$

8. $mmnn = m^2n^2$

9. $(y^7)^3 = y^{7 \cdot 3}$
         $= y^{21}$

10. $(3x)^4 = 3^4 \cdot x^4$
          $= 81x^4$

11. $b^3b^4b^5 = b^{3+4+5}$
            $= b^{12}$

12. $-z^2(z^3y^2) = -y^2z^{2+3}$
              $= -y^2z^5$

13. $(-16s)^2s = (-16)^2s^2s$
             $= 256s^{2+1}$
             $= 256s^3$

14. $(2x^2y)^2 = 2^2 \cdot x^{2 \cdot 2} \cdot y^{1 \cdot 2}$
            $= 4x^4y^2$

15. $(x^2x^3)^3 = (x^{2+3})^3$
            $= (x^5)^3$
            $= x^{5 \cdot 3}$
            $= x^{15}$

16. $\left(\dfrac{x^2y}{xy^2}\right)^2 = \left(x^{2-1}y^{1-2}\right)^2$

    $= \left(xy^{-1}\right)^2$

    $= \left(x^{1 \bullet 2}y^{-1 \bullet 2}\right)$

    $= x^2y^{-2}$

    $= \dfrac{x^2}{y^2}$

17. $\dfrac{(m-25)^{16}}{(m-25)^4} = (m-25)^{16-4}$

    $= (m-25)^{12}$

18. $\dfrac{\left(5y^2z^3\right)^3}{\left(yz\right)^5} = \dfrac{5^3 y^{2 \bullet 3} z^{3 \bullet 3}}{y^{1 \bullet 5} z^{1 \bullet 5}}$

    $= \dfrac{125y^6z^9}{y^5z^5}$

    $= 125y^{6-5}z^{9-5}$

    $= 125yz^4$

19. $V = (4x^4)^3$
       $= 4^3 \cdot x^{4 \cdot 3}$
       $= 64x^{12}$

20. $A = (y^2)^2$
       $= y^{2 \cdot 2}$
       $= y^4$

21. $x^0 = 1$

22. $(3x^2y^2)^0 = 1$

23. $(3x^0)^2 = 3^2 \cdot x^{0 \cdot 2}$
           $= 9 \cdot 1$
           $= 9$

24. $10^{-3} = \dfrac{1}{10^3}$

    $= \dfrac{1}{1,000}$

25. $\left(\dfrac{3}{4}\right)^{-1} = \dfrac{4}{3}$

26. $-5^{-2} = -\dfrac{1}{5^2}$

$\qquad\quad = -\dfrac{1}{25}$

27. $x^{-5} = \dfrac{1}{x^5}$

28. $-6y^4 y^{-5} = -6y^{4+(-5)}$

$\qquad\qquad = -6y^{-1}$

$\qquad\qquad = -\dfrac{6}{y}$

29. $\dfrac{7^{-2}}{2^{-3}} = \dfrac{2^3}{7^2}$

$\qquad\quad = \dfrac{8}{49}$

30. $\left(x^{-3}x^{-4}\right)^{-2} = \left(x^{-3+(-4)}\right)^{-2}$

$\qquad\qquad\quad = \left(x^{-7}\right)^{-2}$

$\qquad\qquad\quad = x^{-7(-2)}$

$\qquad\qquad\quad = x^{14}$

31. $\left(\dfrac{-3r^4 r^{-3}}{r^{-3}r^7}\right)^3 = \left(\dfrac{-3r^{4+(-3)}}{r^{-3+7}}\right)^3$

$\qquad\qquad\quad = \left(\dfrac{-3r}{r^4}\right)^3$

$\qquad\qquad\quad = \left(-3r^{1-4}\right)^3$

$\qquad\qquad\quad = \left(-3r^{-3}\right)^3$

$\qquad\qquad\quad = (-3)^3 r^{-3\bullet 3}$

$\qquad\qquad\quad = -27r^{-9}$

$\qquad\qquad\quad = -\dfrac{27}{r^9}$

32. $\left(\dfrac{4z^4}{z^3}\right)^{-2} = \left(4z^{4-3}\right)^{-2}$

$\qquad\qquad\quad = \left(4z\right)^{-2}$

$\qquad\qquad\quad = \dfrac{1}{\left(4z\right)^2}$

$\qquad\qquad\quad = \dfrac{1}{4^2 z^2}$

$\qquad\qquad\quad = \dfrac{1}{16z^2}$

33. $728 = 7.28 \times 10^2$

34. $9{,}370{,}000{,}000{,}000{,}000 = 9.37 \times 10^{15}$

35. $0.0136 = 1.36 \times 10^{-2}$

36. $0.00942 = 9.42 \times 10^{-3}$

37. $0.018 \times 10^{-2} = 1.8 \times 10^{-4}$

38. $753 \times 10^3 = 7.53 \times 10^{2+3}$

$\qquad\qquad\quad = 7.53 \times 10^5$

39. $7.26 \times 10^5 = 726{,}000$; move the decimal 5 places to the right

40. $3.91 \times 10^{-8} = 0.0000000391$; move the decimal 8 places to the left

41. $2.68 \times 10^0 = 2.68$; the decimal does not move

42. $5.76 \times 10^1 = 57.6$; move the decimal 1 place to the right

43. $\dfrac{(0.00012)(0.00004)}{0.00000016} = \dfrac{\left(1.2\times10^{-4}\right)\left(4\times10^{-5}\right)}{1.6\times10^{-7}}$

$\qquad\qquad\qquad\quad = \dfrac{(1.2\bullet 4)\left(10^{-4+(-5)}\right)}{1.6\times10^{-7}}$

$\qquad\qquad\qquad\quad = \dfrac{4.8\times10^{-9}}{1.6\times10^{-7}}$

$\qquad\qquad\qquad\quad = \dfrac{4.8}{1.6}\times\dfrac{10^{-9}}{10^{-7}}$

$\qquad\qquad\qquad\quad = 3\times10^{-9-(-7)}$

$\qquad\qquad\qquad\quad = 3\times10^{-2}$

$\qquad\qquad\qquad\quad = 0.03$

44. $\dfrac{(4{,}800)(20{,}000)}{600{,}000} = \dfrac{\left(4.8\times10^3\right)\left(2\times10^4\right)}{6\times10^5}$

$\qquad\qquad\qquad\quad = \dfrac{(4.8\bullet 2)\left(10^{3+4}\right)}{6\times10^5}$

$\qquad\qquad\qquad\quad = \dfrac{9.6\times10^7}{6\times10^5}$

$\qquad\qquad\qquad\quad = \dfrac{9.6}{6}\times\dfrac{10^7}{10^5}$

$\qquad\qquad\qquad\quad = 1.6\times10^{7-5}$

$\qquad\qquad\qquad\quad = 1.6\times10^2$

$\qquad\qquad\qquad\quad = 160$

45. 6.31 billion = 6,310,000,000
$$= 6.31 \times 10^9$$

46. $\dfrac{1.0 \times 10^{-8}}{1.0 \times 10^{-13}} = 1.0 \times 10^{-8-(-13)}$

$$= 1.0 \times 10^5$$

$$= 100,000 \text{ nuclei}$$

SECTION 4.4
Polynomials

47. a) 4 terms
    b) $3x^3$
    c) 3, -1, 1, 10
    d) 10

48. a) 7th degree; monomial
    b) 3rd degree; monomial
    c) 2nd degree; binomial
    d) 5th degree; trinomial
    e) 6th degree; binomial
    f) 4th degree; none of these

49. $-x^5 - 3x^4 + 3$
    Let $x = 0$.
    $-(0)^5 - 3(0)^4 + 3 = -0 - 3(0) + 3$
    $$= 0 - 0 + 3$$
    $$= 3$$

    Let $x = -2$.
    $-(-2)^5 - 3(-2)^4 + 3 = -(-32) - 3(16) + 3$
    $$= 32 - 48 + 3$$
    $$= -13$$

50. DIVING

    $0.1875x^2 - 0.0078125x^3$
    $$= 0.1875(8)^2 - 0.0078125(8)^3$$
    $$= 0.1875(8)^2 - 0.0078125(8)^3$$
    $$= 0.1875(64) - 0.0078125(512)$$
    $$= 12 - 4$$
    $$= 8$$
    The amount of deflection is 8 in.

51.

x	$y = x^2$
-3	$(-3)^2 = 9$
-2	$(-2)^2 = 4$
-1	$(-1)^2 = 1$
0	$(0)^2 = 0$
1	$(1)^2 = 1$
2	$(2)^2 = 4$
3	$(3)^2 = 9$

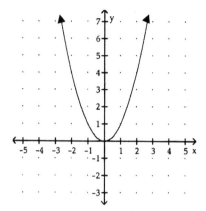

52.

x	$y = x^3 + 1$
-3	$(-3)^3 + 1 = -27 + 1 = -26$
-2	$(-2)^3 + 1 = -8 + 1 = -7$
-1	$(-1)^3 + 1 = -1 + 1 = 0$
0	$(0)^3 + 1 = 0 + 1 = 1$
1	$(1)^3 + 1 = 1 + 1 = 2$
2	$(2)^3 + 1 = 8 + 1 = 9$
3	$(3)^3 + 1 = 27 + 1 = 28$

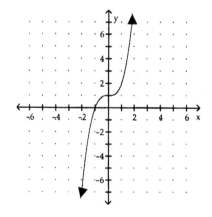

SECTION 4.5
Adding and Subtracting Polynomials

53. $6y^3 + 8y^4 + 7y^3 + (-8y^4)$
    $$= (8y^4 - 8y^4) + (6y^3 + 7y^3)$$
    $$= 0y^4 + 13y^3$$
    $$= 13y^3$$

54. $4a^2b + 5 - 6a^3b - 3a^2b + 2a^3b$
$\quad = (-6a^3b + 2a^3b) + (4a^2b - 3a^2b) + 5$
$\quad = -4a^3b + a^2b + 5$

55. $\dfrac{5}{6}x^2 + \dfrac{1}{3}y^2 - \dfrac{1}{4}x^2 - \dfrac{3}{4}xy + \dfrac{2}{3}y^2$

$\quad = \left(\dfrac{5}{6}x^2 - \dfrac{1}{4}x^2\right) - \dfrac{3}{4}xy + \left(\dfrac{1}{3}y^2 + \dfrac{2}{3}y^2\right)$

$\quad = \left(\dfrac{5}{6}\bullet\dfrac{2}{2}x^2 - \dfrac{1}{4}\bullet\dfrac{3}{3}x^2\right) - \dfrac{3}{4}xy + \left(\dfrac{1}{3}y^2 + \dfrac{2}{3}y^2\right)$

$\quad = \left(\dfrac{10}{12}x^2 - \dfrac{3}{12}x^2\right) - \dfrac{3}{4}xy + \left(\dfrac{1}{3}y^2 + \dfrac{2}{3}y^2\right)$

$\quad = \dfrac{7}{12}x^2 - \dfrac{3}{4}xy + \dfrac{3}{3}y^2$

$\quad = \dfrac{7}{12}x^2 - \dfrac{3}{4}xy + y^2$

56. $-(c^5 + 5c^4 - 12) = -c^5 - 5c^4 + 12$

57. $(2r^6 + 14r^3) + (23r^6 - 5r^3 + 5r)$
$\quad = (2r^6 + 23r^6) + (14r^3 - 5r^3) + 5r$
$\quad = 25r^6 + 9r^3 + 5r$

58. $(7a^2 + 2a - 5) - (3a^2 - 2a + 1)$
$\quad = 7a^2 + 2a - 5 - 3a^2 + 2a - 1$
$\quad = (7a^2 - 3a^2) + (2a + 2a)(-5 - 1)$
$\quad = 4a^2 + 4a - 6$

59. $(3r^3s + r^2s^2 - 3rs^3 - 3s^4) + (r^3s - 8r^2s^2 - 4rs^3 + s^4)$
$\quad = (3r^3s + r^3s) + (r^2s^2 - 8r^2s^2) + (-3rs^3 - 4rs^3) +$
$\quad\quad (-3s^4 + s^4)$
$\quad = 4r^3s - 7r^2s^2 - 7rs^3 - 2s^4$

60. $3(9x^2 + 3x + 7) - 2(11x^2 - 5x + 9)$
$\quad = 3(9x^2) + 3(3x) + 3(7) - 2(11x^2) - 2(5x) - 2(9)$
$\quad = 27x^2 + 9x + 21 - 22x^2 + 10x - 18$
$\quad = (27x^2 - 22x^2) + (9x + 10x) + (21 - 18)$
$\quad = 5x^2 + 19x + 3$

61. $(2z^2 + 3z - 7) + (-4z^3 - 2z - 3) - (-3z^3 - 4z + 7)$
$\quad = 2z^2 + 3z - 7 - 4z^3 - 2z - 3 + 3z^3 + 4z - 7$
$\quad = (-4z^3 + 3z^3) + 2z^2 + (3z - 2z + 4z) + (-7 - 3 - 7)$
$\quad = -z^3 + 2z^2 + 5z - 17$

62. $\quad 3x^2 + 5x + 2$
$+ \quad \underline{x^2 - 3x + 6}$
$\quad 4x^2 + 2x + 8$

63. $\quad \begin{array}{r} 20x^3 \quad\quad + 12x \\ - \quad \underline{12x^3 - 7x^2 - 7x} \end{array}$ $\longrightarrow$ $\quad \begin{array}{r} 20x^3 \quad\quad + 12x \\ + \quad \underline{12x^3 + 7x^2 + 7x} \\ 32x^3 + 7x^2 + 19x \end{array}$

64. GARDENING

$\quad (2x^2 + x + 1) - (x^2 - 2)$
$\quad\quad = 2x^2 + x + 1 - x^2 + 2$
$\quad\quad = (2x^2 - x^2) + x + (1 + 2)$
$\quad\quad = (x^2 + x + 3)$ in.

SECTION 4.6
Multiplying Polynomials

65. $(2x^2)(5x) \quad = (2 \cdot 5)(x^{2+1})$
$\quad\quad\quad\quad\quad = 10x^3$

66. $(-6x^4z^3)(x^6z^2) = -6x^{4+6}z^{3+2}$
$\quad\quad\quad\quad\quad\quad = -6x^{10}z^5$

67. $5b^3 \cdot 6b^2 \cdot 4b^6 = (5\cdot6\cdot4)b^{3+2+6}$
$\quad\quad\quad\quad\quad\quad = 120b^{11}$

68. $\dfrac{2}{3}h^5\left(3h^9 + 12h^6\right) = \dfrac{2}{3}h^5\left(3h^9\right) + \dfrac{2}{3}h^5\left(12h^6\right)$

$\quad\quad\quad\quad = \left(\dfrac{2}{3}\bullet\dfrac{3}{1}\right)h^{5+9} + \left(\dfrac{2}{3}\bullet\dfrac{12}{1}\right)h^{5+6}$

$\quad\quad\quad\quad = 2h^{14} + 8h^{11}$

69. $x^2y(y^2 - xy) = x^2y(y^2) - x^2y(xy)$
$\quad\quad\quad\quad\quad = x^2y^{1+2} - x^{2+1}y^{1+1}$
$\quad\quad\quad\quad\quad = x^2y^3 - x^3y^2$

70. $3n^2(3n^2 - 5n + 2) = 3n^2(3n^2) + 3n^2(-5n) + 3n^2(2)$
$\quad\quad\quad\quad\quad\quad = 9n^4 - 15n^3 + 6n^2$

71. $2x(3x^4)(x + 2) = (2\cdot3\cdot x^{1+4})(x + 2)$
$\quad\quad\quad\quad\quad = 6x^5(x + 2)$
$\quad\quad\quad\quad\quad = 6x^5(x) + 6x^5(2)$
$\quad\quad\quad\quad\quad = 6x^6 + 12x^5$

72. $-a^2b^2(-a^4b^2 + a^3b^3 - ab^4 + 7a)$
$\quad = -a^2b^2(-a^4b^2) - a^2b^2(a^3b^3) - a^2b^2(-ab^4) - a^2b^2(7a)$
$\quad = a^6b^4 - a^5b^5 + a^3b^6 - 7a^3b^2$

73. $(x + 3)(x + 2) = x\cdot x + x\cdot 2 + 3\cdot x + 3\cdot 2$
$\quad\quad\quad\quad\quad = x^2 + 2x + 3x + 6$
$\quad\quad\quad\quad\quad = x^2 + 5x + 6$

74. $(2x + 1)(x - 1) = 2x\cdot x + 2x\cdot(-1) + 1\cdot x + 1\cdot(-1)$
$\quad\quad\quad\quad\quad\quad = 2x^2 - 2x + x - 1$
$\quad\quad\quad\quad\quad\quad = 2x^2 - x - 1$

75. $(3a - 3)(2a + 2) = 3a\cdot 2a + 3a\cdot 2 - 3\cdot 2a - 3\cdot 2$
$\quad\quad\quad\quad\quad\quad = 6a^2 + 6a - 6a - 6$
$\quad\quad\quad\quad\quad\quad = 6a^2 - 6$

76. $6(a-1)(a+1) = 6(a \cdot a + a \cdot 1 - 1 \cdot a - 1 \cdot 1)$
$= 6(a^2 + a - a - 1)$
$= 6(a^2 - 1)$
$= 6a^2 - 6$

77. $(a-b)(2a+b) = a \cdot 2a + a \cdot b - b \cdot 2a - b \cdot b$
$= 2a^2 + ab - 2ab - b^2$
$= 2a^2 - ab - b^2$

78. $(3n^4 - 5n^2)(2n^4 - n^2)$
$= 3n^4 \cdot 2n^4 + 3n^4 \cdot (-n^2) - 5n^2 \cdot 2n^4 - 5n^2 \cdot (-n^2)$
$= 6n^8 - 3n^6 - 10n^6 + 5n^4$
$= 6n^8 - 13n^6 + 5n^4$

79. $(2a-3)(4a^2 + 6a + 9)$
$= 2a(4a^2 + 6a + 9) - 3(4a^2 + 6a + 9)$
$= 2a(4a^2) + 2a(6a) + 2a(9) - 3(4a^2) - 3(6a) - 3(9)$
$= 8a^3 + 12a^2 + 18a - 12a^2 - 18a - 27$
$= 8a^3 + (12a^2 - 12a^2) + (18a - 18a) - 27$
$= 8a^3 - 27$

80. $(8x^2 + x - 2)(7x^2 + x - 1)$
$= 8x^2(7x^2 + x - 1) + x(7x^2 + x - 1) - 2(7x^2 + x - 1)$
$= 56x^4 + 8x^3 - 8x^2 + 7x^3 + x^2 - x - 14x^2 - 2x + 2$
$= 56x^4 + (8x^3 + 7x^3) + (-8x^2 + x^2 - 14x^2) + (-x - 2x) + 2$
$= 56x^4 + 15x^3 - 21x^2 - 3x + 2$

81.
$$\begin{array}{r} 4x^2 - 2x + 1 \\ 2x + 1 \\ \hline 4x^2 - 2x + 1 \\ 8x^3 - 4x^2 + 2x \\ \hline 8x^3 + 1 \end{array}$$

82. APPLIANCE

perimeter:
$(x+6) + (2x-1) = (x+2x) + (6-1)$
$= (3x+5)$ in

area:
$(x+6)(2x-1) = x(2x) + x(-1) + 6(2x) + 6(-1)$
$= 2x^2 - x + 12x - 6$
$= (2x^2 + 11x - 6)$ in^2

volume:
$3x(x+6)(2x-1) = 3x[x(2x) + x(-1) + 6(2x) + 6(-1)]$
$= 3x[2x^2 - x + 12x - 6]$
$= 3x(2x^2 + 11x - 6)$
$= (6x^3 + 33x^2 - 18x)$ in^3

SECTION 4.7
Special Products

83. $(x+3)(x+3) = x(x) + x(3) + 3(x) + 3(3)$
$= x^2 + 3x + 3x + 9$
$= x^2 + 6x + 9$

84. $(2x-0.9)(2x+0.9) = (2x)^2 - (0.9)^2$
$= 4x^2 - 0.81$

85. $(a-3)^2 = a^2 - 2(a)(3) + (-3)^2$
$= a^2 - 6a + 9$

86. $(x+4)^2 = x^2 + 2(x)(4) + 4^2$
$= x^2 + 8x + 16$

87. $(-2y+1)^2 = (-2y)^2 + 2(-2y)(1) + 1^2$
$= 4y^2 - 4y + 1$

88. $(y^2+1)(y^2-1) = (y^2)^2 - 1^2$
$= y^4 - 1$

89. $(6r^2 + 10s)^2 = (6r^2)^2 + 2(6r^2)(10s) + (10s)^2$
$= 36r^4 + 120r^2s + 100s^2$

90. $-(8a-3)^2 = -[(8a)^2 - 2(8a)(3) + (-3)^2]$
$= -[64a^2 - 48a + 9]$
$= -64a^2 + 48a - 9$

91. $80s(r^2 + s^2)(r^2 - s^2) = 80s[(r^2)^2 - (s^2)^2]$
$= 80s(r^4 - s^4)$
$= 80r^4s - 80s^5$

92. $4b(3b-4)^2 = 4b[(3b)^2 - 2(3b)(4) + (-4)^2]$
$= 4b[9b^2 - 24b + 16]$
$= 36b^3 - 96b^2 + 64b$

93. $\left(t - \dfrac{3}{4}\right)^2 = t^2 - 2(t)\left(\dfrac{3}{4}\right) + \left(-\dfrac{3}{4}\right)^2$
$= t^2 - \dfrac{3}{2}t + \dfrac{9}{16}$

94. $(m+2)^3 = (m+2)^2(m+2)$
$= (m^2 + 2 \cdot m \cdot 2 + 2^2)(m+2)$
$= (m^2 + 4m + 4)(m+2)$
$= m^2(m+2) + 4m(m+2) + 4(m+2)$
$= m^3 + 2m^2 + 4m^2 + 8m + 4m + 8$
$= m^3 + 6m^2 + 12m + 8$

95. $(5c-1)^2 - (c+6)(c-6)$
$$= [(5c)^2 - 2(5c)(1) + (-1)^2] - [c^2 - 6^2]$$
$$= (25c^2 - 10c + 1) - (c^2 - 36)$$
$$= 25c^2 - 10c + 1 - c^2 + 36$$
$$= 24c^2 - 10c + 37$$

96. GRAPHIC ARTS

To find the length and width of the picture, subtract the border on each end from the total length and width.

The length of the picture would be
$$(x+3) - 2(\tfrac{1}{2}) = x + 3 - 1$$
$$= x + 2.$$
The width of the picture would be
$$(x-3) - 2(\tfrac{1}{2}) = x - 3 - 1$$
$$= x - 4.$$
The area of the picture would be
$$(x+2)(x-4) = x(x) + x(-4) + 2(x) + 2(-4)$$
$$= x^2 - 4x + 2x - 8$$
$$= (x^2 - 2x - 8) \text{ in.}^2$$

SECTION 4.8
Division of Polynomials

97. $\dfrac{16n^8}{8n^5} = 2n^{8-5}$
$$= 2n^3$$

98. $\dfrac{-14x^2 y}{21xy^3} = -\dfrac{2}{3}x^{2-1}y^{1-3}$
$$= -\dfrac{2}{3}x^1 y^{-2}$$
$$= -\dfrac{2x}{3y^2}$$

99. $\dfrac{a^{15} - 24a^8}{6a^{12}} = \dfrac{a^{15}}{6a^{12}} - \dfrac{24a^8}{6a^{12}}$
$$= \dfrac{1}{6}a^{15-12} - 4a^{8-12}$$
$$= \dfrac{1}{6}a^3 - 4a^{-4}$$
$$= \dfrac{a^3}{6} - \dfrac{4}{a^4}$$

100. $\dfrac{15a^2 b + 20ab^2 - 25ab}{5ab}$
$$= \dfrac{15a^2 b}{5ab} + \dfrac{20ab^2}{5ab} - \dfrac{25ab}{5ab}$$
$$= 3a + 4b - 5$$

101.
$$\begin{array}{r}
x - 5 \\
x-1 \overline{) x^2 - 6x + 5} \\
\underline{x^2 - x} \\
-5x + 5 \\
\underline{-5x + 5} \\
0
\end{array}$$

102. Make sure the dividend is in descending order
$$\begin{array}{r}
2x + 1 \\
x+3 \overline{) 2x^2 + 7x + 3} \\
\underline{2x^2 + 6x} \\
x + 3 \\
\underline{x + 3} \\
0
\end{array}$$

103. Insert $0x$ for the missing x term.
$$\begin{array}{r}
3x^2 + 2x + 1 + \frac{2}{2x-1} \\
2x-1 \overline{) 6x^3 + x^2 + 0x + 1} \\
\underline{6x^3 - 3x^2} \\
4x^2 + 0x \\
\underline{4x^2 - 2x} \\
2x + 1 \\
\underline{2x - 1} \\
2
\end{array}$$

104. Make sure the dividend is in descending order and insert $0x^2$ for the missing x^2 term.
$$\begin{array}{r}
3x^2 - x - 4 \\
3x+1 \overline{) 9x^3 + 0x^2 - 13x - 4} \\
\underline{9x^3 + 3x^2} \\
-3x^2 - 13x \\
\underline{-3x^2 - x} \\
-12x - 4 \\
\underline{-12x + 4} \\
0
\end{array}$$

105. $(y+3)(3y+2) = y\cdot 3y + y\cdot(2) + 3\cdot 3y + 3\cdot(2)$

$$= 3y^2 + 2y + 9y + 6$$

$$= 3y^2 + 11y + 6$$

106.   SAVING BONDS

$$\frac{50x+250}{50} = \frac{50x}{50} + \frac{250}{50}$$

$$= x+5$$

The number would be $(x + 5)$.

CHAPTER 4 TEST

1. $2xxxyyy = 2x^3y^4$

2. $-6^2 = -(6)(6)$
   $= -36$

3. $y^2(yy^3) = y^2(y^{1+3})$
   $\quad = y^2(y^4)$
   $\quad = y^{2+4}$
   $\quad = y^6$

4. $(2x^3)^5(x^2)^3 = 2^5 \cdot x^{3 \cdot 5} \cdot x^{2 \cdot 3}$
   $\quad = 32x^{15} \cdot x^6$
   $\quad = 32x^{15+6}$
   $\quad = 32x^{21}$

5. $3x^0 = 3 \cdot 1$
   $\quad = 3$

6. $2y^{-5}y^2 = 2y^{-5+2}$
   $\quad = 2y^{-3}$
   $\quad = \dfrac{2}{y^3}$

7. $5^{-3} = \dfrac{1}{5^3}$
   $\quad = \dfrac{1}{125}$

8. $\dfrac{(x+1)^{15}}{(x+1)^6} = (x+1)^{15-6}$
   $\quad = (x+1)^9$

9. $\dfrac{y^2}{yy^{-2}} = \dfrac{y^2}{y^{1+(-2)}}$
   $\quad = \dfrac{y^2}{y^{-1}}$
   $\quad = y^{2-(-1)}$
   $\quad = y^3$

10. $\left(\dfrac{a^2b^{-1}}{4a^3b^{-2}}\right)^{-3} = \left(\dfrac{4a^3b^{-2}}{a^2b^{-1}}\right)^3$
    $\quad = \left(4a^{3-2}b^{-2-(-1)}\right)^3$
    $\quad = \left(4ab^{-1}\right)^3$
    $\quad = \left(\dfrac{4a}{b}\right)^3$
    $\quad = \dfrac{4^3 a^3}{b^3}$
    $\quad = \dfrac{64a^3}{b^3}$

11. $\left(10y^4\right)^3 = 10^3 y^{4 \bullet 3}$
    $\quad = 1,000y^{12}$ in.^{3}

12. ELECTRICITY

    $6,250,000,000,000,000,000 = 6.25 \times 10^{18}$

13. $9.3 \times 10^{-5} = 0.000093$

14. $(2.3 \times 10^{18})(4.0 \times 10^{-15}) = (2.3 \cdot 4.0) \times 10^{18 + (-15)}$
    $\quad = 9.2 \times 10^3$
    $\quad = 9,200$

15.

Term	Coefficient	Degree
x^4	1	4
$2x^2$	2	2
-12	-12	0
Degree of the polynomial		4

16. binomial

17. The degree of $3x^2y^3$ is 2+3 = 5.
    The degree of $2x^3y$ is 3 + 1 = 4.
    The degree of $-5x^2y$ is 2 + 1 = 3.

    So, the degree of the polynomial is 5.

18. $y = x^2 + 2$

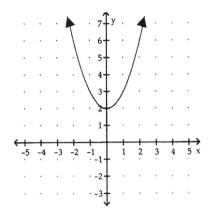

19. FREE FALL

Let $t = 18$.
$-16(\mathbf{18})^2 + 5 = -16(324) + 5{,}184$
$= -5{,}184 + 5{,}184$
$= 0$ ft
The rock hits the canyon floor 18 second after being dropped.

20. $4a^2b + 5 - 6a^3b - 3a^2b + 2a^3b$
$= (-6a^3b + 2a^3b) + (4a^2b - 3a^2b) + 5$
$= -4a^3b + a^2b + 5$

21. $(3a^2 - 4a - 6) + (2a^2 - a + 9)$
$= (3a^2 + 2a^2) + (-4a - a) + (-6 + 9)$
$= 5a^2 - 5a + 3$

22. $(7b^3c - 5bc) - (b^3c - 3bc + 12)$
$= 7b^3c - 5bc - b^3c + 3bc - 12$
$= (7b^3c - b^3c) + (-5bc + 3bc) - 12$
$= 6b^3c - 2bc - 12$

23. $\begin{array}{r} -5y^3 + 4y^2 - 11y + 3 \\ - \underline{-2y^3 - 14y^2 + 17y - 32} \end{array}$

$\begin{array}{r} -5y^3 + 4y^2 - 11y + 3 \\ + \underline{2y^3 + 14y^2 - 17y + 32} \\ -3y^3 + 18y^2 - 28y + 35 \end{array}$

24. $-6(x - y) + 2(x + y)$
$= -6x + 6y + 2x + 2y$
$= (-6x + 2x) + (6y + 2y)$
$= -4x + 8y$

25. $(-2x^3)(2x^2y) = -4x^{3+2}y$
$= -4x^5y$

26. $3y^2(y^2 - 2y + 3)$
$= 3y^2(y^2) + 3y^2(-2y) + 3y^2(3)$
$= 3y^4 - 6y^3 + 9y^2$

27. $(2x - 5)(3x + 4)$
$= 2x(3x) + 2x(4) - 5(3x) - 5(4)$
$= 6x^2 + 8x - 15x - 20$
$= 6x^2 - 7x - 20$

28. $(2x - 3)(x^2 - 2x + 4)$
$= 2x(x^2 - 2x + 4) - 3(x^2 - 2x + 4)$
$= 2x^3 - 4x^2 + 8x - 3x^2 + 6x - 12$
$= 2x^3 + (-4x^2 - 3x^2) + (8x + 6x) - 12$
$= 2x^3 - 7x^2 + 14x - 12$

29. $(1 + 10c)(1 - 10c) = 1^2 - (10c)^2$
$= 1 - 100c^2$

30. $(7b^3 - 3)^2 = (7b^3)^2 - 2(7b^3)(3) + 3^2$
$= 49b^6 - 42b^3 + 9$

31. $(x + y)(x - y) + x(x + y)$
$= (x^2 - y^2) + (x^2 + xy)$
$= 2x^2 + xy - y^2$

32. $\dfrac{6a^2 - 12b^2}{24ab} = \dfrac{6a^2}{24ab} - \dfrac{12b^2}{24ab}$

$= \dfrac{1}{4}a^{2-1}b^{-1} - \dfrac{1}{2}a^{-1}b^{2-1}$

$= \dfrac{1}{4}a^1b^{-1} - \dfrac{1}{2}a^{-1}b^1$

$= \dfrac{a}{4b} - \dfrac{b}{2a}$

33. $\begin{array}{r} x - 2 \\ 2x+3{\overline{\smash{\big)}\,2x^2 - x - 6}} \\ \underline{2x^2 + 3x} \\ -4x - 6 \\ \underline{-4x - 6} \\ 0 \end{array}$

34.

$$2x-1 \overline{)\, 6x^3 + x^2 + 0x + 1 \,}$$

quotient: $3x^2 + 2x + 1 + \frac{2}{2x-1}$

$$
\begin{array}{r}
6x^3 - 3x^2 \\
\hline
3x^2 + 0x \\
3x^2 - 2x \\
\hline
2x + 1 \\
2x - 1 \\
\hline
2
\end{array}
$$

35.

$$x-1 \overline{)\, x^2 - 6x + 5 \,}$$

quotient: $x - 5$

$$
\begin{array}{r}
x^2 - x \\
\hline
-5x + 5 \\
-5x + 5 \\
\hline
0
\end{array}
$$

Width of rectangle is $(x - 5)$ ft.

CHAPTERS 1-4
CUMULATIVE REVIEW EXERCISES

1. a) 1993
    b) 1986
    c) 1996-1997

2. $\dfrac{3}{4} \div \dfrac{6}{5} = \dfrac{3}{4} \bullet \dfrac{5}{6}$

   $= \dfrac{3(5)}{4(6)}$

   $= \dfrac{15}{24}$

   $= \dfrac{5}{8}$

3. $\dfrac{7}{10} - \dfrac{1}{14} = \dfrac{7}{10}\left(\dfrac{7}{7}\right) - \dfrac{1}{14}\left(\dfrac{5}{5}\right)$

   $= \dfrac{49}{70} - \dfrac{5}{70}$

   $= \dfrac{44}{70}$

   $= \dfrac{22}{35}$

4. π is an irrational number.

5. RACING

   $250 - x$

6. CLINICAL TRIALS

   6% of 300 = 0.06(300) = 18

   5% of 320 = 0.05(320) = 16

7. $100 = 2 \cdot 2 \cdot 5 \cdot 5$

   $= 2^2 \cdot 5^2$

8. GRAPH

9. $\dfrac{2}{3} = 3\overline{)2.000}^{\,0.666} = 0.\overline{6}$

10. associative property of multiplication

11. $10d$ cents

12. $13r - 12r = 1r$

   $= r$

13. $27\left(\dfrac{2}{3}x\right) = \dfrac{27}{1}\left(\dfrac{2}{3}\right)x$

   $= \dfrac{54}{3}x$

   $= 18x$

14. $4(d-3) - (d-1) = 4d - 12 - d + 1$

   $= (4d - d) + (-12 + 1)$

   $= 3d - 11$

15. $(13c - 3)(-6) = 13c(-6) - 3(-6)$

   $= -78c + 18$

16. $-3^2 + \left|4^2 - 5^2\right| = -9 + |16 - 25|$

   $= -9 + |-9|$

   $= -9 + 9$

   $= 0$

17. $(4 - 5)^{20} = (-1)^{20}$

   $= 1$

18. $\dfrac{-3 - (-7)}{2^2 - 3} = \dfrac{-3 + 7}{4 - 3}$

   $= \dfrac{4}{1}$

   $= 4$

19. $12 - 2[1 - (-8 + 2)] = 12 - 2[1 - (-6)]$

   $= 12 - 2[1 + 6]$

   $= 12 - 2[7]$

   $= 12 - 14$

   $= -2$

20. $3(x - 5) + 2 = 2x$

   $3x - 15 + 2 = 2x$

   $3x - 13 = 2x$

   $3x - 13 - 3x = 2x - 3x$

   $-13 = -1x$

   $\dfrac{-13}{-1} = \dfrac{-1x}{-1}$

   $\boxed{13 = x}$

21.
$$\frac{x-5}{3}-5=7$$

$$\frac{x-5}{3}-5+5=7+5$$

$$\frac{x-5}{3}=12$$

$$3\left(\frac{x-5}{3}\right)=3(12)$$

$$x-5=36$$

$$x-5+5=36+5$$

$$\boxed{x=41}$$

22.
$$\frac{2}{5}x+1=\frac{1}{3}+x$$

$$15\left(\frac{2}{5}x+1\right)=15\left(\frac{1}{3}+x\right)$$

$$6x+15=5+15x$$

$$6x+15-15=5+15x-15$$

$$6x=15x-10$$

$$6x-15x=15x-10-15x$$

$$-9x=-10$$

$$\frac{-9x}{-9}=\frac{-10}{-9}$$

$$\boxed{x=\frac{10}{9}}$$

23.
$$-\frac{5}{8}h=15$$

$$8\left(-\frac{5}{8}h\right)=8(15)$$

$$-5h=120$$

$$\frac{-5h}{-5}=\frac{120}{-5}$$

$$\boxed{h=-24}$$

24.
$$8(4+x)>10(6+x)$$

$$32+8x>60+10x$$

$$32+8x-10x>60+10x-10x$$

$$32-2x>60$$

$$32-2x-32>60-32$$

$$-2x>28$$

$$\frac{-2x}{-2}>\frac{28}{-2}$$

$$x<-14$$

$$(-\infty,-14)$$

25.
$$A=\frac{1}{2}h(b+B)$$

$$2(A)=2\left[\frac{1}{2}h(b+B)\right]$$

$$2A=h(b+B)$$

$$\frac{2a}{(b+B)}=\frac{h(b+B)}{(b+B)}$$

$$\frac{2A}{b+B}=h$$

$$\boxed{h=\frac{2A}{b+B}}$$

26. CANDY

Let x = the number of pounds of red licorice.

$30-x$ = the number of pounds of lemon gumdrops.

$$\$1.90x + \$2.20(30-x) = \$2(30)$$
$$1.90x + 66 - 2.2x = 60$$
$$-0.3x + 66 = 60$$
$$-0.3x + 66 - 66 = 60 - 66$$
$$-0.3x = -6$$
$$\frac{-0.3x}{-0.3} = \frac{-6}{-0.3}$$
$$x = 20$$
$$30 - x = 30 - 20$$
$$= 10$$

20 lb of $1.90 candy
10 lb of $2.20 candy

27. Graph: $4x - 3y = 12$

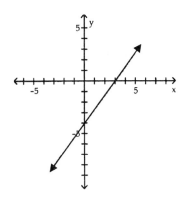

28. Graph: $x = 4$

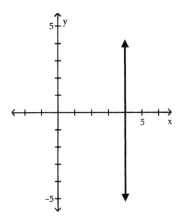

29. $m = \dfrac{y_2 - y_1}{x_2 - x_1}$

$= \dfrac{8 - 4}{6 - (-2)}$

$= \dfrac{4}{8}$

$= \dfrac{1}{2}$

30. The slope of a horizontal line is always <u>0</u>.

31. Solve the equation for y.

$2x - 3y = 12$

$2x - 3y - 2x = 12 - 2x$

$-3y = -2x + 12$

$\dfrac{-3y}{-3} = \dfrac{-2x}{-3} + \dfrac{12}{-3}$

$y = \dfrac{2}{3}x - 4$

$m = \dfrac{2}{3}$

32. $m = \dfrac{2}{3}$ and $b = 5$

$y = mx + b$

$y = \dfrac{2}{3}x + 5$

33. Find the slope:

$m = \dfrac{y_2 - y_1}{x_2 - x_1}$

$= \dfrac{10 - 4}{6 - (-2)}$

$= \dfrac{6}{8}$

$= \dfrac{3}{4}$

Use the point-slope formula:

$y - y_1 = m(x - x_1)$

$y - 10 = \dfrac{3}{4}(x - 6)$

$y - 10 = \dfrac{3}{4}(x) - \dfrac{3}{4}\left(\dfrac{6}{1}\right)$

$y - 10 = \dfrac{3}{4}x - \dfrac{9}{2}$

$y - 10 + 10 = \dfrac{3}{4}x - \dfrac{9}{2} + 10$

$y = \dfrac{3}{4}x - \dfrac{11}{2}$

$4(y) = 4\left(\dfrac{3}{4}x - \dfrac{11}{2}\right)$

$4y = 3x - 22$

$-3x + 4y = 3x - 22 - 3x$

$-3x + 4y = -22$

$\dfrac{-3x}{-1} + \dfrac{4y}{-1} = \dfrac{-22}{-1}$

$3x - 4y = 22$

34. $y = 4$

A horizontal line has the equation $y = b$.

Chapter 4 Cumulative Review

35. Find the slope of both lines.

$$y = -\frac{3}{4}x + \frac{15}{4}$$

so the slope is $-\frac{3}{4}$.

$$4x - 3y = 25$$
$$4x - 3y - 4x = 25 - 4x$$
$$-3y = -4x + 25$$
$$\frac{-3y}{-3} = \frac{-4x}{-3} + \frac{25}{-3}$$
$$y = \frac{4}{3}x - \frac{25}{3}$$

The slope is $\frac{4}{3}$.

Since $-\frac{3}{4}$ and $\frac{4}{3}$ are negative reciprocals,

the two lines are **perpendicular**.

36. $(17x^4 - 3x^2 - 65x - 12) - (23x^4 + 14x^2 + 3x - 23)$
$= 17x^4 - 3x^2 - 65x - 12 - 23x^4 - 14x^2 - 3x + 23$
$= 17x^4 - 23x^4 - 3x^2 - 14x^2 - 65x - 3x - 12 + 23$
$= -6x^4 - 17x^2 - 68x + 11$

37. $(-3x^2 y^4)^2 = (-3)^2 x^{2 \cdot 2} y^{4 \cdot 2}$
$= 9x^4 y^8$

38. $(2y)^{-4} = \dfrac{1}{(2y)^4}$

$= \dfrac{1}{2^4 y^4}$

$= \dfrac{1}{16 y^4}$

39. $(x^3 x^4)^2 = (x^{3+4})^2$
$= (x^7)^2$
$= x^{7 \cdot 2}$
$= x^{14}$

40. $ab^3 c^4 \cdot ab^4 c^2 = a^{1+1} b^{3+4} c^{4+2}$
$= a^2 b^7 c^6$

41. $\dfrac{a^4 b^0}{a^{-3}} = a^{4-(-3)} \cdot b^0$

$= a^7 \cdot 1$

$= a^7$

42. $\left(\dfrac{4t^3 t^4 t^5}{3t^2 t^6}\right)^3 = \left(\dfrac{4t^{3+4+5}}{3t^{2+6}}\right)^3$

$= \left(\dfrac{4t^{12}}{3t^8}\right)^3$

$= \left(\dfrac{4}{3} t^{12-8}\right)^3$

$= \left(\dfrac{4t^4}{3}\right)^3$

$= \dfrac{4^3 t^{4 \cdot 3}}{3^3}$

$= \dfrac{64 t^{12}}{27}$

43. $(4c^2 + 3c - 2) + (3c^2 + 4c + 2)$
$= (4c^2 + 3c^2) + (3c + 4c) + (-2 + 2)$
$= 7c^2 + 7c$

44. $3x(2x + 3)^2 = 3x[(2x)^2 + 2(2x)(3) + 3^2]$
$= 3x(4x^2 + 12x + 9)$
$= 12x^3 + 36x^2 + 27x$

45. $(2t+3s)(3t-s) = 2t(3t) + 2t(-s) + 3s(3t) + 3s(-s)$
$= 6t^2 - 2st + 9st - 3s^2$
$= 6t^2 + 7st - 3s^2$

46. Write the dividend in descending order.

$$\begin{array}{r}
2x + 1 \\
5x+3 \overline{)10x^2 + 11x + 3} \\
\underline{10x^2 + 6x} \\
5x + 3 \\
\underline{5x + 3} \\
0
\end{array}$$

47. Graph: x^2

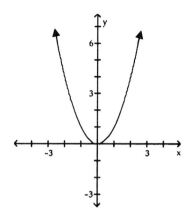

48. $615,000 = 6.15 \times 10^5$; move the decimal 5 places to the left

49. $0.0000013 = 1.3 \times 10^{-6}$; move the decimal 6 places to the right

50. MUSICAL INSTRUMENTS

Let $x = 2$ (half of the distance of 4 feet)

$0.01875(2)^4 - 0.15(2)^3 + 1.2(2)$
$= 0.01875(16) - 0.15(8) + 1.2(2)$
$= 0.3 - 1.2 + 2.4$
$= 1.5$

The amount of deflection is 1.5 in.

Chapter 4 Cumulative Review

SECTION 5.1

VOCABULARY

1. The letters GCF stand for **greatest common factor**.

3. When we write 24 as $2 \cdot 2 \cdot 2 \cdot 3$, we say that 24 has been written as a product of **primes**.

5. To factor $m^3 + 3m^2 + 4m + 12$ by **grouping**, we begin by writing $m^2(m + 3) + 4(m + 3)$.

CONCEPTS

7. a. $15 = 3 \cdot 7$
   b. $8 = 2 \cdot 2 \cdot 2$
   c. $36 = 2 \cdot 2 \cdot 3 \cdot 3$
   d. $98 = 2 \cdot 7 \cdot 7$

9. $12a^2b^2 = 2 \cdot 2 \cdot 3 \cdot a \cdot a \cdot b \cdot b$
   $15a^3b = 3 \cdot 5 \cdot a \cdot a \cdot a \cdot b$
   $75a^4b^2 = 3 \cdot 5 \cdot 5 \cdot a \cdot a \cdot a \cdot a \cdot b \cdot b$
   $GCF = 3 \cdot a \cdot a \cdot b = 3a^2b$

11. a. the distributive property

    b. $2x^2 - 6x = 2x \cdot x - 2x \cdot 3$
       $= \boxed{2x}(x - 3)$

13. a. The GCF is $6a^2$, not $6a$.
    b. The 0 in the first line should be 1.

15. No. The common factor $(c - d)$ still needs to be factored out.

NOTATION

17. $40m^4 - 8m^3 + 32m = \underline{\mathbf{8m}}(5m^3 - m^2 + \underline{\mathbf{4}})$

PRACTICE

19. $12 = 2 \cdot 2 \cdot 3$
    $= 2^2 \cdot 3$

21. $40 = 2 \cdot 2 \cdot 2 \cdot 5$
    $= 2^3 \cdot 5$

23. $98 = 2 \cdot 7 \cdot 7$
    $= 2 \cdot 7^2$

25. $225 = 3 \cdot 3 \cdot 5 \cdot 5$
    $= 3^2 \cdot 5^2$

27. $18 = 2 \cdot 3 \cdot 3$
    $24 = 2 \cdot 2 \cdot 2 \cdot 3$
    $GCF = 2 \cdot 3$
    $\quad = 6$

29. $m^4 = m \cdot m \cdot m \cdot m$
    $m^3 = m \cdot m \cdot m$
    $GCF = m \cdot m \cdot m$
    $\quad = m^3$

31. $20c^2 = 2 \cdot 2 \cdot 5 \cdot c \cdot c$
    $12c = 2 \cdot 2 \cdot 3 \cdot c$
    $GCF = 2 \cdot 2 \cdot c$
    $\quad = 4c$

33. $6m^4n = 2 \cdot 3 \cdot m \cdot m \cdot m \cdot m \cdot n$
    $12m^3n^2 = 2 \cdot 2 \cdot 3 \cdot m \cdot m \cdot m \cdot n \cdot n$
    $9m^3n^3 = 3 \cdot 3 \cdot m \cdot m \cdot m \cdot n \cdot n \cdot n$
    $GCF = 3 \cdot m \cdot m \cdot m \cdot n$
    $\quad = 3m^3n$

35. $6x = 3(2x)$

37. $24y^2 = 8y \cdot 3y$

39. $30t^3 = 15t(2t^2)$

41. $32x^3z^4 = 4xz^2(8x^2z^2)$

43. $3x + 6 = 3 \cdot x + 3 \cdot 2$
    $\quad = 3(x + 2)$

45. $18m - 9 = 9 \cdot 2m - 9 \cdot 1$
    $\quad = 9(2m - 1)$

47. $18x + 24 = 6 \cdot 3x + 6 \cdot 4$
    $\quad = 6(3x + 4)$

49. $d^2 - 7d = d \cdot d - d \cdot 7$
    $\quad = d(d - 7)$

51. $14c^3 + 63 = 7 \cdot 2c^3 + 7 \cdot 9$
    $\quad = 7(2c^3 + 9)$

53. $12x^2 - 6x - 24 = 6 \cdot 2x^2 - 6 \cdot x - 6 \cdot 4$
    $\quad = 6(2x^2 - x - 4)$

55. $t^3 + 2t^2 = t^2 \cdot t + t^2 \cdot 2$
    $\quad = t^2(t + 2)$

57. $ab + ac - ad = a \cdot b + a \cdot c - a \cdot d$
$\qquad = a(b + c - d)$

59. $a^3 - a^2 = a^2 \cdot a - a^2 \cdot 1$
$\qquad = a^2(a - 1)$

61. $24x^2y^3 + 8xy^2 = 8xy^2 \cdot 3xy + 8xy^2 \cdot 1$
$\qquad = 8xy^2(3xy + 1)$

63. $12uvw^3 - 18uv^2w^2 = 6uvw^2 \cdot 2w - 6uvw^2 \cdot 3v$
$\qquad = 6uvw^2(2w - 3v)$

65. $12r^2 - 3rs + 9r^2s^2 = 3r \cdot 4r - 3r \cdot s + 3r \cdot 3rs^2$
$\qquad = 3r(4r - s + 3rs^2)$

67. $\pi R^2 - \pi ab = \pi \cdot R^2 - \pi \cdot ab$
$\qquad = \pi(R^2 - ab)$

69. $3(x + 2) - x(x + 2) = (x + 2)(3 - x)$

71. $h^2(14 + r) + 14 + r = h^2(14 + r) + 1(14 + r)$
$\qquad = (14 + r)(h^2 + 1)$

73. $-a - b = -(a + b)$

75. $-2x + 5y = -(2x - 5y)$

77. $-3r + 2s - 3 = -(3r - 2s + 3)$

79. $-3x^2 - 6x \quad = -3x \cdot x - 3x \cdot 2$
$\qquad = -3x(x + 2)$

81. $-4a^2b^3 + 12a^3b^2 = -4a^2b^2 \cdot b - 4a^2b^2 \cdot -3a$
$\qquad = -4a^2b^2(b - 3a)$

83. $-4a^2b^2c^2 + 14a^2b^2c - 10ab^2c^2$
$\qquad = -2ab^2c \cdot 2ac - 2ab^2c \cdot -7a - 2ab^2c \cdot 5c$
$\qquad = -2ab^2c(2ac - 7a + 5c)$

85. $2x + 2y + ax + ay = (2x + 2y) + (ax + ay)$
$\qquad = 2(x + y) + a(x + y)$
$\qquad = (x + y)(2 + a)$

87. $9p - 9q + mp - mq = (9p - 9q) + (mp - mq)$
$\qquad = 9(p - q) + m(p - q)$
$\qquad = (p - q)(9 + m)$

89. $pm - pn + qm - qn = (pm - pn) + (qm - qn)$
$\qquad = p(m - n) + q(m - n)$
$\qquad = (m - n)(p + q)$

91. $2xy - 3y^2 + 2x - 3y$
$\qquad = (2xy - 3y^2) + (2x - 3y)$
$\qquad = y(2x - 3y) + 1(2x - 3y)$
$\qquad = (2x - 3y)(y + 1)$

93. $ax + bx - a - b = (ax + bx) - (a + b)$
$\qquad = x(a + b) - 1(a + b)$
$\qquad = (a + b)(x - 1)$

95. $ax^3 + bx^3 + 2ax^2y + 2bx^2y$
$\qquad = x^2[ax + bx + 2ay + 2by]$
$\qquad = x^2[(ax + bx) + (2ay + 2by)]$
$\qquad = x^2[x(a + b) + 2y(a + b)]$
$\qquad = x^2(a + b)(x + 2y)$

97. $2x^3z - 4x^2z + 32xz - 64z$
$\qquad = 2z[x^3 - 2x^2 + 16x - 32]$
$\qquad = 2z[(x^3 - 2x^2) + (16x - 32)]$
$\qquad = 2z[x^2(x - 2) + 16(x - 2)]$
$\qquad = 2z(x - 2)(x^2 + 16)$

99. $x(y + 9) - 11(y + 9) = (y + 9)(x - 11)$

101. $9mp + 3mq - 3np - nq$
$\qquad = (9mp + 3mq) - (3np + nq)$
$\qquad = 3m(3p + q) - n(3p + q)$
$\qquad = (3p + q)(3m - n)$

103. $25x^5y^7z^3 - 45x^3y^2z^6$
$\qquad = 5x^3y^2z^3 \cdot 5x^2y^5 - 5x^3y^2z^3 \cdot 9z^3$
$\qquad = 5x^3y^2z^3(5x^2y^5 - 9z^3)$

105. $24m - 12n + 16$
$\qquad = 4 \cdot 6m - 4 \cdot 3n + 4 \cdot 4$
$\qquad = 4(6m - 3n + 4)$

107. $-60P^2 - 80P = -20P \cdot 3P - 20P \cdot 4$
$\qquad = -20P(3P + 4)$

APPLICATIONS

109. REARVIEW MIRRORS
a. Windshield:
$\qquad A = lw$
$\qquad\quad = (2x)(3x^2)$
$\qquad\quad = 6x^3 \text{ cm}^2$

b. Driver's side door:
$\qquad A = lw$
$\qquad\quad = (3x)(4x)$
$\qquad\quad = 12x^2 \text{ cm}^2$
Passenger side door:
$\qquad A = lw$
$\qquad\quad = (3x)(4x)$
$\qquad\quad = 12x^2 \text{ cm}^2$
Area of both mirrors is $24x^2 \text{ cm}^2$

c. Total area:
$\qquad 6x^3 + 24x^2 = 6x^2 \cdot x + 6x^2 \cdot 4$
$\qquad\qquad\quad = 6x^2(x + 4) \text{ cm}^2$

Section 5.1

111. AIRCRAFT CARRIERS
$$x^3 + 4x^2 + 5x + 20$$
$$= (x^3 + 4x^2) + (5x + 20)$$
$$= x^2(x + 4) + 5(x + 4)$$
$$= (x + 4)(x^2 + 5)$$
The length and width are $(x^2 + 5)$ ft. and $(x + 4)$ ft.

WRITING

113. Answers will vary.

115. Answers will vary.

REVIEW

117. INSURANCE COST

Find the difference in the two premiums:
$$\$1,050 - \$925 = \$125$$
$125 is what percent of $1,050

$$125 = x \cdot 1,050$$
$$125 = 1,050x$$
$$\frac{125}{1,050} = \frac{1,050x}{1,050}$$
$$0.119 \approx x$$
$$\boxed{12\% \approx x}$$

CHALLENGE PROBLEMS

119. $6x^{4m}y^n + 21x^{3m}y^{2n} - 15x^{2m}y^{3n}$
$= 3x^{2m}y^n \cdot 2x^{2m} + 3x^{2m}y^n \cdot 7x^m y^n - 3x^{2m}y^n \cdot 5y^{2n}$
$= 3x^{2m}y^n(2x^{2m} + 7x^m y^n - 5y^{2n})$

SECTION 5.2

VOCABULARY

1. A polynomial that has three terms is called a **trinomial**. A polynomial that has two terms is called a **binomial**.

3. Since $10 = (-5)(-2)$, we say that -5 and -2 are **factors** of 10.

5. The **lead** coefficient of the trinomial $x^2 - 3x + 2$ is 1, the **coefficient** of the middle term is -3, and the last term is 2.

CONCEPTS

7. Two factorizations of 4 that involve only positive numbers are **4 · 1** and **2 · 2**. Two factorizations of 4 that involve only negative numbers are **-4(-1)** and **-2(-2)**.

9. To factor $x^2 + x - 56$, we must find two integers who **product** is -56 and whose **sum** is 1.

11. $x^2 + 5x + 3$ cannot be factored because we cannot find two integers whose product is **3** and whose sum is **5**.

13. a) the Last step: $5 \cdot 4 = 20$
    b) the Outer step: $x \cdot 4 = 4x$
       the Inner step: $5 \cdot x = 5x$

15. a) 1
    b) 15, 8
    c) 5 and 3

17. a) They are both positive or both negative.
    b) One will be positive, the other negative.

19. $x(x + 1) + 5(x + 1) = (x + 1)(x + 5)$

21.

Negative factors of 8	Sum of factors of 8
$-1(-8) = 8$	$-1 + (-8) = -9$
$-2(-4) = 8$	$-2 + (-4) = -6$

NOTATION

23. $(x + 3)(x - 2)$

PRACTICE

25. $x^2 + 3x + 2 = (x + 2)(x + 1)$

27. $t^2 - 9t + 14 = (t - 7)(t - 2)$

29. $a^2 + 6a - 16 = (a + 8)(a - 2)$

31. The factors of 11 whose sum is 12 are 11 and 1.
    $z^2 + 12z + 11 = (z + 11)(z + 1)$

33. The factors of 6 whose sum is -5 are -2 and -3.
    $m^2 - 5m + 6 = (m - 3)(m - 2)$

35. The factors of -5 whose sum is -4 are -5 and 1.
    $a^2 - 4a - 5 = (a - 5)(a + 1)$

37. The factors of -24 whose sum is 5 are 8 and -3.
    $x^2 + 5x - 24 = (x + 8)(x - 3)$

39. The factors of -39 whose sum is -10 are -13 and 3.
    $a^2 - 10a - 39 = (a - 13)(a + 3)$

41. Prime. There are no factors of 15 whose sum is 10.

43. Prime. There are no factors of 4 whose sum is -2.

45. The factors of 18 whose sum is -9 are -3 and -6.
    $r^2 - 9r + 18 = (r - 3)(r - 6)$

47. The factors of 4 whose sum is 4 are 2 and 2.
    $x^2 + 4xy + 4y^2 = (x + 2y)(x + 2y)$

49. The factors of -12 whose sum is -4 are -6 and 2.
    $a^2 - 4ab - 12b^2 = (a - 6b)(a + 2b)$

51. Prime. There are no factors of 4 whose sum is 2.

53. $-x^2 - 7x - 10 = -1(x^2 + 7x + 10)$
$= -(x + 5)(x + 2)$

55. $-t^2 - 15t + 34 = -1(t^2 + 15t - 34)$
$= -(t + 17)(t - 2)$

57. $-a^2 - 4ab - 3b^2 = -(a^2 + 4ab + 3b^2)$
$= -(a + 3b)(a + b)$

59. $-x^2 + 6xy + 7y^2 = -1(x^2 - 6xy - 7y^2)$
$= -(x - 7y)(x + y)$

61. $4 - 5x + x^2 = x^2 - 5x + 4$
$= (x - 4)(x - 1)$

63. $10y + 9 + y^2 = y^2 + 10y + 9$
$= (y + 9)(y + 1)$

65. $r^2 - 2 - r = r^2 - r - 2$
$= (r - 2)(r + 1)$

67. $4rx + r^2 + 3x^2 = r^2 + 4rx + 3x^2$
$= (r + 3x)(r + x)$

69. $2x^2 + 10x + 12 = 2(x^2 + 5x + 6)$
$= 2(x + 3)(x + 2)$

71. $-5a^2 + 25a - 30 = -5(a^2 - 5a + 6)$
$= -5(a - 3)(a - 2)$

73. $z^3 - 29z^2 + 100z = z(z^2 - 29z + 100)$
$= z(z - 4)(z - 25)$

75. $12xy + 4x^2y - 72y$
$= 4x^2y + 12xy - 72y$
$= 4y(x^2 + 3x - 18)$
$= 4y(x + 6)(x - 3)$

77. $-r^2 + 14r - 40 = -1(r^2 - 14r + 40)$
$= -(r - 10)(r - 4)$

79. $-13yz + y^2 - 14z^2 = y^2 - 13yz - 14z^2$
$= (y - 14z)(y + z)$

81. $s^2 + 11s - 26 = (s + 13)(s - 2)$

83. $a^2 + 10ab + 9b^2 = (a + 9b)(a + b)$

85. $-x^2 + 21x + 22 = -1(x^2 - 21x - 22)$
$= -(x - 22)(x + 1)$

87. $d^3 - 11d^2 - 26d = d(d^2 - 11d - 26)$
$= d(d - 13)(d + 2)$

APPLICATIONS

89. PETS
$x^3 + 12x^2 + 27x = x(x^2 + 12x + 27)$
$= x(x + 3)(x + 9)$

The length is $(x + 9)$ in.
The width is x in.
The height is $(x + 3)$ in.

WRITING

91. Answers will vary.

93. Answers will vary.

95. Answers will vary.

REVIEW

97. $\dfrac{x^{12}x^{-7}}{x^3x^4} = \dfrac{x^{12+(-7)}}{x^{3+4}}$

$= \dfrac{x^5}{x^7}$

$= x^{5-7}$

$= x^{-2}$

$= \dfrac{1}{x^2}$

99. $\dfrac{a^4a^{-2}}{a^2a^0} = \dfrac{a^{4+(-2)}}{a^{2+0}}$

$= \dfrac{a^2}{a^2}$

$= a^{2-2}$

$= a^0$

$= 1$

CHALLENGE PROBLEMS

101. $x^2 - \dfrac{6}{5}x + \dfrac{9}{25} = \left(x - \dfrac{3}{5}\right)\left(x - \dfrac{3}{5}\right)$

103. $x^{2m} - 12x^m - 45 = (x^m + 3)(x^m - 15)$

105. $c = 5; 5 + 1 = 6$
$c = 8; 2 + 4 = 6$
$c = 9; 3 + 3 = 6$

SECTION 5.3

VOCABULARY

1. The trinomial $3x^2 - x - 12$ has a **lead** coefficient of 3. The middle term is $-x$ and the last **term** is -12.

3. The numbers 6 and -2 are two integers whose **product** is -12 and whose **sum** is 4.

5. To factor a trinomial by grouping, we begin by finding ac, the key number.

CONCEPTS

7. a) $1 \cdot 9 = 9$ or $3 \cdot 3 = 9$
   b) $-2(-8) = 16$, $-1(-16) = 16$, or $-4(-4) = 16$
   c) $-2(5) = -10$, $2(-5) = -10$, $-1(10) = -10$, or $1(-10) = -10$

9. a) $4y \cdot 2 = 8y$
   b) $-8 \cdot 3y = -24y$
   c) $8y + (-24y) = -16y$

11. $10x$ and x, or $5x$ and $2x$

13. a) When factoring a trinomial, we write it in **descending** powers of the variable. Then we factor out any **GCF** (including -1 if that is necessary to make the lead coefficient **positive**).

    b) $3s^2$
    c) $-(2d^2 - 19d + 8)$

15. Since the last term of the trinomial is **positive** and the middle term is **negative**, the integers must be **negative** factors of 6.

17. Since the last term of the trinomial is **negative**, the signs of the integers will be **different**.

19. $3x(5x - 2) + 2(5x - 2) = (5x - 2)(3x + 2)$

21.

Negative factors of 12	Sum of factors of 12
-1(-12) = 12	**-1 + (-12) = -13**
-2(-6) = 12	**-2 + (-6) = -8**
-3(-4) = 12	**-3 + (-4) = -7**

NOTATION

23. a) Answers will vary. $x^2 + 6x + 1$
    b) Answers will vary. $3x^2 + 6x + 1$

25. a) $a = 12$, $b = 20$, $c - 9$
    b) key number $= ac = 12(-9) = -108$

PRACTICE

27. $3a^2 + 13a + 4 = (3a + 1)(a + 4)$

29. $4z^2 - 13z + 3 = (z - 3)(4z - 1)$

31. $2m^2 + 5m - 12 = (2m - 3)(m + 4)$

33. $12t^2 + 17t + 6 = 12t^2 + \underline{9}t + \underline{8}t + 6$
    $= \underline{3t}(4t + 3) + \underline{2}(4t + 3)$
    $= (\underline{4t + 3})(3t + 2)$

35. $3a^2 + 13a + 4 = 3a^2 + 1a + 12a + 4$
    $= a(3a + 1) + 4(3a + 1)$
    $= (3a + 1)(a + 4)$

37. $5x^2 + 11x + 2 = 5x^2 + 10x + 1x + 2$
    $= 5x(x + 2) + 1(x + 2)$
    $= (x + 2)(5x + 1)$

39. $4x^2 + 8x + 3 = 4x^2 + 2x + 6x + 3$
    $= 2x(2x + 1) + 3(2x + 1)$
    $= (2x + 1)(2x + 3)$

41. $6x^2 + 25x + 21 = 6x^2 + 18x + 7x + 21$
    $= 6x(x + 3) + 7(x + 3)$
    $= (x + 3)(6x + 7)$

43. $2x^2 - 3x + 1 = 2x^2 - 2x - x + 1$
    $= 2x(x - 1) - 1(x - 1)$
    $= (x - 1)(2x - 1)$

45. $4t^2 - 4t + 1 = 4t^2 - 2t - 2t + 1$
    $= 2t(2t - 1) - 1(2t - 1)$
    $= (2t - 1)(2t - 1)$

47. $15t^2 - 34t + 8 = 15t^2 - 30t - 4t + 8$
    $= 15t(t - 2) - 4(t - 2)$
    $= (t - 2)(15t - 4)$

49. $2x^2 - 3x - 2 = 2x^2 - 4x + x - 2$
    $= 2x(x - 2) + 1(x - 2)$
    $= (x - 2)(2x + 1)$
    $= (a - 2)(3a + 2)$

51. $12y^2 - y - 1 = 12y^2 - 4y + 3y - 1$
$= 4y(3y - 1) + 1(3y - 1)$
$= (3y - 1)(4y + 1)$

53. $10y^2 - 3y - 1 = 10y^2 - 5y + 2y - 1$
$- 5y(2y - 1) + 1(2y - 1)$
$= (2y - 1)(5y + 1)$

55. $12y^2 - 5y - 2 = 12y^2 - 8y + 3y - 2$
$= 4y(3y - 2) + 1(3y - 2)$
$= (3y - 2)(4y + 1)$

57. $2m^2 + 5m - 10$
prime

59. $-13x + 3x^2 - 10 = 3x^2 - 13x - 10$
$= 3x^2 - 15x + 2x - 10$
$= 3x(x - 5) + 2(x - 5)$
$= (x - 5)(3x + 2)$

61. $6r^2 + rs - 2s^2 = 6r^2 + 4rs - 3rs - 2s^2$
$= 2r(3r + 2s) - s(3r + 2s)$
$= (3r + 2s)(2r - s)$

63. $8n^2 + 91n + 33 = 8n^2 + 88n + 3n + 33$
$= 8n(n + 11) + 3(n + 11)$
$= (n + 11)(8n + 3)$

65. $4a^2 - 15ab + 9b^2$
$= 4a^2 - 3ab - 12ab + 9b^2$
$= a(4a - 3b) - 3b(4a - 3b)$
$= (4a - 3b)(a - 3b)$

67. $130r^2 + 20r - 110$
$= 10(13r^2 + 2r - 11)$
$= 10(13r^2 + 13r - 11r - 11)$
$= 10[13r(r + 1) - 11(r + 1)]$
$= 10(13r - 11)(r + 1)$

69. $-5t^2 - 13t - 6 = -(5t^2 + 13t + 6)$
$= -(5t^2 + 10t + 3t + 6)$
$= -[5t(t + 2) + 3(t + 2)]$
$= -(t + 2)(5t + 3)$

71. $36y^2 - 88 + 32 = 4(9y^2 - 22y + 8)$
$= 4[9y^2 - 18y - 4y + 8]$
$= 4[9y(y - 2) - 4(y - 2)]$
$= 4(y - 2)(9y - 4)$
$= 5(2a - 1)(7a - 6)$

73. $4x^2 + 8xy + 3y^2$
$= 4x^2 + 2xy + 6xy + 3y^2$
$= 2x(2x + y) + 3y(2x + y)$
$= (2x + y)(2x + 3y)$

75. $12y^2 + 12 - 25y = 12y^2 - 25y + 12$
$= 12y^2 - 16y - 9y + 12$
$= 4y(3y - 4) - 3(3y - 4)$
$= (3y - 4)(4y - 3)$

77. $18x^7 + 31x - 10 = 18x^7 + 36x - 5x - 10$
$= 18x(x + 2) - 5(x + 2)$
$= (x + 2)(18x - 5)$

79. $-y^3 - 13y^2 - 12y = -y(y^2 + 13y + 12)$
$= -y(y + 1)(y + 12)$

81. $3x^2 + 6 + x = 3x^2 + x + 6$
prime

83. $30r^5 + 63r^4 - 30r^3$
$= 3r^3(10r^2 + 21r - 10)$
$= 3r^3(10r^2 + 25r - 4r - 10)$
$= 3r^3[5r(2r + 5) - 2(2r + 5)]$
$= 3r^3(2r + 5)(5r - 2)$

85. $2a^2 + 3b^2 + 5ab = 2a^2 + 5ab + 3b^2$
$= 2a^2 + 2ab + 3ab + 3b^2$
$= 2a(a + b) + 3b(a + b)$
$= (a + b)(2a + 3b)$

87. $pq + 6p^2 - q^2 = 6p^2 + pq - q^2$
$= 6p^2 + 3pq - 2pq - q^2$
$= 3p(2p + q) - q(2p + q)$
$= (2p + q)(3p - q)$

89. $6x^3 - 15x^2 - 9x = 3x(2x^2 - 5x - 3)$
$= 3x(2x^2 - 6x + x - 3)$
$= 3x[2x(x - 3) + 1(x - 3)]$
$= 3x(x - 3)(2x + 1)$

91. $15 + 8a^2 - 26a = 8a^2 - 26a + 15$
$= 8a^2 - 20a - 6a + 15$
$= 4a(2a - 5) - 3(2a - 5)$
$= (2a - 5)(4a - 3)$

93. $16m^3n + 20m^2n^2 + 6mn^3$
$= 2mn(8m^2 + 10mn + 3n^2)$
$= 2mn(8m^2 + 4mn + 6mn + 3n^2)$
$= 2mn[4m(2m + n) + 3n(2m + n)]$
$= 2mn(2m + n)(4m + 3n)$

APPLICATIONS

95. FURNITURE

$4x^2 + 20x - 11 = 4x^2 - 2x + 22x - 11$
$$= 2x(2x - 1) + 11(2x - 1)$$
$$= (2x - 1)(2x + 11)$$

The length is $(2x + 11)$ in.
The width is $(2x - 1)$ in.

To find the difference, subtract:
$(2x + 11) - (2x - 1) = 2x + 11 - 2x + 1$
$$= 12 \text{ in.}$$
The difference is 12 in.

WRITING

97. Answers will vary.

99. Answers will vary.

101. Answers will vary.

REVIEW

103. $-7^2 = -(7)(7)$
$$= -49$$

105. $7^0 = 1$

107. $\dfrac{1}{7^{-2}} = 7^2$
$$= 49$$

CHALLENGE PROBLEMS

109. $6a^{10} + 5a^5 - 21$
$$= 6a^{10} + 14a^5 - 9a^5 - 21$$
$$= 2a^5(3a^5 + 7) - 3(3a^5 + 7)$$
$$= (3a^5 + 7)(2a^5 - 3)$$

111. $8x^2(c^2+c-2) - 2x(c^2+c-2) - 1(c^2+c-2)$
$$= (c^2 + c - 2)(8x^2 - 2x - 1)$$
$$= (c + 2)(c - 1)(4x + 1)(2x - 1)$$

SECTION 5.4

VOCABULARY

1. $x^2 + 6x + 9$ is a **perfect** square trinomial because it is the square of the binomial $(x + 3)$.

CONCEPTS

3. a) The first term is the square of **5x**.
   b) The last term is the square of **3**.
   c) The middle term is twice the product of **5x** and **3**.

5. a) $x^2 + 2xy + y^2 = (\underline{x} + \underline{y})^2$
   b) $x^2 - 2xy + y^2 = (x - y)^2$
   c) $x^2 - y^2 = (x + y)(x - y)$

7. 1, 4, 9, 16, 25, 36, 47, 64, 81, 100

9. a) $(x + 6)(x - 6) = x^2 - 6x + 6x - 36$
   $= x^2 - 36$

   $(x + 6)(x + 6) = x^2 + 6x + 6x + 36$
   $= x^2 + 12x + 36$

   $(x - 6)(x - 6) = x^2 - 6x - 6x + 36$
   $= x^2 - 12x + 36$

   b) No

11. $(3x + 4y)(3x - 4y)$
    $= 9x^2 - 12xy + 12xy - 16y^2$
    $= 9x^2 - 16y^2$

NOTATION

13. Answers will vary.
    a) $x^2 - 9$
    b) $(x - 9)^2$
    c) $x^2 + 2xy + y^2$
    d) $x^2 + 25$

15. a) base: $(x + 3)$
    exponent: 2

    b) $(x - 8)(x - 8)$

PRACTICE

17. $a^2 - 6a + 9 = (a - \underline{3})^2$

19. $4x^2 + 4x + 1 = (2x + 1)^2$

21. $x^2 + 6x + 9 = (x + 3)^2$

23. $b^2 + 2b + 1 = (b + 1)^2$

25. $c^2 - 12c + 36 = (c - 6)^2$

27. $y^2 - 8y + 16 = (y - 4)^2$

29. $t^2 + 20t + 100 = (t + 10)^2$

31. $2u^2 - 36u + 162 = 2(u^2 - 18u + 81)$
    $= 2(u - 9)^2$

33. $36x^3 + 12x^2 + x = x(36x^2 + 12x + 1)$
    $= x(6x + 1)^2$

35. $9 + 4x^2 + 12x = 4x^2 + 12x + 9$
    $= (2x + 3)^2$

37. $a^2 + 2ab + b^2 = (a + b)^2$

39. $25m^2 + 70mn + 49n^2 = (5m + 7n)^2$

41. $9x^2y^2 + 30xy + 25 = (3xy + 5)^2$

43. $t^2 - \dfrac{2}{3}t + \dfrac{1}{9} = \left(t - \dfrac{1}{3}\right)^2$

45. $s^2 - 1.2s + 0.36 = (s - 0.6)^2$

47. $y^2 - 49 = (y + \underline{7})(y - \underline{7})$

49. $t^2 - w^2 = (\underline{t} + \underline{w})(t - w)$

51. $x^2 - 16 = (x - 4)(x + 4)$

53. $4y^2 - 1 = (2y + 1)(2y - 1)$

55. $9x^2 - y^2 = (3x + y)(3x - y)$

57. $16a^2 - 25b^2 = (4a + 5b)(4a - 5b)$

59. $36 - y^2 = (6 + y)(6 - y)$

61. $a^2 + b^2$
    prime

63. $a^4 - 144b^2 = (a^2 + 12b)(a^2 - 12b)$

65. $t^2z^2 - 64 = (tz + 8)(tz - 8)$

67. $y^2 - 63$
    prime

69. $8x^2 - 32y^2 = 8(x^2 - 4y^2)$
$= 8(x + 2y)(x - 2y)$

71. $7a^2 - 7 = 7(a^2 - 1)$
$= 7(a + 1)(a - 1)$

73. $-25 + v^2 = v^2 - 25$
$= (v + 5)(v - 5)$

75. $6x^4 - 6x^2y^2 = 6x^2(x^2 - y^2)$
$= 6x^2(x + y)(x - y)$

77. $x^4 - 81 = (x^2 + 9)(x^2 - 9)$
$= (x^2 + 9)(x + 3)(x - 3)$

79. $a^4 - 16 = (a^2 + 4)(a^2 - 4)$
$= (a^2 + 4)(a + 2)(a - 2)$

81. $c^2 - \dfrac{1}{16} = \left(c + \dfrac{1}{4}\right)\left(c - \dfrac{1}{4}\right)$

APPLICATIONS

83. GENETICS
$p^2 + 2pq + q^2 = 1$
$(p + q)^2 = 1$

85. PHYSICS
$0.5gt_1^2 - 0.5gt_2^2 = 0.5g(t_1^2 - t_2^2)$
$= 0.5g(t_1 + t_2)(t_1 - t_2)$
The distance is $0.5g(t_1 + t_2)(t_1 - t_2)$.

WRITING

87. Answers will vary.

89. Answers will vary.

91. Answers will vary.

REVIEW

93. $\dfrac{5x^2 + 10y^2 - 15xy}{5xy} = \dfrac{5x^2}{5xy} + \dfrac{10y^2}{5xy} - \dfrac{15xy}{5xy}$
$= \dfrac{x}{y} + \dfrac{2y}{x} - 3$

95. Write the dividend in descending order of exponents.

$$
\begin{array}{r}
3a + 2 \\
2a - 1 \overline{)6a^2 + a - 2} \\
\underline{6a^2 - 3a} \\
4a - 2 \\
\underline{4a - 2} \\
0
\end{array}
$$

CHALLENGE PROBLEMS

97. $5(4x + 3)(4x - 3) = 5(16x^2 - 9)$
$= 80x^2 - 45$

$c = 45$

99. $81x^6 + 36x^3y^2 + 4y^4 = (9x^3 + 2y^2)^2$

101. $(x + 5)^2 - y^2 = (x + 5 + y)(x + 5 - y)$

103. $(x + 5)^2 - y^2 = [(x + 5) + y][(x + 5) - y]$
$= (x + 5 + y)(x + 5 - y)$

111 Finished Sec 5.4

SECTION 5.5

VOCABULARY

1. $x^3 + 27$ is the **sum** of two cubes.

3. The factorization of $x^3 + 8$ is
$(x + 2)(x^2 - 2x + 4)$. The first factor is a
binomial and the second is a trinomial.

CONCEPTS

5. a) $F^3 + L^3 = (\boxed{F} + \boxed{L})(F^2 - FL + L^2)$

   b) $F^3 - L^3 = (F\boxed{-}L)(\boxed{F^2} + FL + \boxed{L^2})$

   a) $125 = (\boxed{5})^3$

   b) $27m^3 = (\boxed{3m})^3$

   c) $8x^3 - 27 = (\boxed{2x})^3 - (\boxed{3})^3$

   d) $x^3 + 64y^3 = (\boxed{x})^3 + (\boxed{4y})^3$

7. This is $\boxed{m}$ cubed.

   This is $\boxed{4}$ cubed.

9. 1, 8, 27, 64, 125, 216

11. It is factored completely.

NOTATION

13. Answers will vary.
   a) $x^3 + 8$
   b) $(x + 8)^3$

PRACTICE

15. $a^3 + 8 = (a)^3 + (2)^3$
   $= (a + 2)[(a)^2 - a(2) + (2)^2]$
   $= (a + 2)(a^2 - \boxed{2a} + 4)$

17. $b^3 + 27 = (b)^3 + (3)^3$
   $= (b + 3)[(b)^2 - b(3) + (3)^2]$
   $= (\boxed{b + 3})(b^2 - 3b + 9)$

19. $y^3 + 1 = (y)^3 + (1)^3$
   $= (y + 1)[(y)^2 - y(1) + (1)^2]$
   $= (y + 1)(y^2 - y + 1)$

21. $a^3 - 27 = (a)^3 - (3)^3$
   $= (a - 3)[(a)^2 + a(3) + (3)^2]$
   $= (a - 3)(a^2 + 3a + 9)$

23. $8 + x^3 = (2)^3 + (x)^3$
   $= (2 + x)[(2)^2 - 2(x) + (x)^2]$
   $= (2 + x)(4 - 2x + x^2)$

25. $s^3 - t^3 = (s)^3 - (t)^3$
   $- (s - t)[(s)^2 + s(t) + (t)^2]$
   $= (s - t)(s^2 + st + t^2)$

27. $a^3 + 8b^3 = (a)^3 + (2b)^3$
   $= (a + 2b)[(a)^2 - a(2b) + (2b)^2]$
   $= (a + 2b)(a^2 - 2ab + 4b^2)$

29. $64x^3 - 27 = (4x)^3 - (3)^3$
   $= (4x - 3)[(4x)^2 + 4x(3) + (3)^2]$
   $= (4x - 3)(16a^2 + 12x + 9)$

31. $a^6 - b^3 = (a^2)^3 - (b)^3$
   $= (a^2 - b)[(a^2)^2 + a^2(b) + (b)^2]$
   $= (a^2 - b)(a^4 + a^2b + b^2)$

33. $2x^3 + 54 = 2(x^3 + 27)$
   $= 2[(x)^3 + (3)^3]$
   $= 2(x + 3)[(x)^2 - x(3) + (3)^2]$
   $= 2(x + 3)(x^2 - 3x + 9)$

35. $-x^3 + 216 = -(x^3 - 216)$
   $= -(x)^3 - (6)^3$
   $= -(x - 6)[(x)^2 + x(6) + (6)^2]$
   $= -(x - 6)(x^2 + 6x + 36)$

37. $64m^3x - 8n^3x$
   $= 8x(8m^3 - n^3)$
   $= 8x[(2m)^3 - (n)^3]$
   $= 8x(2m-n)[(2m)^2 + 2m(n) + (n)^2]$
   $= 8x(2m - n)(4m^2 + 2mn + n^2)$

APPLICATIONS

39. MAILING BREAKABLES
   Volume of larger cube:
   $V = 10^3$
   $= 1,000$ in.^{3}

   Volume of smaller cube:
   $V = x^3$ in.^{3}

   Volume of empty space:
   $V = (1,000 - x^3)$ in.^{3}
   $= (10)^3 - (x)^3$
   $= (10 - x)(10^2 + 10(x) + x^2)$
   $= (10 - x)(100 + 10x + x^2)$

WRITING

41. Answers will vary.

REVIEW

43. $\dfrac{7}{9} = 0.\overline{7}$, repeating decimal

45. $\{\ldots, -4, -3, -2, -1, 0, 1, 2, 3, 4, \ldots\}$

47. $2x^2 + 5x - 3$, $x = -3$

$$\begin{aligned}
2x^2 + 5x - 3 &= 2(\mathbf{-3})^2 + 5(\mathbf{-3}) - 3 \\
&= 2(9) + 5(-3) - 3 \\
&= 18 - 15 - 3 \\
&= 0
\end{aligned}$$

CHALLENGE PROBLEMS

49. a) $(x^3)^2 - 1^2$
$$\begin{aligned}
&= (x^3 + 1)(x^3 - 1) \\
&= (x + 1)(x^2 - x + 1)(x + 1)(x^2 - x + 1)
\end{aligned}$$

b) $(x^2)^3 - 1^3$
$$\begin{aligned}
&= (x^2 - 1)(x^4 + x^2 + 1) \\
&= (x + 1)(x - 1)(x^4 + x^2 + 1)
\end{aligned}$$

c) They appear to be different because the trinomial factor $(x^4 + x^2 + 1)$ of part "b" has not been factored completely.
$(x^4 + x^2 + 1) = (x^2 - x + 1)(x^2 + x + 1)$

51. $x^{3m} - y^{3n} = (x^m)^3 - (y^n)^3$
$$= (x^m - y^n)(x^{2m} + x^m y^n + y^{2n})$$

VOCABULARY

1. A polynomial is factored **completely** when no factor can be factored further.

3. A polynomial that does not factor using integers is called a **prime** polynomial.

5. $4x^2 - 12x + 9$ is a perfect square **trinomial** because it is the square of $(2x - 3)$.

7. When factoring a polynomial, always factor out the **GCF** first.

CONCEPTS

9. factor out the GCF

11. perfect square trinomial

13. sum of two cubes

15. trinomial factoring

NOTATION

17. $8m^3$

PRACTICE

19. $2ab^2 + 8ab - 24a = 2a(b^2 + 4a - 12)$
$$= 2a(b + 6)(b - 2)$$

21. $-8p^3q^7 - 4p^2q^3 = -4p^2q^3(2pq^4 + 1)$

23. $20m^2 + 100m + 125 = 5(4m^2 + 20m + 25)$
$$= 5(2m + 5)^2$$

25. $x^2 + 7x + 1$
prime

27. $-2x^5 + 128x^2 = -2x^2(x^3 - 64)$
$$= -2x^2[(x)^3 - (4)^3]$$
$$= -2x^2(x - 4)(x^2 - 4x + 16)$$

29. $a^2c + a^2d^2 + bc + bd^2$
$$= (a^2c + a^2d^2) + (bc + bd^2)$$
$$= a^2(c + d^2) + b(c + d^2)$$
$$= (c + d^2)(a^2 + b)$$
$$= 2t^2(3t - 5)(t + 4)$$

31. $-9x^2y^2 + 6xy - 1 = -(9x^2y^2 - 6xy + 1)$
$$= -(3xy - 1)^2$$

33. $5x^3y^3z^4 + 25x^2y^3z^2 - 35x^3y^2z^5$
$$= 5x^2y^2z^2(xyz^2 + 5y - 7xz^3)$$

35. $2c^2 + 5cd - 3d^2 = (2c + d)(c - 3d)$

37. $8a^2x^3y - 2b^2xy = 2xy(4a^2x^2 - b^2)$
$$= 2xy(2ax + b)(2ax - b)$$

39. $a^2(x - a) - b^2(x - a) = (x - a)(a^2 - b^2)$
$$= (x - a)(a + b)(a - b)$$

41. $a^2b^2 - 144 = (ab + 12)(ab - 12)$

43. $2ac + 4ad + bc + 2bd$
$$= (2ac + 4ad) + (bc + 2bd)$$
$$= 2a(c + 2d) + b(c + 2d)$$
$$= (c + 2d)(2a + b)$$

45. $v^2 - 14v + 49 = (v - 7)^2$

47. $39 + 10n - n^2 = -n^2 + 10n + 39$
$$= -(n^2 - 10n - 39)$$
$$= -(n + 3)(n - 13)$$

49. $16y^8 - 81z^4$
$$= (4y^4 + 9z^2)(4y^4 - 9z^2)$$
$$= (4y^4 + 9z^2)(2y^2 + 3z)(2y^2 - 3z)$$

51. $6x^2 - 14x + 8 = 2(3x^2 - 7x + 4)$
$$= 2(3x - 4)(x - 1)$$

53. $4x^2y^2 + 4xy^2 + y^2 = y^2(4x^2 + 4x + 1)$
$$= y^2(2x + 1)^2$$

55. $4m^5n + 500m^2n^4$
$$= 4m^2n(m^3 + 125n^3)$$
$$= 4m^2n[(m)^3 + (5n)^3]$$
$$= 4m^2n(m + 5n)(m^2 - 5mn + 25n^2)$$

57. $a^2x^2 + b^2y^2 + b^2x^2 + a^2y^2$
$$= (a^2x^2 + a^2y^2) + (b^2x^2 + b^2x^2)$$
$$= a^2(x^2 + y^2) + b^2(x^2 + y^2)$$
$$= (x^2 + y^2)(a^2 + b^2)$$

59. $4x^2 + 9y^2$
prime

61. $16a^5 - 54a^2b^3$
$$= 2a^2(8a^3 - 27b^3)$$
$$= 2a^2[(2a)^3 - (3b)^3]$$
$$= 2a^2(2a - 3b)(4a^2 + 6ab + 9b^2)$$

63. $27x - 27y - 27z = 27(x - y - z)$

65. $xy - ty + xs - ts = (xy - ty) + (xs - ts)$
$\qquad\qquad\qquad = y(x - t) + s(x - t)$
$\qquad\qquad\qquad = (x - t)(y + s)$

89. $x^9 + y^6 = [(x^3)^3 + (y^2)^3]$
$\qquad\qquad = (x^3 + y^2)(x^6 - x^3y^2 + y^4)$

67. $35x^8 - 2x^7 - x^6 = x^6(35x^2 - 2x - 1)$
$\qquad\qquad\qquad = x^6(7x + 1)(5x - 1)$

69. $5(x - 2) + 10y(x - 2) = (x - 2)(5 + 10y)$
$\qquad\qquad\qquad\qquad = (x - 2)(5)(1 + 2y)$
$\qquad\qquad\qquad\qquad = 5(x - 2)(1 + 2y)$

71. $49p^2 + 28pq + 4q^2 = (7p + 2q)^2$

73. $4t^2 + 36 = 4(t^2 + 9)$

75. $p^3 - 2p^2 + 3p - 6 = (p^3 - 2p^2) + (3p - 6)$
$\qquad\qquad\qquad\qquad = p^2(p - 2) + 3(p - 2)$
$\qquad\qquad\qquad\qquad = (p - 2)(p^2 + 3)$

WRITING

77. Answers will vary.

79. Answers will vary.

REVIEW

81.

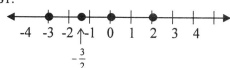

83.

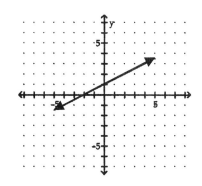

CHALLENGE PROBLEMS

85. $x^4 - 2x^2 - 8 = (x^2 + 2)(x^2 - 4)$
$\qquad\qquad\qquad = (x^2 + 2)(x + 2)(x - 2)$

87. $24 - x^3 + 8x^2 - 3x = 24 - 3x + 8x^2 - x^3$
$\qquad\qquad\qquad\qquad = (24 - 3x) + (8x^2 - x^3)$
$\qquad\qquad\qquad\qquad = 3(8 - x) + x^2(8 - x)$
$\qquad\qquad\qquad\qquad = (8 - x)(3 + x^2)$

Section 5.6

VOCABULARY

1. Any equation that can be written in the form $ax^2 + bx + c = 0$ $(a \neq 0)$ is called a **quadratic** equation.

3. Integers that are 1 unit apart, such as 8 and 9, are called **consecutive** integers.

5. The longest side of a right triangle is the **hypotenuse**. The remaining two sides are the **legs** of the triangle.

CONCEPTS

7. a) yes
   b) no
   c) yes
   d) no

9.
$$5x + 4 = 0$$
$$5x + 4 - 4 = 0 - 4$$
$$5x = -4$$
$$\frac{5x}{5} = \frac{-4}{5}$$
$$\boxed{x = -\frac{4}{5}}$$

11. a) $x^2 + 7x = -6$
    Add 6 to both sides.
    b) $x(x + 7) = 3$
    Distribute the multiplication by x and subtract 3 from both sides.

13. a) $x^2 + 6x - 16 = (x - 2)(x + 8)$

    b) $x^2 + 6x - 16 = 0$
    $$(x - 2)(x + 8) = 0$$
    $$x - 2 = 0$$
    $$x - 2 + 2 = 0 + 2$$
    $$\boxed{x = 2}$$
    $$x + 8 = 0$$
    $$x + 8 - 8 = 0 - 8$$
    $$\boxed{x = -8}$$

17. If the length of the hypotenuse of a right triangle is c and the lengths of the other two legs are a and b, then $\boxed{a^2 + b^2} = c^2$.

NOTATION

19.
$$7y^2 + 14y = 0$$
$$\boxed{7y}(y + 2) = 0$$
$$7y = 0 \quad \text{or} \quad \boxed{y + 2} = 0$$
$$y = \boxed{0} \qquad y = \boxed{-2}$$

PRACTICE

21.
$$(x - 2)(x + 3) = 0$$
$$x - 2 = 0 \quad \text{or} \quad x + 3 = 0$$
$$\boxed{x = 2} \qquad \boxed{x = -3}$$
$$x = 2, \ -3$$
Use this example.

23. $\dfrac{5}{2}, \ -6$

25. 7, -8

27. 1, -2, 3

29. $x(x - 3) = 0$
$$\boxed{x = 0} \quad \text{or} \quad x - 3 = 0$$
$$\boxed{x = 3}$$
$$x = 0, \ 3$$
Use this example.

31. $0, \ \dfrac{5}{2}$

33. 0, 7

35.
$$3x^2 + 8x = 0$$
$$x(3x + 8) = 0$$
$$\boxed{x = 0} \quad \text{or} \quad 3x + 8 = 0$$
$$3x = -8$$
$$\boxed{x = -\frac{8}{3}}$$
$$x = 0, \ -\frac{8}{3}$$
Use this example.

37. 0, 2

39.
$$x^2 - 25 = 0$$
$$(x + 5)(x - 5) = 0$$
$$x + 5 = 0 \quad \text{or} \quad x - 5 = 0$$
$$\boxed{x = -5} \qquad \boxed{x = -5}$$
$$x = 5, \ -5$$
Use this example.

41. $-\dfrac{1}{2}, \ \dfrac{1}{2}$

43. $-\dfrac{2}{3}, \ \dfrac{2}{3}$

45.
$$x^2 = 100$$
$$x^2 - 100 = 0$$
$$(x + 10)(x - 10) = 0$$
$$x + 10 = 0 \quad \text{or} \quad x - 10 = 0$$
$$\boxed{x = -10} \qquad \boxed{x = 10}$$
$$x = -10, \ 10$$
Use this example.

47. $-\dfrac{9}{2}, \ \dfrac{9}{2}$

49.
$$x^2 - 13x + 12 = 0$$
$$(x - 12)(x - 1) = 0$$
$$x - 12 = 0 \quad \text{or} \quad x - 1 = 0$$
$$\boxed{x = 12} \qquad \boxed{x = 1}$$
$$x = 12, \ 1$$
Use this example.

51. $-3, 7$

53. $8, 1$

55.
$$a^2 + 8a = -15$$
$$a^2 + 8a + 15 = 0$$
$$(a + 3)(a + 5) = 0$$
$$a + 3 = 0 \quad \text{or} \quad a + 5 = 0$$
$$\boxed{a = -3} \qquad \boxed{a = -5}$$
$$a = -3, \ -5$$
Use this example.

57. $-2, \ -2$

59. $8, 8$

61. $\dfrac{1}{2}, \ 2$

63. $\dfrac{1}{5}, \ 1$

65. $-\dfrac{1}{2}, \ -\dfrac{1}{2}$

67. $-\dfrac{2}{3}, \ -\dfrac{3}{2}$

69. $\dfrac{2}{3}, \ -\dfrac{1}{5}$

71.
$$x(2x - 3) = 20$$
$$2x^2 - 3x = 20$$
$$2x^2 - 3x - 20 = 0$$
$$(2x + 5)(x - 4) = 0$$
$$2x + 5 = 0 \quad \text{or} \quad x - 4 = 0$$
$$2x = -5 \qquad \boxed{x = 4}$$
$$\boxed{x = -\frac{5}{2}}$$
$$x = -\frac{5}{2}, \ 4$$
Use this example.

73. $(n+8)(n-3) = -30$

$n^2 + 5n - 24 = -30$

$n^2 + 5n + 6 = 0$

$(n+3)(n+2) = 0$

$n+3 = 0 \quad$ or $\quad n+2 = 0$

$\boxed{n = -3} \qquad \boxed{n = -2}$

$x = -3, \ -2$

Use this example.

75. $2, 10$

77. $\dfrac{1}{8}, 1$

79. $(x-2)(x^2 - 8x + 7) = 0$

$(x-2)(x-7)(x-1) = 0$

$x - 2 = 0$ or $x - 7 = 0$ or $x - 1 = 0$

$\boxed{x = 2} \qquad \boxed{x = 7} \qquad \boxed{x = 1}$

$x = 2, \ 7, \ 1$

Use this example.

81. $(x^3 + 3x^2 + 2x) = 0$

$x(x^2 + 3x + 2) = 0$

$x(x+1)(x+2) = 0$

$x = 0$ or $x + 1 = 0$ or $x + 2 = 0$

$\boxed{x = 0} \qquad \boxed{x = -1} \qquad \boxed{x = -2}$

$x = 0, \ -1, \ -2$

Use this example.

83. $0, 9, \ -3$

85. $0, \ -1, 2$

87. OFFICIATING

height is 0 feet (hits the ground)

$h = -16t^2 + 22t + 3$

$0 = -16t^2 + 22t + 3$

$16t^2 - 22t - 3 = 0$

$(8t + 1)(2t - 3) = 0$

$8t + 1 = 0 \quad$ or $\quad 2t - 3 = 0$

$8t = -1 \qquad 2t = 3$

$t = -\dfrac{1}{8} \qquad \boxed{t = \dfrac{3}{2}}$

Time is 1.5 sec.

89. EXHIBITION DIVING

height is 0 feet (enters the water)

$h = -16t^2 + 64$

$0 = -16t^2 + 64$

$16t^2 - 64 = 0$

$16(t^2 - 4) = 0$

$16(t+2)(t-2) = 0$

$t + 2 = 0 \qquad t - 2 = 0$

$t = -2 \qquad \boxed{t = 2}$

Time is 2 sec.

91. CHOREOGRAPHY

# of dancers is 36.

# of rows is unknown.

$d = \dfrac{1}{2}r(r+1)$

$36 = \dfrac{1}{2}r(r+1)$

$72 = r(r+1)$

$72 = r^2 + r$

$0 = r^2 + r - 72$

$0 = (r+9)(r-8)$

$r + 9 = 0 \qquad\qquad r - 8 = 0$

$r = -9 \qquad\qquad \boxed{r = 8}$

# of rows is 8.

93. CUSTOMER SERVICE

x is the current number.

$x + 1$ is the next number.

Their product is 156.

$\quad$ Let $x = $ current #

$\quad x + 1 = $ next #

$\quad\quad x(x+1) = 156$

$\quad\quad\quad x^2 + x = 156$

$\quad\quad x^2 + x - 156 = 0$

$\quad (x+13)(x-12) = 0$

$\quad x + 13 = 0 \quad\quad x + 12 = 0$

$\quad\quad x = -13 \quad\quad \boxed{x = 12}$

$\quad\quad\quad\quad\quad\quad$ Current # is 12.

95. PLOTTING POINTS

x is odd x-cor in QI (all are positive).

$x + 2$ is odd y-cor in QI.

Their product is 143.

$\quad$ Let $x = $ odd x-cor

$\quad\quad x + 2 = $ odd y-cor

$\quad\quad\quad x(x+2) = 143$

$\quad\quad\quad x^2 + 2x = 143$

$\quad\quad x^2 + 2x - 143 = 0$

$\quad (x+13)(x-11) = 0$

$\quad x + 13 = 0 \quad\quad x - 11 = 0$

$\quad\quad x = -13 \quad\quad \boxed{x = 11}$

$\quad\quad\quad$ Corridinate is (11,13).

97. ISULATION

Length is $(2w+1)$ m.

Width is w m.

Area is 36 sq. m.

$A = lw$

$\quad$ Let w = width in m

$\quad\quad 2w + 1 = $ length in ft.

$\quad\quad\quad w(2w+1) = 36$

$\quad\quad\quad 2w^2 + w = 36$

$\quad\quad 2w^2 + w - 36 = 0$

$\quad (2w+9)(w-4) = 0$

$\quad 2w + 9 = 0 \quad\quad w - 4 = 0$

$\quad\quad 2w = -9 \quad\quad \boxed{w = 4}$

$\quad\quad w = -\dfrac{9}{2}$

$\quad\quad$ Dimensions are 4 m by 9 m.

99. DESIGNING A TENT

Height is unknown.

Length(base) is 2 ft more than twice ht.

Area is 30 sq. ft.

$A = \dfrac{1}{2}bh$

$\quad$ Let $x = $ height in ft

$\quad\quad 2x + 2 = $ length (base) in ft

$\quad\quad\quad \dfrac{1}{2}x(2x+2) = 30$

$\quad\quad\quad 2x^2 + 2x = 60$

$\quad\quad 2x^2 + 2x - 60 = 0$

$\quad\quad 2(x^2 + x - 30) = 0$

$\quad\quad 2(x+6)(x-5) = 0$

$\quad x + 6 = 0 \quad\quad x - 5 = 0$

$\quad\quad x = -6 \quad\quad \boxed{x = 5}$

$\quad\quad\quad\quad\quad\quad$ Height is 5 ft.

$\quad\quad\quad\quad\quad\quad$ Base is 12 ft.

101. WIND DAMAGE

Short leg is x ft.

Long leg is $(x + 2)$ ft.

Hypotenuse is $(x + 4)$ ft.

$$a^2 + b^2 = c^2$$

Let x = short leg (trunk) in ft

$x + 2$ = long leg in ft

$x + 4$ = hypotenuse (top) in ft.

$$x^2 + (x + 2)^2 = (x + 4)^2$$

$$x^2 + x^2 + 4x + 4 = x^2 + 8x + 16$$

$$x^2 - 4x - 12 = 0$$

$$(x + 2)(x - 6) = 0$$

$x + 2 = 0 \qquad x - 6 = 0$

$x = -2 \qquad \boxed{x = 6}$

Trunk is 6 ft.

Tree upright is 16 ft.

103. CAR REPAIRS

Back is unknown in ft.

Base is 2 ft longer than back.

Ramp is 1 ft longer than back.

$$a^2 + b^2 = c^2$$

Let x = back (short leg) in ft

$x + 1$ = ramp (long leg) in ft

$x + 2$ = base (hypotenuse) in ft

$$x^2 + (x + 1)^2 = (x + 2)^2$$

$$x^2 + x^2 + 2x + 1 = x^2 + 4x + 4$$

$$x^2 - 2x - 3 = 0$$

$$(x + 1)(x - 3) = 0$$

$x + 1 = 0 \qquad x - 3 = 0$

$x = -1 \qquad \boxed{x = 3}$

Back is 3 feet.

Ramp is 4 feet.

Base is 5 feet.

WRITING

105. Answers will vary.

107. Answers will vary.

REVIEW

109. EXERCISE

$15 \text{ min } \leq t < 30 \text{ min}$

CHALLENGE PROBLEMS

111. $\qquad x^4 - 625 = 0$

$$(x^2 + 25)(x^2 - 25) = 0$$

$x^2 + 25 = 0 \qquad\qquad x^2 - 25 = 0$

$\qquad x^2 = -25 \qquad (x + 5)(x - 5) = 0$

$\qquad$ undefined $\qquad x + 5 = 0 \quad x - 5 = 0$

$\qquad\qquad\qquad\qquad \boxed{x = -5} \quad \boxed{x = 5}$

CHAPTER 5: KEY CONCEPT

FACTORING

1. Start with: $3x + 27$

   Factored: $3(x + 9)$

2. Start with: $x^2 + 12x + 27$

   Factored: $(x + 3)(x + 9)$

3. $-3a^2 + 21a - 36 = -3(a^2 - 7a + 12)$
   $$= -3(a - 3)(a - 4)$$

4. $x^2 - 121y^2 = (x + 11y)(x - 11y)$

5. $rt + 2r + st + 2s = (rt + 2r) + (st + 2s)$
   $$= r(t + 2) + s(t + 2)$$
   $$= (r + s)(t + 2)$$

6. $v^3 - 8 = (v - 2)(v^2 + 2v + 4)$

7. $6t^2 - 19t + 15 = (3t - 5)(2t - 3)$

8. $25y^2 - 20y + 4 = (5y - 2)(5y - 2)$
   $$= (5y - 2)^2$$

9. $2r^3 - 50r = 2r(r^2 - 25)$
   $$= 2r(r + 5)(r - 5)$$

10. $46w - 6 + 16w^2 = 16w^2 + 46w - 6$
    $$= 2(8w^2 + 23w - 3)$$
    $$= 2(8w - 1)(w + 3)$$

CHAPTER 5: REVIEW

SECTION 5.1
The Greatest Common Factor;
Factoring by Grouping

1. $5 \cdot 7$

2. $2^5 \cdot 3$

3. 7

4. $18a^3$

5. $3(x + 3y)$

6. $5a(x^2 + 3)$

7. $7s^3(s^2 + 2)$

8. $\pi\, a(b - c)$

9. $2x(x^2 + 2x - 4)$

10. $xy^2z(x + yz - 1)$

11. $5ab(-b + 2a - 3)$
    or
    $-5ab(b - 2a + 3)$

12. $(x - 2)(4 - x)$

13. $-(a + 7)$

14. $-(4t^2 - 3t + 1)$

15. $2c + 2d + ac + ad = (2c + 2d) + (ac + ad)$
    $= 2(c + d) + a(c + d)$
    $= (c + d)(2 + a)$

16. $3xy + 9x - 2y - 6 = (3xy + 9x) + (-2y - 6)$
    $= 3x(y + 3) + -2(y + 3)$
    $= (y + 3)(3x - 2)$

17. $2a^3 - a + 2a^2 - 1 = (2a^3 - a) + (2a^2 - 1)$
    $= a(2a^2 - 1) + 1(2a^2 - 1)$
    $= (2a^2 - 1)(a + 1)$

18. $4m^2n + 12m^2 - 8mn - 24m$
    $= (4m^2n + 12m^2) + (-8mn - 24m)$
    $= 4m^2(n + 3) + -8m(n + 3)$
    $- (n + 3)(4m^2 - 8m)$
    $= (n + 3)4m(m - 2)$
    $= 4m(n + 3)(m - 2)$

SECTION 5.2
Factoring Trinomials of the Form
$x^2 + bx + c$

19. 1

20. $\boxed{7}$
    $\boxed{5}$
    $\boxed{-7}$
    $\boxed{-5}$

21. $(x + 6)(x - 4)$

22. $(x - 6)(x + 2)$

23. $(x - 5)(x - 2)$

24. prime

25. $-(y - 5)(y - 4)$

26. $(y + 9)(y + 1)$

27. $(c + 5d)(c - 2d)$

28. $(m - 2n)(m - n)$

29. multiply the two binomial factors
    $(x - 4)(x + 5) = x^2 + 5x - 4x - 20$
    $= x^2 + x - 20$

30. There are no two integers whose product is 11 and whose sum is 7.

31. $5a^3(a + 10)(a - 1)$

32. $-4x(x + 3y)(x - 2y)$

SECTION 5.3
Factoring Trinomials of the Form
$ax^2 + bx + c$

33. $(2x+1)(x-3)$

34. $(2y+5)(5y-2)$

35. $-(3x+1)(x-5)$

36. $3p(2p+1)(p-2)$

37. $(4b-c)(b-4c)$

38. prime

39. ENTERTAINING
$12x^2 - x - 1 = (4x+1)(3x-1)$
$(4x+1)$ in , $(3x-1)$ in

40. The middle sign of the second binomial must be negative.

SECTION 5.4
Factoring Perfect Square Trinomials and the Difference of Two Squares

41. $(x+5)^2$

42. $(3y-4)^2$

43. $-(z-1)^2$

44. $(5a+2b)^2$

45. $(x+3)(x-3)$

46. $(7t+5y)(7t-5y)$

47. $(xy+20)(xy-20)$

48. $8a(t+2)(t-2)$

49. $(c^2+16)(c+4)(c-4)$

50. prime

SECTION 5.5
Factoring the Sum and Difference of Two Cubes

51. $(h+1)(h^2-h+1)$

52. $(5p+q)(25p^2-5pq+q^2)$

53. $(x-3)(x^2+3x+9)$

54. $2x^2(2x-3y)(4x^2+6xy+9y^2)$

SECTION 5.6
A Factoring Strategy

55. $2y^2(3y-5)(y+4)$

56. $(t+u^2)(s^2+v)$

57. $(j^2+4)(j+2)(j-2)$

58. $-3(j+2k)(j^2-2jk+4k^2)$

59. $3(2w-3)^2$

60. prime

61. $2(t^3+5)$

62. $(20+m)(20-m)$

63. $(x+8y)^2$

64. $6c^2d(3cd-2c-4)$

SECTION 5.7
Solving Quadratic Equations by Factoring

65. $0, -2$

66. $0, 6$

67. $-3, 3$

68. $3, 4$

69. $-2, -2$

70. $6, -4$

71. $\dfrac{1}{5}$, 1

72. 0, −1, 2

73. CONSTRUCTION

Height is unknown.

Length(base) is 3 ft more than twice ht.

Area is 45 sq m.

$A = \dfrac{1}{2}bh$

Let x = height in ft

$2x + 3$ = length (base) in ft

$\dfrac{1}{2}x(2x + 3) = 45$

$2x^2 + 3x = 90$

$2x^2 + 3x - 90 = 0$

$(2x + 15)(x - 6) = 0$

$2x + 15 = 0 \qquad x - 6 = 0$

$2x = -15 \qquad \boxed{x = 6}$

$x = -\dfrac{15}{2}$

Base is 15 m.

74. ACADEMY AWARDS

x is the # of times for Hepburn (less).

$x + 1$ is the # of times for Streep (more).

Their product is 156.

Let x = Hepburn's #

$x + 1$ = Streep's #

$x(x + 1) = 156$

$x^2 + x = 156$

$x^2 + x - 156 = 0$

$(x + 13)(x - 12) = 0$

$x + 13 = 0 \qquad x + 12 = 0$

$x = -13 \qquad \boxed{x = 12}$

Hepburn was nominated 12 times.

Streep was nominated 13 times.

75. TIGHTROPE WALKERS

Pole is x meters tall.

Ground is $(x + 7)$ meters long.

Rope is $(x + 8)$ meters long.

$a^2 + b^2 = c^2$

Let x = pole (short leg) in m

$x + 7$ = ground (long leg) in m

$x + 8$ = rope (hypotenuse) in m

$x^2 + (x + 7)^2 = (x + 8)^2$

$x^2 + x^2 + 14x + 49 = x^2 + 16x + 64$

$x^2 - 2x - 15 = 0$

$(x + 3)(x - 5) = 0$

$x + 3 = 0 \qquad x - 5 = 0$

$x = -3 \qquad \boxed{x = 5}$

Pole is 5 meters tall.

76. BALLOONING

height is 0 feet (hits the ground)

$h = -16t^2 + 1{,}600$

$0 = -16t^2 + 1{,}600$

$16t^2 - 1{,}600 = 0$

$16t(t - 100) = 0$

$t = 0 \qquad t - 100 = 0$

$\boxed{t = 100}$

Time is 100 sec.

CHAPTER 5 TEST

1. $2^2 \cdot 7^2$

2. $3 \cdot 37$

3. $15x^3y^6$

4. $4(x + 4)$

5. $5ab(6ab^2 - 4a^2b + c)$

6. $(q + 9)(q - 9)$

7. prime

8. $(4x^2 + 9)(2x + 3)(2x - 3)$

9. $(x + 3)(x + 1)$

10. $-(x - 11)(x + 2)$

11. $(a - b)(9 + x)$

12. $(2a - 3)(a + 4)$

13. $2(3x - 5y)^2$

14. $(x + 2)(x^2 - 2x + 4)$

15. $5m^6(4m^2 - 3)$

16. $3(a - 3)(a^2 + 3a + 9)$

17. $r^2(4 - \pi)$

18. $(5x - 4)$ units

19. factors:

$(x - 9)(x + 6)$

product:

$x^2 + 6x - 9x - 54 = x^2 - 3x - 54$

20. $-3, 2$

21. $-5, 5$

22. $0, \dfrac{1}{6}$

23. $-3, 3$

24. $\dfrac{1}{3}, -\dfrac{1}{2}$

25. $9, -2$

26. $0, -1, -6$

27. DRIVING SAFETY

Length is $(w + 3)$ feet.

Width is w feet.

Area is 54 sq. ft.

$A = lw$

Let w = width in ft

$w + 3$ = length in ft

$w(w + 3) = 54$

$w^2 + 3w = 54$

$w^2 + 3w - 54 = 0$

$(w + 9)(w - 6) = 0$

$w + 9 = 0 \qquad w - 6 = 0$

$w = -9 \qquad \boxed{w = 6}$

Dimensions are 6 ft by 9 ft.

28. A quadratic equation is an equation that can be written in the form

$ax^2 + bx + c = 0$.

Example: $x^2 - 2x + 1 = 0$

Answers will vary.

29. Short leg is $(x - 4)$.

Long leg is $(x - 2)$.

Hypotenuse is x.

$$a^2 + b^2 = c^2$$

Let $x - 4 =$ short leg

$x - 2 =$ long leg

$x =$ hypotenuse

$$(x - 4)^2 + (x - 2)^2 = x^2$$

$$x^2 - 8x + 16 + x^2 - 4x + 4 = x^2$$

$$2x^2 - 12x + 20 = x^2$$

$$2x^2 - 12x + 20 - x^2 = x^2 - x^2$$

$$x^2 - 12x + 20 = 0$$

$$(x - 2)(x - 10) = 0$$

$$x - 2 = 0 \qquad x - 10 = 0$$

$$x = 2 \qquad \boxed{x = 10}$$

Hypotenuse is 10.

30. At least one of them is 0.

CHAPTERS 1-5
CUMULATIVE REVIEW EXERCISES

1. 70 year old's heartrate is 150 beats/min

    35 year old's heartrate is 185 beats/min

    Difference is $185 - 150 = 35$ beats/min

2. $2 \cdot 5^3$

3. 0.992

4. a) false

    b) true

    c) true

5. $\dfrac{16}{5} \div \dfrac{10}{3} = \dfrac{16}{5} \cdot \dfrac{3}{10}$

    $= \dfrac{8}{5} \cdot \dfrac{3}{5}$

    $= \dfrac{24}{25}$

6. $-3^3 = -(3)(3)(3)$

    $= -27$

7. $3 + 2[-1 - 4(5)] = 3 + 2[-1 - 20]$

    $= 3 + 2[-21]$

    $= 3 + [-42]$

    $= -39$

8. $\dfrac{|-25| - 2(-5)}{9 - 2^4} = \dfrac{25 + 10}{9 - 16}$

    $= \dfrac{35}{-7}$

    $= -5$

9. $\dfrac{-x - a}{y - b}$, $x = -2$, $y = 1$, $a = 5$, $b = 2$

    $\dfrac{-x - a}{y - b} = \dfrac{-(-2) - 5}{1 - 2}$

    $= \dfrac{2 - 5}{-1}$

    $= \dfrac{-3}{-1}$

    $= 3$

10. $\dfrac{5}{0}$

11. $-8y^2 - 5y^2 + 6 = -13y^2 + 6$

12. $3z + 2(y - z) + y = 3z + 2y - 2z + y$

    $\qquad\qquad\qquad\quad = 3y + z$

13. $\quad -(3a + 1) + a = 2$

    $\qquad -3a - 1 + a = 2$

    $\qquad\quad -2a - 1 = 2$

    $\quad -2a - 1 + 1 = 2 + 1$

    $\qquad\qquad -2a = 3$

    $\qquad\quad \dfrac{-2a}{-2} = \dfrac{3}{-2}$

    $\boxed{a = -\dfrac{3}{2}}$

14. $2 - (4x + 7) = 3 + 2(x + 2)$

    $2 - 4x - 7 = 3 + 2x + 4$

    $\quad -4x - 5 = 2x + 7$

    $-4x - 5 + 5 = 2x + 7 + 5$

    $\qquad -4x = 2x + 12$

    $-4x - 2x = 2x + 12 - 2x$

    $\qquad\quad -6x = 12$

    $\qquad\quad \dfrac{-6x}{-6} = \dfrac{12}{-6}$

    $\boxed{a = -2}$

15. $\qquad \dfrac{3t - 21}{2} = t - 6$

    $2\left(\dfrac{3t - 21}{2}\right) = 2(t - 6)$

    $\quad 3t - 21 = 2t - 12$

    $3t - 21 + 21 = 2t - 12 + 21$

    $\qquad\quad 3t = 2t + 9$

    $\quad 3t - 2t = 2t + 9 - 2t$

    $\boxed{t = 9}$

16. $$-\frac{1}{3} - \frac{x}{5} = \frac{3}{2}$$

$$30\left(-\frac{1}{3} - \frac{x}{5}\right) = 30\left(\frac{3}{2}\right)$$

$$-10 - 6x = 45$$

$$-10 - 6x + 10 = 45 + 10$$

$$-6x = 55$$

$$\frac{-6x}{-6} = \frac{55}{-6}$$

$$\boxed{x = -\frac{55}{6}}$$

17. $A = P + Prt$, solve for t

$$A = P + Prt$$

$$A - P = P + Prt - P$$

$$A - P = Prt$$

$$\frac{A - P}{Pr} = \frac{Prt}{Pr}$$

$$\boxed{\frac{A - P}{Pr} = t}$$

18. $$-\frac{x}{2} + 4 > 5$$

$$2\left(-\frac{x}{2} + 4\right) > 2(5)$$

$$-x + 8 > 10$$

$$-x + 8 - 8 > 10 - 8$$

$$-x > 2$$

$$-(-x) < -(2)$$

$$x < -2$$

$$(-\infty, -2)$$

19. GEOMETRY TOOL

The protractor is a semicircle. Circumference of a circle is found by the formula, $C = 2\pi r$. Radius is 3 inches. $\pi = 3.14$ Divide the formula by 2 and this is the curve part of the protractor.

$$C = \frac{2\pi r}{2}$$

$$= \pi r$$

$$= (3.14)(3)$$

$$= 9.42$$

Curve part is 9.4 inches.

Straight edge is 6 inches.

Total = Curve part + straight edge

$$= 9.4 + 6$$

$$= 15.4$$

Total distance around the outside edge is 15.4 inches.

20. The distance formula is $d = rt$.

$r = 60$ m/h

$t = 5\frac{1}{2}$ hours

$d = ?$

$$d = rt$$

$$= 60(5.5)$$

$$= 330$$

Distance is 330 miles.

21. $I = Prt$

22. $\$20x$

23. PHOTOGRAPHIC CHEMICALS

    6 liter of 5% acetic acid

    ? liter of 10% acetic acid

    ? liter of 7% acetic acid (final)

    Let $x = $ # of liter of 10% needed

    $6(0.05) = $ pure acetic acid

    $x(0.10) = $ pure acetic acid

    $(x+6)(0.07) = $ pure acetic acid (final)

$$6(0.05) + x(0.10) = (x+6)(0.07)$$
$$0.3 + 0.10x = 0.07x + 0.42$$
$$0.3 + 0.10x - 0.3 = 0.07x + 0.42 - 0.3$$
$$0.10x = 0.07x + 0.12$$
$$0.10x - 0.07x = 0.07x + 0.12 - 0.07x$$
$$0.3x = 0.12$$
$$\frac{0.3x}{0.3} = \frac{0.12}{0.3}$$
$$\boxed{x = 4}$$

4 liters of 10% acetic acid is needed.

24. HISTORY

    Let $x = $ number for 1st time

    $x + 2 = $ number for 2nd time

    46 is the sum of the two numbers.

$$x + (x+2) = 46$$
$$2x + 2 = 46$$
$$2x + 2 - 2 = 46 - 2$$
$$2x = 44$$
$$\frac{2x}{2} = \frac{44}{2}$$
$$\boxed{x = 22}$$

He was the 22nd and 24th president.

25. $m = 1$

    y-intercept $= (0, -2)$

26. Given: $(-2, 5)$ and $(-3, -2)$

    (x_1, y_1) and (x_2, y_2)

$$m = \frac{y_2 - y_1}{x_2 - x_1}$$
$$= \frac{-2 - 5}{-3 - (-2)}$$
$$= \frac{-7}{-3 + 2}$$
$$= \frac{-7}{-1}$$
$$= 7$$

Use $(-2, 5)$ and $m = 7$.

$$y - y_1 = m(x - x_1)$$
$$y - 5 = 7(x - (-2))$$
$$y - 5 = 7x + 14$$
$$y - 5 + 5 = 7x + 14 + 5$$
$$\boxed{y = 7x + 19}$$

27.

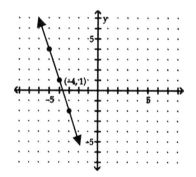

28.

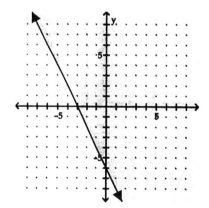

29. They are the same.

30. BEVERAGES

Given: $(1995, 21.6)$ and $(2000, 17.3)$

(x_1, y_1) and (x_2, y_2)

$$m = \frac{y_2 - y_1}{x_2 - x_1}$$

$$= \frac{17.3 - 21.6}{2000 - 1995}$$

$$= \frac{-4.3}{5}$$

$$= -0.86$$

A decrease of 0.86 gal/yr.

31. $0.00009011 = 9.011 \times 10^{-5}$

32. $1,700,000 = 1.7 \times 10^6$

33. $-y^2(4y^3) = -4y^{2+3}$

$$= -4y^5$$

34. $\dfrac{(x^2y^5)^5}{(x^3y)^2} = \dfrac{x^{2(5)}y^{5(5)}}{x^{3(2)}y^2}$

$$= \frac{x^{10}y^{25}}{x^6y^2}$$

$$= x^{10-6}y^{25-2}$$

$$= x^4y^{23}$$

35. $\left(\dfrac{b^5}{b^{-2}}\right)^{-2} = \dfrac{b^{5(-2)}}{b^{-2(-2)}}$

$$= \frac{b^{-10}}{b^4}$$

$$= \frac{1}{b^{4+10}}$$

$$= \frac{1}{b^{14}}$$

36. $2x^0 = 2(1)$

$$= 2$$

37. $(x^2 - 3x + 8) - (3x^2 + x + 3)$

$$= x^2 - 3x + 8 - 3x^2 - x - 3$$

$$= -2x^2 - 4x + 5$$

38. $4b^3(2b^2 - 2b) = 8b^5 - 8b^4$

39. $(y - 6)^2 = (y - 6)(y - 6)$

$$= y^2 - 6y - 6y + 36$$

$$= y^2 - 12y + 36$$

40. $(3x - 2)(x + 4) = 3x^2 + 12x - 2x - 8$

$$= 3x^2 + 10x - 8$$

41. $\dfrac{12a^2b^2 - 8a^2b - 4ab}{4ab}$

$$= \frac{12a^2b^2}{4ab} - \frac{8a^2b}{4ab} - \frac{4ab}{4ab}$$

$$= 3ab - 2a - 1$$

42.

$$\begin{array}{r} 2x \\ x-3 \overline{)2x^2 - 5x - 3} \\ 2x^2 - 6x \end{array}$$

$$\begin{array}{r} 2x \\ x-3 \overline{)2x^2 - 5x - 3} \\ \underline{-2x^2 + 6x} \\ x - 3 \end{array}$$

$$\begin{array}{r} 2x + 1 \\ x-3 \overline{)2x^2 - 5x - 3} \\ \underline{-2x^2 + 6x} \\ x - 3 \\ \underline{x - 3} \end{array}$$

$$\begin{array}{r} \boxed{2x + 1} \\ x-3 \overline{)2x^2 - 5x - 3} \\ \underline{-2x^2 + 6x} \\ x - 3 \\ \underline{-x + 3} \\ 0 \end{array}$$

43. PLAYPENS

    a) $P = 2l + 2w$

$$= 2(x+3) + 2(x+1)$$
$$= 2x + 6 + 2x + 2$$
$$= 4x + 8$$

    Perimeter is $(4x + 8)$ inches.

    b) $A = lw$

$$= (x+3)(x+1)$$
$$= x^2 + x + 3x + 3$$
$$= x^2 + 4x + 3$$

    Area is $(x^2 + 4x + 3)$ in^2.

    c) $V = lwh$

$$= (x+3)(x+1)(x)$$
$$= (x^2 + x + 3x + 3)x$$
$$= (x^2 + 4x + 3)x$$

    Area is $(x^3 + 4x^2 + 3x)$ in^3.

44. 3

45.

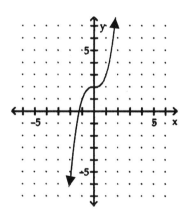

46. $24x^5y^8$ and $54x^6y$

    GCF $= 6x^5y$

47. $b^2(b - 3)$

48. $(u + 3)(u - 1)$

49. $(2x + 1)(x - 2)$

50. $(3z + 1)(3z - 1)$

51. $-5(a - 3)(a - 2)$

52. $(x + y)(a + b)$

53. $(t - 2)(t^2 + 2t + 4)$

54. $(2a - 3)^2$

55. $0, \dfrac{4}{3}$

56. $\dfrac{1}{2}, 2$

SECTION 6.1

VOCABULARY

1. A quotient of two polynomials, such as
    $\dfrac{x^2 + x}{x^2 - 3x}$, is called a **rational** expression.

3. Because of the division by 0, the expression
    $\frac{8}{0}$ is **undefined**.

5. To **simplify** a rational expression, we
    remove factors common to the numerator
    and denominator.

CONCEPTS

7. To find what value of x makes each rational
    expression undefined, set each denominator
    equal to zero and solve for x.

    a) $x = 0$
    b) $x - 6 = 0$
        $x - 6 + 6 = 0 + 6$
        $x = 6$
    c) $x + 6 = 0$
        $x + 6 - 6 = 0 - 6$
        $x = -6$

9. a) $x + 1$
    b) To simplify the rational expression, we
    replace $\dfrac{x+1}{x+1}$ with the equivalent fraction
    $\dfrac{1}{1}$. This removes the factor $\dfrac{x+1}{x+1}$, which is
    equal to 1.

11. a) The factors $x - 2$ and $2 - x$ are opposites.
    b) To simplify the rational expression, we
    replace $\dfrac{x-2}{2-x}$ with the equivalent
    fraction $\dfrac{-1}{1}$. This removes the
    factor $\dfrac{x-2}{2-x}$, which is equal to -1.

13. a) $\dfrac{(x+2)(x-2)}{(x+1)(x+2)} = \dfrac{x-2}{x+1}$

b) $\dfrac{y(y-2)}{9(2-y)} = \dfrac{-y(2-y)}{9(2-y)}$

$= -\dfrac{y}{9}$

c) $\dfrac{(2m+7)(m-5)}{(2m+7)} = \dfrac{m-5}{1}$

$= m - 5$

d) $\dfrac{x \cdot x}{x \cdot x (x-30)} = \dfrac{1}{x-30}$

NOTATION

15. Yes, all three answers are equivalent.

PRACTICE

17. $\dfrac{x-2}{x-5} = \dfrac{6-2}{6-5}$

$= \dfrac{4}{1}$

$= 4$

19. $\dfrac{x^2 - 4x - 12}{x^2 + x - 2} = \dfrac{6^2 - 4(6) - 12}{6^2 + 6 - 2}$

$= \dfrac{36 - 24 - 12}{36 + 6 - 2}$

$= \dfrac{0}{40}$

$= 0$

21. $\dfrac{y+5}{3y-2} = \dfrac{-3+5}{3(-3)-2}$

$= \dfrac{2}{-9-2}$

$= \dfrac{2}{-11}$

$= -\dfrac{2}{11}$

23. $\dfrac{y^3}{3y+1} = \dfrac{(-3)^3}{3(-3)^2+1}$

$\qquad\qquad = \dfrac{-27}{3(9)+1}$

$\qquad\qquad = \dfrac{-27}{27+1}$

$\qquad\qquad = -\dfrac{27}{28}$

25. $8x = 0$

$\dfrac{8x}{8} = \dfrac{0}{8}$

$\boxed{x = 0}$

27. $x - 2 = 0$

$x - 2 + 2 = 0 + 2$

$\boxed{x = 2}$

29. No matter what real number is substituted for x, $x^2 + 6$ will not be 0. Thus, **no real numbers** will make the rational expression undefined.

31. $2x - 1 = 0$

$2x - 1 + 1 = 0 + 1$

$2x = 1$

$\dfrac{2x}{2} = \dfrac{1}{2}$

$\boxed{x = \dfrac{1}{2}}$

33. $x^2 - 36 = 0$

$x^2 - 36 + 36 = 0 + 36$

$x^2 = 36$

$\boxed{x = 6 \text{ and } x = -6}$

Both 6^2 and $(-6)^2 = 36$

35. $x^2 + x - 2 = 0$

$(x + 2)(x - 1) = 0$

$x + 2 = 0 \quad$ and $\quad x - 1 = 0$

$x + 2 - 2 = 0 - 2 \quad x - 1 + 1 = 0 + 1$

$\boxed{x = -2 \qquad \text{and} \qquad x = 1}$

37. $\dfrac{45}{9a} = \dfrac{5 \cdot \cancel{3} \cdot \cancel{3}}{\cancel{3} \cdot \cancel{3} \cdot a}$

$\qquad = \dfrac{5}{a}$

39. $\dfrac{6x^2}{4x^2} = \dfrac{\cancel{2} \cdot 3 \cdot \cancel{x} \cdot \cancel{x}}{\cancel{2} \cdot 2 \cdot \cancel{x} \cdot \cancel{x}}$

$\qquad = \dfrac{3}{2}$

41. $\dfrac{2x^2}{x+2} = $ does not simplify

43. $\dfrac{15x^2 y}{5xy^2} = \dfrac{\cancel{5} \cdot 3 \cdot \cancel{x} \cdot x \cdot \cancel{y}}{\cancel{5} \cdot \cancel{x} \cdot \cancel{y} \cdot y}$

$\qquad = \dfrac{3x}{y}$

45. $\dfrac{6x+3}{3y} = \dfrac{\cancel{3}(2x+1)}{\cancel{3}y}$

$\qquad = \dfrac{2x+1}{y}$

47. $\dfrac{x+3}{3x+9} = \dfrac{\cancel{x+3}}{3(\cancel{x+3})}$

$\qquad = \dfrac{1}{3}$

49. $\dfrac{a^3 - a^2}{a^4 - a^3} = \dfrac{a^2(a-1)}{a^3(a-1)}$

$\qquad = \dfrac{\cancel{a} \cdot \cancel{a}(\cancel{a-1})}{\cancel{a} \cdot \cancel{a} \cdot a(\cancel{a-1})}$

$\qquad = \dfrac{1}{a}$

51. $\dfrac{4x+16}{5x+20} = \dfrac{4(\cancel{x+4})}{5(\cancel{x+4})}$

$\qquad = \dfrac{4}{5}$

53. $\dfrac{4c+4d}{d+c}=\dfrac{4(c+d)}{d+c}$

$\qquad\qquad =\dfrac{4\cancel{(c+d)}}{\cancel{c+d}}$

$\qquad\qquad =\dfrac{4}{1}$

$\qquad\qquad =4$

55. $\dfrac{x-7}{7-x}=\dfrac{\cancel{x-7}}{-1\cancel{(x-7)}}$

$\qquad\qquad =\dfrac{1}{-1}$

$\qquad\qquad =-1$

57. $\dfrac{6x-30}{5-x}=\dfrac{6(x-5)}{5-x}$

$\qquad\qquad =\dfrac{6\cancel{(x-5)}}{-1\cancel{(x-5)}}$

$\qquad\qquad =\dfrac{6}{-1}$

$\qquad\qquad =-6$

59. $\dfrac{x^2-4}{x^2-x-2}=\dfrac{\cancel{(x-2)}(x+2)}{\cancel{(x-2)}(x+1)}$

$\qquad\qquad =\dfrac{x+2}{x+1}$

61. $\dfrac{x^2+3x+2}{x^2+x-2}=\dfrac{\cancel{(x+2)}(x+1)}{\cancel{(x+2)}(x-1)}$

$\qquad\qquad =\dfrac{x+1}{x-1}$

63. $\dfrac{2x^2-8x}{x^2-6x+8}=\dfrac{2x\cancel{(x-4)}}{(x-2)\cancel{(x-4)}}$

$\qquad\qquad =\dfrac{2x}{x-2}$

65. $\dfrac{2-a}{a^2-a-2}=\dfrac{2-a}{(a-2)(a+1)}$

$\qquad\qquad =\dfrac{-1\cancel{(a-2)}}{\cancel{(a-2)}(a+1)}$

$\qquad\qquad =-\dfrac{1}{a+1}$

67. $\dfrac{b^2+2b+1}{(b+1)^3}=\dfrac{(b+1)(b+1)}{\cancel{(b+1)}\cancel{(b+1)}(b+1)}$

$\qquad\qquad =\dfrac{\cancel{1}}{b+1}$

69.

$\dfrac{6x^2-13x+6}{3x^2+x-2}=\dfrac{(2x-3)\cancel{(3x-2)}}{\cancel{(3x-2)}(x+1)}$

$\qquad\qquad =\dfrac{2x-3}{x+1}$

71.

$\dfrac{10(c-3)+10}{3(c-3)+3}=\dfrac{10c-30+10}{3c-9+3}$

$\qquad\qquad =\dfrac{10c-20}{3c-6}$

$\qquad\qquad =\dfrac{10\cancel{(c-2)}}{3\cancel{(c-2)}}$

$\qquad\qquad =\dfrac{10}{3}$

73.

$\dfrac{6a+3(a+2)+12}{a+2}=\dfrac{6a+3a+6+12}{a+2}$

$\qquad\qquad =\dfrac{9a+18}{a+2}$

$\qquad\qquad =\dfrac{9\cancel{(a+2)}}{\cancel{a+2}}$

$\qquad\qquad =\dfrac{9}{1}$

$\qquad\qquad =9$

75.

$\dfrac{3x^2-27}{2x^2-5x-3}=\dfrac{3(x^2-9)}{(2x+1)(x-3)}$

$\qquad\qquad =\dfrac{3(x+3)\cancel{(x-3)}}{(2x+1)\cancel{(x-3)}}$

$\qquad\qquad =\dfrac{3(x+3)}{2x+1}$

77.

$$\frac{-x^2 - 4x + 77}{x^2 - 4x - 21} = \frac{-(x^2 + 4x - 77)}{(x-7)(x+3)}$$

$$= -\frac{(x+11)\cancel{(x-7)}}{\cancel{(x-7)}(x+3)}$$

$$= -\frac{x+11}{x+3}$$

79.

$$\frac{x(x-8)+16}{16-x^2} = \frac{x^2 - 8x + 16}{(4-x)(4+x)}$$

$$= \frac{(x-4)(x-4)}{(4+x)(4-x)}$$

$$= -\frac{(x-4)\cancel{(x-4)}}{(4+x)\cancel{(x-4)}}$$

$$= -\frac{x-4}{x+4}$$

$$\text{or } \frac{4-x}{4+x}$$

81.

$$\frac{m^2 - 2mn + n^2}{2m^2 - 2n^2} = \frac{m^2 - 2mn + n^2}{2(m^2 - n^2)}$$

$$= \frac{(m-n)\cancel{(m-n)}}{2(m+n)\cancel{(m-n)}}$$

$$= \frac{m-n}{2(m+n)}$$

83.

$$\frac{16a^2 - 1}{4a + 4} = \frac{(4a-1)(4a+1)}{4(a+1)}$$

$$= \text{does not simplify}$$

85.

$$\frac{8u^2 - 2u - 15}{4u^4 + 5u^3} = \frac{(2u-3)\cancel{(4u+5)}}{u^3\cancel{(4u+5)}}$$

$$= \frac{2u-3}{u^3}$$

87.

$$\frac{y - xy}{xy - x} = \frac{y(1-x)}{x(y-1)}$$

$$= \text{does not simplify}$$

89.

$$\frac{6a - 6b + 6c}{9a - 9b + 9c} = \frac{6\cancel{(a-b+c)}}{9\cancel{(a-b+c)}}$$

$$= \frac{6}{9}$$

$$= \frac{2}{3}$$

91.

$$\frac{15x - 3x^2}{25y - 5xy} = \frac{3x\cancel{(5-x)}}{5y\cancel{(5-x)}}$$

$$= \frac{3x}{5y}$$

APPLICATIONS

93. ROOFING

$$\frac{rise}{run} = \frac{x^2 + 4x + 4}{x^2 - 4}$$

$$= \frac{\cancel{(x+2)}(x+2)}{\cancel{(x+2)}(x-2)}$$

$$= \frac{x+2}{x-2}$$

95. ORGAN PIPES
Let $L = 6$

$$n = \frac{512}{6}$$

$$= \frac{\cancel{2} \cdot 2 \cdot 2 \cdot 2 \cdot 2 \cdot 2 \cdot 2 \cdot 2 \cdot 2}{\cancel{2} \cdot 3}$$

$$= \frac{256}{3}$$

$$= 85\frac{1}{3}$$

97. MEDICAL SCHOOL

For 1:00, let $c = 1$.
$$c = \frac{4(1)}{(1)^2 + 1}$$
$$= \frac{4}{1+1}$$
$$= \frac{4}{2}$$
$$= 2 \text{ mg per liter}$$

For 2:00, let $c = 2$.
$$c = \frac{4(2)}{(2)^2 + 1}$$
$$= \frac{8}{4+1}$$
$$= \frac{8}{5}$$
$$= 1.6 \text{ mg per liter}$$

For 3:00, let $c = 3$.
$$c = \frac{4(3)}{(3)^2 + 1}$$
$$= \frac{12}{9+1}$$
$$= \frac{12}{10}$$
$$= 1.2 \text{ mg per liter}$$

WRITING

99. Answers will vary.

101. Answers will vary.

REVIEW

103. $(a + b) + c = a + (b + c)$

105. At least one of them is zero.

CHALLENGE PROBLEMS

107.
$$\frac{(x^2 + 2x + 1)(x^2 - 2x + 1)}{(x^2 - 1)^2}$$
$$= \frac{[(x+1)(x+1)][(x-1)(x-1)]}{(x^2-1)(x^2-1)}$$
$$= \frac{\cancel{(x+1)}\,\cancel{(x+1)}\,\cancel{(x-1)}\,\cancel{(x-1)}}{\cancel{(x+1)}\,\cancel{(x-1)}\,\cancel{(x+1)}\,\cancel{(x-1)}}$$
$$= 1$$

109.
$$\frac{x^3 - 27}{x^3 - 9x} = \frac{(x-3)(x^2 + 3x + 9)}{x(x^2 - 9)}$$
$$= \frac{\cancel{(x-3)}(x^2 + 3x + 9)}{x\,\cancel{(x-3)}(x+3)}$$
$$= \frac{x^2 + 3x + 9}{x(x+3)}$$

SECTION 6.2

VOCABULARY

1. A quotient of **opposites** is -1. For example, $\dfrac{x-8}{8-x} = -1$.

3. To find the reciprocal of a rational expression, we **invert** its numerator and denominator.

CONCEPTS

5. To multiply rational expressions, multiply their **numerators** and multiply their **denominators**. In symbols,

$$\frac{A}{B} \bullet \frac{C}{D} = \boxed{\frac{AC}{BD}}$$

7. $\dfrac{(x+7) \bullet 2 \bullet 5}{5(x+1)(x+7)(x-9)} = \dfrac{2}{(x+1)(x-9)}$

9. $6n = \dfrac{6n}{1}$

11. $\dfrac{1}{18x}$

13. $\dfrac{x^2 + 2x + 1}{x^2 + 1}$

15. feet

NOTATION

17. a) m^2
     b) m^3

PRACTICE

19. $\dfrac{3}{y} \cdot \dfrac{y}{2} = \dfrac{3y}{2y}$

$$= \frac{3}{2}$$

21. $\dfrac{35n}{12} \cdot \dfrac{16}{7n^2} = \dfrac{35n \cdot 16}{12 \cdot 7n^2}$

$$= \frac{5 \cdot 7 \cdot n \cdot 2 \cdot 2 \cdot 2 \cdot 2}{3 \cdot 2 \cdot 2 \cdot 7 \cdot n \cdot n}$$

$$= \frac{20}{3n}$$

23. $\dfrac{2x^2 y}{3xy} \cdot \dfrac{3xy^2}{2} = \dfrac{2x^2 y \cdot 3xy^2}{3xy \cdot 2}$

$$= \frac{2 \cdot x \cdot x \cdot y \cdot 3 \cdot x \cdot y \cdot y}{3 \cdot x \cdot y \cdot 2}$$

$$= x^2 y^2$$

25. $\dfrac{10r^2 st^3}{6rs^2} \cdot \dfrac{3r^3 t}{2rst} = \dfrac{10r^2 st^3 \cdot 3r^3 t}{6rs^2 \cdot 2rst}$

$$= \frac{2 \cdot 5 \cdot r \cdot r \cdot s \cdot t \cdot t \cdot t \cdot 3 \cdot r \cdot r \cdot r \cdot t}{2 \cdot 3 \cdot r \cdot s \cdot s \cdot 2 \cdot r \cdot s \cdot t}$$

$$= \frac{5r^3 t^3}{2s^2}$$

27. $\dfrac{z+7}{7} \cdot \dfrac{z+2}{z} = \dfrac{(z+7)(z+2)}{7z}$

29. $\dfrac{x+5}{5} \cdot \dfrac{x}{x+5} = \dfrac{(x+5)x}{5(x+5)}$

$$= \frac{x}{5}$$

31. $\dfrac{x-2}{2} \cdot \dfrac{2x}{2-x} = \dfrac{(x-2) \cdot 2x}{2(2-x)}$

$$= -\frac{(x-2) \cdot 2x}{2(x-2)}$$

$$= -x$$

33. $\dfrac{5}{m} \cdot m = \dfrac{5}{m} \cdot \dfrac{m}{1}$

$$= \frac{5m}{m}$$

$$= 5$$

35.

$$15x\left(\frac{x+1}{15x}\right) = \frac{15x}{1}\left(\frac{x+1}{15x}\right)$$

$$= \frac{\cancel{15x}(x+1)}{\cancel{15x}}$$

$$= x+1$$

37.

$$12y\left(\frac{y+8}{6y}\right) = \frac{12y}{1}\left(\frac{y+8}{6y}\right)$$

$$= \frac{2\cdot\cancel{2}\cdot\cancel{3y}(y+8)}{\cancel{2}\cdot\cancel{3y}}$$

$$= 2(y+8)$$

$$= 2y+16$$

39.

$$x+8\left(\frac{x+5}{x+8}\right) = \frac{x+8}{1}\left(\frac{x+5}{x+8}\right)$$

$$= \frac{\cancel{(x+8)}(x+5)}{\cancel{(x+8)}}$$

$$= x+5$$

41.

$$10(h+9)\frac{h-3}{h+9} = \frac{10\cancel{(h+9)}(h-3)}{\cancel{h+9}}$$

$$= 10(h-3)$$

$$= 10h-30$$

43.

$$\frac{(x+1)^2}{x+1}\cdot\frac{x+2}{x+1} = \frac{\cancel{(x+1)}\cancel{(x+1)}(x+2)}{\cancel{(x+1)}\cancel{(x+1)}}$$

$$= x+2$$

45.

$$\frac{2x+6}{x+3}\cdot\frac{3}{4x} = \frac{2(x+3)}{x+3}\cdot\frac{3}{4x}$$

$$= \frac{2\cancel{(x+3)}\cdot 3}{\cancel{(x+3)}\cdot 2\cdot 2x}$$

$$= \frac{3}{2x}$$

47.

$$\frac{x^2-x}{x}\cdot\frac{3x-6}{3-3x} = \frac{x(x-1)}{x}\cdot\frac{3(x-2)}{3(1-x)}$$

$$= \frac{\cancel{3}\cdot\cancel{x}\cancel{(x-1)}(x-2)}{-\cancel{3}\cdot\cancel{x}\cancel{(x-1)}}$$

$$= \frac{x-2}{-1}$$

$$= -(x-2)$$

49.

$$\frac{x^2+x-6}{5x}\cdot\frac{5x-10}{x+3} = \frac{(x+3)(x-2)}{5x}\cdot\frac{5(x-2)}{x+3}$$

$$= \frac{5\cancel{(x+3)}(x-2)(x-2)}{5x\cancel{(x+3)}}$$

$$= \frac{(x-2)^2}{x}$$

51.

$$\frac{m^2-2m-3}{2m+4}\cdot\frac{m^2-4}{m^2+3m+2}$$

$$= \frac{(m-3)(m+1)}{2(m+2)}\cdot\frac{(m-2)(m+2)}{(m+2)(m+1)}$$

$$= \frac{(m-3)\cancel{(m+1)}(m-2)\cancel{(m+2)}}{2\cancel{(m+2)}(m+2)\cancel{(m+1)}}$$

$$= \frac{(m-2)(m-3)}{2(m+2)}$$

53.

$$\frac{2x^2+17x+21}{x^2+2x-35}\cdot\frac{x^2-25}{2x^2-7x-15}$$

$$= \frac{(2x+3)(x+7)}{(x-5)(x+7)}\cdot\frac{(x-5)(x+5)}{(2x+3)(x-5)}$$

$$= \frac{\cancel{(2x+3)}\cancel{(x+7)}\cancel{(x-5)}(x+5)}{\cancel{(x-5)}\cancel{(x+7)}\cancel{(2x+3)}(x-5)}$$

$$= \frac{x+5}{x-5}$$

55.

$$\frac{4x^2-12xy+9y^2}{x^3y^2}\cdot\frac{x^2y}{9y^2-4x^2}$$

$$=\frac{(2x-3y)(2x-3y)}{x\cdot x\cdot x\cdot y\cdot y}\cdot\frac{x\cdot x\cdot y}{(3y+2x)(3y-2x)}$$

$$=\frac{x\cdot x\cdot y(2x-3y)(2x-3y)}{x\cdot x\cdot x\cdot y\cdot y(3y+2x)(3y-2x)}$$

$$=-\frac{\cancel{x}\cdot\cancel{x}\cdot\cancel{y}(2x-3y)\cancel{(2x-3y)}}{\cancel{x}\cdot x\cdot x\cdot\cancel{y}\cdot y(3y+2x)\cancel{(2x-3y)}}$$

$$=-\frac{2x-3y}{xy(3y+2x)}$$

57.

$$\frac{3x^2+5x+2}{x^2-9}\cdot\frac{x-3}{x^2-4}\cdot\frac{x^2+5x+6}{6x+4}$$

$$=\frac{(3x+2)(x+1)}{(x+3)(x-3)}\cdot\frac{x-3}{(x-2)(x+2)}\cdot\frac{(x+2)(x+3)}{2(3x+2)}$$

$$=\frac{(3x+2)(x+1)\cancel{(x-3)}\cancel{(x+2)}\cancel{(x+3)}}{2\cancel{(x+3)}\cancel{(x-3)}(x-2)\cancel{(x+2)}\cancel{(3x+2)}}$$

$$=\frac{x+1}{2(x-2)}$$

59.

$$\frac{2}{y}\div\frac{4}{3}=\frac{2}{y}\cdot\frac{3}{4}$$

$$=\frac{6}{4y}$$

$$=\frac{\cancel{2}\cdot3}{\cancel{2}\cdot2\cdot y}$$

$$=\frac{3}{2y}$$

61.

$$\frac{3x}{y}\div\frac{2x}{4}=\frac{3x}{y}\cdot\frac{4}{2x}$$

$$=\frac{12x}{2xy}$$

$$=\frac{\cancel{2}\cdot2\cdot3\cdot\cancel{x}}{\cancel{2}\cdot\cancel{x}\cdot y}$$

$$=\frac{6}{y}$$

63.

$$\frac{x^2y}{3xy}\div\frac{xy^2}{6y}=\frac{x^2y}{3xy}\cdot\frac{6y}{xy^2}$$

$$=\frac{6x^2y^2}{3x^2y^3}$$

$$=\frac{2\cdot\cancel{3}\cdot x\cdot x\cdot y\cdot\cancel{y}}{\cancel{3}\cdot\cancel{x}\cdot\cancel{x}\cdot y\cdot y\cdot\cancel{y}}$$

$$=\frac{2}{y}$$

65.

$$24n^2\div\frac{18n^3}{n-1}=\frac{24n^2}{1}\cdot\frac{n-1}{18n^3}$$

$$=\frac{24n^2(n-1)}{18n^3}$$

$$=\frac{\cancel{2}\cdot2\cdot2\cdot\cancel{3}\cdot\cancel{n}\cdot\cancel{n}(n-1)}{\cancel{2}\cdot\cancel{3}\cdot3\cdot\cancel{n}\cdot\cancel{n}\cdot n}$$

$$=\frac{4(n-1)}{3n}$$

67.

$$\frac{x+2}{3x}\div\frac{x+2}{2}=\frac{x+2}{3x}\cdot\frac{2}{x+2}$$

$$=\frac{2\cancel{(x+2)}}{3x\cancel{(x+2)}}$$

$$=\frac{2}{3x}$$

69.

$$\frac{(z-2)^2}{3z^2}\div\frac{z-2}{6z}=\frac{(z-2)^2}{3z^2}\cdot\frac{6z}{z-2}$$

$$=\frac{6z(z-2)^2}{3z^2(z-2)}$$

$$=\frac{2\cdot\cancel{3}\cancel{z}\cancel{(z-2)}(z-2)}{\cancel{3}\cancel{z}\cdot z\cancel{(z-2)}}$$

$$=\frac{2(z-2)}{z}$$

71.

$$\frac{9a-18}{28}\div\frac{9a^3}{35}=\frac{9a-18}{28}\cdot\frac{35}{9a^3}$$

$$=\frac{\cancel{3}\cdot\cancel{3}(a-2)\cdot5\cdot\cancel{7}}{2\cdot2\cdot\cancel{7}\cdot\cancel{3}\cdot\cancel{3}\cdot a\cdot a\cdot a}$$

$$=\frac{5(a-2)}{4a^3}$$

73.

$$\frac{x^2-4}{3x+6} \div \frac{2-x}{x+2} = \frac{x^2-4}{3x+6} \cdot \frac{x+2}{2-x}$$

$$= \frac{(x-2)(x+2)(x+2)}{3(x+2)(2-x)}$$

$$= -\frac{(x-2)(x+2)(x+2)}{3(x+2)(x-2)}$$

$$= -\frac{x+2}{3}$$

75.

$$\frac{x^2-1}{3x-3} \div (x+1) = \frac{x^2-1}{3x-3} \cdot \frac{1}{x+1}$$

$$= \frac{(x-1)(x+1)}{3(x-1)(x+1)}$$

$$= \frac{1}{3}$$

77.

$$\frac{x^2-2x-35}{3x^2+27x} \div \frac{x^2+7x+10}{6x^2+12x}$$

$$= \frac{x^2-2x-35}{3x^2+27x} \cdot \frac{6x^2+12x}{x^2+7x+10}$$

$$= \frac{(x-7)(x+5)2 \cdot 3x(x+2)}{3x(x+9)(x+2)(x+5)}$$

$$= \frac{2(x-7)}{x+9}$$

79.

$$\frac{36c^2-49d^2}{3d^3} \div \frac{12c+14d}{d^4}$$

$$= \frac{36c^2-49d^2}{3d^3} \cdot \frac{d^4}{12c+14d}$$

$$= \frac{(6c+7d)(6c-7d) \cdot d \cdot d \cdot d \cdot d}{3 \cdot d \cdot d \cdot d \cdot 2(6c+7d)}$$

$$= \frac{d(6c-7d)}{6}$$

81.

$$\frac{2d^2+8d-42}{3-d} \div \frac{2d^2+14d}{d^2+5d}$$

$$= \frac{2d^2+8d-42}{3-d} \cdot \frac{d^2+5d}{2d^2+14d}$$

$$= \frac{2(d^2+4d-21)}{3-d} \cdot \frac{d(d+5)}{2d(d+7)}$$

$$= \frac{2(d+7)(d-3)d(d+5)}{(3-d) \cdot 2d(d+7)}$$

$$= -\frac{2(d+7)(d-3)d(d+5)}{(d-3) \cdot 2 \cdot d(d+7)}$$

$$= -(d+5)$$

83.

$$\frac{2r-3s}{12} \div \left(4r^2-12rs+9s^2\right)$$

$$= \frac{2r-3s}{12} \cdot \frac{1}{\left(4r^2-12rs+9s^2\right)}$$

$$= \frac{(2r-3s)}{2 \cdot 2 \cdot 3(2r-3s)(2r-3s)}$$

$$= \frac{1}{12(2r-3s)}$$

85. $\dfrac{150 \text{ yd}}{1} \cdot \dfrac{3 \text{ ft}}{1 \text{ yd}} = 450 \text{ ft}$

87. $\dfrac{30 \text{ meters}}{1 \text{ sec}} \cdot \dfrac{60 \text{ sec}}{1 \text{ min}} = 1{,}800$ meters per min.

APPLICATIONS

89. INTERNATIONAL ALPHABET

Multiply the number of squares (6) times the area of one square:

$$6\left(\frac{2x+1}{2}\right)\left(\frac{2x+1}{2}\right) = \frac{2 \cdot 3(2x+1)(2x+1)}{2 \cdot 2}$$

$$= \frac{3(2x+1)(2x+1)}{2}$$

$$= \frac{3\left((2x)^2+2(2x)(1)+1^2\right)}{2}$$

$$= \frac{3\left(4x^2+4x+1\right)}{2}$$

$$= \frac{12x^2+12x+3}{2} \text{ in.}^2$$

91. TALKING

There are 365 days in 1 year.

$$\frac{12,000}{1\,\text{day}} \cdot \frac{365\,\text{days}}{1\,\text{year}} = 4,380,000 \text{ words per year}$$

93. NATURAL LIGHT

9 square feet = 1 square yard

$$\frac{72\,\text{ft}^2}{1} \cdot \frac{1\,\text{yd}^2}{9\,\text{ft}^2} = \frac{72}{9}$$
$$= 8\,\text{yd}^2$$

95. BEARS

60 minutes = 1 hour

$$\frac{30\,\text{miles}}{1\,\text{hour}} \cdot \frac{1\,\text{hour}}{60\,\text{minutes}} = \frac{30}{60}$$
$$= \frac{1}{2} \text{ mile per minute}$$

97. TV TRIVIA

1 square mile = 640 acres

$$\frac{160\,\text{acres}}{1} \cdot \frac{1\,\text{mi}^2}{640\,\text{acres}} = \frac{160}{640}$$
$$= \frac{1}{4}\,\text{mi}^2$$

WRITING

99. Answers will vary.

101. Answers will vary.

103. Answers will vary.

REVIEW

105. HARDWARE

Let the width = x.
Then the length = $2x - 2$.
The brace, wall, and shelf form a right
triangle, so use the Pythagorean
Theorem to solve the equation.

$$8^2 + x^2 = (2x - 2)^2$$
$$64 + x^2 = (2x)^2 - 2(2x)(2) + (-2)^2$$

$$64 + x^2 = 4x^2 - 8x + 4$$
$$64 + x^2 - 64 - x^2 = 4x^2 - 8x + 4 - 64 - x^2$$
$$0 = 3x^2 - 8x - 60$$
$$0 = (3x + 10)(x - 6)$$

$$3x + 10 = 0 \quad \text{and} \quad x - 6 = 0$$
$$3x + 10 - 10 = 0 - 10 \qquad x - 6 + 6 = 0 + 6$$
$$3x = -10 \qquad\qquad x = 6$$
$$x = -\frac{10}{3}$$

Since the width cannot be negative, the
only valid answer for x is 6.

width = x
  = 6 in.

length = $2x - 2$
  = 2(6) - 2
  = 12 - 2
  = 10 in.

CHALLENGE PROBLEMS

107.
$$\frac{c^3 - 2c^2 + 5c - 10}{c^2 - c - 2} \cdot \frac{c^3 + c^2 - 5c - 5}{c^4 - 25}$$

$$= \frac{c^2(c - 2) + 5(c - 2)}{(c - 2)(c + 1)} \cdot \frac{c^2(c + 1) - 5(c + 1)}{(c^2 - 5)(c^2 + 5)}$$

$$= \frac{(c^2 + 5)(c - 2)(c^2 - 5)(c + 1)}{(c - 2)(c + 1)(c^2 - 5)(c^2 + 5)}$$

$$= 1$$

109.
$$\frac{y^2}{x + 1} \cdot \frac{x^2 + 2x + 1}{x^2 - 1} \div \frac{3y}{xy - y}$$

$$= \frac{y^2}{x + 1} \cdot \frac{x^2 + 2x + 1}{x^2 - 1} \cdot \frac{xy - y}{3y}$$

$$= \frac{y \cdot y \cdot (x + 1)(x + 1) \cdot y(x - 1)}{(x + 1)(x - 1)(x + 1)3y}$$

$$= \frac{y^2}{3}$$

SECTION 6.3

VOCABULARY

1. The rational expressions $\dfrac{7}{6n}$ and $\dfrac{n+1}{6n}$ have a **common** denominator of $6n$.

3. The **least common denominator** of $\dfrac{x-8}{x+6}$ and $\dfrac{6-5x}{x}$ is $x(x+6)$.

5. To simplify $5y + 8 - y + 4$, we combine **like** terms.

CONCEPTS

7. To add or subtract rational expressions that have the same denominator, add or subtract the **numerators**, and write the sum or difference over the common **denominator**. In symbols, if $\dfrac{A}{D}$ and $\dfrac{B}{D}$ are rational expressions,

$$\frac{A}{D} + \frac{B}{D} = \boxed{\frac{A+B}{D}} \quad \text{or} \quad \frac{A}{D} - \frac{B}{D} = \boxed{\frac{A-B}{D}}$$

9. a) $-(6x + 9) = -6x - 9$
    b) $x^2 + 3x - 1 + x^2 - 5x$
        $\quad = (x^2 + x^2) + (3x - 5x) - 1$
        $\quad = 2x^2 - 2x - 1$
    c) $7x - 1 - (5x - 6)$
        $\quad = 7x - 1 - 5x + 6$
        $\quad = (7x - 5x) + (-1 + 6)$
        $\quad = 2x + 5$
    d) $4x^2 - 2x - (x^2 + x)$
        $\quad = 4x^2 - 2x - x^2 - x$
        $\quad = (4x^2 - x^2) + (-2x - x)$
        $\quad = 3x^2 - 3x$

11.
$$\frac{x^2 - 14x + 49}{x^2 - 49} = \frac{(\cancel{x-7})(x-7)}{(\cancel{x-7})(x+7)}$$
$$= \frac{x-7}{x+7}$$

13. a) $40x^2 = 2 \cdot 2 \cdot 2 \cdot 5 \cdot x \cdot x$
    b) $2x^2 - 6x = 2x(x - 3)$
    c) $n^2 - 64 = (n - 8)(n + 8)$

15. a) $\dfrac{5}{21a}$

    b) $\dfrac{a+4}{a+4}$

NOTATION

17. yes

PRACTICE

19.
$$\frac{9}{x} + \frac{2}{x} = \frac{9+2}{x}$$
$$= \frac{11}{x}$$

21.
$$\frac{x}{18} + \frac{5}{18} = \frac{x+5}{18}$$

23.
$$\frac{m-3}{m^3} - \frac{5}{m^3} = \frac{m-3-5}{m^3}$$
$$= \frac{m-8}{m^3}$$

25.
$$\frac{2x}{y} - \frac{x}{y} = \frac{2x-x}{y}$$
$$= \frac{x}{y}$$

27.
$$\frac{13t}{99} - \frac{35t}{99} = \frac{13t - 35t}{99}$$
$$= \frac{-22t}{99}$$
$$= -\frac{\cancel{11} \cdot 2 \cdot t}{\cancel{11} \cdot 9}$$
$$= -\frac{2t}{9}$$

29.
$$\frac{x}{9} + \frac{2x}{9} = \frac{x+2x}{9}$$
$$= \frac{3x}{9}$$
$$= \frac{\cancel{3} \cdot x}{\cancel{3} \cdot 3}$$
$$= \frac{x}{3}$$

31.

$$\frac{50}{r^3 - 25} + \frac{r}{r^3 - 25} = \frac{r + 50}{r^3 - 25}$$

33.

$$\frac{6a}{a+2} - \frac{4a}{a+2} = \frac{6a - 4a}{a+2}$$
$$= \frac{2a}{a+2}$$

35.

$$\frac{7}{t+5} - \frac{9}{t+5} = \frac{7-9}{t+5}$$
$$= \frac{-2}{t+5}$$
$$= -\frac{2}{t+5}$$

37.

$$\frac{3x-5}{x-2} + \frac{6x-13}{x-2} = \frac{3x-5+6x-13}{x-2}$$
$$= \frac{9x-18}{x-2}$$
$$= \frac{9(\cancel{x-2})}{\cancel{x-2}}$$
$$= 9$$

39.

$$\frac{6x-5}{3xy} - \frac{3x-5}{3xy} = \frac{6x-5-(3x-5)}{3xy}$$
$$= \frac{6x-5-3x+5}{3xy}$$
$$= \frac{\cancel{3x}}{\cancel{3x}y}$$
$$= \frac{1}{y}$$

41.

$$\frac{3y-2}{2y+6} - \frac{2y-5}{2y+6} = \frac{3y-2-(2y-5)}{2y+6}$$
$$= \frac{3y-2-2y+5}{2y+6}$$
$$= \frac{\cancel{y+3}}{2(\cancel{y+3})}$$
$$= \frac{1}{2}$$

43.

$$\frac{x+3}{2y} + \frac{x+5}{2y} = \frac{x+3+x+5}{2y}$$
$$= \frac{2x+8}{2y}$$
$$= \frac{\cancel{2}(x+4)}{\cancel{2}y}$$
$$= \frac{x+4}{y}$$

45.

$$\frac{6x^2 - 11x}{3x+2} - \frac{10}{3x+2} = \frac{6x^2 - 11x - 10}{3x+2}$$
$$= \frac{(\cancel{3x+2})(2x-5)}{\cancel{3x+2}}$$
$$= 2x - 5$$

47.

$$\frac{2-p}{p^2 - p} - \frac{-p+2}{p^2 - p} = \frac{2-p-(-p+2)}{p^2 - p}$$
$$= \frac{2-p+p-2}{p^2 - p}$$
$$= \frac{0}{p(p-1)}$$
$$= 0$$

49.

$$\frac{3x^2}{x+1} + \frac{x-2}{x+1} = \frac{3x^2 + x - 2}{x+1}$$
$$= \frac{(3x-2)(\cancel{x+1})}{\cancel{x+1}}$$
$$= 3x - 2$$

51.

$$\frac{11w+1}{3w(w-9)} - \frac{11w}{3w(w-9)} = \frac{11w+1-11w}{3w(w-9)}$$
$$= \frac{1}{3w(w-9)}$$

53.

$$\frac{a}{a^2 + 5a + 6} + \frac{3}{a^2 + 5a + 6} = \frac{a+3}{a^2 + 5a + 6}$$
$$= \frac{\cancel{a+3}}{(\cancel{a+3})(a+2)}$$
$$= \frac{1}{a+2}$$

55.

$$\frac{2c}{c^2 - d^2} - \frac{2d}{c^2 - d^2} = \frac{2c - 2d}{c^2 - d^2}$$

$$= \frac{2(\cancel{c-d})}{(\cancel{c-d})(c+d)}$$

$$= \frac{2}{c+d}$$

57.

$$\frac{11n-1}{(n+4)(n-2)} - \frac{4n}{(n+4)(n-2)} = \frac{11n-1-4n}{(n+4)(n-2)}$$

$$= \frac{7n-1}{(n+4)(n-2)}$$

59.

$$\frac{1}{t^2-2t+1} - \frac{6-t}{t^2-2t+1} = \frac{1+(6-t)}{t^2-2t+1}$$

$$= \frac{1-6+t}{t^2-2t+1}$$

$$= \frac{t-5}{(t-1)(t-1)}$$

$$= \frac{t-5}{t^2-2t+1}$$

61. $2x = 2 \cdot x$
$6x = 2 \cdot 3 \cdot x$
LCD $= 2 \cdot 3 \cdot x = \mathbf{6x}$

63. $3a^2b = 3 \cdot a \cdot a \cdot b$
$a^2b^3 = a \cdot a \cdot b \cdot b \cdot b$
LCD $= 3 \cdot a \cdot a \cdot b \cdot b \cdot b = \mathbf{3a^2b^3}$

65. c cannot be factored
$c + 2$ cannot be factored
LCD $= \mathbf{c(c + 2)}$

67. $15a^3 = 3 \cdot 5 \cdot a \cdot a \cdot a$
$10a = 2 \cdot 5 \cdot a$
LCD $= 2 \cdot 3 \cdot 5 \cdot a \cdot a \cdot a = \mathbf{30a^3}$

69. $4b + 8 = 4(b + 2) = 2 \cdot 2(b + 2)$
$6 = 2 \cdot 3$
LCD $= 2 \cdot 2 \cdot 3(b + 2) = \mathbf{12(b + 2)}$

71. $x^2 - 1 = (x - 1)(x + 1)$
$x + 1$ cannot be factored
LCD $= \mathbf{(x - 1)(x + 1)}$

73. $3x - 1$ cannot be factored
$3x + 1$ cannot be factored
LCD $= \mathbf{(3x + 1)(3x - 1)}$

75. $x^2 - 4x - 5 = (x + 1)(x - 5)$
$x^2 - 25 = (x - 5)(x + 5)$
LCD $= \mathbf{(x + 1)(x - 5)(x + 5)}$

77. $2n^2 + 13n + 20 = (2n + 5)(n + 4)$
$n^2 + 8n + 16 = (n + 4)(n + 4) = (n + 4)^2$
LCD $= \mathbf{(2n + 5)\, (n + 4)^2}$

79. $\dfrac{25}{4} \cdot \dfrac{5x}{5x} = \dfrac{125x}{20x}$

81. $\dfrac{8}{x} \cdot \dfrac{xy}{xy} = \dfrac{8xy}{x^2y}$

83.

$$\frac{3x}{x+1} \cdot \frac{x+1}{x+1} = \frac{3x(x+1)}{(x+1)^2}$$

$$= \frac{3x^2 + 3x}{(x+1)^2}$$

85.

$$\frac{2y}{x} \cdot \frac{x+3}{x+3} = \frac{2y(x+3)}{x(x+3)}$$

$$= \frac{2xy + 6y}{x(x+3)}$$

87.

$$\frac{10}{b-1} \cdot \frac{3}{3} = \frac{10 \cdot 3}{3(b-1)}$$

$$= \frac{30}{3(b-1)}$$

89.

$$\frac{t+5}{4t+8} = \frac{t+5}{4(t+2)}$$

$$= \frac{t+5}{4(t+2)} \cdot \frac{5}{5}$$

$$= \frac{5(t+5)}{4 \cdot 5(t+2)}$$

$$= \frac{5t+25}{20(t+2)}$$

91.

$$\frac{y+3}{y^2-5y+6} = \frac{y+3}{(y-2)(y-3)}$$

$$= \frac{y+3}{(y-2)(y-3)} \cdot \frac{4y}{4y}$$

$$= \frac{4y(y+3)}{4y(y-2)(y-3)}$$

$$= \frac{4y^2+12y}{4y(y-2)(y-3)}$$

93.

$$\frac{12-h}{h^2-81} = \frac{12-h}{(h-9)(h+9)}$$

$$= \frac{12-h}{(h-9)(h+9)} \cdot \frac{3}{3}$$

$$= \frac{3(12-h)}{3(h-9)(h+9)}$$

$$= \frac{36-3h}{3(h-9)(h+9)}$$

95.

$$\frac{6t}{t^2+4t+3} = \frac{6t}{(t+1)(t+3)}$$

$$= \frac{6t}{(t+1)(t+3)} \cdot \frac{t-2}{t-2}$$

$$= \frac{6t(t-2)}{(t+1)(t+3)(t-2)}$$

$$= \frac{6t^2-12t}{(t+1)(t-2)(t+3)}$$

APPLICATIONS

97. DOING LAUNDRY

Find the LCD of 30 and 45.
$30 = 2 \cdot 3 \cdot 5$
$45 = 3 \cdot 3 \cdot 5$
$\text{LCD} = 2 \cdot 3 \cdot 3 \cdot 5 = \mathbf{90}$
In 90 minutes, the washer and dryer will end simultaneously.

WRITING

99. Answers will vary.

101. Answers will vary.

103. Answers will vary.

REVIEW

105. $I = Prt$

107. $A = \frac{1}{2}bh$

109. $d = rt$

CHALLENGE PROBLEMS

111. $\frac{3xy}{x-y} - \frac{x(3y-x)}{x-y} - \frac{x(x-y)}{x-y}$

$$= \frac{3xy - x(3y-x) - x(x-y)}{x-y}$$

$$= \frac{3xy - 3xy + x^2 - x^2 + xy}{x-y}$$

$$= \frac{xy}{x-y}$$

SECTION 6.4

VOCABULARY

1. The rational expressions $\dfrac{x}{x-7}$ and $\dfrac{1}{x-7}$ have like denominators. The rational expressions $\dfrac{x+5}{x-7}$ and $\dfrac{4x}{x+7}$ have **unlike** denominators.

CONCEPTS

3. a) $20x^2 = 2 \cdot 2 \cdot 5 \cdot x \cdot x$
   b) $x^2 + 4x - 12 = (x+6)(x-2)$

5. $\text{LCD} = (x+6)(x+3)$

7. To build $\dfrac{x}{x+2}$ so that it has a denominator of $5(x+2)$, we multiply it by 1 in the form of $\dfrac{5}{5}$.

9. a) $\dfrac{5n}{5n}$

   b) $\dfrac{3}{3}$

11.
$$\dfrac{x-3}{x-4} \cdot \dfrac{8}{8} = \dfrac{8(x-3)}{8(x-4)}$$
$$= \dfrac{8x-24}{8(x-4)}$$

13. $\dfrac{7x}{3x^2(x-5)} + \dfrac{10}{3x^2(x-5)} = \dfrac{7x+10}{3x^2(x-5)}$

15. $-1(x-10) = -x + 10$
    $= 10 - x$

17. You must multiply **-1** by $(y-4)$ to obtain $4 - y$: $-1(y-4) = -y + 4 = 4 - y$

19. $x = \dfrac{x}{1}$

NOTATION

21.
$$\text{Yes. } \dfrac{m^2+2m}{(m-1)(m-4)} = \dfrac{m^2+2m}{(m-4)(m-1)}$$
$$\text{No. } \dfrac{-5x^2-7}{4x(x+3)} = -\dfrac{5x^2+7}{4x(x+3)} \neq -\dfrac{5x^2-7}{4x(x+3)}$$
$$\text{Yes. } \dfrac{-2x}{x-y} = \dfrac{-2x}{-(-x+y)} = \dfrac{2x}{-x+y} = \dfrac{2x}{y-x}$$

PRACTICE

23.
$$\dfrac{x}{3} + \dfrac{2x}{7} = \dfrac{x}{3}\left(\dfrac{7}{7}\right) + \dfrac{2x}{7}\left(\dfrac{3}{3}\right)$$
$$= \dfrac{7x}{21} + \dfrac{6x}{21}$$
$$= \dfrac{7x+6x}{21}$$
$$= \dfrac{13x}{21}$$

25.
$$\dfrac{2y}{9} + \dfrac{y}{3} = \dfrac{2y}{9} + \dfrac{y}{3}\left(\dfrac{3}{3}\right)$$
$$= \dfrac{2y}{9} + \dfrac{3y}{9}$$
$$= \dfrac{2y+3y}{9}$$
$$= \dfrac{5y}{9}$$

27.
$$\dfrac{11}{5x} - \dfrac{5}{6x} = \dfrac{11}{5x} - \dfrac{5}{2 \cdot 3x}$$
$$= \dfrac{11}{5x}\left(\dfrac{2 \cdot 3}{2 \cdot 3}\right) - \dfrac{5}{2 \cdot 3x}\left(\dfrac{5}{5}\right)$$
$$= \dfrac{66}{30x} - \dfrac{25}{30x}$$
$$= \dfrac{66-25}{30x}$$
$$= \dfrac{41}{30x}$$

29.

$$\frac{7}{m^2} - \frac{2}{m} = \frac{7}{m \cdot m} - \frac{2}{m}$$

$$= \frac{7}{m \cdot m} - \frac{2}{m}\left(\frac{m}{m}\right)$$

$$= \frac{7}{m^2} - \frac{2m}{m^2}$$

$$= \frac{7 - 2m}{m^2}$$

31.

$$\frac{4x}{3} + \frac{2x}{y} = \frac{4x}{3}\left(\frac{y}{y}\right) + \frac{2x}{y}\left(\frac{3}{3}\right)$$

$$= \frac{4xy}{3y} + \frac{6x}{3y}$$

$$= \frac{4xy + 6x}{3y}$$

33.

$$\frac{1}{6c^4} + \frac{8}{9c^2} = \frac{1}{2 \cdot 3c \cdot c \cdot c \cdot c} + \frac{8}{3 \cdot 3c \cdot c}$$

$$= \frac{1}{2 \cdot 3c \cdot c \cdot c \cdot c}\left(\frac{3}{3}\right) + \frac{8}{3 \cdot 3c \cdot c}\left(\frac{2c^2}{2c^2}\right)$$

$$= \frac{3}{18c^4} + \frac{16c^2}{18c^4}$$

$$= \frac{3 + 16c^2}{18c^4}$$

$$= \frac{16c^2 + 3}{18c^4}$$

35.

$$\frac{y}{8} + \frac{y-3}{16} = \frac{y}{2 \cdot 2 \cdot 2} + \frac{y-3}{2 \cdot 2 \cdot 2 \cdot 2}$$

$$= \frac{y}{2 \cdot 2 \cdot 2}\left(\frac{2}{2}\right) + \frac{y-3}{2 \cdot 2 \cdot 2 \cdot 2}$$

$$= \frac{2y}{16} + \frac{y-3}{16}$$

$$= \frac{2y + y - 3}{16}$$

$$= \frac{3y - 3}{16}$$

37.

$$\frac{n}{5} - \frac{n-2}{15} = \frac{n}{5} - \frac{n-2}{3 \cdot 5}$$

$$= \frac{n}{5}\left(\frac{3}{3}\right) - \frac{n-2}{3 \cdot 5}$$

$$= \frac{3n}{15} - \frac{n-2}{15}$$

$$= \frac{3n - (n-2)}{15}$$

$$= \frac{3n - n + 2}{15}$$

$$= \frac{2n + 2}{15}$$

39.

$$\frac{2-b}{6} - \frac{b-7}{21} = \frac{2-b}{2 \cdot 3} - \frac{b-7}{3 \cdot 7}$$

$$= \frac{2-b}{2 \cdot 3}\left(\frac{7}{7}\right) - \frac{b-7}{3 \cdot 7}\left(\frac{2}{2}\right)$$

$$= \frac{7(2-b) - 2(b-7)}{42}$$

$$= \frac{14 - 7b - 2b + 14}{42}$$

$$= \frac{-9b + 28}{42}$$

41.

$$\frac{x-1}{9x} - \frac{x-2}{x^3} = \frac{x-1}{3 \cdot 3 \cdot x} - \frac{x-2}{x \cdot x \cdot x}$$

$$= \frac{x-1}{3 \cdot 3 \cdot x}\left(\frac{x \cdot x}{x \cdot x}\right) - \frac{x-2}{x \cdot x \cdot x}\left(\frac{3 \cdot 3}{3 \cdot 3}\right)$$

$$= \frac{x^2(x-1)}{9x^3} - \frac{9(x-2)}{9x^3}$$

$$= \frac{x^2(x-1) - 9(x-2)}{9x^3}$$

$$= \frac{x^3 - x^2 - 9x + 18}{9x^3}$$

43.

$$\frac{y+2}{5y^2} + \frac{y+4}{15y} = \frac{y+2}{5 \cdot y \cdot y} + \frac{y+4}{5 \cdot 3 \cdot y}$$

$$= \frac{y+2}{5 \cdot y \cdot y}\left(\frac{3}{3}\right) + \frac{y+4}{5 \cdot 3 \cdot y}\left(\frac{y}{y}\right)$$

$$= \frac{3(y+2)}{15y^2} + \frac{y(y+4)}{15y^2}$$

$$= \frac{3(y+2) + y(y+4)}{15y^2}$$

$$= \frac{3y+6+y^2+4y}{15y^2}$$

$$= \frac{y^2+7y+6}{15y^2}$$

45.

$$\frac{x+5}{xy} - \frac{x-1}{x^2 y} = \frac{x+5}{x \cdot y} - \frac{x-1}{x \cdot x \cdot y}$$

$$= \frac{x+5}{x \cdot y}\left(\frac{x}{x}\right) - \frac{x-1}{x \cdot x \cdot y}$$

$$= \frac{x(x+5)}{x^2 y} - \frac{x-1}{x^2 y}$$

$$= \frac{x(x+5) - (x-1)}{x^2 y}$$

$$= \frac{x^2 + 5x - x + 1}{x^2 y}$$

$$= \frac{x^2 + 4x + 1}{x^2 y}$$

47.

$$\frac{x-3}{6x} + \frac{x+4}{8x} = \frac{x-3}{2 \cdot 3 \cdot x} + \frac{x+4}{2 \cdot 2 \cdot 2 \cdot x}$$

$$= \frac{x-3}{2 \cdot 3 \cdot x}\left(\frac{2 \cdot 2}{2 \cdot 2}\right) + \frac{x+4}{2 \cdot 2 \cdot 2 \cdot x}\left(\frac{3}{3}\right)$$

$$= \frac{4(x-3)}{24x} + \frac{3(x+4)}{24x}$$

$$= \frac{4(x-3) + 3(x+4)}{24x}$$

$$= \frac{4x - 12 + 3x + 12}{24x}$$

$$= \frac{7x}{24x}$$

$$= \frac{7}{24}$$

49.

$$\frac{a+2}{b} + \frac{b-2}{a} = \frac{a+2}{b}\left(\frac{a}{a}\right) + \frac{b-2}{a}\left(\frac{b}{b}\right)$$

$$= \frac{a(a+2) + b(b-2)}{ab}$$

$$= \frac{a^2 + 2a + b^2 - 2b}{ab}$$

51.

$$\frac{x}{x+1} + \frac{x-1}{x} = \frac{x}{x+1}\left(\frac{x}{x}\right) + \frac{x-1}{x}\left(\frac{x+1}{x+1}\right)$$

$$= \frac{x(x)}{x(x+1)} + \frac{(x-1)(x+1)}{x(x+1)}$$

$$= \frac{x^2}{x(x+1)} + \frac{x^2 - 1}{x(x+1)}$$

$$= \frac{x^2 + x^2 - 1}{x(x+1)}$$

$$= \frac{2x^2 - 1}{x(x+1)}$$

53.

$$\frac{1}{5x} + \frac{7x}{x+5} = \frac{1}{5x}\left(\frac{x+5}{x+5}\right) + \frac{7x}{x+5}\left(\frac{5x}{5x}\right)$$

$$= \frac{(x+5)}{5x(x+5)} + \frac{7x(5x)}{5x(x+5)}$$

$$= \frac{x+5}{5x(x+5)} + \frac{35x^2}{5x(x+5)}$$

$$= \frac{x+5+35x^2}{5x(x+5)}$$

$$= \frac{35x^2 + x + 5}{5x(x+5)}$$

55.

$$\frac{9}{t+3} + \frac{8}{t+2} = \frac{9}{t+3}\left(\frac{t+2}{t+2}\right) + \frac{8}{t+2}\left(\frac{t+3}{t+3}\right)$$

$$= \frac{9(t+2)}{(t+3)(t+2)} + \frac{8(t+3)}{(t+3)(t+2)}$$

$$= \frac{9t+18}{(t+3)(t+2)} + \frac{8t+24}{(t+3)(t+2)}$$

$$= \frac{9t+18+8t+24}{(t+3)(t+2)}$$

$$= \frac{17t+42}{(t+3)(t+2)}$$

57.

$$\frac{3x}{2x-1}-\frac{2x}{2x+3}=\frac{3x}{2x-1}\left(\frac{2x+3}{2x+3}\right)-\frac{2x}{2x+3}\left(\frac{2x-1}{2x-1}\right)$$

$$=\frac{3x(2x+3)}{(2x-1)(2x+3)}-\frac{2x(2x-1)}{(2x-1)(2x+3)}$$

$$=\frac{6x^2+9x}{(2x-1)(2x+3)}-\frac{4x^2-2x}{(2x-1)(2x+3)}$$

$$=\frac{6x^2+9x-(4x^2-2x)}{(2x-1)(2x+3)}$$

$$=\frac{6x^2+9x-4x^2+2x}{(2x-1)(2x+3)}$$

$$=\frac{2x^2+11x}{(2x-1)(2x+3)}$$

59.

$$\frac{4}{a+2}-\frac{7}{(a+2)^2}=\frac{4}{a+2}\left(\frac{a+2}{a+2}\right)-\frac{7}{(a+2)^2}$$

$$=\frac{4(a+2)}{(a+2)^2}-\frac{7}{(a+2)^2}$$

$$=\frac{4a+8}{(a+2)^2}-\frac{7}{(a+2)^2}$$

$$=\frac{4a+8-7}{(a+2)^2}$$

$$=\frac{4a+1}{(a+2)^2}$$

61.

$$\frac{3m}{m-2}+\frac{m-3}{m+5}=\frac{3m}{m-2}\left(\frac{m+5}{m+5}\right)+\frac{m-3}{m+5}\left(\frac{m-2}{m-2}\right)$$

$$=\frac{3m(m+5)}{(m-2)(m+5)}+\frac{(m-3)(m-2)}{(m-2)(m+5)}$$

$$=\frac{3m^2+15m}{(m-2)(m+5)}+\frac{m^2-2m-3m+6}{(m-2)(m+5)}$$

$$=\frac{3m^2+15m+m^2-2m-3m+6}{(m-2)(m+5)}$$

$$=\frac{4m^2+10m+6}{(m-2)(m+5)}$$

63.

$$\frac{s+7}{s+3}-\frac{s-3}{s+7}=\frac{s+7}{s+3}\left(\frac{s+7}{s+7}\right)-\frac{s-3}{s+7}\left(\frac{s+3}{s+3}\right)$$

$$=\frac{(s+7)(s+7)}{(s+3)(s+7)}-\frac{(s-3)(s+3)}{(s+3)(s+7)}$$

$$=\frac{s^2+7s+7s+49}{(s+3)(s+7)}-\frac{s^2-9}{(s+3)(s+7)}$$

$$=\frac{s^2+7s+7s+49-(s^2-9)}{(s+3)(s+7)}$$

$$=\frac{s^2+7s+7s+49-s^2+9}{(s+3)(s+7)}$$

$$=\frac{14s+58}{(s+3)(s+7)}$$

65.

$$\frac{7}{(a+1)(a+3)}+\frac{5}{(a+3)^2}=\frac{7}{(a+1)(a+3)}\left(\frac{a+3}{a+3}\right)+\frac{5}{(a+3)^2}\left(\frac{a+1}{a+1}\right)$$

$$=\frac{7(a+3)}{(a+1)(a+3)^2}+\frac{5(a+1)}{(a+1)(a+3)^2}$$

$$=\frac{7a+21}{(a+1)(a+3)^2}+\frac{5a+5}{(a+1)(a+3)^2}$$

$$=\frac{7a+21+5a+5}{(a+1)(a+3)^2}$$

$$=\frac{12a+26}{(a+1)(a+3)^2}$$

67.

$$\frac{2c-3}{(2c+1)(c-3)}-\frac{3}{c-3}=\frac{2c-3}{(2c+1)(c-3)}-\frac{3}{c-3}\left(\frac{2c+1}{2c+1}\right)$$

$$=\frac{2c-3}{(2c+1)(c-3)}-\frac{6c+3}{(2c+1)(c-3)}$$

$$=\frac{2c-3-(6c+3)}{(2c+1)(c-3)}$$

$$=\frac{2c-3-6c-3}{(2c+1)(c-3)}$$

$$=\frac{-4c-6}{(2c+1)(c-3)}$$

$$=-\frac{4c+6}{(2c+1)(c-3)}$$

69.

$$\frac{b}{b+1}-\frac{b+1}{2b+2}=\frac{b}{b+1}-\frac{b+1}{2(b+1)}$$

$$=\frac{b}{b+1}\left(\frac{2}{2}\right)-\frac{b+1}{2(b+1)}$$

$$=\frac{2b}{2(b+1)}-\frac{b+1}{2(b+1)}$$

$$=\frac{2b-(b+1)}{2(b+1)}$$

$$=\frac{2b-b-1}{2(b+1)}$$

$$=\frac{b-1}{2(b+1)}$$

71.

$$\frac{2}{a^2+4a+3}+\frac{1}{a+3}=\frac{2}{(a+1)(a+3)}+\frac{1}{a+3}$$

$$=\frac{2}{(a+1)(a+3)}+\frac{1}{a+3}\left(\frac{a+1}{a+1}\right)$$

$$=\frac{2}{(a+1)(a+3)}+\frac{a+1}{a+3}$$

$$=\frac{2+a+1}{(a+1)(a+3)}$$

$$=\frac{(a+3)}{(a+1)(a+3)}$$

$$=\frac{1}{a+1}$$

73.

$$\frac{7s}{s^2+s-12}-\frac{4}{s+4}=\frac{7s}{(s+4)(s-3)}-\frac{4}{s+4}$$

$$=\frac{7s}{(s+4)(s-3)}-\frac{4}{s+4}\left(\frac{s-3}{s-3}\right)$$

$$=\frac{7s}{(s+4)(s-3)}-\frac{4s-12}{(s+4)(s-3)}$$

$$=\frac{7s-(4s-12)}{(s+4)(s-3)}$$

$$=\frac{7s-4s+12}{(s+4)(s-3)}$$

$$=\frac{3s+12}{(s+4)(s-3)}$$

$$=\frac{3(s+4)}{(s+4)(s-3)}$$

$$=\frac{3}{s-3}$$

75.

$$\frac{x}{x-2}+\frac{4+2x}{x^2-4}=\frac{x}{x-2}+\frac{4+2x}{(x+2)(x-2)}$$

$$=\frac{x}{x-2}\left(\frac{x+2}{x+2}\right)+\frac{4+2x}{(x+2)(x-2)}$$

$$=\frac{x^2+2x}{(x+2)(x-2)}+\frac{4+2x}{(x+2)(x-2)}$$

$$=\frac{x^2+2x+4+2x}{(x+2)(x-2)}$$

$$=\frac{x^2+4x+4}{(x+2)(x-2)}$$

$$=\frac{(x+2)(x+2)}{(x+2)(x-2)}$$

$$=\frac{x+2}{x-2}$$

77.

$$\frac{x+1}{x+2}-\frac{x^2+1}{x^2-x-6}=\frac{x+1}{x+2}-\frac{x^2+1}{(x+2)(x-3)}$$

$$=\frac{x+1}{x+2}\left(\frac{x-3}{x-3}\right)-\frac{x^2+1}{(x+2)(x-3)}$$

$$=\frac{x^2-3x+x-3}{(x+2)(x-3)}-\frac{x^2+1}{(x+2)(x-3)}$$

$$=\frac{x^2-2x-3-(x^2+1)}{(x+2)(x-3)}$$

$$=\frac{x^2-2x-3-x^2-1}{(x+2)(x-3)}$$

$$=\frac{-2x-4}{(x+2)(x-3)}$$

$$=\frac{-2(x+2)}{(x+2)(x-3)}$$

$$=-\frac{2}{x-3}$$

79.

$$\frac{2}{3h-6}+\frac{3}{4h+8}=\frac{2}{3(h-2)}+\frac{3}{4(h+2)}$$

$$=\frac{2}{3(h-2)}\left(\frac{4(h+2)}{4(h+2)}\right)+\frac{3}{4(h+2)}\left(\frac{3(h-2)}{3(h-2)}\right)$$

$$=\frac{8(h+2)}{12(h-2)(h+2)}+\frac{9(h-2)}{12(h-2)(h+2)}$$

$$=\frac{8h+16}{12(h-2)(h+2)}+\frac{9h-18}{12(h-2)(h+2)}$$

$$=\frac{8h+16+9h-18}{12(h-2)(h+2)}$$

$$=\frac{17h-2}{12(h-2)(h+2)}$$

81.

$$\frac{8}{y^2-16}-\frac{7}{y^2-y-12}=\frac{8}{(y+4)(y-4)}-\frac{7}{(y-4)(y+3)}$$

$$=\frac{8}{(y+4)(y-4)}\left(\frac{y+3}{y+3}\right)-\frac{7}{(y-4)(y+3)}\left(\frac{y+4}{y+4}\right)$$

$$=\frac{8y+24}{(y+4)(y-4)(y+3)}-\frac{7y+28}{(y+4)(y-4)(y+3)}$$

$$=\frac{8y+24-(7y+28)}{(y+4)(y-4)(y+3)}$$

$$=\frac{8y+24-7y-28}{(y+4)(y-4)(y+3)}$$

$$=\frac{y-4}{(y+4)(y-4)(y+3)}$$

$$=\frac{1}{(y+4)(y+3)}$$

83.

$$\frac{4}{s^2+5s+4}+\frac{s}{s^2+2s+1}$$

$$=\frac{4}{(s+1)(s+4)}+\frac{s}{(s+1)^2}$$

$$=\frac{4}{(s+1)(s+4)}\left(\frac{s+1}{s+1}\right)+\frac{s}{(s+1)(s+1)}\left(\frac{s+4}{s+4}\right)$$

$$=\frac{4s+4}{(s+4)(s+1)^2}+\frac{s^2+4s}{(s+4)(s+1)^2}$$

$$=\frac{4s+4+s^2+4s}{(s+4)(s+1)^2}$$

$$=\frac{s^2+8s+4}{(s+4)(s+1)^2}$$

85.

$$\frac{5}{x^2-9x+8}-\frac{3}{x^2-6x-16}=\frac{5}{(x-8)(x-1)}-\frac{3}{(x-8)(x+2)}$$

$$=\frac{5}{(x-8)(x-1)}\left(\frac{x+2}{x+2}\right)-\frac{3}{(x-8)(x+2)}\left(\frac{x-1}{x-1}\right)$$

$$=\frac{5x+10}{(x-8)(x-1)(x+2)}-\frac{3x-3}{(x-8)(x-1)(x+2)}$$

$$=\frac{5x+10-(3x-3)}{(x-8)(x-1)(x+2)}$$

$$=\frac{5x+10-3x+3}{(x-8)(x-1)(x+2)}$$

$$=\frac{2x+13}{(x-8)(x-1)(x+2)}$$

87.

$$\frac{x+1}{2x+4} - \frac{x^2}{2x^2-8} = \frac{x+1}{2(x+2)} - \frac{x^2}{2(x^2-4)}$$

$$= \frac{x+1}{2(x+2)} - \frac{x^2}{2(x+2)(x-2)}$$

$$= \frac{x+1}{2(x+2)}\left(\frac{x-2}{x-2}\right) - \frac{x^2}{2(x+2)(x-2)}$$

$$= \frac{x^2-2x+x-2}{2(x+2)(x-2)} - \frac{x^2}{2(x+2)(x-2)}$$

$$= \frac{x^2-2x+x-2-x^2}{2(x+2)(x-2)}$$

$$= \frac{-x-2}{2(x+2)(x-2)}$$

$$= -\frac{x+2}{2(x+2)(x-2)}$$

$$= -\frac{1}{2(x-2)}$$

89.

$$\frac{8}{x} + 6 = \frac{8}{x} + \frac{6}{1}\left(\frac{x}{x}\right)$$

$$= \frac{8}{x} + \frac{6x}{x}$$

$$= \frac{8+6x}{x}$$

$$= \frac{6x+8}{x}$$

91.

$$b - \frac{3}{a^2} = \frac{b}{1}\left(\frac{a^2}{a^2}\right) - \frac{3}{a^2}$$

$$= \frac{a^2 b}{a^2} - \frac{3}{a^2}$$

$$= \frac{a^2 b - 3}{a^2}$$

93.

$$\frac{9}{x-4} + x = \frac{9}{x-4} + \frac{x}{1}\left(\frac{x-4}{x-4}\right)$$

$$= \frac{9}{x-4} + \frac{x^2-4x}{x-4}$$

$$= \frac{9+x^2-4x}{x-4}$$

$$= \frac{x^2-4x+9}{x-4}$$

95.

$$\frac{x+2}{x+1} - 5 = \frac{x+2}{x+1} - \frac{5}{1}\left(\frac{x+1}{x+1}\right)$$

$$= \frac{x+2}{x+1} - \frac{5x+5}{x+1}$$

$$= \frac{x+2-(5x+5)}{x+1}$$

$$= \frac{x+2-5x-5}{x+1}$$

$$= \frac{-4x-3}{x+1}$$

$$= -\frac{4x+3}{x+1}$$

97.

$$\frac{5}{a-4} + \frac{7}{4-a} = \frac{5}{a-4} + \frac{7}{4-a}\left(\frac{-1}{-1}\right)$$

$$= \frac{5}{a-4} + \frac{-7}{-4+a}$$

$$= \frac{5}{a-4} + \frac{-7}{a-4}$$

$$= \frac{5-7}{a-4}$$

$$= \frac{-2}{a-4}$$

$$= -\frac{2}{a-4}$$

99.

$$\frac{r+2}{r^2-4} + \frac{4}{4-r^2} = \frac{r+2}{r^2-4} + \frac{4}{4-r^2}\left(\frac{-1}{-1}\right)$$

$$= \frac{r+2}{r^2-4} + \frac{-4}{-4+r^2}$$

$$= \frac{r+2}{r^2-4} + \frac{-4}{r^2-4}$$

$$= \frac{r+2-4}{r^2-4}$$

$$= \frac{r-2}{(r-2)(r+2)}$$

$$= \frac{1}{r+2}$$

101.

$$\frac{y+3}{y-1} - \frac{y+4}{1-y} = \frac{y+3}{y-1} - \frac{y+4}{1-y}\left(\frac{-1}{-1}\right)$$

$$= \frac{y+3}{y-1} - \frac{-y-4}{-1+y}$$

$$= \frac{y+3}{y-1} - \frac{-y-4}{y-1}$$

$$= \frac{y+3-(-y-4)}{y-1}$$

$$= \frac{y+3+y+4}{y-1}$$

$$= \frac{2y+7}{y-1}$$

APPLICATIONS

103.

$$\frac{3}{2x^2} + \frac{10}{3x} = \frac{3}{2x^2}\left(\frac{3}{3}\right) + \frac{10}{3x}\left(\frac{2x}{2x}\right)$$

$$= \frac{9}{6x^2} + \frac{20x}{6x^2}$$

$$= \frac{20x+9}{6x^2} \text{ cm}$$

WRITING

105. Answers will vary.

107. Answers will vary.

REVIEW

109. Use $y = mx + b$ for $y = 8x + 2$.
Then, $m = 8$ and $b = (0, 2)$.

111. $y = -2$
$y = 0x - 2$
$m = 0$

CHALLENGE PROBLEMS

113.

$$\frac{a}{a-1} - \frac{2}{a+2} + \frac{3(a-2)}{a^2+a-2} = \frac{a}{a-1} - \frac{2}{a+2} + \frac{3(a-2)}{(a+2)(a-1)}$$

$$= \frac{a}{a-1}\left(\frac{a+2}{a+2}\right) - \frac{2}{a+2}\left(\frac{a-1}{a-1}\right) + \frac{3(a-2)}{(a+2)(a-1)}$$

$$= \frac{a^2+2a}{(a+2)(a-1)} - \frac{2a-2}{(a+2)(a-1)} + \frac{3a-6}{(a+2)(a-1)}$$

$$= \frac{a^2+2a-(2a-2)+(3a-6)}{(a+2)(a-1)}$$

$$= \frac{a^2+2a-2a+2+3a-6}{(a+2)(a-1)}$$

$$= \frac{a^2+3a-4}{(a+2)(a-1)}$$

$$= \frac{(a+4)(a-1)}{(a+2)(a-1)}$$

$$= \frac{a+4}{a+2}$$

115.

$$\frac{1}{a+1} + \frac{a^2-7a+10}{2a^2-2a-4} \cdot \frac{2a^2-50}{a^2+10a+25}$$

$$= \frac{1}{a+1} + \left(\frac{(a-2)(a-5)}{2(a^2-a-2)} \cdot \frac{2(a^2-25)}{(a+5)(a+5)}\right)$$

$$= \frac{1}{a+1} + \left(\frac{(a-2)(a-5)}{2(a-2)(a+1)} \cdot \frac{2(a-5)(a+5)}{(a+5)(a+5)}\right)$$

$$= \frac{1}{a+1} + \left(\frac{(a-2)(a-5)}{2(a-2)(a+1)} \cdot \frac{2(a-5)(a+5)}{(a+5)(a+5)}\right)$$

$$= \frac{1}{a+1} + \frac{(a-5)(a-5)}{(a+1)(a+5)}$$

$$= \frac{1}{a+1}\left(\frac{a+5}{a+5}\right) + \frac{(a-5)(a-5)}{(a+1)(a+5)}$$

$$= \frac{a+5}{(a+1)(a+5)} + \frac{a^2-5a-5a+25}{(a+1)(a+5)}$$

$$= \frac{a+5+a^2-10a+25}{(a+1)(a+5)}$$

$$= \frac{a^2-9a+30}{(a+1)(a+5)}$$

VOCABULARY

1. The expression $\dfrac{\frac{2}{3}-\frac{1}{x}}{\frac{x-3}{4}}$ is called a **complex** rational expression or, more simply, a **complex** fraction.

15. $\dfrac{4x^2}{15} \div \dfrac{16x}{25} = \dfrac{\frac{4x^2}{15}}{\frac{16x}{25}}$

3. To find the **reciprocal** of $\dfrac{x+8}{x+7}$, we invert it.

PRACTICE

CONCEPTS

17.

$\dfrac{\frac{x}{2}}{\frac{6}{5}} = \dfrac{x}{2} \div \dfrac{6}{5}$

5. To simplify a complex fraction, write its numerator and denominator as **single** rational expressions. Then perform the division by multiplying the numerator of the complex fraction by the **reciprocal** of the denominator of the complex fraction. **Simplify** the result, if possible.

$\quad = \dfrac{x}{2} \bullet \dfrac{5}{6}$

$\quad = \dfrac{5x}{12}$

7. a) $\dfrac{x-3}{4}$

19.

$\dfrac{\frac{2}{3}}{\frac{3}{4}} = \dfrac{2}{3} \div \dfrac{3}{4}$

   b) yes

   c) $\dfrac{1}{12} - \dfrac{x}{6}$

$\quad = \dfrac{2}{3} \bullet \dfrac{4}{3}$

   d) no

$\quad = \dfrac{8}{9}$

9. a) y, 3, and 6
   b) $6y$
   c) $\dfrac{6y}{6y}$

21.

$\dfrac{\frac{x}{y}}{\frac{1}{x}} = \dfrac{x}{y} \div \dfrac{1}{x}$

11. $\dfrac{x}{12}(24) = \dfrac{24x}{12}$

$\quad = \dfrac{x}{y} \bullet \dfrac{x}{1}$

$\qquad\qquad = 2x$

$\quad = \dfrac{x^2}{y}$

23.

$\dfrac{\frac{n}{8}}{\frac{1}{n^2}} = \dfrac{n}{8} \div \dfrac{1}{n^2}$

13. $\dfrac{2}{3}\left(6y^2\right) = \dfrac{12y^2}{3}$

$\qquad\qquad = 4y^2$

$\quad = \dfrac{n}{8} \bullet \dfrac{n^2}{1}$

$\quad = \dfrac{n^3}{8}$

25. Multiply by the LCD of x.

$$\frac{\dfrac{1}{x}-3}{\dfrac{5}{x}+2}=\frac{\dfrac{1}{x}-3}{\dfrac{5}{x}+2}\bullet\frac{x}{x}$$

$$=\frac{x\bullet\dfrac{1}{x}-x\bullet3}{x\bullet\dfrac{5}{x}+x\bullet2}$$

$$=\frac{1-3x}{5+2x}$$

27. Multiply by the LCD of 3.

$$\frac{\dfrac{2}{3}+1}{\dfrac{1}{3}+1}=\frac{\dfrac{2}{3}+1}{\dfrac{1}{3}+1}\bullet\frac{3}{3}$$

$$=\frac{3\left(\dfrac{2}{3}+1\right)}{3\left(\dfrac{1}{3}+1\right)}$$

$$=\frac{3\bullet\dfrac{2}{3}+3\bullet1}{3\bullet\dfrac{1}{3}+3\bullet1}$$

$$=\frac{2+3}{1+3}$$

$$=\frac{5}{4}$$

29. Simplify using division.

$$\frac{\dfrac{4a}{11}}{\dfrac{6a}{55}}=\frac{4a}{11}\div\frac{6a}{55}$$

$$=\frac{4a}{11}\bullet\frac{55}{6a}$$

$$=\frac{220a}{66a}$$

$$=\frac{10}{3}$$

31. Simplify using division.

$$\frac{\dfrac{40x^2}{20x}}{9}=\frac{40x^2}{1}\div\frac{20x}{9}$$

$$=\frac{40x^2}{1}\bullet\frac{9}{20x}$$

$$=\frac{360x^2}{20x}$$

$$=18x$$

33. Multiply by the LCD of $2y$.

$$\frac{\dfrac{1}{y}-\dfrac{5}{2}}{\dfrac{3}{y}}=\frac{\dfrac{1}{y}-\dfrac{5}{2}}{\dfrac{3}{y}}\bullet\frac{2y}{2y}$$

$$=\frac{\left(\dfrac{1}{y}-\dfrac{5}{2}\right)2y}{\left(\dfrac{3}{y}\right)2y}$$

$$=\frac{\dfrac{1}{y}(2y)-\dfrac{5}{2}(2y)}{\dfrac{3}{y}(2y)}$$

$$=\frac{2-5y}{6}$$

35. Multiply by the LCD of 12.

$$\frac{\dfrac{d+2}{2}}{\dfrac{d}{3}-\dfrac{d}{4}}=\frac{\dfrac{d+2}{2}}{\dfrac{d}{3}-\dfrac{d}{4}}\bullet\frac{12}{12}$$

$$=\frac{\left(\dfrac{d+2}{2}\right)12}{\left(\dfrac{d}{3}\right)12-\left(\dfrac{d}{4}\right)12}$$

$$=\frac{6(d+2)}{4d-3d}$$

$$=\frac{6d+12}{d}$$

Section 6.5

37. Multiply by the LCD of 16.

$$\frac{\dfrac{s+15}{16}}{\dfrac{s+15}{8}} = \frac{\dfrac{s+15}{16}}{\dfrac{s+15}{8}} \cdot \frac{16}{16}$$

$$= \frac{\left(\dfrac{s+15}{16}\right)16}{\left(\dfrac{s+15}{8}\right)16}$$

$$= \frac{s+15}{2(s+15)}$$

$$= \frac{1}{2}$$

39. Multiply by the LCD of 4.

$$\frac{\dfrac{1}{2}+\dfrac{3}{4}}{\dfrac{3}{2}+\dfrac{1}{4}} = \frac{\dfrac{1}{2}+\dfrac{3}{4}}{\dfrac{3}{2}+\dfrac{1}{4}} \cdot \frac{4}{4}$$

$$= \frac{\left(\dfrac{1}{2}\right)4+\left(\dfrac{3}{4}\right)4}{\left(\dfrac{3}{2}\right)4+\left(\dfrac{1}{4}\right)4}$$

$$= \frac{2+3}{6+1}$$

$$= \frac{5}{7}$$

41. Multiply by the LCD of 30.

$$\frac{\dfrac{x^4}{30}}{\dfrac{7x}{15}} = \frac{\dfrac{x^4}{30}}{\dfrac{7x}{15}} \cdot \frac{30}{30}$$

$$= \frac{\left(\dfrac{x^4}{30}\right)30}{\left(\dfrac{7x}{15}\right)30}$$

$$= \frac{x^4}{14x}$$

$$= \frac{x^3}{14}$$

43. Multiply by the LCD of s^3.

$$\frac{\dfrac{2}{s}-\dfrac{2}{s^2}}{\dfrac{4}{s^3}+\dfrac{4}{s^2}} = \frac{\dfrac{2}{s}-\dfrac{2}{s^2}}{\dfrac{4}{s^3}+\dfrac{4}{s^2}} \cdot \frac{s^3}{s^3}$$

$$= \frac{\left(\dfrac{2}{s}\right)s^3 - \left(\dfrac{2}{s^2}\right)s^3}{\left(\dfrac{4}{s^3}\right)s^3 + \left(\dfrac{4}{s^2}\right)s^3}$$

$$= \frac{2s^2 - 2s}{4 + 4s}$$

$$= \frac{2s(s-1)}{2 \cdot 2(1+s)}$$

$$= \frac{s(s-1)}{2(1+s)}$$

$$= \frac{s^2 - s}{2 + 2s}$$

45. Simplify using division.

$$\frac{-\dfrac{3}{x^3}}{\dfrac{6}{x^5}} = -\frac{3}{x^3} \div \frac{6}{x^5}$$

$$= -\frac{3}{x^3} \cdot \frac{x^5}{6}$$

$$= -\frac{3x^5}{6x^3}$$

$$= -\frac{x^2}{2}$$

47. Multiply by the LCD of x.

$$\frac{\dfrac{2}{x}+2}{\dfrac{4}{x}+2} = \frac{\dfrac{2}{x}+2}{\dfrac{4}{x}+2} \cdot \frac{x}{x}$$

$$= \frac{\dfrac{2}{x}(x)+2(x)}{\dfrac{4}{x}(x)+2(x)}$$

$$= \frac{2+2x}{4+2x}$$

$$= \frac{2(1+x)}{2(2+x)}$$

$$= \frac{1+x}{2+x}$$

49. Simplify using division.

$$\frac{\dfrac{2x-8}{15}}{\dfrac{3x-12}{35x}} = \frac{2x-8}{15} \div \frac{3x-12}{35x}$$

$$= \frac{2x-8}{15} \bullet \frac{35x}{3x-12}$$

$$= \frac{2(x-4)\cdot 5\cdot 7\cdot x}{3\cdot 5\cdot 3(x-4)}$$

$$= \frac{14x}{9}$$

51. Simplify using division.

$$\frac{\dfrac{t-6}{16}}{\dfrac{12-2t}{t}} = \frac{t-6}{16} \div \frac{12-2t}{t}$$

$$= \frac{t-6}{16} \bullet \frac{t}{12-2t}$$

$$= \frac{(t-6)\cdot t}{2\cdot 2\cdot 2\cdot 2\cdot 2(6-t)}$$

$$= -\frac{(t-6)\cdot t}{2\cdot 2\cdot 2\cdot 2\cdot 2(t-6)}$$

$$= -\frac{t}{32}$$

53. Multiply by the LCD of $4c^2$.

$$\frac{\dfrac{2}{c^2}}{\dfrac{1}{c}+\dfrac{5}{4}} = \frac{\dfrac{2}{c^2}}{\dfrac{1}{c}+\dfrac{5}{4}} \bullet \frac{4c^2}{4c^2}$$

$$= \frac{\dfrac{2}{c^2}\left(4c^2\right)}{\dfrac{1}{c}\left(4c^2\right)+\dfrac{5}{4}\left(4c^2\right)}$$

$$= \frac{8}{4c+5c^2}$$

55. Multiply by the LCD of a^2b^2.

$$\frac{\dfrac{1}{a^2b}-\dfrac{5}{ab}}{\dfrac{3}{ab}-\dfrac{7}{ab^2}} = \frac{\dfrac{1}{a^2b}-\dfrac{5}{ab}}{\dfrac{3}{ab}-\dfrac{7}{ab^2}} \bullet \frac{a^2b^2}{a^2b^2}$$

$$= \frac{\dfrac{1}{a^2b}\left(a^2b^2\right)-\dfrac{5}{ab}\left(a^2b^2\right)}{\dfrac{3}{ab}\left(a^2b^2\right)-\dfrac{7}{ab^2}\left(a^2b^2\right)}$$

$$= \frac{b-5ab}{3ab-7a}$$

57. Multiply by the LCD of x.

$$\frac{\dfrac{3y}{x}-y}{y-\dfrac{y}{x}} = \frac{\dfrac{3y}{x}-y}{y-\dfrac{y}{x}} \bullet \frac{x}{x}$$

$$= \frac{\dfrac{3y}{x}(x)-y(x)}{y(x)-\dfrac{y}{x}(x)}$$

$$= \frac{3y-xy}{xy-y}$$

$$= \frac{y(3-x)}{y(x-1)}$$

$$= \frac{3-x}{x-1}$$

59. Simplify using division.

$$\frac{\dfrac{b^2-81}{18a^2}}{\dfrac{4b-36}{9a}} = \frac{b^2-81}{18a^2} \div \frac{4b-36}{9a}$$

$$= \frac{b^2-81}{18a^2} \bullet \frac{9a}{4b-36}$$

$$= \frac{(b-9)(b+9)\bullet 3\cdot 3\cdot a}{2\cdot 3\cdot 3\cdot a\cdot a\bullet 2\cdot 2(b-9)}$$

$$= \frac{b+9}{8a}$$

61. Multiply by the LCD of $8h$.

$$\dfrac{4 - \dfrac{1}{8h}}{12 + \dfrac{3}{4h}} = \dfrac{4 - \dfrac{1}{8h}}{12 + \dfrac{3}{4h}} \bullet \dfrac{8h}{8h}$$

$$= \dfrac{4(8h) - \dfrac{1}{8h}(8h)}{12(8h) + \dfrac{3}{4h}(8h)}$$

$$= \dfrac{32h - 1}{96h + 6}$$

63. Multiply by a LCD of xy.

$$\dfrac{1}{\dfrac{1}{x} + \dfrac{1}{y}} = \dfrac{1}{\dfrac{1}{x} + \dfrac{1}{y}} \bullet \dfrac{xy}{xy}$$

$$= \dfrac{1(xy)}{\dfrac{1}{x}(xy) + \dfrac{1}{y}(xy)}$$

$$= \dfrac{xy}{y + x}$$

65. Multiply by the LCD of $(x + 1)$.

$$\dfrac{\dfrac{1}{x+1}}{1 + \dfrac{1}{x+1}} = \dfrac{\dfrac{1}{x+1}}{1 + \dfrac{1}{x+1}} \bullet \dfrac{x+1}{x+1}$$

$$= \dfrac{\dfrac{1}{x+1}(x+1)}{1(x+1) + \dfrac{1}{x+1}(x+1)}$$

$$= \dfrac{1}{x+1+1}$$

$$= \dfrac{1}{x+2}$$

67. Multiply by the LCD of $(x + 2)$.

$$\dfrac{\dfrac{x}{x+2}}{\dfrac{x}{x+2} + x} = \dfrac{\dfrac{x}{x+2}}{\dfrac{x}{x+2} + x} \bullet \dfrac{x+2}{x+2}$$

$$= \dfrac{\dfrac{x}{x+2}(x+2)}{\dfrac{x}{x+2}(x+2) + x(x+2)}$$

$$= \dfrac{x}{x + x^2 + 2x}$$

$$= \dfrac{x}{x^2 + 3x}$$

$$= \dfrac{x}{x(x+3)}$$

$$= \dfrac{1}{x+3}$$

69. Simplify using division.

$$\dfrac{\dfrac{5t^2}{9x^2}}{\dfrac{3t}{x^2t}} = \dfrac{5t^2}{9x^2} \div \dfrac{3t}{x^2t}$$

$$= \dfrac{5t^2}{9x^2} \bullet \dfrac{x^2t}{3t}$$

$$= \dfrac{5 \cdot t \cdot t \cdot x \cdot x \cdot t}{3 \cdot 3 \cdot x \cdot x \cdot 3 \cdot t}$$

$$= \dfrac{5t^2}{27}$$

71. Simplify using division.

$$\dfrac{\dfrac{m^2 - 4}{3m + 3}}{\dfrac{2m + 4}{m + 1}} = \dfrac{m^2 - 4}{3m + 3} \div \dfrac{2m + 4}{m + 1}$$

$$= \dfrac{m^2 - 4}{3m + 3} \bullet \dfrac{m + 1}{2m + 4}$$

$$= \dfrac{(m+2)(m-2)(m+1)}{3(m+1)2(m+2)}$$

$$= \dfrac{m - 2}{6}$$

73. Multiply by a LCD of xy.

$$\frac{\dfrac{2}{x}}{\dfrac{2}{y}-\dfrac{4}{x}}=\frac{\dfrac{2}{x}}{\dfrac{2}{y}-\dfrac{4}{x}}\bullet\frac{xy}{xy}$$

$$=\frac{\dfrac{2}{x}(xy)}{\dfrac{2}{y}(xy)-\dfrac{4}{x}(xy)}$$

$$=\frac{2y}{2x-4y}$$

$$=\frac{2y}{2(x-2y)}$$

$$=\frac{y}{x-2y}$$

75. Multiply by the LCD of mn.

$$\frac{\dfrac{m}{n}+\dfrac{n}{m}}{\dfrac{m}{n}-\dfrac{n}{m}}=\frac{\dfrac{m}{n}+\dfrac{n}{m}}{\dfrac{m}{n}-\dfrac{n}{m}}\bullet\frac{mn}{mn}$$

$$=\frac{\dfrac{m}{n}(mn)+\dfrac{n}{m}(mn)}{\dfrac{m}{n}(mn)-\dfrac{n}{m}(mn)}$$

$$=\frac{m^2+n^2}{m^2-n^2}$$

77. Multiply by the LCD of $(x-1)$.

$$\frac{3+\dfrac{3}{x-1}}{3-\dfrac{3}{x-1}}=\frac{3+\dfrac{3}{x-1}}{3-\dfrac{3}{x-1}}\bullet\frac{x-1}{x-1}$$

$$=\frac{3(x-1)+\dfrac{3}{x-1}(x-1)}{3(x-1)-\dfrac{3}{x-1}(x-1)}$$

$$=\frac{3x-3+3}{3x-3-3}$$

$$=\frac{3x}{3x-6}$$

$$=\frac{3x}{3(x-2)}$$

$$=\frac{x}{x-2}$$

APPLICATIONS

79. GARDENING TOOLS

$$\frac{\dfrac{x}{2}}{\dfrac{7x}{3}}=\frac{x}{2}\div\frac{7x}{3}$$

$$=\frac{x}{2}\bullet\frac{3}{7x}$$

$$=\frac{3x}{14x}$$

$$=\frac{3}{14}\text{ in.}$$

81. ELECTRONICS

Multiply by the LCD of R_1R_2.

$$\frac{1}{\dfrac{1}{R_1}+\dfrac{1}{R_2}}=\frac{1}{\dfrac{1}{R_1}+\dfrac{1}{R_2}}\bullet\frac{R_1R_2}{R_1R_2}$$

$$=\frac{1(R_1R_2)}{\dfrac{1}{R_1}(R_1R_2)+\dfrac{1}{R_2}(R_1R_2)}$$

$$=\frac{R_1R_2}{R_2+R_1}$$

WRITING

83. Answers will vary.

85. Answers will vary.

REVIEW

87. $(8x)^0=1$

89. $\left(\dfrac{3r}{4r^3}\right)^4=\left(\dfrac{3}{4}r^{1-3}\right)^4$

$$=\left(\dfrac{3}{4}r^{-2}\right)^4$$

$$=\left(\dfrac{3}{4r^2}\right)^4$$

$$=\dfrac{3^4}{4^4r^{2\bullet4}}$$

$$=\dfrac{81}{256r^8}$$

91.

$$\left(\frac{6r^{-2}}{2r^3}\right)^{-2} = \left(3r^{-2-3}\right)^{-2}$$

$$= \left(3r^{-5}\right)^{-2}$$

$$= 3^{-2}r^{-5\bullet-2}$$

$$= \frac{1}{3^2}\bullet r^{10}$$

$$= \frac{1}{9}\bullet r^{10}$$

$$= \frac{r^{10}}{9}$$

CHALLENGE PROBLEMS

93. Multiply by the LCD of $(h+1)(h+2)$.

$$\frac{\dfrac{h}{h^2+3h+2}}{\dfrac{4}{h+2}-\dfrac{4}{h+1}} = \frac{\dfrac{h}{(h+1)(h+2)}}{\dfrac{4}{h+2}-\dfrac{4}{h+1}}$$

$$= \frac{\dfrac{h}{(h+1)(h+2)}}{\dfrac{4}{h+2}-\dfrac{4}{h+1}}\bullet\frac{(h+1)(h+2)}{(h+1)(h+2)}$$

$$= \frac{\dfrac{h}{(h+1)(h+2)}\bullet(h+1)(h+2)}{\dfrac{4}{h+2}\bullet(h+1)(h+2)-\dfrac{4}{h+1}\bullet(h+1)(h+2)}$$

$$= \frac{h}{4(h+1)-4(h+2)}$$

$$= \frac{h}{4h+4-4h-8}$$

$$= \frac{h}{-4}$$

$$= -\frac{h}{4}$$

95. Multiply the fraction by the LCD of $(a+1)$. Then add the two rational expressions by finding a common denominator.

$$a+\frac{a}{1+\dfrac{a}{a+1}} = a+\left(\frac{a}{1+\dfrac{a}{a+1}}\bullet\frac{a+1}{a+1}\right)$$

$$= a+\left(\frac{a(a+1)}{1(a+1)+\dfrac{a}{a+1}(a+1)}\right)$$

$$= a+\left(\frac{a^2+a}{a+1+a}\right)$$

$$= a+\frac{a^2+a}{2a+1}$$

$$= \frac{a}{1}\left(\frac{2a+1}{2a+1}\right)+\frac{a^2+a}{2a+1}$$

$$= \frac{2a^2+a}{2a+1}+\frac{a^2+a}{2a+1}$$

$$= \frac{2a^2+a+a^2+a}{2a+1}$$

$$= \frac{3a^2+2a}{2a+1}$$

SECTION 6.6

VOCABULARY

1. Equations that contain one or more rational expressions, such as $\dfrac{x}{x+2}=4+\dfrac{10}{x+2}$, are called **rational** equations.

3. To **clear** a rational equation of fractions, multiply both sides by the LCD of all rational expressions in the equation.

5. $x^2 - x + 2 = 0$ is a **quadratic** equation.

CONCEPTS

7. a) Yes.
$$\frac{1}{5-1} \overset{?}{=} 1 - \frac{3}{5-1}$$
$$\frac{1}{4} \overset{?}{=} 1 - \frac{3}{4}$$
$$\frac{1}{4} \overset{?}{=} \frac{4}{4} - \frac{3}{4}$$
$$\frac{1}{4} = \frac{1}{4}$$

    b) No, 5 is an extraneous solution because it makes the denominator 0.
$$\frac{5}{5-5} \overset{?}{=} 3 + \frac{5}{5-5}$$
$$\frac{5}{0} \overset{?}{=} 3 + \frac{5}{0}$$

9. a) 3 and 0
    b) 3 and 0
    c) 3 and 0

11. y

13. $(x + 8)(x - 8)$

15. $(x + 2)(x - 2)$

17. $12 = 1 + 10x$

NOTATION

19.
$$\frac{2}{a} + \frac{1}{2} = \frac{7}{2a}$$
$$\boxed{2a}\left(\frac{2}{a} + \frac{1}{2}\right) = \boxed{2a}\left(\frac{7}{2a}\right)$$
$$\boxed{2a}\left(\frac{2}{a}\right) + \boxed{2a}\left(\frac{1}{2}\right) = \boxed{2a}\left(\frac{7}{2a}\right)$$
$$\boxed{4} + a = \boxed{7}$$
$$4 + a - 4 = 7 - 4$$
$$\boxed{a = 3}$$

21. multiplication: $fpq = f \cdot p \cdot q$

PRACTICE

23. Multiply both sides of the equation by 6.
$$\frac{2}{3} = \frac{1}{2} + \frac{x}{6}$$
$$6\left(\frac{2}{3}\right) = 6\left(\frac{1}{2} + \frac{x}{6}\right)$$
$$6\left(\frac{2}{3}\right) = 6\left(\frac{1}{2}\right) + 6\left(\frac{x}{6}\right)$$
$$4 = 3 + x$$
$$4 - 3 = 3 + x - 3$$
$$1 = x$$
$$\boxed{x = 1}$$

Check :
$$\frac{2}{3} \overset{?}{=} \frac{1}{2} + \frac{1}{6}$$
$$\frac{2}{3} \overset{?}{=} \frac{3}{6} + \frac{1}{6}$$
$$\frac{2}{3} \overset{?}{=} \frac{4}{6}$$
$$\frac{2}{3} = \frac{2}{3}$$

25. Multiply both sides of the equation by 18.

$$\frac{x}{18} = \frac{1}{3} - \frac{x}{2}$$

$$18\left(\frac{x}{18}\right) = 18\left(\frac{1}{3} - \frac{x}{2}\right)$$

$$18\left(\frac{x}{18}\right) = 18\left(\frac{1}{3}\right) - 18\left(\frac{x}{2}\right)$$

$$x = 6 - 9x$$

$$x + 9x = 6 - 9x + 9x$$

$$10x = 6$$

$$\frac{10x}{10} = \frac{6}{10}$$

$$\boxed{x = \frac{3}{5}}$$

Check:

$$\frac{\frac{3}{5}}{18} \overset{?}{=} \frac{1}{3} - \frac{\frac{3}{5}}{2}$$

$$\frac{3}{5} \div 18 \overset{?}{=} \frac{1}{3} - \frac{3}{5} \div 2$$

$$\frac{3}{5} \bullet \frac{1}{18} \overset{?}{=} \frac{1}{3} - \frac{3}{5} \bullet \frac{1}{2}$$

$$\frac{1}{30} \overset{?}{=} \frac{1}{3} - \frac{3}{10}$$

$$\frac{1}{30} \overset{?}{=} \frac{10}{30} - \frac{9}{30}$$

$$\frac{1}{30} = \frac{1}{30}$$

27. Multiply both sides of the equation by 14.

$$\frac{a-1}{7} - \frac{a-2}{14} = \frac{1}{2}$$

$$14\left(\frac{a-1}{7} - \frac{a-2}{14}\right) = 14\left(\frac{1}{2}\right)$$

$$14\left(\frac{a-1}{7}\right) - 14\left(\frac{a-2}{14}\right) = 14\left(\frac{1}{2}\right)$$

$$2(a-1) - (a-2) = 7$$

$$2a - 2 - a + 2 = 7$$

$$(2a - a) + (-2 + 2) = 7$$

$$\boxed{a = 7}$$

Check:

$$\frac{7-1}{7} - \frac{7-2}{14} \overset{?}{=} \frac{1}{2}$$

$$\frac{6}{7} - \frac{5}{14} \overset{?}{=} \frac{1}{2}$$

$$\frac{12}{14} - \frac{5}{14} \overset{?}{=} \frac{1}{2}$$

$$\frac{7}{14} \overset{?}{=} \frac{1}{2}$$

$$\frac{1}{2} = \frac{1}{2}$$

29. Multiply both sides of the equation by x.

$$\frac{3}{x} + 2 = 3$$

$$x\left(\frac{3}{x} + 2\right) = x(3)$$

$$x\left(\frac{3}{x}\right) + x(2) = x(3)$$

$$3 + 2x = 3x$$

$$3 + 2x - 2x = 3x - 2x$$

$$3 = x$$

$$\boxed{x = 3}$$

Check:

$$\frac{3}{3} + 2 \overset{?}{=} 3$$

$$1 + 2 \overset{?}{=} 3$$

$$3 = 3$$

31. Multiply both sides of the equation by $(x-5)$.

$$\frac{x}{x-5} - \frac{5}{x-5} = 3$$

$$(x-5)\left(\frac{x}{x-5} - \frac{5}{x-5}\right) = (x-5)(3)$$

$$(x-5)\frac{x}{x-5} - (x-5)\frac{5}{x-5} = 3(x-5)$$

$$x - 5 = 3x - 15$$

$$x - 5 + 5 = 3x - 15 + 5$$

$$x = 3x - 10$$

$$x - 3x = 3x - 10 - 3x$$

$$-2x = -10$$

$$\frac{-2x}{-2} = \frac{-10}{-2}$$

$$\boxed{x = 5}$$

$$\boxed{\text{No solution; 5 is extraneous}}$$

Check:

$$\frac{5}{5-5} - \frac{5}{5-5} \overset{?}{=} 3$$

$$\frac{5}{0} - \frac{5}{0} \overset{?}{=} 3$$

If $x = 5$, the denominators of both fractions are 0 and the expressions are undefined; so 5 is extraneous and there are no solutions.

33. Multiply both sides of the equation by $4a$.

$$\frac{a}{4} - \frac{4}{a} = 0$$

$$4a\left(\frac{a}{4} - \frac{4}{a}\right) = 4a(0)$$

$$4a\left(\frac{a}{4}\right) - 4a\left(\frac{4}{a}\right) = 4a(0)$$

$$a^2 - 16 = 0$$

$$(a-4)(a+4) = 0$$

$$a - 4 = 0 \quad \text{and} \quad a + 4 = 0$$

$$a - 4 + 4 = 0 + 4 \quad a + 4 - 4 = 0 - 4$$

$$\boxed{a = 4 \qquad\qquad a = -4}$$

Check:

$$\frac{4}{4} - \frac{4}{4} \overset{?}{=} 0 \qquad\qquad \frac{-4}{4} - \frac{4}{-4} \overset{?}{=} 0$$

$$1 - 1 \overset{?}{=} 0 \qquad\qquad -1 - (-1) \overset{?}{=} 0$$

$$0 = 0 \qquad\qquad 0 = 0$$

35. Multiply both sides of the equation by $(y+1)$.

$$\frac{2}{y+1} + 5 = \frac{12}{y+1}$$

$$(y+1)\left(\frac{2}{y+1} + 5\right) = (y+1)\left(\frac{12}{y+1}\right)$$

$$(y+1)\bullet\frac{2}{y+1} + (y+1)\bullet 5 = (y+1)\bullet\frac{12}{y+1}$$

$$2 + 5(y+1) = 12$$

$$2 + 5y + 5 = 12$$

$$5y + 7 = 12$$

$$5y + 7 - 7 = 12 - 7$$

$$5y = 5$$

$$\frac{5y}{5} = \frac{5}{5}$$

$$\boxed{y = 1}$$

Check:

$$\frac{2}{1+1} + 5 \overset{?}{=} \frac{12}{1+1}$$

$$\frac{2}{2} + 5 \overset{?}{=} \frac{12}{2}$$

$$1 + 5 \overset{?}{=} 6$$

$$6 = 6$$

- 263 -

Section 6.6

37. Multiply both sides of the equation by b.

$$-\frac{1}{b} = \frac{5}{b} - 9 - \frac{6}{b}$$

$$b\left(-\frac{1}{b}\right) = b\left(\frac{5}{b} - 9 - \frac{6}{b}\right)$$

$$b\left(-\frac{1}{b}\right) = b\left(\frac{5}{b}\right) - b(9) - b\left(\frac{6}{b}\right)$$

$$-1 = 5 - 9b - 6$$

$$-1 = -1 - 9b$$

$$-1 + 1 = -1 - 9b + 1$$

$$0 = 9b$$

$$\frac{0}{9} = \frac{9b}{9}$$

$$\boxed{0 = b}$$

$$\boxed{\text{No solution; } 0 \text{ is extraneous}}$$

Check:

$$-\frac{1}{0} \overset{?}{=} \frac{5}{0} - 9 - \frac{6}{0}$$

39. Multiply both sides of the equation by $(t + 2)(t + 7)$.

$$\frac{1}{t+2} = \frac{t-1}{t+7}$$

$$(t+2)(t+7)\frac{1}{t+2} = (t+2)(t+7)\frac{t-1}{t+7}$$

$$1(t+7) = (t+2)(t-1)$$

$$t + 7 = t^2 - t + 2t - 2$$

$$t + 7 = t^2 + t - 2$$

$$t + 7 - t - 7 = t^2 + t - 2 - t - 7$$

$$0 = t^2 - 9$$

$$0 = (t-3)(t+3)$$

$$t - 3 = 0 \quad \text{and} \quad t + 3 = 0$$

$$\boxed{t = 3 \quad \text{and} \quad t = -3}$$

Check:

$$\frac{1}{3+2} \overset{?}{=} \frac{3-1}{3+7} \qquad \frac{1}{-3+2} \overset{?}{=} \frac{-3-1}{-3+7}$$

$$\frac{1}{5} \overset{?}{=} \frac{2}{10} \qquad \frac{1}{-1} \overset{?}{=} \frac{-4}{-4}$$

$$\frac{1}{5} = \frac{1}{5} \qquad -1 = -1$$

41. Multiply both sides of the equation by $12n$.

$$\frac{5}{n} + \frac{5}{12} = 0$$

$$12n\left(\frac{5}{n} + \frac{5}{12}\right) = 12n(0)$$

$$12n\left(\frac{5}{n}\right) + 12n\left(\frac{5}{12}\right) = 12n(0)$$

$$60 + 5n = 0$$

$$60 + 5n - 60 = 0 - 60$$

$$5n = -60$$

$$\frac{5n}{5} = \frac{-60}{5}$$

$$\boxed{n = -12}$$

Check:

$$\frac{5}{-12} + \frac{5}{12} \overset{?}{=} 0$$

$$0 = 0$$

43. Multiply both sides of the equation by $40y$.

$$\frac{1}{8} + \frac{2}{y} = \frac{1}{y} + \frac{1}{10}$$

$$40y\left(\frac{1}{8} + \frac{2}{y}\right) = 40y\left(\frac{1}{y} + \frac{1}{10}\right)$$

$$40y\left(\frac{1}{8}\right) + 40y\left(\frac{2}{y}\right) = 40y\left(\frac{1}{y}\right) + 40y\left(\frac{1}{10}\right)$$

$$5y + 80 = 40 + 4y$$

$$5y + 80 - 80 = 40 + 4y - 80$$

$$5y = -40 + 4y$$

$$5y - 4y = -40 + 4y - 4y$$

$$\boxed{y = -40}$$

Check:

$$\frac{1}{8} + \frac{2}{-40} \overset{?}{=} \frac{1}{-40} + \frac{1}{10}$$

$$\frac{5}{40} - \frac{2}{40} \overset{?}{=} -\frac{1}{40} + \frac{4}{40}$$

$$\frac{3}{40} = \frac{3}{40}$$

45. Multiply both sides of the equation by $24b$.

$$\frac{1}{8} + \frac{2}{b} - \frac{1}{12} = 0$$

$$24b\left(\frac{1}{8} + \frac{2}{b} - \frac{1}{12}\right) = 24b(0)$$

$$24b\left(\frac{1}{8}\right) + 24b\left(\frac{2}{b}\right) - 24b\left(\frac{1}{12}\right) = 24b(0)$$

$$3b + 48 - 2b = 0$$

$$b + 48 = 0$$

$$b + 48 - 48 = 0 - 48$$

$$\boxed{b = -48}$$

Check:

$$\frac{1}{8} + \frac{2}{-48} - \frac{1}{12} \overset{?}{=} 0$$

$$\frac{6}{48} - \frac{2}{48} - \frac{4}{48} \overset{?}{=} 0$$

$$0 = 0$$

47. Multiply both sides of the equation by $5x(x-5)$.

$$\frac{4}{5x} = \frac{8}{x-5}$$

$$5x(x-5) \bullet \frac{4}{5x} = 5x(x-5) \bullet \frac{8}{x-5}$$

$$4(x-5) = 5x(8)$$

$$4x - 20 = 40x$$

$$4x - 20 - 4x = 40x - 4x$$

$$-20 = 36x$$

$$\frac{-20}{36} = x$$

$$-\frac{5}{9} = x$$

$$\boxed{x = -\frac{5}{9}}$$

Check:

$$\frac{4}{5\left(-\frac{5}{9}\right)} \overset{?}{=} \frac{8}{-\frac{5}{9} - 5}$$

$$\frac{4}{-\frac{25}{9}} \overset{?}{=} \frac{8}{-\frac{50}{9}}$$

$$4 \div -\frac{25}{9} \overset{?}{=} 8 \div -\frac{50}{9}$$

$$-\frac{36}{25} = -\frac{36}{25}$$

49. Multiply both sides of the equation by $12a$.

$$\frac{1}{4} - \frac{5}{6} = \frac{1}{a}$$

$$12a\left(\frac{1}{4} - \frac{5}{6}\right) = 12a\left(\frac{1}{a}\right)$$

$$12a \bullet \frac{1}{4} - 12a \bullet \frac{5}{6} = 12a \bullet \frac{1}{a}$$

$$3a - 10a = 12$$

$$-7a = 12$$

$$\boxed{a = -\frac{12}{7}}$$

Check:

$$\frac{1}{4} - \frac{5}{6} \overset{?}{=} \frac{1}{-\frac{12}{7}}$$

$$\frac{3}{12} - \frac{10}{12} \overset{?}{=} 1 \div -\frac{12}{7}$$

$$-\frac{7}{12} = -\frac{7}{12}$$

51. Multiply both sides of the equation by $4h$.

$$\frac{3}{4h} + \frac{2}{h} = 1$$

$$4h\left(\frac{3}{4h} + \frac{2}{h}\right) = 4h(1)$$

$$4h \bullet \frac{3}{4h} + 4h \bullet \frac{2}{h} = 4h \bullet 1$$

$$3 + 8 = 4h$$

$$11 = 4h$$

$$\boxed{\frac{11}{4} = h}$$

Check:

$$\frac{3}{4\left(\frac{11}{4}\right)} + \frac{2}{\frac{11}{4}} \overset{?}{=} 1$$

$$\frac{3}{11} + \frac{8}{11} \overset{?}{=} 1$$

$$\frac{11}{11} = 1$$

53. Multiply both sides of the equation by $2r$.

$$\frac{3r}{2} - \frac{3}{r} = \frac{3r}{2} + 3$$

$$2r\left(\frac{3r}{2} - \frac{3}{r}\right) = 2r\left(\frac{3r}{2} + 3\right)$$

$$2r \bullet \frac{3r}{2} - 2r \bullet \frac{3}{r} = 2r \bullet \frac{3r}{2} + 2r \bullet 3$$

$$3r^2 - 6 = 3r^2 + 6r$$

$$3r^2 - 6 - 3r^2 = 3r^2 + 6r - 3r^2$$

$$-6 = 6r$$

$$\frac{-6}{6} = \frac{6r}{6}$$

$$-1 = r$$

$$\boxed{r = -1}$$

Check:

$$\frac{3(-1)}{2} - \frac{3}{-1} \overset{?}{=} \frac{3(-1)}{2} + 3$$

$$\frac{-3}{2} + \frac{3}{1} \overset{?}{=} \frac{-3}{2} + 3$$

$$\frac{-3}{2} + \frac{6}{2} \overset{?}{=} \frac{-3}{2} + \frac{6}{2}$$

$$\frac{3}{2} = \frac{3}{2}$$

55. Multiply both sides of the equation by $3(x - 3)$.

$$\frac{1}{3} + \frac{2}{x - 3} = 1$$

$$3(x - 3)\left(\frac{1}{3} + \frac{2}{x - 3}\right) = 3(x - 3)(1)$$

$$3(x - 3) \bullet \frac{1}{3} + 3(x - 3) \bullet \frac{2}{x - 3} = 3(x - 3) \bullet 1$$

$$x - 3 + 3 \bullet 2 = 3(x - 3)$$

$$x - 3 + 6 = 3x - 9$$

$$x + 3 = 3x - 9$$

$$x + 3 - 3 = 3x - 9 - 3$$

$$x = 3x - 12$$

$$x - 3x = 3x - 12 - 3x$$

$$-2x = -12$$

$$\frac{-2x}{-2} = \frac{-12}{-2}$$

$$\boxed{x = 6}$$

Check:

$$\frac{1}{3} + \frac{2}{6 - 3} \overset{?}{=} 1$$

$$\frac{1}{3} + \frac{2}{3} \overset{?}{=} 1$$

$$\frac{3}{3} = 1$$

57. Multiply both sides of the equation by $(z - 3)(z + 1)$.

$$\frac{z - 4}{z - 3} = \frac{z + 2}{x + 1}$$

$$(z - 3)(z + 1)\left(\frac{z - 4}{z - 3}\right) = (z - 3)(z + 1)\left(\frac{z + 2}{x + 1}\right)$$

$$(z + 1)(z - 4) = (z - 3)(z + 2)$$

$$z^2 - 4z + z - 4 = z^2 + 2z - 3z - 6$$

$$z^2 - 3z - 4 = z^2 - z - 6$$

$$z^2 - 3z - 4 - z^2 = z^2 - z - 6 - z^2$$

$$-3z - 4 = -z - 6$$

$$-3z - 4 + z = -z - 6 + z$$

$$-2z - 4 = -6$$

$$-2z - 4 + 4 = -6 + 4$$

$$-2z = -2$$

$$z = 1$$

Check:

$$\frac{1 - 4}{1 - 3} \overset{?}{=} \frac{1 + 2}{1 + 1}$$

$$\frac{-3}{-2} \overset{?}{=} \frac{3}{2}$$

$$\frac{3}{2} = \frac{3}{2}$$

59. Multiply both sides of the equation by $(v+2)(v-1)$.

$$\frac{v}{v+2}+\frac{1}{v-1}=1$$

$$(v+2)(v-1)\left(\frac{v}{v+2}+\frac{1}{v-1}\right)=(v+2)(v-1)(1)$$

$$v(v-1)+1(v+2)=(v+2)(v-1)$$

$$v^2-v+v+2=v^2-v+2v-2$$

$$v^2+2=v^2+v-2$$

$$v^2+2-v^2=v^2+v-2-v^2$$

$$2=v-2$$

$$2+2=v-2+2$$

$$\boxed{4=v}$$

Check:

$$\frac{4}{4+2}+\frac{1}{4-1}\overset{?}{=}1$$

$$\frac{4}{6}+\frac{1}{3}\overset{?}{=}1$$

$$\frac{4}{6}+\frac{2}{6}\overset{?}{=}1$$

$$\frac{6}{6}=1$$

61. Multiply both sides of the equation by $(a+2)$

$$\frac{a^2}{a+2}-\frac{4}{a+2}=a$$

$$(a+2)\left(\frac{a^2}{a+2}-\frac{4}{a+2}\right)=(a+2)(a)$$

$$(a+2)\left(\frac{a^2}{a+2}\right)-(a+2)\left(\frac{4}{a+2}\right)=(a+2)(a)$$

$$a^2-4=a^2+2a$$

$$a^2-4-a^2=a^2+2a-a^2$$

$$-4=2a$$

$$\frac{-4}{2}=\frac{2a}{2}$$

$$\boxed{-2=a}$$

$$\boxed{\text{No solution; -2 is extraneous}}$$

Check:

$$\frac{(-2)^2}{-2+2}-\frac{4}{-2+2}\overset{?}{=}-2$$

$$\frac{4}{0}-\frac{4}{0}\overset{?}{=}-2$$

63. Multiply both sides of the equation by $(x+4)$.

$$\frac{5}{x+4}+\frac{1}{x+4}=x-1$$

$$(x+4)\left(\frac{5}{x+4}+\frac{1}{x+4}\right)=(x+4)(x-1)$$

$$(x+4)\left(\frac{5}{x+4}\right)+(x+4)\left(\frac{1}{x+4}\right)=(x+4)(x-1)$$

$$5+1=x^2-x+4x-4$$

$$6=x^2+3x-4$$

$$6-6=x^2+3x-4-6$$

$$0=x^2+3x-10$$

$$0=(x+5)(x-2)$$

$$0=x+5 \quad \text{and} \quad 0=x-2$$

$$0-5=x+5-5 \quad 0+2=x-2+2$$

$$\boxed{-5=x \quad \text{and} \quad x=2}$$

Check:

$$\frac{5}{-5+4}+\frac{1}{-5+4}\overset{?}{=}-5-1 \qquad \frac{5}{2+4}+\frac{1}{2+4}\overset{?}{=}2-1$$

$$\frac{5}{-1}+\frac{1}{-1}\overset{?}{=}-6 \qquad \frac{5}{6}+\frac{1}{6}\overset{?}{=}1$$

$$-5-1\overset{?}{=}-6 \qquad \frac{6}{6}=1$$

$$-6=-6$$

65. Multiply both sides of the equation by $2(x+1)$.

$$\frac{3}{x+1} - \frac{x-2}{2} = \frac{x-2}{x+1}$$

$$2(x+1)\left(\frac{3}{x+1} - \frac{x-2}{2}\right) = 2(x+1)\left(\frac{x-2}{x+1}\right)$$

$$2(3) - (x+1)(x-2) = 2(x-2)$$

$$6 - \left(x^2 - 2x + x - 2\right) = 2x - 4$$

$$6 - x^2 + 2x - x + 2 = 2x - 4$$

$$-x^2 + x + 8 = 2x - 4$$

$$-x^2 + x + 8 + x^2 - x - 8 = 2x - 4 + x^2 - x - 8$$

$$0 = x^2 + x - 12$$

$$0 = (x-3)(x+4)$$

$$0 = x-3 \text{ and } 0 = x+4$$

$$\boxed{x=3 \quad \text{and} \quad x=-4}$$

Check:

$$\frac{3}{3+1} - \frac{3-2}{2} \overset{?}{=} \frac{3-2}{3+1} \qquad \frac{3}{-4+1} - \frac{-4-2}{2} \overset{?}{=} \frac{-4-2}{-4+1}$$

$$\frac{3}{4} - \frac{1}{2} \overset{?}{=} \frac{1}{4} \qquad \frac{3}{-3} - \frac{-6}{2} \overset{?}{=} \frac{-6}{-3}$$

$$\frac{1}{4} = \frac{1}{4} \qquad -1+3 \overset{?}{=} 2$$

$$2 = 2$$

67. Multiply both sides of the equation by $(b+3)(b-5)$.

$$\frac{b+2}{b+3} + 1 = \frac{-7}{b-5}$$

$$(b+3)(b-5)\left(\frac{b+2}{b+3}+1\right) = (b+3)(b-5)\left(\frac{-7}{b-5}\right)$$

$$(b-5)(b+2) + (b+3)(b-5) = -7(b+3)$$

$$b^2 + 2b - 5b - 10 + b^2 - 5b + 3b - 15 = -7b - 21$$

$$2b^2 - 5b - 25 = -7b - 21$$

$$2b^2 - 5b - 25 + 7b + 21 = -7b - 21 + 7b + 21$$

$$2b^2 + 2b - 4 = 0$$

$$2(b^2 + b - 2) = 0$$

$$2(b+2)(b-1) = 0$$

$$b+2=0 \text{ and } b-1=0$$

$$\boxed{b=-2 \text{ and } b=1}$$

Check:

$$\frac{-2+2}{-2+3} + 1 \overset{?}{=} \frac{-7}{-2-5} \qquad \frac{1+2}{1+3} + 1 \overset{?}{=} \frac{-7}{1-5}$$

$$\frac{0}{1} + 1 \overset{?}{=} \frac{-7}{-7} \qquad \frac{3}{4} + 1 \overset{?}{=} \frac{-7}{-4}$$

$$0 + 1 \overset{?}{=} 1 \qquad \frac{3}{4} + \frac{4}{4} \overset{?}{=} \frac{7}{4}$$

$$1 = 1 \qquad \frac{7}{4} = \frac{7}{4}$$

69. Multiply both sides of the equation by $u(u-1)$.

$$\frac{u}{u-1} + \frac{1}{u} = \frac{u^2+1}{u^2-u}$$

$$\frac{u}{u-1} + \frac{1}{u} = \frac{u^2+1}{u(u-1)}$$

$$u(u-1)\left(\frac{u}{u-1} + \frac{1}{u}\right) = u(u-1)\left(\frac{u^2+1}{u(u-1)}\right)$$

$$u(u) + 1(u-1) = u^2 + 1$$

$$u^2 + u - 1 = u^2 + 1$$

$$u^2 + u - 1 - u^2 = u^2 + 1 - u^2$$

$$u - 1 = 1$$

$$u - 1 + 1 = 1 + 1$$

$$\boxed{u = 2}$$

Check:

$$\frac{2}{2-1} + \frac{1}{2} \overset{?}{=} \frac{2^2+1}{2^2-2}$$

$$\frac{2}{1} + \frac{1}{2} \overset{?}{=} \frac{4+1}{4-2}$$

$$\frac{4}{2} + \frac{1}{2} \overset{?}{=} \frac{5}{2}$$

$$\frac{5}{2} = \frac{5}{2}$$

71. Multiply both sides of the equation by $(n-3)(n+3)$.

$$\frac{n}{n^2-9}+\frac{n+8}{n+3}=\frac{n-8}{n-3}$$

$$\frac{n}{(n-3)(n+3)}+\frac{n+8}{n+3}=\frac{n-8}{n-3}$$

$$(n+3)(n-3)\left(\frac{n}{(n+3)(n-3)}+\frac{n+8}{n+3}\right)=(n+3)(n-3)\left(\frac{n-8}{n-3}\right)$$

$$n+(n-3)(n+8)=(n+3)(n-8)$$

$$n+n^2+8n-3n-24=n^2-8n+3n-24$$

$$n^2+6n-24=n^2-5n-24$$

$$n^2+6n-24-n^2=n^2-5n-24-n^2$$

$$6n-24=-5n-24$$

$$6n-24+5n=-5n-24+5n$$

$$11n-24=-24$$

$$11n-24+24=-24+24$$

$$11n=0$$

$$\frac{11n}{11}=\frac{0}{11}$$

$$\boxed{n=0}$$

Check:

$$\frac{0}{0^2-9}+\frac{0+8}{0+3}\stackrel{?}{=}\frac{0-8}{0-3}$$

$$\frac{0}{-9}+\frac{8}{3}\stackrel{?}{=}\frac{-8}{-3}$$

$$\frac{8}{3}=\frac{8}{3}$$

73. Multiply both sides of the equation by $x(x-1)$.

$$\frac{x}{x-1}-\frac{12}{x^2-x}=\frac{-1}{x-1}$$

$$\frac{x}{x-1}-\frac{12}{x(x-1)}=\frac{-1}{x-1}$$

$$x(x-1)\left(\frac{x}{x-1}-\frac{12}{x(x-1)}\right)=x(x-1)\left(\frac{-1}{x-1}\right)$$

$$x(x)-12=x(-1)$$

$$x^2-12=-x$$

$$x^2-12+x=-x+x$$

$$x^2+x-12=0$$

$$(x+4)(x-3)=0$$

$$x+4=0 \quad \text{and} \quad x-3=0$$

$$x+4-4=0-4 \quad x-3+3=0+3$$

$$\boxed{x=-4 \quad \text{and} \quad x=3}$$

Check:

$$\frac{-4}{-4-1}-\frac{12}{(-4)^2-(-4)}\stackrel{?}{=}\frac{-1}{-4-1} \qquad \frac{3}{3-1}-\frac{12}{(3)^2-3}\stackrel{?}{=}\frac{-1}{3-1}$$

$$\frac{-4}{-5}-\frac{12}{16+4}\stackrel{?}{=}\frac{-1}{-5} \qquad \frac{3}{2}-\frac{12}{9-3}\stackrel{?}{=}\frac{-1}{2}$$

$$\frac{4}{5}-\frac{12}{20}\stackrel{?}{=}\frac{1}{5} \qquad \frac{3}{2}-\frac{12}{6}\stackrel{?}{=}-\frac{1}{2}$$

$$\frac{16}{20}-\frac{12}{20}\stackrel{?}{=}\frac{1}{5} \qquad \frac{9}{6}-\frac{12}{6}\stackrel{?}{=}-\frac{1}{2}$$

$$\frac{4}{20}=\frac{1}{5} \qquad -\frac{3}{6}=-\frac{1}{2}$$

75. Multiply both sides of the equation by $b(b+3)$.

$$1-\frac{3}{b}=\frac{-8b}{b^2+3b}$$

$$1-\frac{3}{b}=\frac{-8b}{b(b+3)}$$

$$b(b+3)\left(1-\frac{3}{b}\right)=b(b+3)\left(\frac{-8b}{b(b+3)}\right)$$

$$b(b+3)-3(b+3)=-8b$$

$$b^2+3b-3b-9=-8b$$

$$b^2-9=-8b$$

$$b^2-9+8b=-8b+8b$$

$$b^2+8b-9=0$$

$$(b+9)(b-1)=0$$

$$b+9=0 \quad \text{and} \quad b-1=0$$

$$b+9-9=0-9 \quad \text{and} \quad b-1+1=0+1$$

$$\boxed{b=-9 \quad \text{and} \quad b=1}$$

Check:

$$1-\frac{3}{-9}\stackrel{?}{=}\frac{-8(-9)}{(-9)^2+3(-9)} \qquad 1-\frac{3}{1}\stackrel{?}{=}\frac{-8(1)}{(1)^2+3(1)}$$

$$1+\frac{1}{3}\stackrel{?}{=}\frac{72}{81-27} \qquad 1-3\stackrel{?}{=}\frac{-8}{1+3}$$

$$\frac{4}{3}\stackrel{?}{=}\frac{72}{54} \qquad -2\stackrel{?}{=}\frac{-8}{4}$$

$$\frac{4}{3}=\frac{4}{3} \qquad -2=-2$$

77. Multiply both sides of the equation by $(x-1)(x+9)$.

$$\frac{3}{x-1} - \frac{1}{x+9} = \frac{18}{x^2+8x-9}$$

$$\frac{3}{x-1} - \frac{1}{x+9} = \frac{18}{(x-1)(x+9)}$$

$$(x-1)(x+9)\left(\frac{3}{x-1} - \frac{1}{x+9}\right) = (x-1)(x+9)\left(\frac{18}{(x-1)(x+9)}\right)$$

$$3(x+9) - (x-1) = 18$$

$$3x + 27 - x + 1 = 18$$

$$2x + 28 = 18$$

$$2x + 28 - 28 = 18 - 28$$

$$2x = -10$$

$$\frac{2x}{2} = \frac{-10}{2}$$

$$\boxed{x = -5}$$

Check:

$$\frac{3}{-5-1} - \frac{1}{-5+9} \overset{?}{=} \frac{18}{(-5)^2 + 8(-5) - 9}$$

$$\frac{3}{-6} - \frac{1}{4} \overset{?}{=} \frac{18}{25 - 40 - 9}$$

$$-\frac{6}{12} - \frac{3}{12} \overset{?}{=} \frac{18}{-24}$$

$$-\frac{9}{12} \overset{?}{=} -\frac{18}{24}$$

$$-\frac{3}{4} = -\frac{3}{4}$$

79. Muliply both sides of the equation by n.

$$\frac{P}{n} = rt$$

$$n\left(\frac{P}{n}\right) = n(rt)$$

$$\boxed{P = nrt}$$

81. Multiply both sides of the equation by bd.

$$\frac{a}{b} = \frac{c}{d}$$

$$bd\left(\frac{a}{b}\right) = bd\left(\frac{c}{d}\right)$$

$$da = bc$$

$$\frac{da}{a} = \frac{bc}{a}$$

$$\boxed{d = \frac{bc}{a}}$$

83. Multiply both sides of the equation by $(b+d)$.

$$h = \frac{2A}{b+d}$$

$$(b+d)h = (b+d)\left(\frac{2A}{b+d}\right)$$

$$h(b+d) = 2A$$

$$\frac{h(b+d)}{2} = \frac{2A}{2}$$

$$\boxed{\frac{h(b+d)}{2} = A}$$

85. Multiply both sides of the equation by ab.

$$\frac{1}{a} + \frac{1}{b} = 1$$

$$(ab)\frac{1}{a} + (ab)\frac{1}{b} = (ab)1$$

$$b + a = ab$$

$$b + a - a = ab - a$$

$$b = ab - a$$

$$b = a(b-1)$$

$$\frac{b}{b-1} = \frac{a(b-1)}{b-1}$$

$$\boxed{\frac{b}{b-1} = a}$$

87. Multiply both sides of the equation by $(R + r)$.

$$I = \frac{E}{R + r}$$

$$(R + r)I = (R + r)\frac{E}{R + r}$$

$$I(R + r) = E$$

$$IR + Ir = E$$

$$IR + Ir - IR = E - IR$$

$$Ir = E - IR$$

$$\frac{Ir}{I} = \frac{E - IR}{I}$$

$$\boxed{r = \frac{E - IR}{I}}$$

89. Multiply both sides of the equation by rst.

$$\frac{1}{r} + \frac{1}{s} = \frac{1}{t}$$

$$(rst)\frac{1}{r} + (rst)\frac{1}{s} = (rst)\frac{1}{t}$$

$$st + rt = rs$$

$$st + rt - rt = rs - rt$$

$$st = rs - rt$$

$$st = r(s - t)$$

$$\frac{st}{s - t} = \frac{r(s - t)}{s - t}$$

$$\boxed{\frac{st}{s - t} = r}$$

91. Multiply both sides of the equation by $6d$.

$$F = \frac{L^2}{6d} + \frac{d}{2}$$

$$(6d)F = (6d)\frac{L^2}{6d} + (6d)\frac{d}{2}$$

$$6dF = L^2 + 3d^2$$

$$6dF - 3d^2 = L^2 + 3d^2 - 3d^2$$

$$\boxed{6dF - 3d^2 = L^2}$$

APPLICATIONS

93. MEDICINE

Multiply both sides of the equation by $(R + B)$.

$$H = \frac{RB}{R + B}$$

$$(R + B)H = (R + B)\frac{RB}{R + B}$$

$$HR + HB = RB$$

$$HR + HB - HR = RB - HR$$

$$HB = RB - HR$$

$$HB = R(B - H)$$

$$\boxed{\frac{HB}{B - H} = R}$$

95. ELECTRONICS

Multiply both sides of the equation by (rr_1r_2).

$$\frac{1}{r} = \frac{1}{r_1} + \frac{1}{r_2}$$

$$(rr_1r_2)\frac{1}{r} = (rr_1r_2)\frac{1}{r_1} + (rr_1r_2)\frac{1}{r_2}$$

$$r_1r_2 = rr_2 + rr_1$$

$$r_1r_2 = r(r_2 + r_1)$$

$$\frac{r_1r_2}{r_2 + r_1} = \frac{r(r_2 + r_1)}{r_2 + r_1}$$

$$\boxed{\frac{r_1r_2}{r_2 + r_1} = r}$$

WRITING

97. Answers will vary.

99. Answers will vary.

REVIEW

101. Factor out the GCF.
$$x^2 + 4x = x(x + 4)$$

103. Factor using FOIL
$$2x^2 + x - 3 = (2x + 3)(x - 1)$$

Section 6.6

105. Factor Difference of Two Squares

$$x^4 - 81 = (x^2 + 9)(x^2 - 9)$$
$$= (x^2 + 9)(x + 3)(x - 3)$$

CHALLENGE PROBLEMS

107. Rewrite the equation without negative exponents. Then multiply both sides of the equation by x^2.

$$x^{-2} + 2x^{-1} + 1 = 0$$

$$\frac{1}{x^2} + \frac{2}{x} + 1 = 0$$

$$\left(x^2\right)\frac{1}{x^2} + \left(x^2\right)\frac{2}{x} + \left(x^2\right) \cdot 1 = \left(x^2\right)0$$

$$1 + 2x + x^2 = 0$$

$$x^2 + 2x + 1 = 0$$

$$(x + 1)(x + 1) = 0$$

$$x + 1 = 0$$

$$x + 1 - 1 = 0 - 1$$

$$\boxed{x = -1}$$

VOCABULARY

1. In the formula $d = rt$, the variable d stands for the **distance** traveled, r is the **rate**, and t is the **time**.

3. In the formula $I = Prt$, the variable I stands for the amount of **interest** earned, P is the **principal**, r stands for the annual interest **rate**, and t is the **time**.

CONCEPTS

5. iii. $\dfrac{5+x}{8+x} = \dfrac{2}{3}$

7. Divide both sides by r.
$$d = rt$$
$$\frac{d}{r} = \frac{rt}{r}$$
$$\frac{d}{r} = t$$
$$t = \frac{d}{r}$$

9. a) $9\% = 0.09$
    b) $0.035 = 3.5\%$

11. $\dfrac{1}{3}$

13. $\dfrac{x}{4}$

15. City savings: $\dfrac{50}{r} \cdot r \cdot 1 = 50$

    Credit union: $\left(\dfrac{75}{r - 0.02}\right)(r - 0.02)(1) = 75$

NOTATION

17. $\dfrac{55}{9} = 6\dfrac{2}{9}$ days

PRACTICE

19. NUMBER PROBLEM

Let the number be x.

$$\frac{2+x}{5+x} = \frac{2}{3}$$
$$3(5+x)\frac{2+x}{5+x} = 3(5+x)\frac{2}{3}$$
$$3(2+x) = 2(5+x)$$
$$6 + 3x = 10 + 2x$$
$$6 + 3x - 2x = 10 + 2x - 2x$$
$$6 + x = 10$$
$$6 + x - 6 = 10 - 6$$
$$\boxed{x = 4}$$

21. NUMBER PROBLEM

Let the number be x.

$$\frac{3 \bullet 2}{4 + x} = 1$$
$$(4 + x)\frac{6}{4 + x} = 1(4 + x)$$
$$6 = 4 + x$$
$$6 - 4 = 4 + x - 4$$
$$\boxed{2 = x}$$

23. NUMBER PROBLEM

Let the number be x.

$$\frac{3 + x}{4 + 2x} = \frac{4}{7}$$
$$7(4 + 2x)\frac{3 + x}{4 + 2x} = 7(4 + 2x)\frac{4}{7}$$
$$7(3 + x) = 4(4 + 2x)$$
$$21 + 7x = 16 + 8x$$
$$21 + 7x - 7x = 16 + 8x - 7x$$
$$\boxed{21} = 16 + x$$
$$21 - 16 = 16 + x - 16$$
$$\boxed{5 = x}$$

25. NUMBER PROBLEM

Let the number be x.

$$x + \frac{1}{x} = \frac{13}{6}$$

$$(6x)\left(x + \frac{1}{x}\right) = (6x)\left(\frac{13}{6}\right)$$

$$(6x)(x) + (6x)\left(\frac{1}{x}\right) = (6x)\left(\frac{13}{6}\right)$$

$$6x^2 + 6 = 13x$$

$$6x^2 + 6 - 13x = 13x - 13x$$

$$6x^2 - 13x + 6 = 0$$

$$(2x - 3)(3x - 2) = 0$$

$$2x - 3 = 0 \quad \text{and} \quad 3x - 2 = 0$$

$$2x = 3 \quad \text{and} \quad 3x = 2$$

$$\boxed{x = \frac{3}{2} \text{ and } x = \frac{2}{3}}$$

APPLICATIONS

27. COOKING

$$\frac{1+x}{4+x} = \frac{3}{4}$$

$$4(4+x)\left(\frac{1+x}{4+x}\right) = 4(4+x)\left(\frac{3}{4}\right)$$

$$4(1+x) = 3(4+x)$$

$$4 + 4x = 12 + 3x$$

$$4 + 4x - 3x = 12 + 3x - 3x$$

$$4 + x = 12$$

$$4 + x - 4 = 12 - 4$$

$$\boxed{x = 8}$$

29. FILLING A POOL

Let x be the number of hours it takes both pipes to fill the pool working together.

Amount of work 1st pipe in 1 hour = $\frac{1}{5}$

Amount of work 2nd pipe in 1 hour = $\frac{1}{4}$

Amount of work together in 1 hour = $\frac{1}{x}$

1st pipe + 2nd pipe = work together

$$\frac{1}{5} + \frac{1}{4} = \frac{1}{x}$$

$$20x\left(\frac{1}{5} + \frac{1}{4}\right) = 20x\left(\frac{1}{x}\right)$$

$$4x + 5x = 20$$

$$9x = 20$$

$$x = \frac{20}{9}$$

$$\boxed{x = 2\frac{2}{9}}$$

It will take both $2\frac{2}{9}$ hours.

31. HOLIDAY DECORATING

Let x be the number of hours it takes both crews to put up the decorations working together.

1st Crew's work in 1 hour = $\dfrac{1}{8}$

2nd Crew's work in 1 hour = $\dfrac{1}{10}$

Amount of work together in 1 hour = $\dfrac{1}{x}$

1st crew + 2nd crew = work together

$$\frac{1}{8} + \frac{1}{10} = \frac{1}{x}$$

$$40x\left(\frac{1}{8} + \frac{1}{10}\right) = 40x\left(\frac{1}{x}\right)$$

$$5x + 4x = 40$$

$$9x = 40$$

$$x = \frac{40}{9}$$

$$\boxed{x = 4\frac{4}{9}}$$

It will take both $4\dfrac{4}{9}$ hours.

33. FILLING A POOL

Let x be the number of hours it takes both pipes to fill the pool working together. Since the drain is EMPTYING the pool, subtract the work of the drain from the work of the pipe.

Amount of work pipe in 1 hour = $\dfrac{1}{4}$

Amount of work drain in 1 hour = $\dfrac{1}{8}$

Amount of work together in 1 hour = $\dfrac{1}{x}$

pipe's work - drain's work = work together

$$\frac{1}{4} - \frac{1}{8} = \frac{1}{x}$$

$$8x\left(\frac{1}{4} - \frac{1}{8}\right) = 8x\left(\frac{1}{x}\right)$$

$$2x - x = 8$$

$$\boxed{x = 8}$$

It will take the pipe 8 hours.

35. GRADING PAPERS

Let x be the number of minutes it takes the teacher and the aide to grade the papers working together.

Teacher's work in 1 minute = $\dfrac{1}{30}$

Aide's work in 1 minute = $\dfrac{1}{2(30)} = \dfrac{1}{60}$

Amount of work together in 1 minute = $\dfrac{1}{x}$

$$\frac{1}{30} + \frac{1}{60} = \frac{1}{x}$$

$$60x\left(\frac{1}{30} + \frac{1}{60}\right) = 60x\left(\frac{1}{x}\right)$$

$$2x + x = 60$$

$$3x = 60$$

$$\frac{3x}{3} = \frac{60}{3}$$

$$\boxed{x = 20}$$

It will take both 20 minutes.

37. PRINTERS

Let x be the number of hours it takes both printers to print the schedules working together.

Slow printer's work in 1 hour = $\dfrac{1}{6}$

Old printer's work in 1 hour = $\dfrac{1}{4}$

$\dfrac{3}{4}$ of work together in 1 hour = $\dfrac{3}{4} \bullet \dfrac{1}{x} = \dfrac{3}{4x}$

$$\frac{1}{6} + \frac{1}{4} = \frac{3}{4x}$$

$$12x\left(\frac{1}{6} + \frac{1}{4}\right) = 12x\left(\frac{3}{4x}\right)$$

$$2x + 3x = 9$$

$$5x = 9$$

$$x = \frac{9}{5}$$

$$x = 1\frac{4}{5}$$

$$\boxed{x = 1.8}$$

It will take both 1.8 hours.

39. PHYSICAL FITNESS

Use $t = \dfrac{d}{r}$ to find the time for both riding and walking. Let the rate walking = r. Then the rate riding would be $(r + 10)$ since it is 10 mph faster.

Riding:
$$t = \frac{d}{r}$$
$$= \frac{28}{r + 10}$$

Walking:
$$t = \frac{d}{r}$$
$$= \frac{8}{r}$$

Since the time is the same for both riding and walking, set both times equal to each other and solve the equation for r.

$$\frac{28}{r + 10} = \frac{8}{r}$$
$$r(r + 10)\left(\frac{28}{r + 10}\right) = r(r + 10)\left(\frac{8}{r}\right)$$
$$28r = 8(r + 10)$$
$$28r = 8r + 80$$
$$28r - 8r = 8r + 80 - 8r$$
$$20r = 80$$
$$\frac{20r}{20} = \frac{80}{20}$$
$$\boxed{r = 4}$$

Her rate walking is 4 mph.

41. PACKING FRUIT

Use $t = \dfrac{d}{r}$ to find the time for both conveyor belts. Let the rate of the 1st belt be r, then the rate of the 2nd belt (1 ft per second slower) would be $r - 1$.

1st belt:
$$t = \frac{d}{r}$$
$$= \frac{300}{r}$$

2nd belt:
$$t = \frac{d}{r}$$
$$= \frac{10}{r - 1}$$

Set both times equal to each other and solve the equation for r.

$$\frac{300}{r} = \frac{100}{r - 1}$$
$$r(r - 1)\left(\frac{300}{r}\right) = r(r - 1)\left(\frac{100}{r - 1}\right)$$
$$300(r - 1) = 100r$$
$$300r - 300 = 100r$$
$$300r - 300 - 300r = 100r - 300r$$
$$-300 = -200r$$
$$\frac{-300}{-200} = \frac{-200r}{-200}$$
$$\frac{3}{2} = r$$
$$r = \frac{3}{2}$$
$$\boxed{\begin{array}{l} r = 1.5 \\ r - 1 = 1.5 - 1 \\ \quad = 0.5 \end{array}}$$

The rate of the 1st conveyor belt is 1.5 mph or $1\frac{1}{2}$ mph, and the rate of the 2nd conveyor belt would be $1.5 - 1 = 0.5$ mph or $\frac{1}{2}$ mph.

43. WIND SPEED

Use $t = \dfrac{d}{r}$ to find the time for both planes.

Let x be the rate of the wind. The rate downwind would be $255 + x$ and the rate of the plane upwind would be $255 - x$.

Downwind: $t = \dfrac{d}{r}$

$$= \dfrac{300}{255 + x}$$

Upwind: $t = \dfrac{d}{r}$

$$= \dfrac{210}{255 - x}$$

Since the flying time is the same for both planes, set both times equal to each other and solve the equation for x.

$$\frac{300}{255+x} = \frac{210}{255-x}$$

$$(255+x)(255-x)\left(\frac{300}{255+x}\right) = (255+x)(255-x)\left(\frac{210}{255-x}\right)$$

$$300(255-x) = 210(255+x)$$

$$76{,}500 - 300x = 53{,}550 + 210x$$

$$76{,}500 - 300x - 210x = 53{,}550 + 210x - 210x$$

$$76{,}500 - 510x = 53{,}550$$

$$76{,}500 - 510x - 76{,}500 = 53{,}550 - 76{,}500$$

$$-510x = -22{,}950$$

$$\frac{-510x}{-510} = \frac{-22{,}950}{-510}$$

$$\boxed{x = 45}$$

The rate of the wind is 45 mph.

45. COMPARING INVESTMENTS

Use $P = \dfrac{I}{rt}$ to find the rate of interest paid by each investment. Let x be the rate of the interest paid by the tax-free bonds. Then the rate of the interest paid by the credit union would be $x - 0.02$ (2% less than). Let $t = 1$ year.

Bonds: $P = \dfrac{I}{rt}$

$$= \dfrac{300}{x(1)}$$

$$= \dfrac{300}{x}$$

Credit Union: $P = \dfrac{I}{rt}$

$$= \dfrac{200}{(x-0.02)(1)}$$

$$= \dfrac{200}{x-0.02}$$

Since the principal is the same for both investments, set both rational expressions equal to each other and solve the equation for x.

$$\frac{300}{x} = \frac{200}{x - 0.02}$$

$$x(x-0.02)\left(\frac{300}{x}\right) = x(x-0.02)\left(\frac{200}{x-0.02}\right)$$

$$300(x - 0.02) = 200x$$

$$300x - 6 = 200x$$

$$300x - 6 - 300x = 200x - 300x$$

$$-6 = -100x$$

$$\frac{-6}{-100} = x$$

$$0.06 = x$$

$$\boxed{x = 6\%}$$

rate of interest of bonds = 6%
rate of interest of credit union = 6–2
= 4%

use $P = \dfrac{I}{rt}$ to find the rate of interest paid each investment. Let x be the rate of the interest paid by the 1st CD. Then the rate of the interest paid by the 2nd CD would be $x + 0.01$ (differ by 1%). Let $t = 1$ year.

1st CD: $\quad P = \dfrac{I}{rt}$

$\qquad = \dfrac{175}{x(1)}$

$\qquad = \dfrac{175}{x}$

2nd CD: $\quad P = \dfrac{I}{rt}$

$\qquad = \dfrac{200}{(x+0.01)(1)}$

$\qquad = \dfrac{200}{x+0.01}$

Since the principal is the same for both accounts, set both rational expressions equal to each other and solve the equation for x.

$$\dfrac{175}{x} = \dfrac{200}{x+0.01}$$

$$x(x+0.01)\left(\dfrac{175}{x}\right) = x(x+0.01)\left(\dfrac{200}{x+0.01}\right)$$

$$175(x + 0.01) = 200x$$

$$175x + 1.75 = 200x$$

$$175x + 1.75 - 175x = 200x - 175x$$

$$1.75 = 25x$$

$$\dfrac{1.75}{25} = x$$

$$0.07 = x$$

$$\boxed{7\% = x}$$

The rate of interest of the 1st CD is 7%.
The rate of interest of 2nd CD is $7 + 1 = 8\%$.

WRITING

49. Answers will vary.

REVIEW

51. repeating; $\quad \dfrac{7}{9} = 9\overline{)7.000}^{\,0.777} = 0.\overline{7}$

53. $\{\ldots, -4, -3, -2, -1, 0, 1, 2, 3, 4, \ldots\}$

55. $2(\mathbf{-3})^2 + 5(\mathbf{-3}) - 3 \quad = 2(9) + 5(-3) - 3$
$\qquad\qquad\qquad\qquad\quad = 18 - 15 - 3$
$\qquad\qquad\qquad\qquad\quad = 3 - 3$
$\qquad\qquad\qquad\qquad\quad = 0$

CHALLENGE PROBLEMS

57. RIVER TOURS

Use $t = \dfrac{d}{r}$ to find the time for both trips.

Let x equal the rate of the boat in still water. The rate of the boat downstream would be $x + 5$ and the rate of the boat upstream would be $x - 5$. $d = 60$.

Downstream: $t = \dfrac{d}{r}$

$= \dfrac{60}{x+5}$

Upstream: $t = \dfrac{d}{r}$

$= \dfrac{60}{x-5}$

Since the total time is 5, find the sum of the times and set it equal to 5.

$$\frac{60}{x+5} + \frac{60}{x-5} = 5$$

$$(x+5)(x-5)\left(\frac{60}{x+5} + \frac{60}{x-5}\right) = (x+5)(x-5)(5)$$

$$60(x-5) + 60(x+5) = 5(x^2 - 25)$$

$$60x - 300 + 60x + 300 = 5x^2 - 125$$

$$120x = 5x^2 - 125$$

$$120x - 120x = 5x^2 - 125 - 120x$$

$$0 = 5x^2 - 120x - 125$$

$$\frac{0}{5} = \frac{5x^2}{5} - \frac{120x}{5} - \frac{125}{5}$$

$$0 = x^2 - 24x - 25$$

$$0 = (x-25)(x+1)$$

$$x - 25 = 0 \quad \text{and} \quad x + 1 = 0$$

$$x = 25 \quad \text{and} \quad x = -1$$

The rate cannot be negative, so $\boxed{x = 25}$.
The still-water rate of the boat is 25 mph.

59. SALES

Let x = the number of radios bought.

Cost of each radio $= \dfrac{1{,}200}{x}$

$x - 6$ = the number of radios sold

Cost of each radio $+ \$10 = \dfrac{1{,}200}{x-6}$

Cost of each radio$+\$10-\$10 = \dfrac{1{,}200}{x-6} - \$10$

Cost of each radio $= \dfrac{1{,}200}{x-6} - 10$

Set the two expressions equal to each other and solve for x.

$$\frac{1{,}200}{x} = \frac{1{,}200}{x-6} - 10$$

$$x(x-6)\left(\frac{1{,}200}{x}\right) = x(x-6)\left(\frac{1{,}200}{x-6} - 10\right)$$

$$1{,}200(x-6) = 1{,}200x - 10x(x-6)$$

$$1{,}200x - 7{,}200 = 1{,}200x - 10x^2 + 60x$$

$$1{,}200x - 7{,}200 - 1{,}200x = 1{,}200x - 10x^2 + 60x - 1{,}200x$$

$$-7{,}200 = -10x^2 + 60x$$

$$-7{,}200 + 10x^2 - 60x = -10x^2 + 60x + 10x^2 - 60x$$

$$10x^2 - 60x - 7{,}200 = 0$$

$$10(x^2 - 6x - 720) = 0$$

$$10(x-30)(x+24) = 0$$

$$x - 30 = 0 \quad \text{and} \quad x + 24 = 0$$

$$\boxed{x = 30} \text{ and } \quad x = -24$$

Since x cannot be negative, $x = 30$.
She bought 30 radios.

SECTION 6.8

VOCABULARY

1. A **ratio** is the quotient of two numbers or the quotient of two quantities with the same units. A **rate** is a quotient of two quantities that have different units.

3. The **terms** of the proportion $\frac{2}{x} = \frac{16}{40}$ are 2, x, 16, and 40.

5. The product of the extremes and the product of the means of a proportion are also known as **cross** products.

7. Examples of **unit** prices are $1.65 per gallon, 17¢ per day, and $50 per foot.

CONCEPTS

9. WEST AFRICA

$$\frac{\text{red stripes}}{\text{white stripes}} = \frac{6}{5}$$

11. a) $6(10) = 60$
    $5(12) = 60$
    b) $15(x) = 15x$
    $2(45) = 90$

13. SNACK FOODS

$$\frac{\boxed{25}}{\boxed{2}} = \frac{\boxed{1,000}}{\boxed{x}}$$

15. GROCERY SHOPPING

The 24 twelve ounce bottles would be the better buy.

12 - eight oz bottles =

$$\frac{\text{price}}{\text{number of ounces}} = \frac{1.79}{12(8)}$$

$$= \frac{1.79}{96}$$

$$= 1.9 \text{ cents per oz}$$

24 - twelve oz bottles =

$$\frac{\text{price}}{\text{number of ounces}} = \frac{4.49}{24(12)}$$

$$= \frac{4.49}{288}$$

$$= 1.6 \text{ cents per oz}$$

17. $\dfrac{AB}{DE} = \dfrac{\boxed{BC}}{EF}$

$\dfrac{BC}{\boxed{EF}} = \dfrac{CA}{FD}$

$\dfrac{CA}{FD} = \dfrac{AB}{\boxed{DE}}$

NOTATION

19. $\dfrac{12}{18} = \dfrac{x}{24}$

$12 \cdot 24 = 18 \cdot \boxed{x}$

$\boxed{288} = 18x$

$\dfrac{288}{\boxed{18}} = \dfrac{18x}{\boxed{18}}$

$\boxed{16} = x$

21. a) $\dfrac{12}{15} = \dfrac{4}{5}$
    b) $\dfrac{9 \text{ crates}}{7 \text{ crates}} = \dfrac{9}{7}$

23. We read $\triangle XYZ \sim \triangle MNO$ as: triangle XYZ is **similar** to triangle MNO.

PRACTICE

25. $\dfrac{4}{15}$

27. $\dfrac{30}{24} = \dfrac{5}{4}$

29. 60 minutes = 1 hour

$$\dfrac{90 \text{ minutes}}{3 \text{ hours}} = \dfrac{90 \text{ minutes}}{3(60) \text{ minutes}}$$
$$= \dfrac{90}{180}$$
$$= \dfrac{1}{2}$$

31. 4 quarts = 1 gallon

$$\dfrac{13 \text{ quarts}}{2 \text{ gallons}} = \dfrac{13 \text{ quarts}}{2(4) \text{ quarts}}$$
$$= \dfrac{13}{8}$$

33. no; $9(70) \overset{?}{=} 7(81)$
$630 \neq 567$

35. yes; $7(6) \overset{?}{=} 3(14)$
$42 = 42$

37. no; $9(80) \overset{?}{=} 19(38)$
$720 \neq 722$

39. $2(6) = 3 \cdot x$
$12 = 3x$
$\dfrac{12}{3} = \dfrac{3x}{3}$
$\boxed{4 = x}$

41. $5 \cdot c = 10(3)$
$5c = 30$
$\dfrac{5c}{5} = \dfrac{30}{5}$
$\boxed{c = 6}$

43. $6(4) = 8 \cdot x$
$24 = 8x$
$\dfrac{24}{8} = \dfrac{8x}{8}$
$\boxed{3 = x}$

45. $(x + 1) \cdot 15 = 5 \cdot 3$
$15(x + 1) = 15$
$15x + 15 = 15$
$15x + 15 - 15 = 15 - 15$
$15x = 0$
$\dfrac{15x}{15} = \dfrac{0}{15}$
$\boxed{x = 0}$

47. $(x + 7) \cdot 4 = -4 \cdot 1$
$4(x + 7) = -4$
$4x + 28 = -4$
$4x + 28 - 28 = -4 - 28$
$4x = -32$
$\dfrac{4x}{4} = \dfrac{-32}{4}$
$\boxed{x = -8}$

49. $(5 - x) \cdot 34 = 17 \cdot 13$
$34(5 - x) = 221$
$170 - 34x = 221$
$170 - 34x - 170 = 221 - 170$
$-34x = 51$
$\dfrac{-34x}{-34} = \dfrac{51}{-34}$
$\boxed{x = -\dfrac{3}{2}}$

51. $(2x - 1) \cdot 54 = 18 \cdot 9$
$54(2x - 1) = 162$
$108x - 54 = 162$
$108x - 54 + 54 = 162 + 54$
$108x = 216$
$\dfrac{108x}{108} = \dfrac{216}{108}$
$\boxed{x = 2}$

53. $(x - 1) \cdot 3 = 9 \cdot 2x$
$3(x - 1) = 18x$
$3x - 3 = 18x$
$3x - 3 - 3x = 18x - 3x$
$-3 = 15x$
$\dfrac{-3}{15} = \dfrac{15x}{15}$
$\boxed{-\dfrac{1}{5} = x}$

55.
$$8x \cdot 4 = 3 \cdot (11x + 9)$$
$$32x = 3(11x + 9)$$
$$32x = 33x + 27$$
$$32x - 33x = 33x + 27 - 33x$$
$$-x = 27$$
$$\frac{-x}{-1} = \frac{27}{-1}$$
$$\boxed{x = -27}$$

57.
$$2 \cdot 6 = 3x \cdot x$$
$$12 = 3x^2$$
$$12 - 12 = 3x^2 - 12$$
$$0 = 3x^2 - 12$$
$$0 = 3(x^2 - 4)$$
$$0 = 3(x - 2)(x + 2)$$

$$x - 2 = 0 \qquad \text{and} \qquad x + 2 = 0$$
$$x - 2 + 2 = 0 + 2 \qquad x + 2 - 2 = 0 - 2$$
$$x = 2 \qquad\qquad x = -2$$
$$\boxed{x = 2, -2}$$

59.
$$(b - 5) \cdot b = 3 \cdot 2$$
$$b(b - 5) = 6$$
$$b^2 - 5b = 6$$
$$b^2 - 5b - 6 = 6 - 6$$
$$b^2 - 5b - 6 = 0$$
$$(b - 6)(b + 1) = 0$$

$$b - 6 = 0 \qquad \text{and} \qquad b + 1 = 0$$
$$b - 6 + 6 = 0 + 6 \qquad b + 1 - 1 = 0 - 1$$
$$b = 6 \qquad\qquad b = -1$$
$$\boxed{b = 6, -1}$$

61.
$$(x - 1) \cdot 3x = (x + 1) \cdot 2$$
$$3x(x - 1) = 2(x + 1)$$
$$3x^2 - 3x = 2x + 2$$
$$3x^2 - 3x - 2x - 2 = 2x + 2 - 2x - 2$$
$$3x^2 - 5x - 2 = 0$$
$$(3x + 1)(x - 2) = 0$$

$$3x + 1 = 0 \qquad \text{and} \qquad x - 2 = 0$$
$$3x + 1 - 1 = 0 - 1 \qquad x - 2 + 2 = 0 + 2$$
$$3x = -1 \qquad\qquad x = 2$$
$$\frac{3x}{3} = \frac{-1}{3}$$
$$x = -\frac{1}{3}$$
$$\boxed{x = -\frac{1}{3}, 2}$$

63.
$$\frac{4}{12} = \frac{5}{x}$$
$$4(x) = 12(5)$$
$$4x = 60$$
$$\frac{4x}{4} = \frac{60}{4}$$
$$\boxed{x = 15}$$

65.
$$\frac{x}{12} = \frac{6}{9}$$
$$x(9) = 12(6)$$
$$9x = 72$$
$$\frac{9x}{9} = \frac{72}{9}$$
$$\boxed{x = 8}$$

APPLICATIONS

67. GEAR RATIOS

a) $\dfrac{18}{12} = \dfrac{3}{2}$, 3:2

b) $\dfrac{12}{18} = \dfrac{2}{3}$, 2:3

69. SHOPPING FOR CLOTHES

$$\frac{2 \text{ shirts}}{\$25} = \frac{5 \text{ shirts}}{x}$$
$$2(x) = 25(5)$$
$$2x = 125$$
$$\frac{2x}{2} = \frac{125}{2}$$
$$\boxed{x = \$62.50}$$

71. COOKING

$$\frac{\text{bottles}}{\text{gallons}}$$
$$\frac{4}{2} = \frac{x}{10}$$
$$4(10) = 2(x)$$
$$40 = 2x$$
$$\frac{40}{2} = \frac{2x}{2}$$
$$\boxed{20 = x}$$

73. CPR

$$\frac{\text{compressions}}{\text{breaths}}$$

$$\frac{5}{2} = \frac{210}{x}$$

$$5(x) = 2(210)$$
$$5x = 420$$
$$\frac{5x}{5} = \frac{420}{5}$$
$$\boxed{x = 84}$$

75. NUTRITION

calories:

$$\frac{10\text{-oz}}{16\text{-oz}} = \frac{355}{x}$$

$$10(x) = 16(355)$$
$$10x = 5,680$$
$$\frac{10x}{10} = \frac{5,680}{10}$$
$$\boxed{x = 568}$$

fat:

$$\frac{10\text{-oz}}{16\text{-oz}} = \frac{8}{x}$$

$$10(x) = 16(8)$$
$$10x = 128$$
$$\frac{10x}{10} = \frac{128}{10}$$
$$x = 12.8$$
$$\boxed{x \approx 13}$$

protein:

$$\frac{10\text{-oz}}{16\text{-oz}} = \frac{9}{x}$$

$$10(x) = 16(9)$$
$$10x = 144$$
$$\frac{10x}{10} = \frac{144}{10}$$
$$x = 14.4$$
$$\boxed{x \approx 14}$$

77. QUALITY CONTROL

$$\frac{17}{500} = \frac{x}{15,000}$$

$$17(15,000) = 500(x)$$
$$255,000 = 500x$$
$$\frac{255,000}{500} = \frac{500x}{500}$$
$$\boxed{510 = x}$$

79. MIXING FUELS

$$\frac{50}{1} = \frac{6(128)}{x}$$

$$\frac{50}{1} = \frac{768}{x}$$

$$50x = 768$$
$$\frac{50x}{50} = \frac{768}{50}$$
$$\boxed{x = 15.36}$$

The instructions are not exactly right, but close.

81. CROP DAMAGE

$$\frac{35}{3} = \frac{x}{12}$$

$$35(12) = 3(x)$$
$$420 = 3x$$
$$\frac{420}{3} = \frac{3x}{3}$$
$$\boxed{140 = x}$$

83. MODEL RAILROADS

Change 9 inches to $\frac{3}{4}$ foot.

$$\frac{87}{1} = \frac{x}{9 \text{ inches}}$$

$$\frac{87}{1} = \frac{x}{\frac{9}{12} \text{ foot}}$$

$$\frac{87}{1} = \frac{x}{\frac{3}{4}}$$

$$87\left(\frac{3}{4}\right) = 1(x)$$

$$\frac{261}{4} = x$$

$$64\frac{1}{4} = x$$

$$\boxed{\begin{array}{l} x = 65\frac{1}{4} \text{ feet} \\ x = 65 \text{ feet, 3 inches} \end{array}}$$

85. BLUEPRINTS

Change 1 foot to 12 inches.

$$\frac{\frac{1}{4}}{12} = \frac{2\frac{1}{2}}{x}$$

$$\frac{1}{4}x = 12\left(2\frac{1}{2}\right)$$

$$\frac{1}{4}x = 12\left(\frac{5}{2}\right)$$

$$\frac{1}{4}x = 30$$

$$4 \cdot \frac{1}{4}x = 4 \cdot 30$$

$$x = 120 \quad \text{inches}$$

$$x = \frac{120}{12}$$

$$\boxed{x = 10 \quad \text{feet}}$$

87. $\dfrac{\$25}{45 \text{ minutes}} \approx \$0.56$ per minute

$\dfrac{\$35}{60 \text{ minutes}} \approx \$0.58$ per minute

45 minutes for \$25 is the better buy.

89. $\dfrac{\$9.99}{100 \text{ cards}} \approx \$0.10$ per card

$\dfrac{\$12.99}{150 \text{ cards}} \approx \$0.09$ per card

150 cards for \$12.99 is the better buy.

91. $\dfrac{\$1.50}{6 \text{ drinks}} = \$0.25$ per drink

$\dfrac{\$6.25}{24 \text{ drinks}} \approx \$0.26$ per drink

6-pack for \$1.50 is the better buy.

93. HEIGHT OF A TREE

$$\frac{6}{h} = \frac{4}{26}$$

$$6(26) = h(4)$$

$$156 = 4h$$

$$\frac{156}{4} = \frac{4h}{4}$$

$$\boxed{39 \text{ ft} = h}$$

95. SURVEYING

$$\frac{20}{32} = \frac{w}{75}$$

$$32w = 20(75)$$

$$32w = 1,500$$

$$\frac{32w}{32} = \frac{1,500}{32}$$

$$w = 46.875$$

$$\boxed{w = 46\frac{7}{8} \text{ ft.}}$$

WRITING

97. Answers will vary.

99. Answers will vary.

REVIEW

101. $\dfrac{9}{10} = 10\overline{)9.0}\,^{0.9} = 0.9 = 90\%$

103. $0.30(1,600) = 480$

CHALLENGE PROBLEMS

105. $\dfrac{b}{a} = \dfrac{d}{c}, \quad \dfrac{c}{d} = \dfrac{a}{b}, \quad \dfrac{c}{a} = \dfrac{d}{b}$

CHAPTER 6: KEY CONCEPTS

1. a)
$$\frac{2x^2 - 8x}{x^2 - 6x + 8} = \frac{2x(x-4)}{(x-2)(x-4)}$$
$$= \frac{2x}{x-2}$$

   b) $x - 4$

2. a)
$$\frac{x^2 + 2x + 1}{x} \cdot \frac{x^2 - x}{x^2 - 1} = \frac{(x+1)(x+1) \cdot x(x-1)}{x \cdot (x+1)(x-1)}$$
$$= x + 1$$

   b) $x, x - 1, x + 1$

3. a)
$$\frac{x}{x+1} + \frac{x-1}{x} = \frac{x}{x+1} \cdot \frac{x}{x} + \frac{x-1}{x} \cdot \frac{x+1}{x+1}$$
$$= \frac{x^2 + (x-1)(x+1)}{x(x+1)}$$
$$= \frac{x^2 + x^2 - 1}{x(x+1)}$$
$$= \frac{2x^2 - 1}{x(x+1)}$$

   b) $\dfrac{x}{x}, \dfrac{x+1}{x+1}$

4. a)
$$\frac{n - 1 - \dfrac{2}{n}}{\dfrac{n}{3}} = \frac{n - 1 - \dfrac{2}{n}}{\dfrac{n}{3}} \cdot \frac{3n}{3n}$$
$$= \frac{3n\left(n - 1 - \dfrac{2}{n}\right)}{3n\left(\dfrac{n}{3}\right)}$$
$$= \frac{3n^2 - 3n - 6}{n^2}$$
$$= \frac{3(n^2 - n - 2)}{n^2}$$

   b) $3n$

5. a)
$$\frac{11}{b} + \frac{13}{b} = 12$$
$$b\left(\frac{11}{b} + \frac{13}{b}\right) = b(12)$$
$$b\left(\frac{11}{b}\right) + b\left(\frac{13}{b}\right) = b(12)$$
$$11 + 13 = 12b$$
$$24 = 12b$$
$$\frac{24}{12} = \frac{12b}{12}$$
$$\boxed{2 = b}$$

   b) b

6. a)

$$\frac{3}{s+2} - \frac{5}{s^2+s-2} = \frac{1}{s-1}$$

$$\frac{3}{s+2} - \frac{5}{(s+2)(s-1)} = \frac{1}{s-1}$$

$$(s+2)(s-1)\left(\frac{3}{s+2} - \frac{5}{(s+2)(s-1)}\right) = (s+2)(s-1)\left(\frac{1}{s-1}\right)$$

$$3(s-1) - 5 = s+2$$

$$3s - 3 - 5 = s+2$$

$$3s - 8 = s+2$$

$$3s - 8 - s = s+2-s$$

$$2s - 8 = 2$$

$$2s - 8 + 8 = 2+8$$

$$2s = 10$$

$$\frac{2s}{2} = \frac{10}{2}$$

$$\boxed{s = 5}$$

b) $(s+2)(s-1)$

7. a)

$$y + \frac{3}{4} = \frac{3y-50}{4y-24}$$

$$y + \frac{3}{4} = \frac{3y-50}{4(y-6)}$$

$$4(y-6)\left(y + \frac{3}{4}\right) = 4(y-6)\left(\frac{3y-50}{4(y-6)}\right)$$

$$4y(y-6) + 3(y-6) = 3y-50$$

$$4y^2 - 24y + 3y - 18 = 3y-50$$

$$4y^2 - 21y - 18 = 3y-50$$

$$4y^2 - 21y - 18 - 3y + 50 = 3y - 50 - 3y + 50$$

$$4y^2 - 24y + 32 = 0$$

$$4(y^2 - 6y + 8) = 0$$

$$4(y-2)(y-4) = 0$$

$$y - 2 = 0 \quad \text{and} \quad y - 4 = 0$$

$$y - 2 + 2 = 0 + 2 \quad y - 4 + 4 = 0 + 4$$

$$\boxed{y = 2 \quad \text{and} \quad y = 4}$$

b) $4(y-6)$

8. a)

$$\frac{1}{a} - \frac{1}{b} = 1$$

$$ab\left(\frac{1}{a} - \frac{1}{b}\right) = ab \bullet 1$$

$$b - a = ab$$

$$b - a - b = ab - b$$

$$-a = b(a-1)$$

$$-\frac{a}{a-1} = \frac{b(a-1)}{a-1}$$

$$\frac{a}{1-a} = b$$

$$b = \frac{a}{1-a}$$

b) ab

CHAPTER 6 REVIEW

SECTION 6.1
Simplifying Rational Expressions

1. To find the values of x for which the rational expression is undefined, set the denominator equal to zero and solve for x.
$$x^2 - 16 = 0$$
$$(x - 4)(x + 4) = 0$$
$$x - 4 = 0 \quad \text{and} \quad x + 4 = 0$$
$$\boxed{x = 4 \quad \text{and} \quad x = -4}$$

2.
$$\frac{(-2)^2 - 1}{-2 - 5} = \frac{4 - 1}{-2 - 5}$$
$$= \frac{3}{-7}$$
$$= -\frac{3}{7}$$

3.
$$\frac{3x^2}{6x^3} = \frac{1}{2}x^{2-3}$$
$$= \frac{1}{2}x^{-1}$$
$$= \frac{1}{2x}$$

4.
$$\frac{5xy^2}{2x^2y^2} = \frac{5}{2}x^{1-2}y^{2-2}$$
$$= \frac{5}{2}x^{-1}y^0$$
$$= \frac{5}{2x}$$

5.
$$\frac{x^2}{x^2 + x} = \frac{\cancel{x} \bullet x}{\cancel{x}(x + 1)}$$
$$= \frac{x}{x + 1}$$

6.
$$\frac{a^2 - 4}{a + 2} = \frac{(a + 2)(a - 2)}{a + 2}$$
$$= a - 2$$

7.
$$\frac{3p - 2}{2 - 3p} = -\frac{\cancel{3p - 2}}{\cancel{3p - 2}}$$
$$= -1$$

8.
$$\frac{8 - x}{x^2 - 5x - 24} = \frac{8 - x}{(x - 8)(x + 3)}$$
$$= -\frac{\cancel{x - 8}}{(\cancel{x - 8})(x + 3)}$$
$$= -\frac{1}{x + 3}$$

9.
$$\frac{2x^2 - 16x}{2x^2 - 18x + 16} = \frac{2x(x - 8)}{2(x^2 - 9x + 8)}$$
$$= \frac{\cancel{2}x\cancel{(x - 8)}}{\cancel{2}\cancel{(x - 8)}(x - 1)}$$
$$= \frac{x}{x - 1}$$

10. $\dfrac{x^2 + x - 2}{x^2 - x - 2} = \dfrac{(x + 2)(x - 1)}{(x - 2)(x + 1)}$

   It does not simplify

11.
$$\frac{4(t + 3) + 8}{3(t + 3) + 6} = \frac{4t + 12 + 8}{3t + 9 + 6}$$
$$= \frac{4t + 20}{3t + 15}$$
$$= \frac{4(t + 5)}{3(t + 5)}$$
$$= \frac{4}{3}$$

12. x is not a common factor of the numerator and the denominator; x is a term of the numerator.

13. DOSAGES

    Let $A = 11$ and $D = 300$.

$$C = \frac{300(11+1)}{24}$$
$$= \frac{300(12)}{24}$$
$$= \frac{3{,}600}{24}$$
$$= 150 \text{ mg}$$

SECTION 6.2
Multiplying and Dividing Rational Expressions

14.
$$\frac{3xy}{2x} \bullet \frac{4x}{2y^2} = \frac{3 \cdot \cancel{x} \cdot \cancel{y} \cdot 2 \cdot 2 \cdot x}{2 \cdot \cancel{x} \cdot 2 \cdot \cancel{y} \cdot y}$$
$$= \frac{3x}{y}$$

15.
$$56x\left(\frac{12}{7x}\right) = \frac{56x}{1}\left(\frac{12}{7x}\right)$$
$$= \frac{2 \cdot 2 \cdot 2 \cdot \cancel{7} \cdot \cancel{x} \cdot 2 \cdot 2 \cdot 3}{\cancel{7} \cdot \cancel{x}}$$
$$= 96$$

16.
$$\frac{x^2-1}{x^2+2x} \bullet \frac{x}{x+1} = \frac{(x+1)(x-1)\cancel{x}}{\cancel{x}(x+2)(\cancel{x+1})}$$
$$= \frac{x-1}{x+2}$$

17.
$$\frac{x^2+x}{3x-15} \bullet \frac{6x-30}{x^2+2x+1} = \frac{x(\cancel{x+1}) \cdot 2 \cdot \cancel{3}(\cancel{x-5})}{\cancel{3}(\cancel{x-5})(\cancel{x+1})(x+1)}$$
$$= \frac{2x}{x+1}$$

18.
$$\frac{3x^2}{5x^2y} \div \frac{6x}{15xy^2} = \frac{3x^2}{5x^2y} \bullet \frac{15xy^2}{6x}$$
$$= \frac{\cancel{3} \cdot \cancel{x} \cdot x \cdot 3 \cdot \cancel{5} \cdot \cancel{x} \cdot \cancel{y} \cdot y}{\cancel{5} \cdot \cancel{x} \cdot \cancel{x} \cdot \cancel{y} \cdot 2 \cdot \cancel{3} \cdot \cancel{x}}$$
$$= \frac{3y}{2}$$

19.
$$\frac{x^2+5x}{x^2+4x-5} \div x^2 = \frac{x^2+5x}{x^2+4x-5} \bullet \frac{1}{x^2}$$
$$= \frac{x(\cancel{x+5})}{(\cancel{x+5})(x-1) \cdot \cancel{x} \cdot x}$$
$$= \frac{1}{x(x-1)}$$

20.
$$\frac{x^2-x-6}{1-2x} \div \frac{x^2-2x-3}{2x^2+x-1} = \frac{x^2-x-6}{1-2x} \bullet \frac{2x^2+x-1}{x^2-2x-3}$$
$$= \frac{(x-3)(x+2)(2x-1)(x+1)}{(1-2x)(x-3)(x+1)}$$
$$= \frac{(\cancel{x-3})(x+2)(\cancel{2x-1})(\cancel{x+1})}{(\cancel{2x-1})(\cancel{x-3})(\cancel{x+1})}$$
$$= -(x+2)$$
$$= -x-2$$

21. a) Yes, 1 ft = 12 in.
    b) No, 60 min ≠ 1 day
    c) Yes, 2,000 lbs = 1 ton
    d) Yes, 1 gal = 4 qt.

22. TRAFFIC SIGNS

$$\frac{20 \text{ miles}}{\text{hour}} \bullet \frac{1 \text{ hour}}{60 \text{ minues}} = \frac{20 \text{ miles}}{60 \text{ minutes}}$$
$$= \frac{1}{3} \text{ mile per minute}$$

SECTION 6.3
Addition and Subtraction with Like Denominator; Least Common Denominators

23. $\dfrac{13c}{15d} - \dfrac{8}{15d} = \dfrac{13c-8}{15d}$

24.

$$\frac{x}{x+y} + \frac{y}{x+y} = \frac{\cancel{x+y}}{\cancel{x+y}}$$
$$= 1$$

25.

$$\frac{3x}{x-7} - \frac{x-2}{x-7} = \frac{3x-(x-2)}{x-7}$$
$$= \frac{3x-x+2}{x-7}$$
$$= \frac{2x+2}{x-7}$$

26.

$$\frac{a}{a^2-2a-8} + \frac{2}{a^2-2a-8} = \frac{a+2}{a^2-2a-8}$$
$$= \frac{\cancel{a+2}}{(a-4)\cancel{(a+2)}}$$
$$= \frac{1}{a-4}$$

27. x = does not simplify
  $9 = 3 \cdot 3$
  LCD $= 3 \cdot 3 \cdot x = \mathbf{9x}$

28. $2x^3 = 2 \cdot x \cdot x \cdot x$
  $8x = 2 \cdot 2 \cdot 2 \cdot x$
  LCD $= 2 \cdot 2 \cdot 2 \cdot x \cdot x \cdot x = \mathbf{8x^3}$

29. m = does not simplify
  $m - 8$ = does not simplify
  LCD $= m(m-8)$

30. $5x + 1$ = does not simplify
  $5x - 1$ = does not simplify
  LCD $= (5x+1)(5x-1)$

31. $a^2 - 25 = (a-5)(a+5)$
  $a - 5$ = does not simplify
  LCD $= (a-5)(a+5)$

32. $t^2 + 10t + 25 = (t+5)(t+5)$
  $2t^2 + 17t + 35 = (t+5)(2t+7)$
  LCD $= (t+5)(t+5)(2t+7)$
    $= (2t+7)(t+5)^2$

33.

$$\frac{9}{a} = \frac{9}{a} \cdot \frac{7}{7}$$
$$= \frac{63}{7a}$$

34.

$$\frac{2y+1}{x-9} = \frac{2y+1}{x-9}\left(\frac{x}{x}\right)$$
$$= \frac{x(2y+1)}{x(x-9)}$$
$$= \frac{2xy+x}{x(x-9)}$$

35.

$$\frac{b+7}{3b-15} = \frac{b+7}{3(b-5)}$$
$$= \frac{b+7}{3(b-5)}\left(\frac{2}{2}\right)$$
$$= \frac{2b+14}{6(b-5)}$$

36.

$$\frac{9r}{r^2+6r+5} = \frac{9r}{(r+1)(r+5)}\left(\frac{r-4}{r-4}\right)$$
$$= \frac{9r^2-36r}{(r+1)(r+5)(r-4)}$$

SECTION 6.4
Addition and Subtraction with Unlike Denominator

37.

$$\frac{1}{7} - \frac{1}{a} = \frac{1}{7}\left(\frac{a}{a}\right) - \frac{1}{a}\left(\frac{7}{7}\right)$$
$$= \frac{a}{7a} - \frac{7}{7a}$$
$$= \frac{a-7}{7a}$$

38.

$$\frac{x}{x-1} + \frac{1}{x} = \frac{x}{x-1}\left(\frac{x}{x}\right) + \frac{1}{x}\left(\frac{x-1}{x-1}\right)$$
$$= \frac{x^2}{x(x-1)} + \frac{x-1}{x(x-1)}$$
$$= \frac{x^2+x-1}{x(x-1)}$$

39.

$$\frac{2t+2}{t^2+2t+1} - \frac{1}{t+1} = \frac{2t+2}{(t+1)(t+1)} - \frac{1}{t+1}$$

$$= \frac{2t+2}{(t+1)(t+1)} - \frac{1}{t+1}\left(\frac{t+1}{t+1}\right)$$

$$= \frac{2t+2}{(t+1)(t+1)} - \frac{t+1}{(t+1)(t+1)}$$

$$= \frac{2t+2-(t+1)}{(t+1)(t+1)}$$

$$= \frac{2t+2-t-1}{(t+1)(t+1)}$$

$$= \frac{\cancel{t+1}}{(\cancel{t+1})(t+1)}$$

$$= \frac{1}{t+1}$$

40.

$$\frac{x+2}{2x} - \frac{2-x}{x^2} = \frac{x+2}{2x}\left(\frac{x}{x}\right) - \frac{2-x}{x^2}\left(\frac{2}{2}\right)$$

$$= \frac{x^2+2x}{2x^2} - \frac{4-2x}{2x^2}$$

$$= \frac{x^2+2x-(4-2x)}{2x^2}$$

$$= \frac{x^2+2x-4+2x}{2x^2}$$

$$= \frac{x^2+4x-4}{2x^2}$$

41.

$$\frac{6}{b-1} - \frac{b}{1-b} = \frac{6}{b-1} - \frac{-b}{b-1}$$

$$= \frac{6+b}{b-1}$$

$$= \frac{b+6}{b-1}$$

42.

$$\frac{8}{c} + 6 = \frac{8}{c} + \frac{6}{1}$$

$$= \frac{8}{c} + \frac{6}{1}\left(\frac{c}{c}\right)$$

$$= \frac{8}{c} + \frac{6c}{c}$$

$$= \frac{8+6c}{c}$$

$$= \frac{6c+8}{c}$$

43.

$$\frac{n+7}{n+3} - \frac{n-3}{n+7} = \frac{n+7}{n+3}\left(\frac{n+7}{n+7}\right) - \frac{n-3}{n+7}\left(\frac{n+3}{n+3}\right)$$

$$= \frac{n^2+14n+49}{(n+3)(n+7)} - \frac{n^2-9}{(n+3)(n+7)}$$

$$= \frac{n^2+14n+49-(n^2-9)}{(n+3)(n+7)}$$

$$= \frac{n^2+14n+49-n^2+9}{(n+3)(n+7)}$$

$$= \frac{14n+58}{(n+3)(n+7)}$$

44.

$$\frac{4}{t+2} - \frac{7}{(t+2)^2} = \frac{4}{t+2}\left(\frac{t+2}{t+2}\right) - \frac{7}{(t+2)^2}$$

$$= \frac{4t+8}{(t+2)^2} - \frac{7}{(t+2)^2}$$

$$= \frac{4t+8-7}{(t+2)^2}$$

$$= \frac{4t+1}{(t+2)^2}$$

45.

$$\frac{6}{a^2-9}-\frac{5}{a^2-a-6}=\frac{6}{(a-3)(a+3)}-\frac{5}{(a-3)(a+2)}$$

$$=\frac{6}{(a-3)(a+3)}\left(\frac{a+2}{a+2}\right)-\frac{5}{(a-3)(a+2)}\left(\frac{a+3}{a+3}\right)$$

$$=\frac{6a+12}{(a-3)(a+3)(a+2)}-\frac{5a+15}{(a-3)(a+3)(a+2)}$$

$$=\frac{6a+12-(5a+15)}{(a-3)(a+3)(a+2)}$$

$$=\frac{6a+12-5a-15}{(a-3)(a+3)(a+2)}$$

$$=\frac{\cancel{a-3}}{\cancel{(a-3)}(a+3)(a+2)}$$

$$=\frac{1}{(a+3)(a+2)}$$

46.

$$\frac{2}{3y-6}+\frac{3}{4y+8}=\frac{2}{3(y-2)}+\frac{3}{4(y+2)}$$

$$=\frac{2}{3(y-2)}\left(\frac{4(y+2)}{4(y+2)}\right)+\frac{3}{4(y+2)}\left(\frac{3(y-2)}{3(y-2)}\right)$$

$$=\frac{8(y+2)}{12(y-2)(y+2)}+\frac{9(y-2)}{12(y-2)(y+2)}$$

$$=\frac{8y+16+9y-18}{12(y-2)(y+2)}$$

$$=\frac{17y-2}{12(y-2)(y+2)}$$

47. Yes. Factor a -1 out of the numerator and the answers would be the same.

48. VIDEO CAMERA

Perimeter = $2l + 2w$

$$2\left(\frac{3}{x-1}\right)+2\left(\frac{4}{x+6}\right)=\frac{6}{x-1}+\frac{8}{x+6}$$

$$=\frac{6}{x-1}\left(\frac{x+6}{x+6}\right)+\frac{8}{x+6}\left(\frac{x-1}{x-1}\right)$$

$$=\frac{6x+36+8x-8}{(x+6)(x-1)}$$

$$=\frac{14x+28}{(x+6)(x-1)}\text{ units}$$

Area = $l \cdot w$

$$\frac{3}{x-1}\bullet\frac{4}{x+6}=\frac{12}{(x-1)(x+6)}\text{ sq units}$$

SECTION 6.5
Simplifying Complex Fractions

49.

$$\frac{\dfrac{n^4}{30}}{\dfrac{7n}{15}}=\frac{n^4}{30}\div\frac{7n}{15}$$

$$=\frac{n^4}{30}\bullet\frac{15}{7n}$$

$$=\frac{15n^4}{210n}$$

$$=\frac{n^3}{14}$$

50.

$$\frac{\dfrac{r^2-81}{18s^2}}{\dfrac{4r-36}{9s}}=\frac{\dfrac{r^2-81}{18s^2}}{\dfrac{4r-36}{9s}}\bullet\frac{18s^2}{18s^2}$$

$$=\frac{18s^2\left(\dfrac{r^2-81}{18s^2}\right)}{18s^2\left(\dfrac{4r-36}{9s}\right)}$$

$$=\frac{r^2-81}{2s(4r-36)}$$

$$=\frac{r^2-81}{8rs-72}$$

$$=\frac{\cancel{(r-9)}(r+9)}{8s\cancel{(r-9)}}$$

$$=\frac{r+9}{8s}$$

51.

$$\frac{\frac{1}{y}+1}{\frac{1}{y}-1} = \frac{\frac{1}{y}+1}{\frac{1}{y}-1} \bullet \frac{y}{y}$$

$$= \frac{y\left(\frac{1}{y}+1\right)}{y\left(\frac{1}{y}-1\right)}$$

$$= \frac{1+y}{1-y}$$

52.

$$\frac{\frac{7}{a^2}}{\frac{1}{a}+\frac{10}{3}} = \frac{\frac{7}{a^2}}{\frac{1}{a}+\frac{10}{3}} \bullet \frac{3a^2}{3a^2}$$

$$= \frac{3a^2\left(\frac{7}{a^2}\right)}{3a^2\left(\frac{1}{a}\right)+3a^2\left(\frac{10}{3}\right)}$$

$$= \frac{21}{3a+10a^2}$$

53.

$$\frac{\frac{2}{x-1}+\frac{x-1}{x+1}}{\frac{1}{x^2-1}} = \frac{\frac{2}{x-1}+\frac{x-1}{x+1}}{\frac{1}{(x-1)(x+1)}}$$

$$= \frac{\frac{2}{x-1}+\frac{x-1}{x+1}}{\frac{1}{(x-1)(x+1)}} \bullet \frac{(x-1)(x+1)}{(x-1)(x+1)}$$

$$= \frac{(x-1)(x+1)\left(\frac{2}{x-1}\right)+(x-1)(x+1)\left(\frac{x-1}{x+1}\right)}{(x-1)(x+1)\left(\frac{1}{(x-1)(x+1)}\right)}$$

$$= \frac{2(x+1)+(x-1)(x-1)}{1}$$

$$= 2x+2+x^2-x-x+1$$

$$= x^2+3$$

54.

$$\frac{\frac{1}{x^2y}-\frac{5}{xy}}{\frac{3}{xy}-\frac{7}{xy^2}} = \frac{\frac{1}{x^2y}-\frac{5}{xy}}{\frac{3}{xy}-\frac{7}{xy^2}} \bullet \frac{x^2y^2}{x^2y^2}$$

$$= \frac{x^2y^2\left(\frac{1}{x^2y}\right)-x^2y^2\left(\frac{5}{xy}\right)}{x^2y^2\left(\frac{3}{xy}\right)-x^2y^2\left(\frac{7}{xy^2}\right)}$$

$$= \frac{y-5xy}{3xy-7x}$$

SECTION 6.6
Solving Rational Equations

55.

$$\frac{3}{x} = \frac{2}{x-1}$$

$$x(x-1)\left(\frac{3}{x}\right) = x(x-1)\left(\frac{2}{x-1}\right)$$

$$3(x-1) = 2x$$

$$3x-3 = 2x$$

$$3x-3-3x = 2x-3x$$

$$-3 = -x$$

$$\frac{-3}{-1} = \frac{-x}{-1}$$

$$\boxed{3 = x}$$

Check:

$$\frac{3}{3} \overset{?}{=} \frac{2}{3-1}$$

$$\frac{3}{3} \overset{?}{=} \frac{2}{2}$$

$$1 = 1$$

56.

$$\frac{a}{a-5} = 3 + \frac{5}{a-5}$$

$$(a-5)\left(\frac{a}{a-5}\right) = (a-5)\left(3 + \frac{5}{a-5}\right)$$

$$a = 3(a-5) + 5$$

$$a = 3a - 15 + 5$$

$$a = 3a - 10$$

$$a - 3a = 3a - 10 - 3a$$

$$-2a = -10$$

$$\frac{-2a}{-2} = \frac{-10}{-2}$$

$$a = 5$$

No solution; 5 is extraneous since it makes the denominator zero

Check:

$$\frac{5}{5-5} \overset{?}{=} 3 + \frac{5}{5-5}$$

$$\frac{5}{0} = 3 + \frac{5}{0}$$

57.

$$\frac{2}{3t} + \frac{1}{t} = \frac{5}{9}$$

$$9t\left(\frac{2}{3t} + \frac{1}{t}\right) = 9t\left(\frac{5}{9}\right)$$

$$3(2) + 9(1) = t(5)$$

$$6 + 9 = 5t$$

$$15 = 5t$$

$$\frac{15}{5} = \frac{5t}{5}$$

$$\boxed{3 = t}$$

Check:

$$\frac{2}{3(3)} + \frac{1}{3} \overset{?}{=} \frac{5}{9}$$

$$\frac{2}{9} + \frac{1}{3} \overset{?}{=} \frac{5}{9}$$

$$\frac{2}{9} + \frac{3}{9} \overset{?}{=} \frac{5}{9}$$

$$\frac{5}{9} = \frac{5}{9}$$

58.

$$a = \frac{3a - 50}{4a - 24} - \frac{3}{4}$$

$$a = \frac{3a - 50}{4(a-6)} - \frac{3}{4}$$

$$4(a-6)(a) = 4(a-6)\left(\frac{3a-50}{4(a-6)} - \frac{3}{4}\right)$$

$$4a(a-6) = 3a - 50 - 3(a-6)$$

$$4a^2 - 24a = 3a - 50 - 3a + 18$$

$$4a^2 - 24a = -32$$

$$4a^2 - 24a + 32 = -32 + 32$$

$$4a^2 - 24a + 32 = 0$$

$$4(a^2 - 6a + 8) = 0$$

$$4(a-2)(a-4) = 0$$

$$a - 2 = 0 \quad \text{and} \quad a - 4 = 0$$

$$a - 2 + 2 = 0 + 2 \quad a - 4 + 4 = 0 + 4$$

$$\boxed{a = 2 \quad \text{and} \quad a = 4}$$

Check:

$$2 \overset{?}{=} \frac{3(2) - 50}{4(2) - 24} - \frac{3}{4} \qquad 4 \overset{?}{=} \frac{3(4) - 50}{4(4) - 24} - \frac{3}{4}$$

$$2 \overset{?}{=} \frac{6 - 50}{8 - 24} - \frac{3}{4} \qquad 4 \overset{?}{=} \frac{12 - 50}{16 - 24} - \frac{3}{4}$$

$$2 \overset{?}{=} \frac{-44}{-16} - \frac{3}{4} \qquad 4 \overset{?}{=} \frac{-38}{-8} - \frac{3}{4}$$

$$2 \overset{?}{=} \frac{11}{4} - \frac{3}{4} \qquad 4 \overset{?}{=} \frac{19}{4} - \frac{3}{4}$$

$$2 \overset{?}{=} \frac{8}{4} \qquad 4 \overset{?}{=} \frac{16}{4}$$

$$2 = 2 \qquad 4 = 4$$

59.

$$\frac{4}{x+2} - \frac{3}{x+3} = \frac{6}{x^2+5x+6}$$

$$\frac{4}{x+2} - \frac{3}{x+3} = \frac{6}{(x+2)(x+3)}$$

$$(x+2)(x+3)\left(\frac{4}{x+2} - \frac{3}{x+3}\right) = (x+2)(x+3)\left(\frac{6}{(x+2)(x+3)}\right)$$

$$4(x+3) - 3(x+2) = 6$$

$$4x+12-3x-6 = 6$$

$$x+6 = 6$$

$$x+6-6 = 6-6$$

$$\boxed{x=0}$$

Check:

$$\frac{4}{0+2} - \frac{3}{0+3} \stackrel{?}{=} \frac{6}{0^2+5(0)+6}$$

$$\frac{4}{2} - \frac{3}{3} \stackrel{?}{=} \frac{6}{6}$$

$$2 - 1 \stackrel{?}{=} 1$$

$$1 = 1$$

60.

$$\frac{3}{x+1} - \frac{x-2}{2} = \frac{x-2}{x+1}$$

$$2(x+1)\left(\frac{3}{x+1} - \frac{x-2}{2}\right) = 2(x+1)\left(\frac{x-2}{x+1}\right)$$

$$2(3) - (x+1)(x-2) = 2(x-2)$$

$$6 - (x^2-2x+x-2) = 2x-4$$

$$6 - x^2+2x-x+2 = 2x-4$$

$$-x^2+x+8 = 2x-4$$

$$-x^2+x+8+x^2-x-8 = 2x-4+x^2-x-8$$

$$0 = x^2+x-12$$

$$0 = (x+4)(x-3)$$

$$x+4=0 \quad \text{and} \quad x-3=0$$

$$x+4-4=0-4 \quad x-3+3=0+3$$

$$\boxed{x=-4 \quad \text{and} \quad x=3}$$

Check:

$$\frac{3}{-4+1} - \frac{-4-2}{2} \stackrel{?}{=} \frac{-4-2}{-4+1}$$

$$\frac{3}{-3} - \frac{-6}{2} \stackrel{?}{=} \frac{-6}{-3}$$

$$-1+3 \stackrel{?}{=} 2$$

$$2 = 2$$

$$\frac{3}{3+1} - \frac{3-2}{2} \stackrel{?}{=} \frac{3-2}{3+1}$$

$$\frac{3}{4} - \frac{1}{2} \stackrel{?}{=} \frac{1}{4}$$

$$\frac{3}{4} - \frac{2}{4} \stackrel{?}{=} \frac{1}{4}$$

$$\frac{1}{4} = \frac{1}{4}$$

61. ENGINEERING

$$E = 1 - \frac{T_2}{T_1}$$

$$T_1(E) = T_1\left(1 - \frac{T_2}{T_1}\right)$$

$$T_1 E = T_1 - T_2$$

$$T_1 E - T_1 = T_1 - T_2 - T_1$$

$$T_1 E - T_1 = -T_2$$

$$T_1(E-1) = -T_2$$

$$\frac{T_1(E-1)}{E-1} = \frac{-T_2}{E-1}$$

$$T_1 = \frac{-T_2}{E-1}$$

$$\boxed{T_1 = \frac{T_2}{1-E}}$$

62.

$$\frac{1}{x} = \frac{1}{y} + \frac{1}{z}$$

$$xyz\left(\frac{1}{x}\right) = xyz\left(\frac{1}{y} + \frac{1}{z}\right)$$

$$yz = xz + xy$$

$$yz - xy = xz + xy - xy$$

$$yz - xy = xz$$

$$y(z-x) = xz$$

$$\frac{y(z-x)}{z-x} = \frac{xz}{z-x}$$

$$\boxed{y = \frac{xz}{z-x}}$$

63. NUMBER PROBLEMS

$$\frac{4+2x}{5-x} = 5$$

$$(5-x)\left(\frac{4+2x}{5-x}\right) = (5-x)(5)$$

$$4+2x = 25-5x$$

$$4+2x+5x = 25-5x+5x$$

$$4+7x = 25$$

$$4+7x-4 = 25-4$$

$$7x = 21$$

$$\frac{7x}{7} = \frac{21}{7}$$

$$\boxed{x = 3}$$

The number is 3.

64. EXERCISE

Use the formula $t = \dfrac{d}{r}$.

Biking: $t = \dfrac{30}{x+10}$

Jogging: $t = \dfrac{10}{x}$

Since the times are the same, set the times equal and solve the equation for x.

$$\frac{30}{x+10} = \frac{10}{x}$$

$$x(x+10)\left(\frac{30}{x+10}\right) = x(x+10)\left(\frac{10}{x}\right)$$

$$30x = 10(x+10)$$

$$30x = 10x+100$$

$$30x-10x = 10x+100-10x$$

$$20x = 100$$

$$\frac{20x}{20} = \frac{100}{20}$$

$$\boxed{x = 5}$$

She can jog 5 mph.

65. HOUSE CLEANING

$$r = \frac{W}{t}$$

$$= \frac{1}{4} \text{ of the job per hour}$$

66. HOUSE PAINTING

rate of homeowner $= \dfrac{1}{14}$

rate of professional $= \dfrac{1}{10}$

rate together $= \dfrac{1}{x}$

$$\frac{1}{14} + \frac{1}{10} = \frac{1}{x}$$

$$70x\left(\frac{1}{14} + \frac{1}{10}\right) = 70x\left(\frac{1}{x}\right)$$

$$5x+7x = 70$$

$$12x = 70$$

$$\frac{12x}{12} = \frac{70}{12}$$

$$x = 5\frac{10}{12}$$

$$\boxed{x = 5\frac{5}{6} \text{ days}}$$

67. INVESTMENTS

Use the formula $P = \dfrac{I}{rt}$.

Savings: $P = \dfrac{100}{x(1)}$

$= \dfrac{100}{x}$

Credit Union: $P = \dfrac{120}{(x+0.01)1}$

$= \dfrac{120}{x+0.01}$

Since the principal is the same for both investments, set them equal and solve the equation for x.

$$\frac{100}{x} = \frac{120}{x+0.01}$$

$$x(x+0.01)\left(\frac{100}{x}\right) = x(x+0.01)\left(\frac{120}{x+0.01}\right)$$

$$100(x+0.01) = 120x$$

$$100x + 1 = 120x$$

$$100x + 1 - 100x = 120x - 100x$$

$$1 = 20x$$

$$\frac{1}{20} = \frac{20x}{20}$$

$$\frac{1}{20} = x$$

$$\boxed{x = 0.05 = 5\%}$$

68. WIND SPEED

Use the formula $t = \dfrac{d}{r}$.

Downwind: $t = \dfrac{400}{360+x}$

Upwind: $t = \dfrac{320}{360-x}$

Since the times are the same, set them equal and solve the equation for x.

$$\frac{400}{360+x} = \frac{320}{360-x}$$

$$(360+x)(360-x)\left(\frac{400}{360+x}\right) = (360+x)(360-x)\left(\frac{320}{360-x}\right)$$

$$400(360-x) = 320(360+x)$$

$$144{,}000 - 400x = 115{,}200 + 320x$$

$$144{,}000 - 400x + 400x = 115{,}200 + 320x + 400x$$

$$144{,}000 = 115{,}200 + 720x$$

$$144{,}000 - 115{,}200 = 115{,}200 + 720x - 115{,}200$$

$$28{,}800 = 720x$$

$$\frac{28{,}800}{720} = \frac{720x}{720}$$

$$\boxed{40 = x}$$

The rate of the wind is 40 mph.

SECTION 6.8
Proportions and Similar Triangles

69. No.

$4(34) \overset{?}{=} 7(20)$
$156 \ne 140$

70. Yes.

$5(42) \overset{?}{=} 7(30)$
$210 = 210$

71. $3(9) = x(6)$
$27 = 6x$
$\dfrac{27}{6} = \dfrac{6x}{6}$
$\boxed{\dfrac{9}{2} = x}$

72. $x(5) = 3(x)$
$$5x = 3x$$
$$5x - 3x = 3x - 3x$$
$$2x = 0$$
$$\frac{2x}{2} = \frac{0}{2}$$
$$\boxed{x = 0}$$

73. $(x - 2) \cdot 7 = 5(x)$
$$7x - 14 = 5x$$
$$7x - 14 - 7x = 5x - 7x$$
$$-14 = -2x$$
$$\frac{-14}{-2} = \frac{-2x}{-2}$$
$$\boxed{7 = x}$$

74. $2x(x - 1) = (x + 4) \cdot 3$
$$2x^2 - 2x = 3x + 12$$
$$2x^2 - 2x - 3x - 12 = 3x + 12 - 3x - 12$$
$$2x^2 - 5x - 12 = 0$$
$$(2x + 3)(x - 4) = 0$$
$$2x + 3 = 0 \quad \text{and} \quad x - 4 = 0$$
$$2x = -3 \quad \text{and} \quad x = 4$$
$$x = -\frac{3}{2}$$
$$\boxed{x = -\frac{3}{2}, \, 4}$$

75. DENTISTRY

$$\frac{3}{4} = \frac{x}{340}$$
$$3(340) = 4(x)$$
$$1{,}020 = 4x$$
$$\frac{1{,}020}{4} = \frac{4x}{4}$$
$$\boxed{255 = x}$$

225 will develop gum disease.

76. $\dfrac{x}{12} = \dfrac{6}{3.6}$
$$x(3.6) = 12(6)$$
$$3.6x = 72$$
$$\frac{3.6x}{3.6} = \frac{72}{3.6}$$
$$\boxed{x = 20}$$

The pole is 20 ft. tall.

77. PORCELAIN FIGURINE

$$\frac{1}{12} = \frac{5.5}{x}$$
$$1(x) = 12(5.5)$$
$$\boxed{x = 66}$$

The flutist is 66 in. or 5 ft. 6 in. tall.

78. COMPARISON SHOPPING

$$\frac{60}{150} = \$0.40 \text{ per compact disc}$$
$$\frac{98}{250} = \$0.392 \text{ per compact disc}$$

250 for \$98 is the better buy.

CHAPTER 6 TEST

1. $5x = 0$

$\dfrac{5x}{5} = \dfrac{0}{5}$

$\boxed{x = 0}$

2. $x^2 + x - 6 = 0$

$(x + 3)(x - 2) = 0$

$x + 3 = 0 \quad$ and $\quad x - 2 = 0$

$x + 3 - 3 = 0 - 3 \quad x - 2 + 2 = 0 + 2$

$\boxed{x = -3 \qquad \text{and} \qquad x = 2}$

3. MEMORY

Let $d = 7$ days in one week.

$n = \dfrac{35 + 5(7)}{7}$

$= \dfrac{35 + 35}{7}$

$= \dfrac{70}{7}$

$= 10$

A person can remember 10 words.

4. U.S. SCHOOL CONSTRUCTION

$\dfrac{109{,}512\,\text{ft}^2}{1} \bullet \dfrac{1\,\text{yd}^2}{9\,\text{ft}^2} = \dfrac{109{,}512}{9}$

$= 12{,}168\,\text{yd}^2$

5. $\dfrac{48x^2y}{54xy^2} = \dfrac{\cancel{2} \cdot 2 \cdot 2 \cdot 2 \cdot \cancel{3} \cdot \cancel{x} \cdot x \cdot \cancel{y}}{\cancel{2} \cdot 3 \cdot 3 \cdot \cancel{3} \cdot \cancel{x} \cdot y \cdot \cancel{y}}$

$= \dfrac{8x}{9y}$

6. $\dfrac{7m - 49}{7 - m} = \dfrac{7(m - 7)}{-(-7 + m)}$

$= -\dfrac{7\cancel{(m - 7)}}{\cancel{m - 7}}$

$= -7$

7. $\dfrac{2x^2 - x - 3}{4x^2 - 9} = \dfrac{\cancel{(2x - 3)}(x + 1)}{\cancel{(2x - 3)}(2x + 3)}$

$= \dfrac{x + 1}{2x + 3}$

8. $\dfrac{3(x + 2) - 3}{6x + 5 - (3x + 2)} = \dfrac{3x + 6 - 3}{6x + 5 - 3x - 2}$

$= \dfrac{\cancel{3x + 3}}{\cancel{3x + 3}}$

$= 1$

9. $3c^2d = 3 \cdot c \cdot c \cdot d$

$c^2d^3 = c \cdot c \cdot d \cdot d \cdot d$

LCD $= 3 \cdot c \cdot c \cdot d \cdot d \cdot d$

$= 3c^2d^3$

10. $n^2 - 4n - 5 = (n - 5)(n + 1)$

$n^2 - 25 = (n - 5)(n + 5)$

LCD $= (n + 1)\,(n + 5)(n - 5)$

11. $\dfrac{12x^2y}{15xy} \bullet \dfrac{25y^2}{16x} = \dfrac{\cancel{2} \cdot \cancel{2} \cdot \cancel{3} \cdot \cancel{x} \cdot \cancel{x} \cdot \cancel{y} \cdot \cancel{3} \cdot 5 \cdot y \cdot y}{\cancel{3} \cdot \cancel{3} \cdot \cancel{x} \cdot \cancel{y} \cdot \cancel{2} \cdot \cancel{2} \cdot 2 \cdot 2 \cdot \cancel{x}}$

$= \dfrac{5y^2}{4}$

12. $\dfrac{x^2 + 3x + 2}{3x + 9} \bullet \dfrac{x + 3}{x^2 - 4} = \dfrac{(x + 1)\cancel{(x + 2)}\cancel{(x + 3)}}{3\cancel{(x + 3)}\cancel{(x + 2)}(x - 2)}$

$= \dfrac{x + 1}{3(x - 2)}$

13. $\dfrac{x - x^2}{3x^2 + 6x} \div \dfrac{3x - 3}{3x^3 + 6x^2} = \dfrac{x - x^2}{3x^2 + 6x} \bullet \dfrac{3x^3 + 6x^2}{3x - 3}$

$= \dfrac{x(1 - x) \cdot 3x^2(x + 2)}{3x(x + 2) \cdot 3(x - 1)}$

$= -\dfrac{x\cancel{(x - 1)} \cdot \cancel{3}x \cdot x\cancel{(x + 2)}}{\cancel{3}x\cancel{(x + 2)} \cdot 3\cancel{(x - 1)}}$

$= -\dfrac{x^2}{3}$

14. $\dfrac{a^2 - 16}{a - 4} \div (6a + 24) = \dfrac{a^2 - 16}{a - 4} \bullet \dfrac{1}{6a + 24}$

$= \dfrac{\cancel{(a - 4)}(a + 4)}{\cancel{(a - 4)} \cdot 6\cancel{(a + 4)}}$

$= \dfrac{1}{6}$

15.

$$\frac{3y+7}{2y+3} - \frac{3y-6}{2y+3} = \frac{3y+7-(3y-6)}{2y+3}$$
$$= \frac{3y+7-3y+6}{2y+3}$$
$$= \frac{13}{2y+3}$$

16.

$$\frac{2n}{5m} - \frac{n}{2} = \frac{2n}{5m}\left(\frac{2}{2}\right) - \frac{n}{2}\left(\frac{5m}{5m}\right)$$
$$= \frac{4n}{10m} - \frac{5mn}{10m}$$
$$= \frac{4n-5mn}{10m}$$

17.

$$\frac{x+1}{x} + \frac{x-1}{x+1} = \frac{x+1}{x}\left(\frac{x+1}{x+1}\right) + \frac{x-1}{x+1}\left(\frac{x}{x}\right)$$
$$= \frac{x^2+x+x+1}{x(x+1)} + \frac{x^2-x}{x(x+1)}$$
$$= \frac{x^2+x+x+1+x^2-x}{x(x+1)}$$
$$= \frac{2x^2+x+1}{x(x+1)}$$

18.

$$\frac{a+3}{a-1} - \frac{a+4}{1-a} = \frac{a+3}{a-1} + \frac{a+4}{a-1}$$
$$= \frac{a+3+a+4}{a-1}$$
$$= \frac{2a+7}{a-1}$$

19.

$$\frac{9}{c-4} + c = \frac{9}{c-4} + \frac{c}{1}\left(\frac{c-4}{c-4}\right)$$
$$= \frac{9}{c-4} + \frac{c^2-4c}{c-4}$$
$$= \frac{c^2-4c+9}{c-4}$$

20.

$$\frac{6}{t^2-9} - \frac{5}{t^2-t-6} = \frac{6}{(t-3)(t+3)} - \frac{5}{(t-3)(t+2)}$$
$$= \frac{6}{(t-3)(t+3)}\left(\frac{t+2}{t+2}\right) - \frac{5}{(t-3)(t+2)}\left(\frac{t+3}{t+3}\right)$$
$$= \frac{6t+12}{(t-3)(t+3)(t+2)} - \frac{5t+15}{(t-3)(t+3)(t+2)}$$
$$= \frac{6t+12-(5t+15)}{(t-3)(t+3)(t+2)}$$
$$= \frac{6t+12-5t-15}{(t-3)(t+3)(t+2)}$$
$$= \frac{\cancel{t-3}}{\cancel{(t-3)}(t+3)(t+2)}$$
$$= \frac{1}{(t+3)(t+2)}$$

21.

$$\frac{\dfrac{3m-9}{8m}}{\dfrac{5m-15}{32}} = \frac{\dfrac{3m-9}{8m}}{\dfrac{5m-15}{32}} \bullet \frac{32m}{32m}$$
$$= \frac{32m\left(\dfrac{3m-9}{8m}\right)}{32m\left(\dfrac{5m-15}{32}\right)}$$
$$= \frac{4(3m-9)}{m(5m-15)}$$
$$= \frac{12m-36}{5m^2-15m}$$
$$= \frac{12\cancel{(m-3)}}{5m\cancel{(m-3)}}$$
$$= \frac{12}{5m}$$

22.

$$\frac{\dfrac{3}{as^2}+\dfrac{6}{a^2 s}}{\dfrac{6}{a}-\dfrac{9}{s^2}}=\frac{\dfrac{3}{as^2}+\dfrac{6}{a^2 s}}{\dfrac{6}{a}-\dfrac{9}{s^2}}\bullet\frac{a^2 s^2}{a^2 s^2}$$

$$=\frac{a^2 s^2\left(\dfrac{3}{as^2}\right)+a^2 s^2\left(\dfrac{6}{a^2 s}\right)}{a^2 s^2\left(\dfrac{6}{a}\right)-a^2 s^2\left(\dfrac{9}{s^2}\right)}$$

$$=\frac{3a+6s}{6as^2-9a^2}$$

$$=\frac{\cancel{3}(a+2s)}{\cancel{3}(2as^2-3a^2)}$$

$$=\frac{a+2s}{2as^2-3a^2}$$

23.

$$\frac{1}{3}+\frac{4}{3y}=\frac{5}{y}$$

$$3y\left(\frac{1}{3}+\frac{4}{3y}\right)=3y\left(\frac{5}{y}\right)$$

$$y+4=15$$

$$y+4-4=15-4$$

$$\boxed{y=11}$$

Check:

$$\frac{1}{3}+\frac{4}{3(11)}\overset{?}{=}\frac{5}{11}$$

$$\frac{1}{3}+\frac{4}{33}\overset{?}{=}\frac{5}{11}$$

$$\frac{11}{33}+\frac{4}{33}\overset{?}{=}\frac{5}{11}$$

$$\frac{15}{33}\overset{?}{=}\frac{5}{11}$$

$$\frac{5}{11}=\frac{5}{11}$$

24.

$$\frac{9n}{n-6}=3+\frac{54}{n-6}$$

$$(n-6)\left(\frac{9n}{n-6}\right)=(n-6)\left(3+\frac{54}{n-6}\right)$$

$$9n=3(n-6)+54$$

$$9n=3n-18+54$$

$$9n=3n+36$$

$$9n-3n=3n+36-3n$$

$$6n=36$$

$$\frac{6n}{6}=\frac{36}{6}$$

$$n=6$$

$$\boxed{\text{No Solution; 6 is extraneous}}$$

Check:

$$\frac{9(6)}{6-6}\overset{?}{=}3+\frac{54}{6-6}$$

$$\frac{54}{0}\overset{?}{=}3+\frac{54}{0}$$

25.

$$\frac{7}{q^2-q-2}+\frac{1}{q+1}=\frac{3}{q-2}$$

$$\frac{7}{(q-2)(q+1)}+\frac{1}{q+1}=\frac{3}{q-2}$$

$$(q-2)(q+1)\left(\frac{7}{(q-2)(q+1)}+\frac{1}{q+1}\right)=(q-2)(q+1)\left(\frac{3}{q-2}\right)$$

$$(q-2)(q+1)\frac{7}{(q-2)(q+1)}+(q-2)(q+1)\frac{1}{q+1}=(q-2)(q+1)\frac{3}{q-2}$$

$$7+q-2=3(q+1)$$

$$q+5=3q+3$$

$$q+5-q=3q+3-q$$

$$5=2q+3$$

$$5-3=2q+3-3$$

$$2=2q$$

$$\frac{2}{2}=\frac{2q}{2}$$

$$\boxed{1=q}$$

Check:

$$\frac{7}{1^2-1-2}+\frac{1}{1+1}\overset{?}{=}\frac{3}{1-2}$$

$$\frac{7}{-2}+\frac{1}{2}\overset{?}{=}\frac{3}{-1}$$

$$-\frac{7}{2}+\frac{1}{2}\overset{?}{=}-3$$

$$-\frac{6}{2}\overset{?}{=}-3$$

$$-3=-3$$

26.

$$\frac{2}{3} = \frac{2c-12}{3c-9} - c$$

$$\frac{2}{3} = \frac{2c-12}{3(c-3)} - \frac{c}{1}$$

$$3(c-3)\frac{2}{3} = 3(c-3)\frac{2c-12}{3(c-3)} - 3(c-3)\frac{c}{1}$$

$$2(c-3) = 2c-12-3c(c-3)$$

$$2c-6 = 2c-12-3c^2+9c$$

$$2c-6 = -3c^2+11c-12$$

$$2c-6+3c^2-11c+12 = -3c^2+11c-12+3c^2-11c+12$$

$$3c^2-9c+6 = 0$$

$$3(c^2-3c+2) = 0$$

$$3(c-2)(c-1) = 0$$

$$c-2 = 0 \quad \text{and} \quad c-1 = 0$$

$$c-2+2 = 0+2 \quad c-1+1 = 0+1$$

$$\boxed{c=2 \qquad c=1}$$

Check:

$$\frac{2}{3} \overset{?}{=} \frac{2(1)-12}{3(1)-9} - 1 \qquad \frac{2}{3} \overset{?}{=} \frac{2(2)-12}{3(2)-9} - 2$$

$$\frac{2}{3} \overset{?}{=} \frac{-10}{-6} - 1 \qquad \frac{2}{3} \overset{?}{=} \frac{-8}{-3} - 2$$

$$\frac{2}{3} \overset{?}{=} \frac{10}{6} - \frac{6}{6} \qquad \frac{2}{3} \overset{?}{=} \frac{8}{3} - \frac{6}{3}$$

$$\frac{2}{3} \overset{?}{=} \frac{4}{6} \qquad \frac{2}{3} = \frac{2}{3}$$

$$\frac{2}{3} = \frac{2}{3}$$

27.

$$\frac{y}{y-1} = \frac{y-2}{y}$$

$$y(y-1)\left(\frac{y}{y-1}\right) = y(y-1)\left(\frac{y-2}{y}\right)$$

$$y(y) = (y-1)(y-2)$$

$$y^2 = y^2-2y-y+2$$

$$y^2 = y^2-3y+2$$

$$y^2-y^2 = y^2-3y+2-y^2$$

$$0 = -3y+2$$

$$0+3y = -3y+2+3y$$

$$3y = 2$$

$$\boxed{y = \frac{2}{3}}$$

Check:

$$\frac{\frac{2}{3}}{\frac{2}{3}-1} \overset{?}{=} \frac{\frac{2}{3}-2}{\frac{2}{3}}$$

$$\frac{\frac{2}{3}}{\frac{2}{3}-\frac{3}{3}} \overset{?}{=} \frac{\frac{2}{3}-\frac{6}{3}}{\frac{2}{3}}$$

$$\frac{\frac{2}{3}}{-\frac{1}{3}} \overset{?}{=} \frac{-\frac{4}{3}}{\frac{2}{3}}$$

$$-2 = -2$$

28.

$$H = \frac{RB}{R+B}$$

$$(R+B)(H) = (R+B)\left(\frac{RB}{R+B}\right)$$

$$HR + HB = RB$$

$$HR + HB - HB = RB - HB$$

$$HR = RB - HB$$

$$HR = B(R-H)$$

$$\frac{HR}{R-H} = \frac{B\cancel{(R-H)}}{\cancel{R-H}}$$

$$\boxed{\frac{HR}{R-H} = B}$$

29. HEALTH RISK

Yes.

$$\frac{114}{120} = \frac{\cancel{2}\cdot\cancel{3}\cdot 19}{\cancel{2}\cdot 2\cdot 2\cdot\cancel{3}\cdot 5}$$

$$= \frac{19}{20}$$

30. CURRENCY EXCHANGE RATE

$$\frac{5}{3} = \frac{1{,}500}{x}$$

$$5(x) = 3(1{,}500)$$

$$5x = 4{,}500$$

$$\frac{5x}{5} = \frac{4{,}500}{5}$$

$$\boxed{x = 900}$$

The traveler received 900 British pounds.

31. TV TOWERS

$$\frac{x}{114} = \frac{6}{4}$$

$$x(4) = 114(6)$$

$$4x = 684$$

$$\frac{4x}{4} = \frac{684}{4}$$

$$\boxed{x = 171}$$

The tower is 171 ft. tall.

32. COMPARISON SHOPPING

$$\frac{\$3.89}{80} = \$0.048625 \text{ per sheet}$$

$$\frac{\$6.19}{120} = \$0.051583 \text{ per sheet}$$

80 sheets for $3.89 is the better buy.

33. CLEANING HIGHWAYS

amount of work of 1st worker = $\dfrac{1}{7}$

amount of work of 2nd worker = $\dfrac{1}{9}$

amount of work together = $\dfrac{1}{x}$

$$\frac{1}{7} + \frac{1}{9} = \frac{1}{x}$$

$$63x\left(\frac{1}{7} + \frac{1}{9}\right) = 63x\left(\frac{1}{x}\right)$$

$$9x + 7x = 63$$

$$16x = 63$$

$$\boxed{\frac{16x}{16} = \frac{63}{16}}$$
$$\boxed{x = 3\frac{15}{16}}$$

It will take both of them $3\dfrac{15}{16}$ hrs.

34. PHYSICAL FITNESS

Use the formula $t = \dfrac{d}{r}$.

roller blades: $t = \dfrac{d}{r}$

$$= \dfrac{5}{x+6}$$

jogging: $t = \dfrac{d}{r}$

$$= \dfrac{2}{x}$$

Since the traveling time is the same, set the times equal and solve for x.

$$\frac{5}{x+6} = \frac{2}{x}$$

$$x(x+6)\frac{5}{x+6} = x(x+6)\frac{2}{x}$$

$$5x = 2(x+6)$$

$$5x = 2x + 12$$

$$5x - 2x = 2x + 12 - 2x$$

$$3x = 12$$

$$\frac{3x}{3} = \frac{12}{3}$$

$$\boxed{x = 4}$$

He jogs 4 mph.

35. 5 is not a common factor of the numerator and denominator.

36. We multiply both sides of the equation by the LCD of the rational expressions appearing in the equation. The resulting equation is easier to solve.

CHAPTERS 1 – 6

CUMULATIVE REVIEW

1. $9^2 - 3[45 - 3(6 + 4)] = 81 - 3[45 - 3(10)]$
$= 81 - 3[45 - 30]$
$= 81 - 3[15]$
$= 81 - 45$
$= 36$

2. GRAND KING SIZE BED

$$\frac{7,840 - 6,240}{6,240} = \frac{1,600}{6,240}$$
$$\approx 0.26$$
$$\approx 26\%$$

3. $|2 - 4| \boxed{<} - (-6)$
$|-2| < 6$
$2 < 6$

4.
$$\frac{80 + 73 + 61 + 73 + 98}{5} = \frac{385}{5}$$
$$= 77$$

5. $F = \dfrac{9}{5}C + 32$

$F = \dfrac{9}{5}(\mathbf{40}) + 32$

$= 72 + 32$

$= 104° \text{ C}$

6.
$V = \dfrac{1}{3}Bh$

$= \dfrac{1}{3}(\mathbf{6^2})(\mathbf{20})$

$= \dfrac{1}{3}(36)(20)$

$= 240 \text{ ft}^3$

7. a) false
   b) false
   c) true
   d) true

8. $2 - 3(x - 5) = 4(x - 1)$
$2 - 3x + 15 = 4x - 4$
$17 - 3x = 4x - 4$
$17 - 3x - 4x = 4x - 4 - 4x$
$17 - 7x = -4$
$17 - 7x - 17 = -4 - 17$
$-7x = -21$
$\dfrac{-7x}{-7} = \dfrac{-21}{-7}$
$\boxed{x = 3}$

9. $8(c + 7) - 2(c - 3) = 8c + 56 - 2c + 6$
$= 6c + 62$

10.
$$A - c = 2B + r$$
$$A - c - r = 2B + r - r$$
$$A - c - r = 2B$$
$$\frac{A - c - r}{2} = \frac{2B}{2}$$
$$\boxed{\frac{A - c - r}{2} = B}$$

11. $7x + 2 \geq 4x - 1$
$7x + 2 - 4x \geq 4x - 1 - 4x$
$3x + 2 \geq -1$
$3x + 2 - 2 \geq -1 - 2$
$3x \geq -3$
$\dfrac{3x}{3} \geq \dfrac{-3}{3}$
$x \geq -1$

$[-1, \infty)$

-1

12.
$$\frac{4}{5}d = -4$$
$$\frac{5}{4} \bullet \frac{4}{5}d = -4 \bullet \frac{5}{4}$$
$$\boxed{d = -5}$$

13. BLENDING TEA

Let x = pounds of \$3.20 tea and
$(20 - x)$ = pounds of \$2 tea.
$$3.20x + 2(20 - x) = 2.72(20)$$
$$3.2x + 40 - 2x = 54.4$$
$$10(3.2x + 40 - 2x) = 10(54.4)$$
$$32x + 400 - 20x = 544$$
$$12x + 400 = 544$$
$$12x + 400 - 400 = 544 - 400$$
$$12x = 144$$
$$\frac{12x}{12} = \frac{144}{12}$$

$$\boxed{\begin{array}{c} x = 12 \\ 20 - x = 20 - 12 = 8 \end{array}}$$

12 lbs of \$3.20 tea and 8 lbs of \$2 tea

14. SPEED OF A PLANE

$d = rt$
Let x = rate of plane 1 and
$x + 200$ = rate of plane 2.
The time for both planes is 5 hours.
d for plane 1 = $5x$
d for plane 2 = $5(x + 200)$
Plane 1's dis + plane 2's dis = 6,000 miles.
$$5x + 5(x + 200) = 6,000$$
$$5x + 5x + 1,000 = 6,000$$
$$10x + 1,000 = 6,000$$
$$10x + 1,000 - 1,000 = 6,000 - 1,000$$
$$10x = 5,000$$
$$\frac{10x}{10} = \frac{5,000}{10}$$
$$\boxed{x = 500}$$

15. $y = 2x - 3$

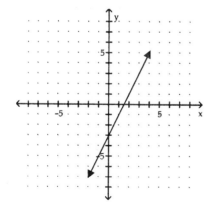

16.

$$3x - 2y \le 6$$
$$3x - 2y - 3x \le 6 - 3x$$
$$-2y \le -3x + 6$$
$$\frac{-2y}{-2} \le \frac{-3x}{-2} + \frac{6}{-2}$$
$$y \ge \frac{3}{2}x - 3$$

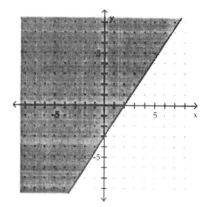

17.
Perpendicular lines have slopes that are opposite reciprocals. The slope of the given line is $-\frac{7}{8}$, so the slope of the line perpendicular to it is $\frac{8}{7}$.

18.
$$m = \frac{y_2 - y_1}{x_2 - x_1}$$
$$= \frac{-1 - 3}{3 - (-1)}$$
$$= \frac{-4}{4}$$
$$= -1$$

19.
$$y - y_1 = m(x - x_1)$$
$$y - 5 = 3(x - 1)$$
$$y - 5 = 3x - 3$$
$$y - 5 + 5 = 3x - 3 + 5$$
$$y = 3x + 2$$

20. CUTTING STEEL

$$m = \frac{y_2 - y_1}{x_2 - x_1}$$
$$= \frac{0.4 - 0}{50 - 0}$$
$$= \frac{0.4}{50}$$
$$= 0.008 \text{ mm/m}$$

21. $x^4 x^3 = x^{4+3}$
    $= x^7$

22. $(x^2 x^3)^5 = (x^{2+3})^5$
    $= (x^5)^5$
    $= x^{5 \cdot 5}$
    $= x^{25}$

23.
$$\left(\frac{y^3 y}{2yy^2}\right)^3 = \left(\frac{y^4}{2y^3}\right)^3$$
$$= \left(\frac{1}{2} y^{4-3}\right)^3$$
$$= \left(\frac{y}{2}\right)^3$$
$$= \frac{y^3}{8}$$

24.
$$\left(\frac{-2a}{b}\right)^5 = \frac{(-2)^5 a^5}{b^5}$$
$$= -\frac{32a^5}{b^5}$$

25.
$$\left(a^{-2} b^3\right)^{-4} = a^{-2 \bullet -4} b^{3 \bullet -4}$$
$$= a^8 b^{-12}$$
$$= \frac{a^8}{b^{12}}$$

26.
$$\frac{9b^0 b^3}{3b^{-3} b^4} = \frac{9b^{0+3}}{3b^{-3+4}}$$
$$= \frac{9b^3}{3b^1}$$
$$= 3b^{3-1}$$
$$= 3b^2$$

27. $290{,}000 = 2.9 \times 10^5$

28. 3

29. CONCENTRIC CIRCLES

$$A = \pi(R + r)(R - r)$$
$$= \pi(R^2 - r^2)$$
$$= \pi R^2 - \pi r^2$$

30. $y = -x^3$

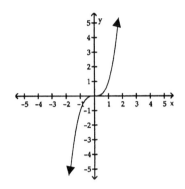

31. $(3x^2 - 3x - 2) + (3x^2 + 4x - 3)$
    $= (3x^2 + 3x^2) + (-3x + 4x) + (-2 - 3)$
    $= 6x^2 + x - 5$

32. $(2x^2 y^3)(3x^3 y^2) = (2 \cdot 3)x^{2+3} y^{3+2}$
    $= 6x^5 y^5$

33. $(2y - 5)(3y + 7) = 6y^2 + 14y - 15y - 35$
    $= 6y^2 - y - 35$

34. $-4x^2 z(3x^2 - z) = -12x^4 z + 4x^2 z^2$

35.
$$\frac{6x + 9}{3} = \frac{6x}{3} + \frac{9}{3}$$
$$= 2x + 3$$

36. $x^2 + 2x - 1$

$$\require{enclose}
\begin{array}{r}
x^2 + 2x - 1 \\
2x + 3 \enclose{longdiv}{2x^3 + 7x^2 + 4x - 3} \\
\underline{2x^3 + 3x^2} \\
4x^2 + 4x \\
\underline{4x^2 + 6x} \\
-2x - 3 \\
\underline{-2x - 3} \\
0
\end{array}$$

37. $k^3 t - 3k^2 t = k^2 t(k - 3)$

38. $2ab + 2ac + 3b + 3c = (2ab+2ac)+(3b+3c)$
$$= 2a(b + c) + 3(b + c)$$
$$= (b + c)(2a + 3)$$

39. $2a^2 - 200b^2 = 2(a^2 - 100b^2)$
$$= 2(a + 10b)(a - 10b)$$

40. $b^3 + 125 = (b + 5)(b^2 - b \cdot 5 + 5^2)$
$$= (b + 5)(b^2 - 5b + 25)$$

41. $u^2 - 18u + 81 = (u - 9)(u - 9)$
$$= (u - 9)^2$$

42. $6x^2 - 13x - 63 = (2x - 9)(3x + 7)$

43. $-r^2 + r + 2 = -(r^2 - r - 2)$
$$= -(r - 2)(r + 1)$$

44. $u^2 + 10u + 15$
prime

45. $5x^2 + x = 0$
$x(5x + 1) = 0$
$x = 0$ or $5x + 1 = 0$
$5x = -1$
$x = -\dfrac{1}{5}$

$\boxed{x = 0, \ -\dfrac{1}{5}}$

46. $6x^2 - 5x = -1$
$6x^2 - 5x + 1 = -1 + 1$
$6x^2 - 5x + 1 = 0$
$(2x - 1)(3x - 1) = 0$
$2x - 1 = 0$ or $3x - 1 = 0$
$2x = 1$ $\qquad$ $3x = 1$
$x = \dfrac{1}{2}$ $\qquad$ $x = \dfrac{1}{3}$

$\boxed{x = \dfrac{1}{2}, \ \dfrac{1}{3}}$

47. COOKING

$w(w + 6) = 160$
$w^2 + 6w = 160$
$w^2 + 6w - 160 = 160 - 160$
$w^2 + 6w - 160 = 0$
$(w + 16)(w - 10) = 0$
$w + 16 = 0$ or $w - 10 = 0$
$w = -16$ $\qquad$ $w = 10$

The width cannot be negative.
$\boxed{\begin{array}{l} w = 10 \text{ in.} \\ l = w + 6 = 10 + 6 = 16 \text{ in.} \end{array}}$

48. $x^2 - 25 = 0$
$(x - 5)(x + 5) = 0$
$x - 5 = 0$ or $x + 5 = 0$
$x = 5$ $\qquad$ $x = -5$
$\boxed{x = 5, -5}$

49.
$$\frac{x^2 - 16}{x - 4} \div \frac{3x + 12}{x} = \frac{x^2 - 16}{x - 4} \bullet \frac{x}{3x + 12}$$
$$= \frac{\cancel{(x-4)}(x+4)}{\cancel{x-4}} \bullet \frac{x}{3\cancel{(x+4)}}$$
$$= \frac{x}{3}$$

50.
$$\frac{4}{x - 3} + \frac{5}{3 - x} = \frac{4}{x - 3} + \frac{-5}{x - 3}$$
$$= \frac{4 + (-5)}{x - 3}$$
$$= \frac{-1}{x - 3}$$
$$= -\frac{1}{x - 3}$$

51.

$$\frac{m}{m^2+5m+6} - \frac{2}{m^2+3m+2}$$

$$= \frac{m}{(m+2)(m+3)} - \frac{2}{(m+1)(m+2)}$$

$$= \frac{m}{(m+2)(m+3)}\left(\frac{m+1}{m+1}\right) - \frac{2}{(m+1)(m+2)}\left(\frac{m+3}{m+3}\right)$$

$$= \frac{m^2+m}{(m+1)(m+2)(m+3)} - \frac{2m+6}{(m+1)(m+2)(m+3)}$$

$$= \frac{m^2+m-(2m+6)}{(m+1)(m+2)(m+3)}$$

$$= \frac{m^2+m-2m-6}{(m+1)(m+2)(m+3)}$$

$$= \frac{m^2-m-6}{(m+1)(m+2)(m+3)}$$

$$= \frac{(m-3)\cancel{(m+2)}}{(m+1)\cancel{(m+2)}(m+3)}$$

$$= \frac{m-3}{(m+1)(m+3)}$$

52.

$$\frac{2-\dfrac{2}{x+1}}{2+\dfrac{2}{x}} = \frac{2-\dfrac{2}{x+1}}{2+\dfrac{2}{x}} \bullet \frac{x(x+1)}{x(x+1)}$$

$$= \frac{2x(x+1) - \dfrac{2}{x+1}x(x+1)}{2x(x+1) + \dfrac{2}{x}x(x+1)}$$

$$= \frac{2x^2+2x-2x}{2x^2+2x+2x+2}$$

$$= \frac{2x^2}{2x^2+4x+2}$$

$$= \frac{\cancel{2}x^2}{\cancel{2}(x^2+2x+1)}$$

$$= \frac{x^2}{(x+1)^2}$$

53.

$$\frac{7}{5x} - \frac{1}{2} = \frac{5}{6x} + \frac{1}{3}$$

$$30x\left(\frac{7}{5x} - \frac{1}{2}\right) = 30x\left(\frac{5}{6x} + \frac{1}{3}\right)$$

$$42 - 15x = 25 + 10x$$

$$42 - 15x - 10x = 25 + 10x - 10x$$

$$42 - 25x = 25$$

$$42 - 25x - 42 = 25 - 42$$

$$-25x = -17$$

$$\frac{-25x}{-25} = \frac{-17}{-25}$$

$$\boxed{x = \frac{17}{25}}$$

54.

$$\frac{u}{u-1} + \frac{1}{u} = \frac{u^2+1}{u^2-u}$$

$$\frac{u}{u-1} + \frac{1}{u} = \frac{u^2+1}{u(u-1)}$$

$$u(u-1)\left(\frac{u}{u-1} + \frac{1}{u}\right) = u(u-1)\left(\frac{u^2+1}{u(u-1)}\right)$$

$$u^2 + u - 1 = u^2 + 1$$

$$u^2 + u - 1 - u^2 = u^2 + 1 - u^2$$

$$u - 1 = 1$$

$$u - 1 + 1 = 1 + 1$$

$$\boxed{u = 2}$$

55. DRAINING A TANK

$$\frac{1}{24} + \frac{1}{36} = \frac{1}{x}$$

$$72x\left(\frac{1}{24} + \frac{1}{36}\right) = 72x\left(\frac{1}{x}\right)$$

$$3x + 2x = 72$$

$$5x = 72$$

$$\frac{5x}{5} = \frac{72}{5}$$

$$\boxed{x = 14\frac{2}{5}}$$

It will take both pipes $14\frac{2}{5}$ hours.

56. HEIGHT OF A TREE

$$\frac{3}{2.5} = \frac{x}{29}$$

$$3(29) = 2.5x$$

$$87 = 2.5x$$

$$\frac{87}{2.5} = \frac{2.5x}{2.5}$$

$$\boxed{34.8 = x}$$

Height of tree is 34.8 ft.

SECTION 7.1

VOCABULARY

1. A pair of equations $\begin{cases} x - y = -1 \\ 2x - y = 1 \end{cases}$ is called

    a **system** of equations.

3. The x-coordinate of the ordered pairs
    (-4, 7) is -4 and a **y-coordinate** is 7.

5. The point of **intersection** of the lines
    graphed in part (a) of the following figure
    is (1, 2).

7. A system of equations that has at least one
    solution is called a **consistent** system. A
    system with no solution is called an
    inconsistent system.

CONCEPTS

9. The solution is the pair of coordinates that
    are the same in both tables. **(-4, -1)**.

11. true

13. false

15.
$$\begin{cases} y = 5x - 1 \\ 30x - 6 = 6y \end{cases}$$

$$30x - 6 = 6y$$
$$\frac{30x}{6} - \frac{6}{6} = \frac{6y}{6}$$
$$5x - 1 = y$$

17. No solution; independent

NOTATION

19. is possibly equal to

21.   solution (1,1)
$$x + y = 2$$
$$1 + 1 \overset{?}{=} 2$$
$$2 = 2$$
$$\text{yes}$$
$$2x - y = 1$$
$$2(1) - 1 \overset{?}{=} 1$$
$$2 - 1 \overset{?}{=} 1$$
$$1 = 1$$
$$\text{yes}$$

23.   solution $(3, -2)$
$$2x + y = 4$$
$$2(3) + (-2) \overset{?}{=} 4$$
$$6 - 2 \overset{?}{=} 4$$
$$4 = 4$$
$$\text{yes}$$
$$y = 1 - x$$
$$-2 \overset{?}{=} 1 - 3$$
$$-2 = -2$$
$$\text{yes}$$

25.   solution $(-2, -4)$
$$4x + 5y = \text{-23}$$
$$4(\text{-2}) + 5(\text{-4}) \overset{?}{=} \text{-23}$$
$$\text{-8} - 20 \overset{?}{=} \text{-23}$$
$$\text{-28} \neq \text{-23}$$
$$\text{no}$$
There is no need to check

the second equation.

27.

solution $\left(\frac{1}{2}, 3\right)$

$2x + y = 4$

$2\left(\frac{1}{2}\right) + 3 \overset{?}{=} 4$

$1 + 3 \overset{?}{=} 4$

$4 = 4$

yes

$4x - 11 = 3y$

$4\left(\frac{1}{2}\right) - 11 \overset{?}{=} 3(3)$

$2 - 11 \overset{?}{=} 9$

$-9 \neq 9$

no

29.

solution $(2.5, 3.5)$

$4x - 3 = 2y$

$4(2.5) - 3 \overset{?}{=} 2(3.5)$

$10 - 3 \overset{?}{=} 7$

$7 = 7$

yes

$4y + 1 = 6x$

$4(3.5) + 1 \overset{?}{=} 6(2.5)$

$14 + 1 \overset{?}{=} 15$

$15 = 15$

yes

31. (3, 2)

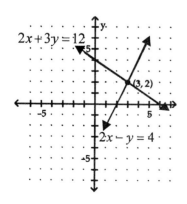

33. (-1, 5)

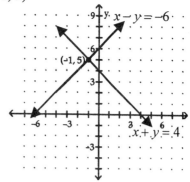

35. (-2, 0)

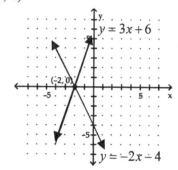

37.

The graphs are the same line.

Infinitely many solutions.

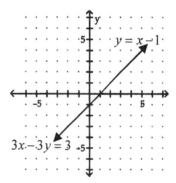

39.

The lines are parallel.

There are no solutions.

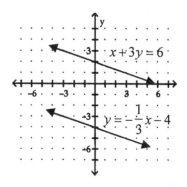

Section 7.1

41. (4, -6)

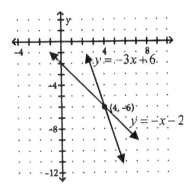

43. (5, -2)

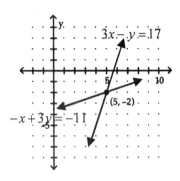

45. (1, 1)

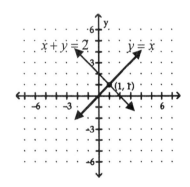

47. (-4, 0)

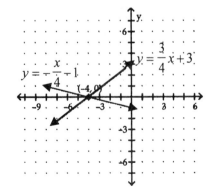

49. The lines are parallel.
There are no solutions.

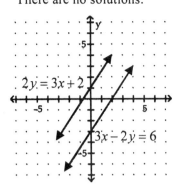

51. The graphs are the same line.
Infinitely many solutions.

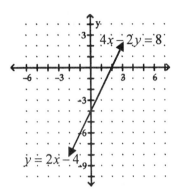

53. (3, -1)

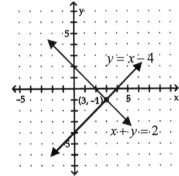

55. (-6, 1)

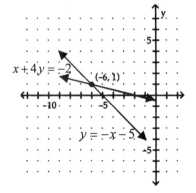

57. (3, 0)

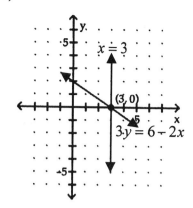

59. (-2, -3)

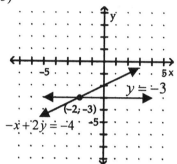

61. (4, -4)

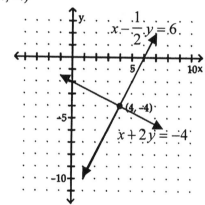

63.

$$y = 6x - 7$$
$$m = 6$$
$$y\text{-int} = (0, -7)$$

$$y = -2x + 1$$
$$m = -2$$
$$y\text{-int} = (0, 1)$$
1 solution

65.

$$3x - y = -3$$
$$y = 3x + 3$$
$$m = 3$$
$$y\text{-int} = (0, 3)$$

$$y - 3x = 3$$
$$y = 3x + 3$$
$$m = 3$$
$$y\text{-int} = (0, 3)$$
same line
infinitely many solutions

67.

$$y = -x + 6$$
$$m = -1$$
$$y\text{-int} = (0, 6)$$

$$x + y = 8$$
$$y = -x + 8$$
$$m = -1$$
$$y\text{-int} = (0, 8)$$
parallel lines
no solution

69.

$$6x + y = 0$$
$$y = -6x + 0$$
$$m = -6$$
$$y\text{-int} = (0, 0)$$

$$2x + 2y = 0$$
$$y = -x + 0$$
$$m = -1$$
$$y\text{-int} = (0, 0)$$
one solution

71. (1, 3)

73. no solution

APPLICATIONS

75. TRANSPLANTS

   1994, 4,100

77. LATITUDE AND LONGITUDE

   a) Houston, New Orleans, St. Augustine
   b) St. Louis, Memphis, New Orleans
   c) New Orleans

80. DAILY TRACKING POLL
   a) the incumbent, 7%
   b) November 2
   c) the challenger, 3

81. TV COVERAGE
   10 miles

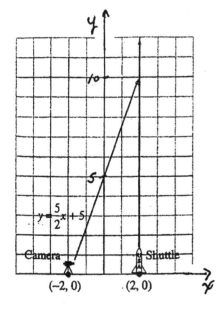

WRITING

83. Answers will vary.

85. Answers will vary.

87. $\dfrac{x+3}{x^2}+\dfrac{x+5}{2x}=\dfrac{2}{2}\left(\dfrac{x+3}{x^2}\right)+\dfrac{x}{x}\left(\dfrac{x+5}{2x}\right)$

$$=\dfrac{2x+6}{2x^2}+\dfrac{x^2+5x}{2x^2}$$

$$=\dfrac{x^2+7x+6}{2x^2}$$

89. $\dfrac{z^2+4z-5}{5z-5}\cdot\dfrac{5z}{z+5}=\dfrac{(z+5)(z-1)}{5(z-1)}\cdot\dfrac{5z}{z+5}$

$$= z$$

CHALLENGE PROBLEMS

91. No. Suppose the equations are dependent. Then they have the same graph, and would intersect at an infinite number of points. Therefore, the system would have at least one solution, and be consistent.

93. Answers will vary.
   Given (2, 3) is a solution.

$$x+y=5$$
$$x-y=10$$

   There are many more systems

   than these two.

SECTION 7.2

VOCABULARY

1. $\begin{cases} y = x + 3 \\ 3x - y = -1 \end{cases}$ is called a **system** of equations.

3. When checking a proposed solution of a system of equations, always use the **original** equations.

5. In mathematics, "to **substitute**" means to replace an expression with one that has the same value.

7. With the substitution method, the basic objective is to use an appropriate substitution to obtain one equation in **one** variable.

9. When the graphs of the equations of a system are identical lines, the equations are called dependent and the system has an **infinitely** many solutions.

CONCEPTS

11. a) $y = -3x$

    b) $x = 2y + 1$

13. a) $x = 2 + 2y$

    b) $y = -11 + 2x$

15. $x + 3x - 4 = 8$

    $x + 3(x - 4) = 8$

17. a) LCD = 15

    b) $\dfrac{x}{5} + \dfrac{2y}{3} = 1$

    $15\left(\dfrac{x}{5}\right) + 15\left(\dfrac{2y}{3}\right) = 15(1)$

    $3x + 10y = 15$

19. a) no

    b) ii

NOTATION

21. $\begin{cases} y = 3x \\ x - y = 4 \end{cases}$

    $x - y = 4$ The 2nd equation

    $x - (\boxed{3x}) = 4$

    $-2x = \boxed{4}$

    $x = \boxed{-2}$

    $y = 3x$ The 1st equation

    $y = 3(\boxed{-2})$

    $y = \boxed{-6}$

    The solution is $(\boxed{-2}, \boxed{-6})$.

PRACTICE

23. $\begin{cases} y = 2x \\ x + y = 6 \end{cases}$

    $x + y = 6$ The 2nd equation

    $x + (2x) = 6$

    $3x = 6$

    $\boxed{x = 2}$

    $y = 2x$ The 1st equation

    $y = 2(2)$

    $\boxed{y = 4}$

    The solution is $(2, 4)$.

25. $\begin{cases} y = 2x - 6 \\ 2x + y = 6 \end{cases}$

    $2x + y = 6$ The 2nd equation

    $2x + (2x - 6) = 6$

    $4x - 6 = 6$

    $4x - 6 + 6 = 6 + 6$

    $4x = 12$

    $\dfrac{4x}{4} = \dfrac{12}{4}$

    $\boxed{x = 3}$

    $y = 2x - 6$ The 1st equation

    $y = 2(3) - 6$

    $\boxed{y = 0}$

    The solution is $(3, 0)$.

27. $\begin{cases} y = 2x + 5 \\ x + 2y = -5 \end{cases}$

$\qquad x + 2y = -5 \quad$ The 2nd equation

$\quad x + 2(2x + 5) = -5$

$\qquad x + 4x + 10 = -5$

$\qquad\quad 5x + 10 = -5$

$\quad 5x + 10 - 10 = -5 - 10$

$\qquad\qquad 5x = -15$

$\qquad\qquad \dfrac{5x}{5} = \dfrac{-15}{5}$

$\qquad\qquad \boxed{x = -3}$

$\qquad y = 2x + 5 \quad$ The 1st equation

$\qquad y = 2(-3) + 5$

$\qquad \boxed{y = -1}$

The solution is $(-3, -1)$.

29. $\begin{cases} 3x + y = -4 \\ x = y \end{cases}$

$\qquad\qquad x = y \quad$ The 2nd equation

$\qquad 3x + x = -4$

$\qquad\quad 4x = -4$

$\qquad\quad \dfrac{4x}{4} = \dfrac{-4}{4}$

$\qquad\quad \boxed{x = -1}$

$\qquad 3x + y = -4 \quad$ The 1st equation

$\qquad 3(-1) + y = -4$

$\qquad -3 + y + 3 = -4 + 3$

$\qquad\qquad \boxed{y = -1}$

The solution is $(-1, -1)$.

31. $\begin{cases} 2a + 4b = -24 \\ a = 20 - 2b \end{cases}$

$\qquad 2a + 4b = -24 \quad$ The 1st equation

$\quad 2(20 - 2b) + 4b = -24$

$\qquad 40 - 4b + 4b = -24$

$\qquad\qquad 40 = -24$

$\qquad\qquad$ false

$\qquad$ No solution.

33. $\begin{cases} 2a - 3b = -13 \\ -b = -2a - 7 \end{cases}$

solve for b in 2nd equation

$\begin{cases} 2a - 3b = -13 \\ b = 2a + 7 \end{cases}$

$\qquad 2a - 3b = -13 \quad$ The 1st equation

$\quad 2a - 3(2a + 7) = -13$

$\qquad 2a - 6a - 21 = -13$

$\qquad\quad -4a - 21 = -13$

$\quad -4a - 21 + 21 = -13 + 21$

$\qquad\qquad -4a = 8$

$\qquad\qquad \dfrac{-4a}{-4} = \dfrac{8}{-4}$

$\qquad\qquad \boxed{a = -2}$

$\qquad b = 2a + 7 \quad$ The 2nd equation

$\qquad b = 2(-2) + 7$

$\qquad b = -4 + 7$

$\qquad \boxed{b = 3}$

The solution is $(-2, 3)$.

35. $\begin{cases} -y = 11 - 3x \\ 2x + 5y = -4 \end{cases}$

solve for y in 1st equation

$\begin{cases} y = 3x - 11 \\ 2x + 5y = -4 \end{cases}$

$\qquad 2x + 5y = -4 \quad$ The 2nd equation

$\quad 2x + 5(3x - 11) = -4$

$\qquad 2x + 15x - 55 = -4$

$\qquad\quad 17x - 55 = -4$

$\quad 17x - 55 + 55 = -4 + 55$

$\qquad\qquad 17x = 51$

$\qquad\qquad \dfrac{17x}{17} = \dfrac{51}{17}$

$\qquad\qquad \boxed{x = 3}$

$\qquad y = 3x - 11 \quad$ The 1st equation

$\qquad y = 3(3) - 11$

$\qquad y = 9 - 11$

$\qquad \boxed{y = -2}$

The solution is $(3, -2)$.

37. $\begin{cases} 2x+3=-4y \\ x-6=-8y \end{cases}$

solve for x in 2nd equation

$\begin{cases} 2x+3=-4y \\ x=6-8y \end{cases}$

$2x+3=-4y$ The 1st equation

$2(6-8y)+3=-4y$

$12-16y+3=-4y$

$-16y+15=-4y$

$-16y+15+16y=-4y+16y$

$15=12y$

$\dfrac{15}{12}=\dfrac{12y}{12}$

$\boxed{\dfrac{5}{4}=y}$

$x=6-8y$ The 2nd equation

$x=6-8\left(\dfrac{5}{4}\right)$

$x=6-10$

$\boxed{x=-4}$

The solution is $\left(-4,\dfrac{5}{4}\right)$.

39. $\begin{cases} r+3s=9 \\ 3r+2s=13 \end{cases}$

solve for r in 1st equation

$\begin{cases} r=9-3s \\ 3r+2s=13 \end{cases}$

$3r+2s=13$ The 2nd equation

$3(9-3s)+2s=13$

$27-9s+2s=13$

$27-7s=13$

$27-7s-27=13-27$

$-7s=-14$

$\dfrac{-7s}{-7}=\dfrac{-14}{-7}$

$\boxed{s=2}$

$r=9-3s$ The 1st equation

$r=9-3(2)$

$r=9-6$

$\boxed{r=3}$

The solution is $(3,2)$.

41. $\begin{cases} 6x-3y=5 \\ 2y+x=0 \end{cases}$

solve for x in 2nd equation

$\begin{cases} 6x-3y=5 \\ x=-2y \end{cases}$

$6x-3y=5$ The 1st equation

$6(-2y)-3y=5$

$-12y-3y=5$

$-15y=5$

$\dfrac{-15y}{-15}=\dfrac{5}{-15}$

$\boxed{y=-\dfrac{1}{3}}$

$x=-2y$ The 2nd equation

$x=-2\left(-\dfrac{1}{3}\right)$

$\boxed{x=\dfrac{2}{3}}$

The solution is $(\dfrac{2}{3},\ -\dfrac{1}{3})$.

43. $\begin{cases} y-3x=-5 \\ 21x=7y+35 \end{cases}$

splve for y in 1st equation

$\begin{cases} y=-5+3x \\ 21x=7y+35 \end{cases}$

$21x=7y+35$ The 2nd equation

$21x=7(-5+3x)+35$

$21x=-35+21x+35$

$21x=21x$

$21x-21x=21x-21x$

$\boxed{0=0}$

true

Infinitely many solutions.

45. $\begin{cases} y = 3x + 6 \\ y = -2x - 4 \end{cases}$

$y = 3x + 6$ The 1st equation

$-2x - 4 = 3x + 6$

$-2x - 4 + 4 = 3x + 6 + 4$

$-2x = 3x + 10$

$-2x - 3x = 3x + 10 - 3x$

$-5x = 10$

$\dfrac{-5x}{-5} = \dfrac{10}{-5}$

$\boxed{x = -2}$

$y = -2x - 4$ The 2nd equation

$y = -2(-2) - 4$

$y = 4 - 4$

$\boxed{y = 0}$

The solution is (-2, 0).

47. $\begin{cases} x = \dfrac{1}{3}y - 1 \\ x = y + 1 \end{cases}$

$x = \dfrac{1}{3}y - 1$ The 1st equation

$y + 1 = \dfrac{1}{3}y - 1$

$3(y) + 3(1) = 3\left(\dfrac{1}{3}y\right) - 3(1)$

$3y + 3 = y - 3$

$3y + 3 - 3 = y - 3 - 3$

$3y = y - 6$

$3y - y = y - 6 - y$

$2y = -6$

$\dfrac{2y}{2} = \dfrac{-6}{2}$

$\boxed{y = -3}$

$x = y + 1$ The 2nd equation

$x = -3 + 1$

$\boxed{x = -2}$

The solution is (-2, -3).

49. $\begin{cases} 4x + 5y = 2 \\ 3x - y = 11 \end{cases}$

solve for y in 2nd equation

$\begin{cases} 4x + 5y = 2 \\ 3x - 11 = y \end{cases}$

$4x + 5y = 2$ The 1st equation

$4x + 5(3x - 11) = 2$

$4x + 15x - 55 = 2$

$19x - 55 = 2$

$19x - 55 + 55 = 2 + 55$

$19x = 57$

$\dfrac{19x}{19} = \dfrac{57}{19}$

$\boxed{x = 3}$

$3x - 11 = y$ The 2nd equation

$3(3) - 11 = y$

$9 - 11 = y$

$\boxed{-2 = y}$

The solution is (3, -2).

51. $\begin{cases} 3x + 4y = -7 \\ 2y - x = -1 \end{cases}$

solve for x in 2nd equation

$\begin{cases} 3x + 4y = -7 \\ 2y + 1 = x \end{cases}$

$3x + 4y = -7$ The 1st equation

$3(2y + 1) + 4y = -7$

$6y + 3 + 4y = -7$

$10y + 3 = -7$

$10y + 3 - 3 = -7 - 3$

$10y = -10$

$\dfrac{10y}{10} = \dfrac{-10}{10}$

$\boxed{y = -1}$

$2y + 1 = x$ The 2nd equation

$2(-1) + 1 = x$

$-2 + 1 = x$

$\boxed{-1 = x}$

The solution is (-1, -1).

53.

$$\begin{cases} 6 - y = 4x \\ 2y = -8x - 20 \end{cases}$$

solve for y in 1ˢᵗ equation

$$\begin{cases} 6 - 4x = y \\ 2y = -8x - 20 \end{cases}$$

$2y = -8x - 20$ The 2ⁿᵈ equation

$2(6 - 4x) = -8x - 20$

$12 - 8x = -8x - 20$

$12 - 8x - 12 = -8x - 20 - 12$

$-8x = -8x - 32$

$-8x + 8x = -8x - 32 + 8x$

$\boxed{0 = -32}$

false

No solution.

57.

$$\begin{cases} 2x + 5y = -2 \\ -\dfrac{x}{2} = y \end{cases}$$

re-arrange 2ⁿᵈ equation

$$\begin{cases} 2x + 5y = -2 \\ x = -2y \end{cases}$$

$2x + 5y = -2$ The 1ˢᵗ equation

$2(-2y) + 5y = -2$

$-4y + 5y = -2$

$\boxed{y = -2}$

$x = -2y$ The 2ⁿᵈ equation

$x = -2(-2)$

$\boxed{x = 4}$

The solution is $(4, -2)$.

55.

$$\begin{cases} b = \dfrac{2}{3}a \\ 8a - 3b = 3 \end{cases}$$

$8a - 3b = 3$ The 2ⁿᵈ equation

$8a - 3\left(\dfrac{2}{3}a\right) = 3$

$8a - 2a = 3$

$6a = 3$

$\dfrac{6a}{6} = \dfrac{3}{6}$

$\boxed{a = \dfrac{1}{2}}$

$b = \dfrac{2}{3}a$ The 1ˢᵗ equation

$b = \dfrac{2}{3}\left(\dfrac{1}{2}\right)$

$\boxed{b = \dfrac{1}{3}}$

The solution is $\left(\dfrac{1}{2}, \dfrac{1}{3}\right)$.

59. $\begin{cases} \dfrac{x}{2} + \dfrac{y}{2} = -1 \\ \dfrac{x}{3} - \dfrac{y}{2} = -4 \end{cases}$

clear both equations of fractions

$\begin{cases} 2\left(\dfrac{x}{2}\right) + 2\left(\dfrac{y}{2}\right) = 2(-1) \\ 6\left(\dfrac{x}{3}\right) - 6\left(\dfrac{y}{2}\right) = 6(-4) \end{cases}$

$\begin{cases} x + y = -2 \\ 2x - 3y = -24 \end{cases}$

solve for x in 1st equation

$\begin{cases} x = -2 - y \\ 2x - 3y = -24 \end{cases}$

$2x - 3y = -24$ The 2nd equation

$2(-2-y) - 3y = -24$

$-4 - 2y - 3y = -24$

$-5y - 4 = -24$

$-5y - 4 + 4 = -24 + 4$

$-5y = -20$

$\dfrac{-5y}{-5} = \dfrac{-20}{-5}$

$\boxed{y = 4}$

$x = -2 - y$ The 1st equation

$x = -2 - 4$

$x = -6$

$\boxed{x = -6}$

The solution is (-6, 4).

61. $\begin{cases} 5x = \dfrac{1}{2}y - 1 \\ \dfrac{1}{4}y = 10x - 1 \end{cases}$

clear both equations of fractions

$\begin{cases} 2(5x) = 2\left(\dfrac{1}{2}y\right) - 2(1) \\ 4\left(\dfrac{1}{4}y\right) = 4(10x) - 4(1) \end{cases}$

$\begin{cases} 10x = y - 2 \\ y = 40x - 4 \end{cases}$

$10x = y - 2$ The 1st equation

$10x = (40x - 4) - 2$

$10x = 40x - 6$

$10x - 40x = 40x - 6 - 40x$

$-30x = -6$

$\dfrac{-30x}{-30} = \dfrac{-6}{-30}$

$\boxed{x = \dfrac{1}{5}}$

$10x = y - 2$ The 1st equation

$10\left(\dfrac{1}{5}\right) = y - 2$

$2 = y - 2$

$2 + 2 = y - 2 + 2$

$\boxed{4 = y}$

The solution is $\left(\dfrac{1}{5}, 4\right)$.

63.
$$\begin{cases} x - \dfrac{4y}{5} = 4 \\ \dfrac{y}{3} = \dfrac{x}{2} - \dfrac{5}{2} \end{cases}$$

clear both equations of fractions

$$\begin{cases} 5(x) - 5\left(\dfrac{4y}{5}\right) = 5(4) \\ 6\left(\dfrac{y}{3}\right) = 6\left(\dfrac{x}{2}\right) - 6\left(\dfrac{5}{2}\right) \end{cases}$$

$$\begin{cases} 5x - 4y = 20 \\ 2y = 3x - 15 \end{cases}$$

solve for y in 2nd equation

$$\begin{cases} 5x - 4y = 20 \\ y = \dfrac{3}{2}x - \dfrac{15}{2} \end{cases}$$

$5x - 4y = 20$ The 1st equation

$$5x - 4\left(\dfrac{3}{2}x - \dfrac{15}{2}\right) = 20$$

$$5x - 6x + 30 = 20$$

$$-x + 30 = 20$$

$$-x + 30 - 30 = 20 - 30$$

$$-x = -10$$

$$\dfrac{-x}{-1} = \dfrac{-10}{-1}$$

$$\boxed{x = 10}$$

$y = \dfrac{3}{2}x - \dfrac{15}{2}$ The 2nd equation

$$y = \dfrac{3}{2}(10) - \dfrac{15}{2}$$

$$y = \dfrac{30}{2} - \dfrac{15}{2}$$

$$\boxed{y = \dfrac{15}{2}}$$

The solution is $\left(10, \dfrac{15}{2}\right)$.

65.
$$\begin{cases} y + x = 2x + 2 \\ 6x - 4y = 21 - y \end{cases}$$

put both equations in general form

$$\begin{cases} y - x = 2 \\ 6x - 3y = 21 \end{cases}$$

solve for y in 1st equation

$$\begin{cases} y = 2 + x \\ 6x - 3y = 21 \end{cases}$$

$6x - 3y = 21$ The 2nd equation

$$6x - 3(2 + x) = 21$$

$$6x - 6 - 3x = 21$$

$$3x - 6 = 21$$

$$3x - 6 + 6 = 21 + 6$$

$$3x = 27$$

$$\dfrac{3x}{3} = \dfrac{27}{3}$$

$$\boxed{x = 9}$$

$y = 2 + x$ The 1st equation

$$y = 2 + 9$$

$$\boxed{y = 11}$$

The solution is $(9, 11)$.

67.
$$\begin{cases} x = -3y + 6 \\ 2x + 4y = 6 + x + y \end{cases}$$

put 2nd equation in general form

$$\begin{cases} x = -3y + 6 \\ x + 3y = 6 \end{cases}$$

$x + 3y = 6$ The 2nd equation

$$-3y + 6 + 3y = 6$$

$$\boxed{6 = 6}$$

true

Infinitely many solutions.

69.

$$\begin{cases} 4x + 5y + 1 = -12 + 2x \\ x - 3y + 2 = -3 - x \end{cases}$$

put both equations in general form

$$\begin{cases} 2x + 5y = -13 \\ 2x - 3y = -5 \end{cases}$$

solve for $2x$ in 1ˢᵗ equation

$$\begin{cases} 2x = -13 - 5y \\ 2x - 3y = -5 \end{cases}$$

$2x - 3y = -5$ The 2ⁿᵈ equation

$$(-13 - 5y) - 3y = -5$$
$$-8y - 13 = -5$$
$$-8y - 13 + 13 = -5 + 13$$
$$-8y = 8$$
$$\frac{-8y}{-8} = \frac{8}{-8}$$
$$\boxed{y = -1}$$

$2x = -13 - 5y$ The 1ˢᵗ equation

$$2x = -13 - 5(-1)$$
$$2x = -13 + 5$$
$$2x = -8$$
$$\frac{2x}{2} = \frac{-8}{2}$$
$$\boxed{x = -4}$$

The solution is (-4, -1).

71.

$$\begin{cases} 3(x-1) + 3 = 8 + 2y \\ 2(x+1) = 8 + y \end{cases}$$

distribute both equations

$$\begin{cases} 3x - 3 + 3 = 8 + 2y \\ 2x + 2 = 8 + y \end{cases}$$

put 1ˢᵗ equation in general form
solve for y in 2ⁿᵈ equation

$$\begin{cases} 3x - 2y = 8 \\ 2x - 6 = y \end{cases}$$

$3x - 2y = 8$ The 1ˢᵗ equation

$$3x - 2(2x - 6) = 8$$
$$3x - 4x + 12 = 8$$
continued

71. continued

$$-x + 12 = 8$$
$$-x + 12 - 12 = 8 - 12$$
$$-x = -4$$
$$\frac{-x}{-1} = \frac{-4}{-1}$$
$$\boxed{x = 4}$$

$2x - 6 = y$ The 2ⁿᵈ equation

$$2(4) - 6 = y$$
$$8 - 6 = y$$
$$\boxed{2 = y}$$

The solution is (4, 2).

APPLICATIONS

73. DINING
   He can pick melon, because it's the same price as the hash browns.

WRITING

75. Answers will vary.

77. Answers will vary.

REVIEW

79. $3x - 8 = 1$; check 3

$$3(3) - 8 \overset{?}{=} 1$$
$$9 - 8 \overset{?}{=} 1$$
$$1 = 1 \text{ true}$$
$$\text{yes}$$

81. $3(x+8) + 5x = 2(12 + 4x)$; check 3

$$3(3+8) + 5(3) \overset{?}{=} 2[12 + 4(3)]$$
$$3(11) + 15 \overset{?}{=} 2(12 + 12)$$
$$33 + 15 \overset{?}{=} 2(24)$$
$$48 = 48 \text{ true}$$
$$\text{yes}$$

83. $x^3 + 7x^2 = x^2 - 9x$; check 3

$(3)^3 + 7(3)^2 \overset{?}{=} (3)^2 - 9(3)$

$27 + 7(9) \overset{?}{=} 9 - 27$

$27 + 63 \overset{?}{=} -18$

$90 = -18$ false

no

CHALLENGE PROBLEMS

85. $\begin{cases} \dfrac{6x-1}{3} - \dfrac{5}{3} = \dfrac{3y+1}{2} \\ \dfrac{1+5y}{4} + \dfrac{x+3}{4} = \dfrac{17}{2} \end{cases}$

clear both equations of fractions

$\begin{cases} 6\left(\dfrac{6x-1}{3}\right) - 6\left(\dfrac{5}{3}\right) = 6\left(\dfrac{3y+1}{2}\right) \\ 4\left(\dfrac{1+5y}{4}\right) + 4\left(\dfrac{x+3}{4}\right) = 4\left(\dfrac{17}{2}\right) \end{cases}$

$\begin{cases} 12x - 2 - 10 = 9y + 3 \\ 1 + 5y + x + 3 = 34 \end{cases}$

put 1st equation in general form

solve for x in 2nd equation

$\begin{cases} 12x - 9y = 15 \\ x = 30 - 5y \end{cases}$

$12x - 9y = 15$ The 1st equation

$12(30 - 5y) - 9y = 15$

$360 - 60y - 9y = 15$

$360 - 69y = 15$

$360 - 69y - 360 = 15 - 360$

$-69y = 345$

$\dfrac{-69y}{-69} = \dfrac{345}{-69}$

$\boxed{y = 5}$

$x = 30 - 5y$ The 2nd equation

$x = 30 - 5(5)$

$x = 30 - 25$

$\boxed{x = 5}$

The solution is $(5, 5)$.

87. $\begin{cases} \dfrac{1}{2}x = y + 3 \\ x - 2y = 6 \end{cases}$

solve 2nd equation for x

$x = 2y + 6$

To find one solution, let $y = -3$.

Substitute that value into

2nd equation for y.

let $y = -3$

$x = 2y + 6$

$x = 2(-3) + 6$

$x = -6 + 6$

$x = 0$

$\boxed{1^{st} \text{ solution is } (0, \text{-}3)}$

solve 2nd equation for x

$x = 2y + 6$

To find one solution let $y = 0$.

Substitute that value into

2nd equation for y.

let $y = 0$

$x = 2y + 6$

$x = 2(0) + 6$

$x = 0 + 6$

$x = 6$

$\boxed{2^{nd} \text{ solution is } (6, 0)}$

solve 2nd equation for x

$x = 2y + 6$

To find one solution let $y = -2$.

Substitute that value into

2nd equation for y.

let $y = -2$

$x = 2y + 6$

$x = 2(-2) + 6$

$x = -4 + 6$

$x = 2$

$\boxed{3^{rd} \text{ solution is } (2, \text{-}2)}$

You can check each of these solutions

by substituting them into the 1st equation.

Section 7.2

SECTION 7.3

VOCABULARY

1. The coefficient of $3x$ and $-3x$ are **opposites**.

3. When the following equations are added, the variable y will be **eliminated**.

5. The elimination method for solving a system is based on the **addition** property of equality: *When equal quantities are added to both sides of an equation, the results are equal.*

CONCEPTS

7. $7y$ and $-7y$

9. a) $\quad 4x + y = 2$

    multiply both sides by 3

    $3(4x + y) = 3(2)$

    simplify

    $\boxed{12x + 3y = 6}$

  b) $\quad x - 3y = 1$

    multiply both sides by -2

    $-2(x - 3y) = -2(1)$

    simplify

    $\boxed{-2x + 6y = -2}$

11. $\begin{cases} 4x + 3y = 11 \\ 3x - 2y = 4 \end{cases}$

    $x = 2$

    pick either equation to work with to solve for y

    $x = 2$

    $4x + 3y = 11$

    $4(2) + 3y = 11$

    $8 + 3y = 11$

    $8 + 3y - 8 = 11 - 8$

    $3y = 3$

    $\dfrac{3y}{3} = \dfrac{3}{3}$

    $\boxed{y = 1}$

13. $\begin{cases} 2x + 3y = -1 \\ 3x + 5y = -2 \end{cases}$

$2x + 3y = -1$

Is $(1, -1)$ is solution?

$2x + 3y = -1$

$2(1) + 3(-1) \stackrel{?}{=} -1$

$2 - 3 \stackrel{?}{=} -1$

$-1 = -1$

true

$3x + 5y = -2$

Is $(1, -1)$ is solution?

$3x + 5y = -2$

$3(1) + 5(-1) \stackrel{?}{=} -2$

$3 - 5 \stackrel{?}{=} -2$

$-2 = -2$

true

NOTATION

15. Solve $\begin{cases} x + y = 5 \\ x - y = -3 \end{cases}$

$x + y = 5$

$\underline{x - y = -3}$

$\boxed{2x} \quad = 2$

$\dfrac{2x}{2} = \dfrac{2}{2}$

$x = \boxed{1}$

$x + y = 5$ The 1st equation

$\boxed{1} + y = 5$

$y = 4$

The solution is $\boxed{(1, 4)}$.

17. $\begin{cases} x + y = 5 \\ x - y = 1 \end{cases}$

eliminate the y

$x + y = 5$

$\underline{x - y = 1}$

$2x \quad = 6$

$\dfrac{2x}{2} = \dfrac{6}{2}$

$\boxed{x = 3}$

$x + y = 5 \quad$ 1st equation

$3 + y = 5$

$3 + y - 3 = 5 - 3$

$\boxed{y = 2}$

solution is (3, 2)

19.

$\begin{cases} x + y = 1 \\ x - y = 5 \end{cases}$

eliminate the y

$x + y = 1$

$\underline{x - y = 5}$

$2x \quad = 6$

$\dfrac{2x}{2} = \dfrac{6}{2}$

$\boxed{x = 3}$

$x + y = 1 \quad$ 1st equation

$3 + y = 1$

$3 + y - 3 = 1 - 3$

$\boxed{y = -2}$

solution is (3, -2)

21. $\begin{cases} x + y = -5 \\ -x + y = -1 \end{cases}$

eliminate the x

$x + y = -5$

$\underline{\text{-}x + y = -1}$

$2y = -6$

$\dfrac{2y}{2} = \dfrac{-6}{2}$

$\boxed{y = -3}$

$x + y = -5 \quad$ 1st equation

$x + (-3) = -5$

$x - 3 + 3 = -5 + 3$

$\boxed{x = -2}$

solution is (-2, -3)

23.

$\begin{cases} 4x + 3y = 24 \\ 4x - 3y = -24 \end{cases}$

eliminate the y

$4x + 3y = 24$

$\underline{4x - 3y = -24}$

$8x \quad = 0$

$\dfrac{8x}{8} = \dfrac{0}{8}$

$\boxed{x = 0}$

$4x + 3y = 24 \quad$ 1st equation

$4(0) + 3y = 24$

$3y = 24$

$\dfrac{3y}{3} = \dfrac{24}{3}$

$\boxed{y = 8}$

solution is (0, 8)

25.

$$\begin{cases} 2s+t=-2 \\ -2s-3t=-6 \end{cases}$$

eliminate the s

$$2s+\ t=-2$$
$$\underline{-2s-3t=-6}$$
$$-2t=-8$$

$$\frac{-2t}{-2}=\frac{-8}{-2}$$

$$\boxed{t=4}$$

$$2s+t=-2 \quad 1^{\text{st}} \text{ equation}$$
$$2s+4=-2$$
$$2s+4-4=-2-4$$
$$2s=-6$$

$$\frac{2s}{2}=\frac{-6}{2}$$

$$\boxed{s=-3}$$

solution is (-3, 4)

27.

$$\begin{cases} 5x-4y=8 \\ -5x-4y=8 \end{cases}$$

eliminate the x

$$5x-4y=8$$
$$\underline{-5x-4y=8}$$
$$-8y=16$$

$$\frac{-8y}{-8}=\frac{16}{-8}$$

$$\boxed{y=-2}$$

$$5x-4y=8 \quad 1^{\text{st}} \text{ equation}$$
$$5x-4(-2)=8$$
$$5x+8=8$$
$$5x+8-8=8-8$$
$$5x=0$$

$$\frac{5x}{5}=\frac{0}{5}$$

$$\boxed{x=0}$$

solution is (0, -2)

29.

$$\begin{cases} 4x-7y=-19 \\ -4x+7y=19 \end{cases}$$

eliminate the x or y

$$4x-7y=-19$$
$$\underline{-4x+7y=19}$$
$$0=0$$

true

infinitely many solutions

31.

$$\begin{cases} x+3y=-9 \\ x+8y=-4 \end{cases}$$

multiply 1st equation by -1

$$-x-3y=9$$
$$\underline{x+8y=-4}$$
$$5y=5$$

$$\frac{5y}{5}=\frac{5}{5}$$

$$\boxed{y=1}$$

$$x+3y=-9 \quad 1^{\text{st}} \text{ equation}$$
$$x+3(1)=-9$$
$$x+3-3=-9-3$$

$$\boxed{x=-12}$$

solution is (-12, 1)

33.

$$\begin{cases} 5c+d=-15 \\ 6c+d=-20 \end{cases}$$

multiply 1st equation by -1

$$-5c-d=15$$
$$\underline{6c+d=-20}$$
$$\boxed{c=-5}$$

$$5c+d=-15 \quad 1^{\text{st}} \text{ equation}$$
$$5(-5)+d=-15$$
$$-25+d=-15$$
$$-25+d+25=-15+25$$

$$\boxed{d=10}$$

solution is (-5, 10)

35.

$$\begin{cases} 7x - y = 10 \\ 8x - y = 13 \end{cases}$$

multiply 1st equation by -1

$-7x + y = -10$

$\underline{8x - y = 13}$

$\boxed{x = 3}$

$7x - y = 10 \quad$ 1st equation

$7(3) - y = 10$

$21 - y = 10$

$21 - y + y = 10 + y$

$21 - 10 = 10 + y - 10$

$\boxed{11 = y}$

solution is (3, 11)

37.

$$\begin{cases} 3x - 5y = -29 \\ 3x - 5y = 15 \end{cases}$$

multiply 1st equation by -1

$-3x + 5y = 29$

$\underline{3x - 5y = 15}$

$\boxed{0 = 44}$

false

no solution

39.

$$\begin{cases} 6x - 3y = -7 \\ 9x + y = 6 \end{cases}$$

multiply 2nd equation by 3

$6x - 3y = -7$

$\underline{27x + 3y = 18}$

$33x = 11$

$\dfrac{33x}{33} = \dfrac{11}{33}$

$\boxed{x = \dfrac{1}{3}}$

$6x - 3y = -7 \quad$ 1st equation

$6\left(\dfrac{1}{3}\right) - 3y = -7$

$2 - 3y = -7$

$2 - 3y - 2 = -7 - 2$

$-3y = -9$

$\dfrac{-3y}{-3} = \dfrac{-9}{-3}$

$\boxed{y = 3}$

solution is $\left(\dfrac{1}{3}, 3\right)$

41.

$$\begin{cases} 9x + 4y = 31 \\ 6x - y = -5 \end{cases}$$

multiply 2nd equation by 4

$$9x + 4y = 31$$
$$\underline{24x - 4y = -20}$$
$$33x = 11$$

$$\frac{33x}{33} = \frac{11}{33}$$

$$\boxed{x = \frac{1}{3}}$$

$$6x - y = -5 \quad 2^{nd} \text{ equation}$$

$$6\left(\frac{1}{3}\right) - y = -5$$

$$2 - y = -5$$
$$2 - y + y = -5 + y$$
$$2 = -5 + y$$
$$2 + 5 = -5 + y + 5$$

$$\boxed{7 = y}$$

solution is $\left(\dfrac{1}{3}, 7\right)$

43.

$$\begin{cases} 8x + 8y = -16 \\ 3x + y = -4 \end{cases}$$

divide 1st equation by -8

$$-x - y = 2$$
$$\underline{3x + y = -4}$$
$$2x = -2$$

$$\frac{2x}{2} = \frac{-2}{2}$$

$$\boxed{x = -1}$$

$$3x + y = -4 \quad 2^{nd} \text{ equation}$$
$$3(-1) + y = -4$$
$$-3 + y = -4$$
$$-3 + y + 3 = -4 + 3$$

$$\boxed{y = -1}$$

solution is (-1, -1)

45.

$$\begin{cases} 7x - 50y = -43 \\ x + 3y = 4 \end{cases}$$

multiply 2nd equation by -7

$$7x - 50y = -43$$
$$\underline{-7x - 21y = -28}$$
$$-71y = -71$$

$$\frac{-71y}{-71} = \frac{-71}{-71}$$

$$\boxed{y = 1}$$

$$x + 3y = 4 \quad 2^{nd} \text{ equation}$$
$$x + 3(1) = 4$$
$$x + 3 - 3 = 4 - 3$$

$$\boxed{x = 1}$$

solution is (1, 1)

47.

$$\begin{cases} 8x - 4y = 18 \\ 3x - 2y = 8 \end{cases}$$

multiply 2nd equation by -2

$$8x - 4y = 18$$
$$\underline{-6x + 4y = -16}$$
$$2x = 2$$
$$\frac{2x}{2} = \frac{2}{2}$$
$$\boxed{x = 1}$$

$3x - 2y = 8 \quad$ 2nd equation

$$3(1) - 2y = 8$$
$$3 - 2y - 3 = 8 - 3$$
$$-2y = 5$$
$$\frac{-2y}{-2} = \frac{5}{-2}$$
$$\boxed{y = -\frac{5}{2}}$$

solution is $\left(1, -\dfrac{5}{2}\right)$

49.

$$\begin{cases} 4x + 3y = 7 \\ 3x - 2y = -16 \end{cases}$$

multiply 1st equation by 2

multiply 2nd equation by 3

$$8x + 6y = 14$$
$$\underline{9x - 6y = -48}$$
$$17x = -34$$
$$\frac{17x}{17} = \frac{-34}{17}$$
$$\boxed{x = -2}$$

$4x + 3y = 7 \quad$ 1st equation

$$4(-2) + 3y = 7$$
$$-8 + 3y + 8 = 7 + 8$$
$$3y = 15$$
$$\frac{3y}{3} = \frac{15}{3}$$
$$\boxed{y = 5}$$

solution is (-2, 5)

51.

$$\begin{cases} 3x + 4y = 12 \\ 4x + 5y = 17 \end{cases}$$

multiply 1st equation by 5

multiply 2nd equation by -4

$$15x + 20y = 60$$
$$\underline{-16x - 20y = -68}$$
$$-x = -8$$
$$\frac{-x}{-1} = \frac{-8}{-1}$$
$$\boxed{x = 8}$$

$3x + 4y = 12 \quad$ 1st equation

$$3(8) + 4y = 12$$
$$24 + 4y - 24 = 12 - 24$$
$$4y = -12$$
$$\frac{4y}{4} = \frac{-12}{4}$$
$$\boxed{y = -3}$$

solution is (8, -3)

53.

$$\begin{cases} -3x + 6y = -9 \\ -5x + 4y = -15 \end{cases}$$

multiply 1st equation by 2

multiply 2nd equation by -3

$$-6x + 12y = -18$$
$$\underline{15x - 12y = 45}$$
$$9x = 27$$
$$\frac{9x}{9} = \frac{27}{9}$$
$$\boxed{x = 3}$$

$-3x + 6y = -9 \quad$ 1st equation

$$-3(3) + 6y = -9$$
$$-9 + 6y + 9 = -9 + 9$$
$$6y = 0$$
$$\frac{6y}{6} = \frac{0}{6}$$
$$\boxed{y = 0}$$

solution is (3, 0)

55.

$$\begin{cases} 4a + 7b = 2 \\ 9a - 3b = 1 \end{cases}$$

multiply 1st equation by 3

multiply 2nd equation by 7

$$12a + 21b = 6$$
$$\underline{63a - 21b = 7}$$
$$75a = 13$$
$$\frac{75a}{75} = \frac{13}{75}$$
$$\boxed{a = \frac{13}{75}}$$

$$4a + 7b = 2 \quad 1^{st} \text{ equation}$$

$$4\left(\frac{13}{75}\right) + 7b = 2$$

$$\frac{52}{75} + 7b = 2$$

$$\frac{52}{75} + 7b - \frac{52}{75} = 2 - \frac{52}{75}$$

$$7b = \frac{150}{75} - \frac{52}{75}$$

$$7b = \frac{98}{75}$$

$$7b\left(\frac{1}{7}\right) = \frac{98}{75}\left(\frac{1}{7}\right)$$

$$\boxed{b = \frac{14}{75}}$$

solution is $\left(\dfrac{13}{75}, \dfrac{14}{75}\right)$

57.

$$\begin{cases} 9x = 10y \\ 3x - 2y = 12 \end{cases}$$

put 1st equation in general form

$$\begin{cases} 9x - 10y = 0 \\ 3x - 2y = 12 \end{cases}$$

multiply 2nd equation by -3

$$9x - 10y = 0$$
$$\underline{-9x + 6y = -36}$$
$$-4y = -36$$
$$\frac{-4y}{-4} = \frac{-36}{-4}$$
$$\boxed{y = 9}$$

$$9x = 10y \quad 1^{st} \text{ equation}$$
$$9x = 10(9)$$
$$9x = 90$$
$$\frac{9x}{9} = \frac{90}{9}$$
$$\boxed{x = 10}$$

solution is (10, 9)

59.

$$\begin{cases} 2x + 5y + 13 = 0 \\ 2x + 5 = 3y \end{cases}$$

put both equations in general form

$$\begin{cases} 2x + 5y = -13 \\ 2x - 3y = -5 \end{cases}$$

multiply 2nd equation by -1

$$2x + 5y = -13$$
$$\underline{-2x + 3y = 5}$$
$$8y = -8$$
$$\frac{8y}{8} = \frac{-8}{8}$$
$$\boxed{y = -1}$$

$$2x + 5 = 3y \quad 2^{nd} \text{ equation}$$
$$2x + 5 = 3(-1)$$
$$2x + 5 - 5 = -3 - 5$$
$$2x = -8$$
$$\frac{2x}{2} = \frac{-8}{2}$$
$$\boxed{x = -4}$$

solution is (-4, -1)

61.

$$\begin{cases} 0 = 4x - 3y \\ 5x = 4y - 2 \end{cases}$$

put 2nd equation in general form

$$\begin{cases} 4x - 3y = 0 \\ 5x - 4y = -2 \end{cases}$$

multiply 1st equation by 4

multiply 2nd equation by -3

$$16x - 12y = 0$$
$$\underline{-15x + 12y = 6}$$
$$\boxed{x = 6}$$

$$4x - 3y = 0 \quad 1^{st} \text{ equation}$$
$$4(6) - 3y = 0$$
$$24 - 3y - 24 = 0 - 24$$
$$-3y = -24$$
$$\frac{-3y}{-3} = \frac{-24}{-3}$$
$$\boxed{y = 8}$$

solution is (6, 8)

63.

$$\begin{cases} 3x - 16 = 5y \\ -3x + 5y - 33 = 0 \end{cases}$$

put both equations in general form

$$\begin{cases} 3x - 5y = 16 \\ -3x + 5y = 33 \end{cases}$$

$$3x - 5y = 16$$
$$\underline{-3x + 5y = 33}$$
$$\boxed{0 = 49}$$

false

no solution

65.

$$\begin{cases} \dfrac{3}{5}s + \dfrac{4}{5}t = 1 \\ -\dfrac{1}{4}s + \dfrac{3}{8}t = 1 \end{cases}$$

clear both equations of fractions

$$\begin{cases} 5\left(\dfrac{3}{5}s\right) + 5\left(\dfrac{4}{5}t\right) = 5(1) \\ 8\left(-\dfrac{1}{4}s\right) + 8\left(\dfrac{3}{8}t\right) = 8(1) \end{cases}$$

$$\begin{cases} 3s + 4t = 5 \\ -2s + 3t = 8 \end{cases}$$

multiply 1st equation by 2

multiply 2nd equation by 3

$$6s + 8t = 10$$
$$\underline{-6s + 9t = 24}$$
$$17t = 34$$
$$\frac{17t}{17} = \frac{34}{17}$$
$$\boxed{t = 2}$$

$$3s + 4t = 5 \quad 1^{st} \text{ equation}$$
$$3s + 4(2) = 5$$
$$3s + 8 - 8 = 5 - 8$$
$$3s = -3$$
$$\frac{3s}{3} = \frac{-3}{3}$$
$$\boxed{s = -1}$$

solution is (-1, 2)

67.

$$\begin{cases} \dfrac{1}{2}s - \dfrac{1}{4}t = 1 \\ \dfrac{1}{3}s + t = 3 \end{cases}$$

clear both equations of fractions

$$\begin{cases} 4\left(\dfrac{1}{2}s\right) - 4\left(\dfrac{1}{4}t\right) = 4(1) \\ 3\left(\dfrac{1}{3}s\right) + 3(t) = 3(3) \end{cases}$$

$$\begin{cases} 2s - t = 4 \\ s + 3t = 9 \end{cases}$$

multiply 2nd equation by -2

$$2s - t = 4$$
$$\underline{-2s - 6t = -18}$$
$$-7t = -14$$
$$\dfrac{-7t}{-7} = \dfrac{-14}{-7}$$
$$\boxed{t = 2}$$

$$\dfrac{1}{3}s + t = 3 \quad 2^{nd} \text{ equation}$$
$$\dfrac{1}{3}s + (2) = 3$$
$$\dfrac{1}{3}s + 2 - 2 = 3 - 2$$
$$\dfrac{1}{3}s = 1$$
$$\dfrac{3}{1}\left(\dfrac{1}{3}s\right) = 3(1)$$
$$\boxed{s = 3}$$

solution is (3, 2)

69.

$$\begin{cases} -\dfrac{m}{4} - \dfrac{n}{3} = \dfrac{1}{12} \\ \dfrac{m}{2} - \dfrac{5n}{4} = \dfrac{7}{4} \end{cases}$$

clear both equations of fractions

$$\begin{cases} 12\left(-\dfrac{m}{4}\right) - 12\left(\dfrac{n}{3}\right) = 12\left(\dfrac{1}{12}\right) \\ 4\left(\dfrac{m}{2}\right) - 4\left(\dfrac{5n}{4}\right) = 4\left(\dfrac{7}{4}\right) \end{cases}$$

$$\begin{cases} -3m - 4n = 1 \\ 2m - 5n = 7 \end{cases}$$

multiply 1st equation by 2
multiply 2nd equation by 3

$$-6m - 8n = 2$$
$$\underline{6m - 15n = 21}$$
$$-23n = 23$$
$$\dfrac{-23n}{-23} = \dfrac{23}{-23}$$
$$\boxed{n = -1}$$

$$2m - 5n = 7 \quad 2^{nd} \text{ equation}$$
$$2m - 5(-1) = 7$$
$$2m + 5 - 5 = 7 - 5$$
$$2m = 2$$
$$\dfrac{2m}{2} = \dfrac{2}{2}$$
$$\boxed{m = 1}$$

solution is (1, -1)

71.

$$\begin{cases} \dfrac{x}{2} - 3y = 7 \\ -x + 6y = -14 \end{cases}$$

clear 1st equation of fractions

$$\begin{cases} 2\left(\dfrac{x}{2}\right) - 2(3y) = 2(7) \\ -x + 6y = -14 \end{cases}$$

$$\begin{cases} x - 6y = 14 \\ -x + 6y = -14 \end{cases}$$

$$\begin{array}{r} x - 6y = 14 \\ -x + 6y = -14 \\ \hline \boxed{0 = 0} \end{array}$$

true

infinitely many solutions

73.

$$\begin{cases} 2x + y = 10 \\ 0.1x + 0.2y = 1.0 \end{cases}$$

multiply 2nd equation by 10

$$\begin{cases} 2x + y = 10 \\ 1x + 2y = 10 \end{cases}$$

multiply 2nd equation by -2

$$\begin{array}{r} 2x + y = 10 \\ -2x - 4y = -20 \\ \hline -3y = -10 \end{array}$$

$$\dfrac{-3y}{-3} = \dfrac{-10}{-3}$$

$$\boxed{y = \dfrac{10}{3}}$$

$x + 2y = 10$ 2nd equation

$$x + 2\left(\dfrac{10}{3}\right) = 10$$

$$x + \dfrac{20}{3} = 10$$

$$x + \dfrac{20}{3} - \dfrac{20}{3} = 10 - \dfrac{20}{3}$$

$$x = \dfrac{30}{3} - \dfrac{20}{3}$$

$$\boxed{x = \dfrac{10}{3}}$$

solution is $\left(\dfrac{10}{3}, \dfrac{10}{3}\right)$

75.

$$\begin{cases} 2x - y = 16 \\ 0.03x + 0.02y = 0.03 \end{cases}$$

multiply 2nd equation by 100

$$\begin{cases} 2x - y = 16 \\ 3x + 2y = 3 \end{cases}$$

multiply 1st equation by 2

$$\begin{array}{r} 4x - 2y = 32 \\ 3x + 2y = 3 \\ \hline 7x = 35 \end{array}$$

$$\dfrac{7x}{7} = \dfrac{35}{7}$$

$$\boxed{x = 5}$$

$2x - y = 16$ 1st equation

$$2(5) - y = 16$$

$$10 - y - 10 = 16 - 10$$

$$-y = 6$$

$$\dfrac{-y}{-1} = \dfrac{6}{-1}$$

$$\boxed{y = -6}$$

solution is (5, -6)

APPLICATIONS

77.

$$\begin{cases} 9x + 11y = 352 \\ 5x - 11y = -198 \end{cases}$$

$$\begin{array}{r} 9x + 11y = 352 \\ 5x - 11y = -198 \\ \hline 14x = 154 \end{array}$$

$$\dfrac{14x}{14} = \dfrac{154}{14}$$

$$\boxed{x = 11}$$

$x = 0$ implies 1980

$x = 11$ implies $1980 + 11$

The year when the percents are equal is 1991.

WRITING

79. Answers will vary.

81. Answers will vary.

83. 10 is less than x

$$\boxed{x-10}$$

85. $x - x$

$$\boxed{0}$$

87. $A = \dfrac{1}{2}bh$

$b = 4$ feet

$h = 3.75$ feet

$A = ??$

$A = 0.5(4 \text{ ft})(3.75 \text{ ft})$

$A = 7.5 \text{ ft}^2$

89.

$$\begin{cases} \dfrac{x-3}{2} + \dfrac{y+5}{3} = \dfrac{11}{6} \\ \dfrac{x+3}{3} - \dfrac{5}{12} = \dfrac{y+3}{4} \end{cases}$$

clear both equations of fractions

$$\begin{cases} 6\left(\dfrac{x-3}{2}\right) + 6\left(\dfrac{y+5}{3}\right) = 6\left(\dfrac{11}{6}\right) \\ 12\left(\dfrac{x+3}{3}\right) - 12\left(\dfrac{5}{12}\right) = 12\left(\dfrac{y+3}{4}\right) \end{cases}$$

$$\begin{cases} 3(x-3) + 2(y+5) = 11 \\ 4(x+3) - 1(5) = 3(y+3) \end{cases}$$

$$\begin{cases} 3x - 9 + 2y + 10 = 11 \\ 4x + 12 - 5 = 3y + 9 \end{cases}$$

$$\begin{cases} 3x + 2y + 1 = 11 \\ 4x + 7 = 3y + 9 \end{cases}$$

put both equations in general form

$$\begin{cases} 3x + 2y = 10 \\ 4x - 3y = 2 \end{cases}$$

multiply 1st equation by 3

multiply 2nd equation by 2

$9x + 6y = 30$

$\underline{8x - 6y = 4}$

$\qquad 17x = 34$

$$\dfrac{17x}{17} = \dfrac{34}{17}$$

$$\boxed{x = 2}$$

$3x + 2y = 10 \quad$ 1st equation

$3(2) + 2y = 10$

$6 + 2y - 6 = 10 - 6$

$\qquad 2y = 4$

$$\dfrac{2y}{2} = \dfrac{4}{2}$$

$$\boxed{y = 2}$$

solution is $(2, 2)$

VOCABULARY

1. A **variable** is a letter that stands for a number.

3. $\begin{cases} a+b = 20 \\ a = 2b+4 \end{cases}$ is a **system** of equations.

5. Two angles are said to be **complementary** if the sum of their measures is 90°.

CONCEPTS

7.
   Let x = length of shorter piece

   y = length of longer piece

   20 in. = total length

   longer = 1 in. less than twice shorter

   $\boxed{\begin{array}{l} x+y = 20 \text{ , } 1^{st} \text{ equation} \\ y = 2x-1 \text{ , } 2^{nd} \text{ equation} \end{array}}$

9.
   Let x = larger $\angle$

   y = smaller $\angle$

   sum of supplementary $\angle = 180^0$

   smaller $\angle = 25°$ less than larger $\angle$

   $\boxed{\begin{array}{l} x+y = 180 \text{ , } 1^{st} \text{ equation} \\ y = x-25 \text{ , } 2^{nd} \text{ equation} \end{array}}$

11.
    let x = cost of a chicken taco

    y = cost of a beef taco

    $\boxed{5x+2y = 10}$

13.
    a) Downstream: $x+c$
    b) Upstream: $x-c$

15.
    a) x ml of 30% acid solution

    y ml of 40% acid solution

    $(x+y)$ ml

    b) 33% acid solution

17.
    Let x = measure of one $\angle$

    y = measure of other $\angle$

    sum of complementary $\angle = 90^0$

    one $\angle$ is 10° more than three times other $\angle$

    $\begin{cases} x+y = 90 \\ x = 10+3y \end{cases}$

    $\begin{cases} x+y = 90 \\ x-3y = 10 \end{cases}$

    $\begin{array}{l} x+y = 90 \\ \underline{-x+3y = -10} \\ \quad\quad 4y = 80 \end{array}$

    $\dfrac{4y}{4} = \dfrac{80}{4}$

    $\boxed{y = 20}$

    $x = 10+3y$

    $x = 10+3(20)$

    $\boxed{x = 70}$

    One $\angle$ is 20°.

    Other $\angle$ is 70°.

19.
    Let x = large $\angle$

    y = small $\angle$

    sum of supplementary $\angle = 180^0$

    large $\angle$ – small $\angle = 80°$

    $\begin{cases} x+y = 180 \\ x-y = 80 \end{cases}$

    $\begin{array}{l} x+y = 180 \\ \underline{x-y = 80} \\ 2x = 260 \end{array}$

    $\dfrac{2x}{2} = \dfrac{260}{2}$

    $\boxed{x = 130}$

    $x+y = 180$

    $130+y = 180$

    $130+y-130 = 180-130$

    $\boxed{y = 50}$

    Large $\angle$ is 130°.

    Small $\angle$ is 50°.

APPLICATIONS

21. TREE TRIMMING

Let x = length of upper arm

y – length of lower arm

total length of both arms = 51 ft.

upper arm = 7 ft. shorter than lower arm

$$\begin{cases} x+y=51 \\ x=y-7 \end{cases}$$

$$\begin{cases} x+y=51 \\ x-y=-7 \end{cases}$$

$$x+y=51$$
$$\underline{x-y=-7}$$
$$2x=44$$

$$\frac{2x}{2}=\frac{44}{2}$$

$$\boxed{x=22}$$

$$x+y=51$$
$$22+y=51$$
$$22+y-22=51-22$$

$$\boxed{y=29}$$

Upper arm is 22 ft.

Lower arm is 29 ft.

23. EXECUTIVE BRANCH

Let x = president's salary (large)

y = vice-president salary (small)

total of both salaries = $592,600

president makes $207,400 more than VP

$$\begin{cases} x+y=592,600 \\ x=y+207,400 \end{cases}$$

$$\begin{cases} x+y=592,600 \\ x-y=207,400 \end{cases}$$

$$x+y=592,600$$
$$\underline{x-y=207,400}$$
$$2x=800,000$$

$$\frac{2x}{2}=\frac{800,000}{2}$$

$$\boxed{x=400,000}$$

$$x+y=592,600$$
$$400,000+y=592,600$$

subtract 400,000 from both sides

$$-400,000 \ = \ -400,000$$

$$\boxed{y=192,600}$$

President's salary is $400,000.

Vice-President salary is $192,600.

25. MARINE CORPS

Let $x = \angle 1$

$y = \angle 2$

sum of supplementary $\angle = 180^0$

$\angle 2 = 15°$ less than twice $\angle 1$

$$\begin{cases} x+y=180 \\ y=2x-15 \end{cases}$$

$$\begin{cases} x+y=180 \\ -2x+y=-15 \end{cases}$$

$$x+y=180$$
$$\underline{2x-y=15}$$
$$3x=195$$

$$\frac{3x}{3}=\frac{195}{3}$$

$$\boxed{x=65}$$

$$y=2x-15$$
$$y=2(65)-15$$
$$y=130-15$$

$$\boxed{y=115}$$

$\angle 1$ is 65°.

$\angle 2$ is 115°.

27. THEATER SCREEN

Let l = length in ft.

w = width in ft.

width = 26 ft. less than length

perimeter = 332 ft.

perimeter = $2l + 2w$

$$\begin{cases} 2l+2w=332 \\ w=l-26 \end{cases}$$

$$2l+2(l-26)=332$$
$$2l+2l-52=332$$
$$4l-52+52=332+52$$
$$4l=384$$

$$\frac{4l}{4}=\frac{384}{4}$$

$$\boxed{l=96}$$

$$w=l-26$$
$$w=96-26$$

$$\boxed{w=70}$$

Length is 96 ft.

Width is 70 ft.

29. GEOMETRY

Let l = length in meters (large)

w = width in meters (small)

perimeter of path is 50 meters

$2l + 2w$ = perimeter

width is two-thirds its length

$$\begin{cases} 2l + 2w = 50 \\ w = \dfrac{2}{3}l \end{cases}$$

$$2l + 2\left(\dfrac{2}{3}l\right) = 50$$

$$3(2l) + 3\left(\dfrac{4}{3}l\right) = 3(50)$$

$$6l + 4l = 150$$

$$10l = 150$$

$$\dfrac{10l}{10} = \dfrac{150}{10}$$

$$\boxed{l = 15}$$

$$2l + 2w = 50$$

$$2(15) + 2w = 50$$

$$30 + 2w - 30 = 50 - 30$$

$$2w = 20$$

$$\dfrac{2w}{2} = \dfrac{20}{2}$$

$$\boxed{w = 10}$$

Length is 15 m.

Width is 10 m.

31. BUYING PAINTING SUPPLIES

Let x = price of one gallon of paint

y = price of one brush

cost of 8 gal and 3 brushes is $270

cost of 6 gal and 2 brushes is $200

$$\begin{cases} 8x + 3y = 270 \\ 6x + 2y = 200 \end{cases}$$

$$\begin{cases} -2(8x + 3y) = -2(270) \\ 3(6x + 2y) = 3(200) \end{cases}$$

$$\begin{array}{r} -16x - 6y = -540 \\ 18x + 6y = 600 \\ \hline 2x = 60 \end{array}$$

$$\dfrac{2x}{2} = \dfrac{60}{2}$$

$$\boxed{x = 30}$$

$$6x + 2y = 200$$

$$6(30) + 2y = 200$$

$$180 + 2y - 180 = 200 - 180$$

$$2y = 20$$

$$\dfrac{2y}{2} = \dfrac{20}{2}$$

$$\boxed{y = 10}$$

A gallon of paint costs $30.

One brush costs $10.

33. SELLING ICE CREAM

Let x = # of cones sold

y = # of sundaes sold

a cone cost 0.90 , a sundae cost 1.65

# of cones and # of sundaes = 148

total receipts = 180.45

$$\begin{cases} x + y = 148 \\ 0.90x + 1.65y = 180.45 \end{cases}$$

$$\begin{cases} x = 148 - y \\ 0.90x + 1.65y = 180.45 \end{cases}$$

$$0.90x + 1.65y = 180.45$$

$$0.90(148 - y) + 1.65y = 180.45$$

$$133.20 - 0.90y + 1.65y = 180.45$$

$$133.20 + 0.75y - 133.20 = 180.45 - 133.20$$

$$0.75y = 47.25$$

$$\frac{0.75y}{0.75} = \frac{47.25}{0.75}$$

$$\boxed{y = 63}$$

$$x + y = 148$$

$$x + 63 = 148$$

$$x + 63 - 63 = 148 - 63$$

$$\boxed{x = 85}$$

85 cones were sold.

63 sundaes were sold.

35. MAKING TIRES

a) cost of 1st mold = $1,000 + 15t$

cost of 2nd mold = $3,000 + 10t$

Let t = break point for tires

$$\begin{cases} c = 1,000 + 15t \\ c = 3,000 + 10t \end{cases}$$

cost of 1st mold = cost of 2nd mold

$$1,000 + 15t = 3,000 + 10t$$

$$1,000 + 15t - 1,000 = 3,000 + 10t - 1,000$$

$$15t = 2,000 + 10t$$

$$15t - 10t = 2,000 + 10t - 10t$$

$$5t = 2,000$$

$$\frac{5t}{5} = \frac{2,000}{5}$$

$$\boxed{t = 400}$$

If 400 tires are made, the cost will be the same on either machine.

b)

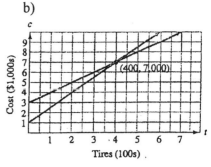

c) cost of 1st mold = $1,000 + 15t$

let $t = 500$

$$c = 1,000 + 15t$$

$$c = 1,000 + 15(500)$$

$$c = 1,000 + 7,500$$

$$\boxed{c = 8,500}$$

cost of 2nd mold = $3,000 + 10t$

let $t = 500$

$$c = 3,000 + 10t$$

$$c = 3,000 + 10(500)$$

$$c = 3,000 + 5,000$$

$$\boxed{c = 8,000}$$

The second mold will cost less to make 500 tires.

37. TELEPHONE RATES

Plan 1 = $10 + 0.08m$

Plan 2 = $15 + 0.04m$

Let m = break point in minutes

$$\begin{cases} c = 10 + 0.08m \\ c = 15 + 0.04m \end{cases}$$

$$10 + 0.08m = 15 + 0.04m$$

$$10 + 0.08m - 10 = 15 + 0.04m - 10$$

$$0.08m = 5 + 0.04m$$

$$0.08m - 0.04m = 5 + 0.04m - 0.04m$$

$$0.04m = 5$$

$$\frac{0.04m}{0.04} = \frac{5}{0.04}$$

$$\boxed{m = 125}$$

It will take 125 minutes for each cost to be the same for either plan.

39. GULF STREAM

Let s = speed of boat in still water (mph)

c = speed of current (mph)

rate • time = distance

north: 300 mi. in 10 hr. with current

south: 300 mi. in 15 hr. against current

$$\begin{cases} 10(s+c) = 300 \\ 15(s-c) = 300 \end{cases}$$

$$\begin{cases} \dfrac{10(s+c)}{10} = \dfrac{300}{10} \\ \dfrac{15(s-c)}{15} = \dfrac{300}{15} \end{cases}$$

$$\begin{cases} s+c = 30 \\ s-c = 20 \end{cases}$$

$$\begin{array}{r} s+c = 30 \\ s-c = 20 \\ \hline 2s = 50 \end{array}$$

$$\frac{2s}{2} = \frac{50}{2}$$

$$\boxed{s = 25}$$

$$s+c = 30$$

$$25+c = 30$$

$$25+c-25 = 30-25$$

$$\boxed{c = 5}$$

Speed of boat is 25 mph.

Speed of current is 5 mph.

41. AVIATION

Let s = speed of plane in still air (mph)

c = speed of wind (mph)

rate • time = distance

with wind: 800 mi. in 4 hr.

against wind: 800 mi. in 5 hr.

$$\begin{cases} 4(s+c) = 800 \\ 5(s-c) = 800 \end{cases}$$

$$\begin{cases} \dfrac{4(s+c)}{4} = \dfrac{800}{4} \\ \dfrac{5(s-c)}{5} = \dfrac{800}{5} \end{cases}$$

$$\begin{cases} s+c = 200 \\ s-c = 160 \end{cases}$$

$$\begin{array}{r} s+c = 200 \\ s-c = 160 \\ \hline 2s = 360 \end{array}$$

$$\frac{2s}{2} = \frac{360}{2}$$

$$\boxed{s = 180}$$

$$s+c = 200$$

$$180+c = 200$$

$$180+c-180 = 200-180$$

$$\boxed{c = 20}$$

Speed of plane is 180 mph.

Speed of jet stream is 20 mph.

43. STUDENT LOANS

Let x = amt loaned to nursing student

y = amt loaned to business student

principal • rate • time = interest

total gift: $5,000

interest rate for nurse: 5%

interest rate for business: 7%

collected interest first year: $310

$$\begin{cases} x + y = 5,000 \\ 0.05x + 0.07y = 310 \end{cases}$$

$$\begin{cases} x = 5,000 - y \\ 0.05x + 0.07y = 310 \end{cases}$$

$$0.05x + 0.07y = 310$$

$$0.05(5,000 - y) + 0.07y = 310$$

$$250 - 0.05y + 0.07y = 310$$

$$250 + 0.02y - 250 = 310 - 250$$

$$0.02y = 60$$

$$\frac{0.02y}{0.02} = \frac{60}{0.02}$$

$$\boxed{y = 3,000}$$

$$x + y = 5,000$$

$$x + 3,000 = 5,000$$

$$x + 3,000 - 3,000 = 5,000 - 3,000$$

$$\boxed{x = 2,000}$$

Nursing student loan is $2,000.

Business student loan is $3,000.

45. MARINE BIOLOGY

Let x = # of liters of 6% water

y = # of liters of 2% water

percent • amt of salt water = pure salt

total water: 16 liters

new percent of salt water: 3%

$$\begin{cases} x + y = 16 \\ 0.06x + 0.02y = 0.03(16) \end{cases}$$

$$\begin{cases} x = 16 - y \\ 0.06x + 0.02y = 0.48 \end{cases}$$

$$0.06x + 0.02y = 0.48$$

$$0.06(16 - y) + 0.02y = 0.48$$

$$0.96 - 0.06y + 0.02y = 0.48$$

$$0.96 - 0.04y - 0.96 = 0.48 - 0.96$$

$$-0.04y = -0.48$$

$$\frac{-0.04y}{-0.04} = \frac{-0.48}{-0.04}$$

$$\boxed{y = 12}$$

$$x + y = 16$$

$$x + 12 = 16$$

$$x + 12 - 12 = 16 - 12$$

$$\boxed{x = 4}$$

4 liters of 6% salt water needed.

12 liters of 2% salt water needed.

46.

47. COFFEE SALES

Let x = # of lb of Columbian coffee

y = # of lb of Brazilian coffee

cost/lb • amt = value

Columbian coffee: $8.75/lb

Brazilian coffee: $3.75/lb

blend: $6.35/lb

total lb.: 100 lb

$$\begin{cases} x + y = 100 \\ 8.75x + 3.75y = 6.35(100) \end{cases}$$

$$\begin{cases} x = 100 - y \\ 8.75x + 3.75y = 635 \end{cases}$$

$$8.75x + 3.75y = 635$$

$$8.75(100 - y) + 3.75y = 635$$

$$875 - 8.75y + 3.75y = 635$$

$$875 - 5y - 875 = 635 - 875$$

$$-5y = -240$$

$$\frac{-5y}{-5} = \frac{-240}{-5}$$

$$\boxed{y = 48}$$

$$x + y = 100$$

$$x + 48 = 100$$

$$x + 48 - 48 = 100 - 48$$

$$\boxed{x = 52}$$

52 lb of Columbian ($8.75/lb) coffee needed.

48 lb of Brazilian ($3.75/lb) coffee needed.

WRITING

49. Answers will vary.

REVIEW

51. $x < 4$

4

$(-\infty, 4)$

53. $-1 < x \leq 2$

$-1 \quad 2$

$(-1, 2]$

CHALLENGE PROBLEMS

55.

$$\begin{cases} 3 \text{ nails} + 1 \text{ bolt} = 3 \text{ nuts} \\ 1 \text{ bolt} + 1 \text{ nut} = 5 \text{ nails} \end{cases}$$

$$\begin{cases} 3 \text{ nails} + 1 \text{ bolt} = 3 \text{ nuts} \\ 1 \text{ bolt} = 5 \text{ nails} - 1 \text{ nut} \end{cases}$$

$$3 \text{ nails} + 1 \text{ bolt} = 3 \text{ nuts}$$

$$3 \text{ nails} + (5 \text{ nails} - 1 \text{ nut}) = 3 \text{ nuts}$$

$$8 \text{ nails} - 1 \text{ nut} = 3 \text{ nuts}$$

$$8 \text{ nails} - 1 \text{ nut} + 1 \text{ nut} = 3 \text{ nuts} + 1 \text{ nut}$$

$$8 \text{ nails} = 4 \text{ nuts}$$

$$\frac{8 \text{ nails}}{4} = \frac{4 \text{ nuts}}{4}$$

$$\boxed{2 \text{ nails} = 1 \text{ nut}}$$

The weight of 1 nut is the same as 2 nails.

VOCABULARY

1. $\begin{cases} x + y > 2 \\ x + y < 4 \end{cases}$ is a system of linear

   inequalities.

3. To **solve** a system of linear inequalities means to find all of its solutions.

5. To find the solutions of a system of two linear inequalities graphically, look for the **intersection**, or overlap, of the two shaded regions.

CONCEPTS

7. a) $3x - y = 5$
   b) dashed

9. slope: $4 = \dfrac{4}{1}$; y-intercept: $(0, -3)$

11.
   a) Does $(0, 0)$ make $2x + y > 4$ true?

   $$2x + y > 4$$

   $$2(0) + (0) \overset{?}{>} 4$$

   $$\boxed{2 > 4} \text{ false}$$

   b) above

13.
   a) $(4, -2)$; yes
   b) $(1, 3)$; no
   c) the origin ; no

15.
   a) $x + y = 2$, ii
   b) $x + y \ge 2$, iii
   c) $\begin{cases} x + y = 2 \\ x - y = 2 \end{cases}$, iv
   d) $\begin{cases} x + y = 2 \\ x - y = 2 \end{cases}$, i

NOTATION

17. **ABC**

PRACTICE

Use the next three detailed exercises as guides for this set of PRACTICE exercises.

19. $\begin{cases} x + 2y \le 3 \\ 2x - y \ge 1 \end{cases}$

   Step 1, graph the boundary of 1st inequality, as $x + 2y = 3$. Shade the region.

   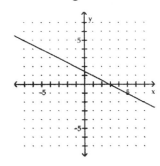

   Step 2, graph the boundary of 2nd inequality, as $2x - y = 1$. Shade the region.

   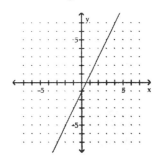

   Step 3, identify the intersection of the two regions.

   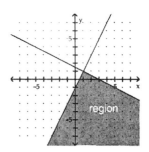

21.
$$\begin{cases} x+y < -1 \\ x-y > -1 \end{cases}$$

Step 1, graph the boundary of 1st inequality, as $x+y=-1$. Shade the region.

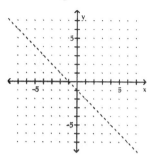

Step 2, graph the boundary of 2nd inequality, as $x-y=-1$. Shade the region.

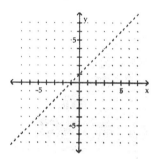

Step 3, identify the intersection of the two regions.

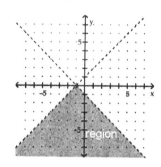

23.
$$\begin{cases} 2x-3y \le 0 \\ y \ge x-1 \end{cases}$$

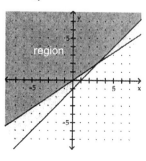

25.
$$\begin{cases} x+y < 2 \\ x+y \le 1 \end{cases}$$

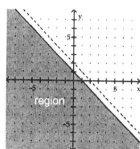

27.
$$\begin{cases} x \ge 2 \\ y \le 3 \end{cases}$$

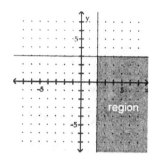

29.
$$\begin{cases} x > 0 \\ y > 0 \end{cases}$$

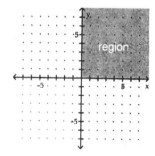

31.
$$\begin{cases} 3x+4y \ge -7 \\ 2x-3y \ge 1 \end{cases}$$

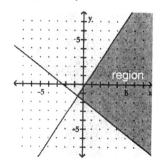

Section 7.5

33. $\begin{cases} 2x + y < 7 \\ y > 2 - 2x \end{cases}$

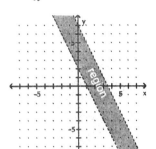

35. $\begin{cases} 2x - 4y > -6 \\ 3x + y \geq 5 \end{cases}$

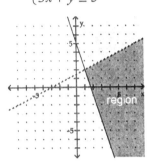

37. $\begin{cases} 3x - y + 4 \leq 0 \\ 3y > -2x - 10 \end{cases}$

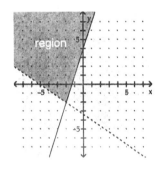

39. $\begin{cases} y \geq x \\ y \leq \dfrac{1}{3}x + 1 \end{cases}$

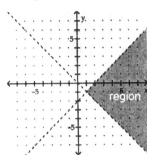

41. $\begin{cases} x + y > 0 \\ y - x < -2 \end{cases}$

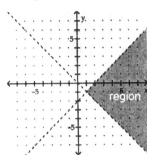

43. $\begin{cases} x \geq 0 \\ y \geq 0 \\ x + y \leq 3 \end{cases}$

With this system, one extra boundary is set up before identifying the intersection of the system.

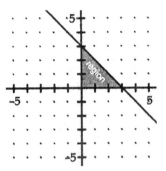

45. $\begin{cases} x - y < 4 \\ y \leq 0 \\ x \geq 0 \end{cases}$

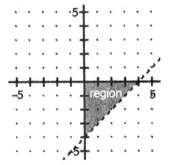

APPLICATIONS

47. BIRDS OF PREY

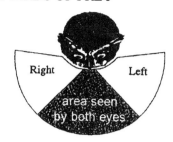

49. BUYING COMPACT DISCS

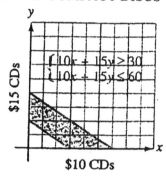

1 $10 CD and 2 $15 CD's
1 $10 CD and 3 $15 CD's
2 $10 CD's and 1 $15 CD
2 $10 CD's and 2 $15 CD's
3 $10 CD's and 1 $15 CD
3 $10 CD's and 2 $15 CD's
4 $10 CD's and 1 $15 CD

51. BUYING FURNITURE

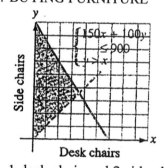

1 desk chair and 2 side chairs
1 desk chair and 3 side chairs
1 desk chair and 4 side chairs
1 desk chair and 5 side chairs
1 desk chair and 6 side chairs
1 desk chair and 7 side chairs
2 desk chair and 3 side chairs
2 desk chair and 4 side chairs
2 desk chair and 5 side chairs
2 desk chair and 6 side chairs
3 desk chair and 4 side chairs

53. PESTICIDES

$$\begin{cases} y \geq -2x + 1 \\ y \geq \dfrac{1}{4}x - 4 \end{cases}$$

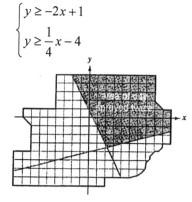

WRITING

55. Answers will vary.

57. Answers will vary.

REVIEW

59. $2.3 \times 10^2 = 230$

61. $9.76 \times 10^{-4} = 0.000976$

63. $290,000 = 2.9 \times 10^5$

65. $0.0000051 = 5.1 \times 10^{-6}$

CHALLENGE PROBLEMS

67.
$$\begin{cases} \dfrac{x}{3} - \dfrac{y}{2} < -3 \\ \dfrac{x}{3} + \dfrac{y}{2} > -1 \end{cases}$$

Clear both of fractions

$$\begin{cases} 2x - 3y < -18 \\ 2x + 3y > -6 \end{cases}$$

Graph the boundaries of both.
Identify the intersection.

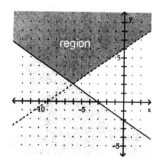

Section 7.5

69.
$$\begin{cases} 2x + 3y \le 6 \\ 3x + y \le 1 \\ x \le 0 \end{cases}$$

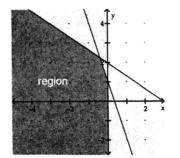

CHAPTER 7 KEY CONCEPTS

Solving Systems of Equations by Graphing

1. FOOD SERVICES

Caterer	Setup Fee	Cost per Meal	Equation
Sunshine	$1,000	$4	$y = 4x + 1,000$
Lucy's	$500	$5	$y = 5x + 500$

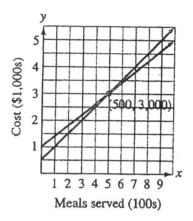

Meals served (100s)

Point of intersection: (500, 3,000)
500 meals, $3,000

Solving Systems of Equations by Substitution

2.
$$\begin{cases} y = 2x - 9 \\ x + 3y = 8 \end{cases}$$
$$x + 3(2x - 9) = 8$$
$$x + 6x - 27 = 8$$
$$7x - 27 + 27 = 8 + 27$$
$$7x = 35$$
$$\frac{7x}{7} = \frac{35}{7}$$
$$\boxed{x = 5}$$
$$y = 2x - 9$$
$$y = 2(5) - 9$$
$$y = 10 - 9$$
$$\boxed{y = 1}$$
solution is (5, 1)

3.
$$\begin{cases} 3x + 4y = -7 \\ 2y - x = -1 \end{cases}$$
$$\begin{cases} 3x + 4y = -7 \\ 2y + 1 = x \end{cases}$$
$$3x + 4y = -7$$
$$3(2y + 1) + 4y = -7$$
$$6y + 3 + 4y = -7$$
$$10y + 3 - 3 = -7 - 3$$
$$10y = -10$$
$$\frac{10y}{10} = \frac{-10}{10}$$
$$\boxed{y = -1}$$
$$x = 2y + 1$$
$$x = 2(-1) + 1$$
$$x = -2 + 1$$
$$\boxed{x = -1}$$
solution is (-1, -1)

Solving Systems of Equations by Elimination

4.
$$\begin{cases} x + y = 1 \\ x - y = 5 \end{cases}$$
eliminate the y
$$x + y = 1$$
$$\underline{x - y = 5}$$
$$2x = 6$$
$$\frac{2x}{2} = \frac{6}{2}$$
$$\boxed{x = 3}$$
$$x + y = 1$$
$$3 + y = 1$$
$$3 + y - 3 = 1 - 3$$
$$\boxed{y = -2}$$
solution is (3, -2)

5.

$$\begin{cases} 2x - 3y = -18 \\ 3x + 2y = -1 \end{cases}$$

multiply 1st equation by 2

multiply 2nd equation by 3

$$\begin{cases} 4x - 6y = -36 \\ 9x + 6y = -3 \end{cases}$$

eliminate the y

$$4x - 6y = -36$$
$$\underline{9x + 6y = -3}$$
$$13x = -39$$

$$\frac{13x}{13} = \frac{-39}{13}$$

$$\boxed{x = -3}$$

$$3x + 2y = -1$$

$$3(-3) + 2y = -1$$

$$-9 + 2y + 9 = -1 + 9$$

$$2y = 8$$

$$\frac{2y}{2} = \frac{8}{2}$$

$$\boxed{y = 4}$$

solution is (-3, 4)

SECTION 7.1
Solving Systems of Equations by Graphing

1.

$$(2, -3) \begin{cases} 3x - 2y = 12 \\ 2x + 3y = -5 \end{cases}$$

substitute $(2, -3)$ into

$$3x - 2y = 12$$

$$3(2) - 2(-3) \overset{?}{=} 12$$

$$6 + 6 \overset{?}{=} 12$$

$$\boxed{12 = 12} \quad \text{yes}$$

substitute $(2, -3)$ into

$$2x + 3y = -5$$

$$2(2) + 3(-3) \overset{?}{=} -5$$

$$4 - 9 \overset{?}{=} -5$$

$$\boxed{-5 = -5} \quad \text{yes}$$

$(2, -3)$ is a solution.

2.

$$\left(\frac{7}{2}, \frac{-2}{3}\right) \begin{cases} 3y = 2x - 9 \\ 2x + 3y = 5 \end{cases}$$

substitute $\left(\frac{7}{2}, \frac{-2}{3}\right)$ into

$$3y = 2x - 9$$

$$3\left(\frac{-2}{3}\right) \overset{?}{=} 2\left(\frac{7}{2}\right) - 9$$

$$-2 \overset{?}{=} 7 - 9$$

$$\boxed{-2 = -2} \quad \text{yes}$$

substitute $\left(\frac{7}{2}, \frac{-2}{3}\right)$ into

$$2x + 3y = 5$$

$$2\left(\frac{7}{2}\right) + 3\left(\frac{-2}{3}\right) \overset{?}{=} 5$$

$$7 - 2 \overset{?}{=} 5$$

$$\boxed{5 = 5} \quad \text{yes}$$

$\left(\frac{7}{2}, \frac{-2}{3}\right)$ is a solution.

3. COLLEGE ENROLLMENT

(1978, 5.6)

In 1978, the same number of men as women were enrolled in college, about 5.6 million of each.

4.

$$\begin{cases} x + y = 7 \\ 2x - y = 5 \end{cases}$$

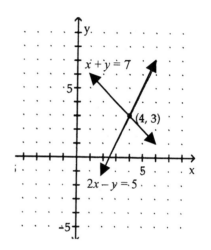

5.

$$\begin{cases} 2x + y = 5 \\ y = -\dfrac{x}{3} \end{cases}$$

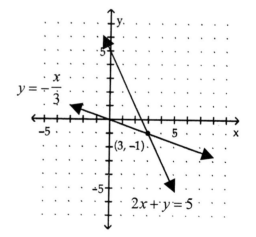

6.

$$\begin{cases} 3x + 6y = 6 \\ x + 2y - 2 = 0 \end{cases}$$

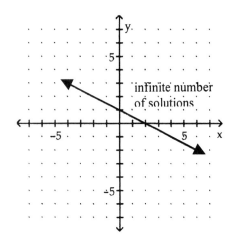

infinite number of solutions

7.

$$\begin{cases} 6x + 3y = 12 \\ y = -2x + 2 \end{cases}$$

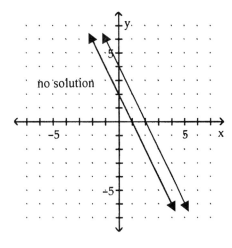

no solution

SECTION 7.2
Solving Systems of Equations by Substitution

8.
$$\begin{cases} y = 15 - 3x \\ 7y + 3x = 15 \end{cases}$$

$$7y + 3x = 15$$
$$7(15 - 3x) + 3x = 15$$
$$105 - 21x + 3x = 15$$
$$105 - 18x - 105 = 15 - 105$$
$$-18x = -90$$
$$\frac{-18x}{-18} = \frac{-90}{-18}$$
$$\boxed{x = 5}$$

continued

8. continued

$$y = 15 - 3x$$
$$y = 15 - 3(5)$$
$$y = 15 - 15$$
$$\boxed{y = 0}$$

solution is (5, 0)

9.
$$\begin{cases} x = y \\ 5x - 4y = 3 \end{cases}$$

$$5x - 4y = 3$$
$$5(y) - 4y = 3$$
$$\boxed{y = 3}$$

$$x = y$$
$$\boxed{x = 3}$$

solution is (3, 3)

10.
$$\begin{cases} 6x + 2y = 8 - y + x \\ 3x = 2 - y \end{cases}$$

$$\begin{cases} 5x + 3y = 8 \\ y = 2 - 3x \end{cases}$$

$$5x + 3y = 8$$
$$5x + 3(2 - 3x) = 8$$
$$5x + 6 - 9x = 8$$
$$-4x + 6 - 6 = 8 - 6$$
$$-4x = 2$$
$$\frac{-4x}{-4} = \frac{2}{-4}$$
$$\boxed{x = -\frac{1}{2}}$$

$$y = 2 - 3x$$
$$y = 2 - 3\left(-\frac{1}{2}\right)$$
$$y = 2 + \frac{3}{2}$$
$$\boxed{y = \frac{7}{2}}$$

solution is $\left(-\frac{1}{2}, \frac{7}{2}\right)$

11.
$$\begin{cases} r = 3s + 7 \\ r = 2s + 5 \end{cases}$$

$$3s + 7 = 2s + 5$$
$$3s + 7 - 7 = 2s + 5 - 7$$
$$3s = 2s - 2$$
$$3s - 2s = 2s - 2 - 2s$$
$$\boxed{s = -2}$$
$$r = 2(-2) + 5$$
$$\boxed{r = 1}$$
solution is (1, -2)

12.
$$\begin{cases} 9x + 3y - 5 = 0 \\ 3x + y = \dfrac{5}{3} \end{cases}$$

$$\begin{cases} 9x + 3y - 5 = 0 \\ y = \dfrac{5}{3} - 3x \end{cases}$$

$$9x + 3y - 5 = 0$$
$$9x + 3\left(\dfrac{5}{3} - 3x\right) - 5 = 0$$
$$9x + 5 - 9x - 5 = 0$$
$$\boxed{0 = 0}$$
infinitely many solutions

13.
$$\begin{cases} \dfrac{x}{2} + \dfrac{y}{2} = 11 \\ \dfrac{5x}{16} - \dfrac{3y}{16} = \dfrac{15}{8} \end{cases}$$
clear both equations of fractions

$$\begin{cases} 2\left(\dfrac{x}{2}\right) + 2\left(\dfrac{y}{2}\right) = 2(11) \\ 16\left(\dfrac{5x}{16}\right) - 16\left(\dfrac{3y}{16}\right) = 16\left(\dfrac{15}{8}\right) \end{cases}$$

$$\begin{cases} x + y = 22 \\ 5x - 3y = 30 \end{cases}$$

$$\begin{cases} x = 22 - y \\ 5x - 3y = 30 \end{cases}$$
continued

13. continued
$$5x - 3y = 30$$
$$5(22 - y) - 3y = 30$$
$$110 - 5y - 3y = 30$$
$$-8y + 110 - 110 = 30 - 110$$
$$-8y = -80$$
$$\dfrac{-8y}{-8} = \dfrac{-80}{-8}$$
$$\boxed{y = 10}$$
$$x = 22 - y$$
$$x = 22 - 10$$
$$\boxed{x = 12}$$
solution is (12, 10)

14.　　a.　no solutions

　　　　b.　two parallel lines

　　　　c.　inconsistent system

SECTION 7.3
Solving Systems of Equations by Elimination

15.
$$\begin{cases} 2x + y = 1 \\ 5x - y = 20 \end{cases}$$
eliminate the y
$$2x + y = 1$$
$$\underline{5x - y = 20}$$
$$7x = 21$$
$$\dfrac{7x}{7} = \dfrac{21}{7}$$
$$\boxed{x = 3}$$
$$2x + y = 1 \quad 1^{st} \text{ equation}$$
$$2(3) + y = 1$$
$$6 + y - 6 = 1 - 6$$
$$\boxed{y = -5}$$
solution is (3, -5)

- 351 -

16.
$$\begin{cases} x+8y=7 \\ x-4y=1 \end{cases}$$

multiply 2nd equation by -1

eliminate the x

$x+8y=7$

$\underline{-x+4y=-1}$

$12y=6$

$\dfrac{12y}{12}=\dfrac{6}{12}$

$\boxed{y=\dfrac{1}{2}}$

$x+8y=7 \quad 1^{st}$ equation

$x+8\left(\dfrac{1}{2}\right)=7$

$x+4-4=7-4$

$\boxed{x=3}$

solution is $\left(3, \dfrac{1}{2}\right)$

17.
$$\begin{cases} 5a+b=2 \\ 3a+2b=11 \end{cases}$$

multiply 1st equation by -2

to eliminate the b

-10a - 2b = -4

$\underline{3a+2b=11}$

-7a = 7

$\dfrac{-7a}{-7}=\dfrac{7}{-7}$

$\boxed{a=-1}$

$5a+b=2 \quad 1^{st}$ equation

$5(-1)+b=2$

$-5+b+5=2+5$

$\boxed{b=7}$

solution is (-1, 7)

18.
$$\begin{cases} 11x+3y=27 \\ 8x+4y=36 \end{cases}$$

multiply 1st equation by 4

multiply 2nd equation by -3

eliminate the y

$44x+12y=108$

$\underline{-24x-12y=-108}$

$20x=0$

$\dfrac{20x}{20}=\dfrac{0}{20}$

$\boxed{x=0}$

$11x+3y=27 \quad 1^{st}$ equation

$11(0)+3y=27$

$\dfrac{3y}{3}=\dfrac{27}{3}$

$\boxed{y=9}$

solution is (0, 9)

19.
$$\begin{cases} 9x+3y=15 \\ 3x=5-y \end{cases}$$

divide 1st equation by 3

put 2nd equation in standard form

$$\begin{cases} 3x+y=5 \\ 3x+y=5 \end{cases}$$

both equations are exactly the same

infinitely many solutions

20.
$$\begin{cases} -\dfrac{a}{4} - \dfrac{b}{3} = \dfrac{1}{12} \\ \dfrac{a}{2} - \dfrac{5b}{4} = \dfrac{7}{4} \end{cases}$$

remove fractions from both

multiply 1st equation by 12

multiply 2nd equation by 4

$$\begin{cases} -3a - 4b = 1 \\ 2a - 5b = 7 \end{cases}$$

multiply 1st equation by 2

multiply 2nd equation by 3

to eliminate the a

$-6a - 8b = 2$

$\underline{6a - 15b = 21}$

$-23b = 23$

$\dfrac{-23b}{-23} = \dfrac{23}{-23}$

$\boxed{b = -1}$

$2a - 5b = 7$ 2nd equation

$2a - 5(-1) = 7$

$2a + 5 - 5 = 7 - 5$

$2a = 2$

$\dfrac{2a}{2} = \dfrac{2}{2}$

$\boxed{a = 1}$

solution is $(1, -1)$

21.
$$\begin{cases} 0.02x + 0.05y = 0 \\ 0.3x - 0.2y = -1.9 \end{cases}$$

remove the decimals

multiply 1st equation by 100

multiply 2nd equation by 10

$$\begin{cases} 2x + 5y = 0 \\ 3x - 2y = -19 \end{cases}$$

multiply 1st equation by 2

multiply 2nd equation by 5

eliminate the y

$4x + 10y = 0$

$\underline{15x - 5y = -95}$

$19x = -95$

$\dfrac{19x}{19} = \dfrac{-95}{19}$

$\boxed{x = -5}$

$2x + 5y = 0$ 1st equation

$2(-5) + 5y = 0$

$-10 + 5y + 10 = 0 + 10$

$5y = 10$

$\dfrac{5y}{5} = \dfrac{10}{5}$

$\boxed{y = 2}$

solution is $(-5, 2)$

22.
$$\begin{cases} -\dfrac{1}{4}x = 1 - \dfrac{2}{3}y \\ 6x - 18y = 5 - 2y \end{cases}$$

remove the fractions from 1st equation

by multiplying by 12

put both equations in standard form

$$\begin{cases} -3x + 8y = 12 \\ 6x - 16y = 5 \end{cases}$$

multiply 1st equation by 2

eliminate the y

$-6x + 16y = 24$

$\underline{6x - 16y = 5}$

$0 = 29$ false

no solution

23. Elimination; no variables have a coefficient of 1 or -1

24. Substitution; equation 1 is solved for x

SECTION 7.4
Problem Solving Using Systems of Equations

25. ELEVATION

Let x = Las Vegas' elevation (higher)

y = Baltimore's elevation (lower)

total of both elevations = 2,100 ft.

LV ele = 20 times greater than Balt ele

$$\begin{cases} x + y = 2{,}100 \\ x = 20y \end{cases}$$

$$x + y = 2{,}100$$

$$20y + y = 2{,}100$$

$$21y = 2{,}100$$

$$\frac{21y}{21} = \frac{2{,}100}{21}$$

$$\boxed{y = 100}$$

$$x + y = 2{,}100$$

$$x + 100 = 2{,}100$$

$$x + 100 - 100 = 2{,}100 - 100$$

$$\boxed{x = 2{,}000}$$

Las Vegas' elevation is 2,000 ft.

Balimore's elevation is 100 ft.

26. PAINTING EQUIPMENT

Let x = length of base part of ladder (longer)

y = length of extension (shorter)

fully extended = 35 ft.

extension = 7 ft. shorter than base

$$\begin{cases} x + y = 35 \\ y = x - 7 \end{cases}$$

$$x + y = 35$$

$$x + x - 7 = 35$$

$$2x - 7 + 7 = 35 + 7$$

$$2x = 42$$

$$\frac{2x}{2} = \frac{42}{2}$$

$$\boxed{x = 21}$$

$$x + y = 35$$

$$21 + y = 35$$

$$21 + y - 21 = 35 - 21$$

$$\boxed{y = 14}$$

Base part of ladder is 21 ft.

Extension part of ladder is 14 ft.

27. CRASH INVESTIGATION

Let l = length in yards (large)

w = width in yards (small)

perimeter of scene is 420 yards

$2l + 2w$ = perimeter

if width is $\frac{3}{4}$ of the length

find the area to be searched

$$\begin{cases} 2l + 2w = 420 \\ w = \frac{3}{4}l \end{cases}$$

$$2l + 2(0.75l) = 420$$

$$2l + 1.50l = 420$$

$$\frac{3.5l}{3.5} = \frac{420}{3.5}$$

$$\boxed{l = 120}$$

$$w = 0.75l$$

$$w = 0.75(120)$$

$$\boxed{w = 90}$$

area to be searched = (120 yds)(90 yds)

$$= 10{,}800 \text{ yd}^2$$

28. CELEBRITY ENDORSEMENT

a) Athlete's equation:
$$y = 5x + 30,000$$
Actor's equation:
$$y = 10x + 20,000$$

b) earned by Athlete: $y = 5x + 30,000$
earned by Actor: $y = 10x + 20,000$
Let x = break point for juice machine
$$\begin{cases} y = 5x + 30,000 \\ y = 10x + 20,000 \end{cases}$$
earned by Athlete = earned by Actor
$$5x + 30,000 = 10x + 20,000$$
subtract 20,000 from both sides
$$20,000 = 20,000$$
$$5x + 10,000 = 10x$$
$$5x + 10,000 - 5x = 10x - 5x$$
$$10,000 = 5x$$
$$\frac{10,000}{5} = \frac{5x}{5}$$
$$\boxed{2,000 = x}$$

When 2,000 juice machine are
sold, both would earn the same.

c)

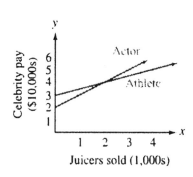

Celebrity pay ($10,000s) / Juicers sold (1,000s)

29. CANDY OUTLET STORE

Let x = # of lb of gummy worms
y = # of lb of gummy bears
cost/lb • amt = value
gummy worms: $3.00/lb
gummy bears: $1.50/lb
mixture: $2.10/lb
total pounds: 30
$$\begin{cases} x + y = 30 \\ 3x + 1.5y = 2.1(30) \end{cases}$$
$$\begin{cases} x = 30 - y \\ 3x + 1.5y = 63 \end{cases}$$
$$3x + 1.5y = 63$$
$$3(30 - y) + 1.5y = 63$$
$$90 - 3y + 1.5y = 63$$
$$90 - 1.5y - 90 = 63 - 90$$
$$-1.5y = -27$$
$$\frac{-1.5y}{-1.5} = \frac{-27}{-1.5}$$
$$\boxed{y = 18}$$
$$x + y = 30$$
$$x + 18 = 30$$
$$x + 18 - 18 = 30 - 18$$
$$\boxed{x = 12}$$

12 lb of g. worms ($3/lb) is needed.
18 lb of g. bears ($6/lb) is needed.

Chapter 7 Review

30. BOATING

Let s = speed of boat in still water (mph)

c = speed of current (mph)

rate • time = distance

with current: 56 mi. in 4 hr.

against current: 56 mi. in $4+3$ (hr. longer)

$$\begin{cases} 4(s+c) = 56 \\ 7(s-c) = 56 \end{cases}$$

$$\begin{cases} \dfrac{4(s+c)}{4} = \dfrac{56}{4} \\ \dfrac{7(s-c)}{7} = \dfrac{56}{7} \end{cases}$$

$$\begin{cases} s+c = 14 \\ s-c = 8 \end{cases}$$

$$s+c = 14$$
$$\underline{s-c = 8}$$
$$2s = 22$$

$$\dfrac{2s}{2} = \dfrac{22}{2}$$

$$\boxed{s = 11}$$

$$s+c = 14$$
$$11+c = 14$$
$$11+c-11 = 14-11$$
$$\boxed{c = 3}$$

Speed of boat is 11 mph.

Speed of current is 3 mph.

31. SHOPPING

Let x = price of one bottle of cleaner

y = price of one bottle of soaking

2 cleaners and 3 soaking is \$63.40

3 cleaners and 2 soaking is \$69.60

$$\begin{cases} 2x+3y = 63.40 \\ 3x+2y = 69.60 \end{cases}$$

$$-2(2x+3y) = -2(63.40)$$
$$3(3x+2y) = 3(69.60)$$

$$\underline{\qquad\qquad\qquad}$$

$$-4x-6y = -126.80$$
$$9x+6y = 208.80$$

$$\underline{\qquad\qquad\qquad}$$

$$5x = 82$$

$$\dfrac{5x}{5} = \dfrac{82}{5}$$

$$\boxed{x = 16.40}$$

$$2x+3y = 63.40$$
$$2(16.4)+3y = 63.40$$
$$32.8+3y-32.8 = 63.40-32.8$$
$$3y = 30.60$$

$$\dfrac{3y}{3} = \dfrac{30.60}{3}$$

$$\boxed{y = 10.20}$$

A bottle of cleaner cost \$16.40.

A bottle of soaking cost \$10.20.

32. INVESTING

Let x = amt invested at 10%

y = amt invested at 6%

principal • rate • time = interest

total investment: \$3,000

annual interest for both: \$270

$$\begin{cases} x+y = 3,000 \\ 0.10x+0.06y = 270 \end{cases}$$

$$\begin{cases} x = 3,000-y \\ 0.10x+0.06y = 270 \end{cases}$$

$$0.10x+0.06y = 270$$
$$0.10(3,000-y)+0.06y = 270$$
$$300-0.10y+0.06y = 270$$
$$300-0.04y-300 = 270-300$$
$$-0.04y = -30$$

$$\dfrac{-0.04y}{-0.04} = \dfrac{-30}{-0.04}$$

$$\boxed{y = 750}$$

\$750 invested at 6%.

33. ANTIFREEZE

Let x = # of gal of 40% antifreeze

y = # of gal of 70% antifreeze

percent • amt of antifreeze = pure antifreeze

total gallons: 20

final mixture percent: 50%

$$\begin{cases} x + y = 20 \\ 0.4x + 0.7y = 0.5(20) \end{cases}$$

$$\begin{cases} x = 20 - y \\ 0.4x + 0.7y = 10 \end{cases}$$

$$0.4x + 0.7y = 10$$

$$0.4(20 - y) + 0.7y = 10$$

$$8 - 0.4y + 0.7y = 10$$

$$8 + 0.3y - 8 = 10 - 8$$

$$0.3y = 2$$

$$\frac{0.3y}{0.3} = \frac{2}{0.3}$$

$$\boxed{\begin{array}{c} y = 6.666 \\ \text{or} \\ y = 6\dfrac{2}{3} \end{array}}$$

$$x + y = 20$$

$$x + 6\frac{2}{3} = 20$$

$$x + 6\frac{2}{3} - 6\frac{2}{3} = 20 - 6\frac{2}{3}$$

$$\boxed{x = 13\frac{1}{3}}$$

$13\dfrac{1}{3}$ gal of 40% antifreeze is needed.

$6\dfrac{2}{3}$ gal of 70% antifreeze is needed.

34. $\begin{cases} 5x + 3y < 15 \\ 3x - y > 3 \end{cases}$

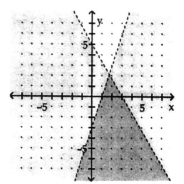

35. $\begin{cases} 3y \le x \\ y > 3x \end{cases}$

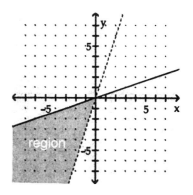

36. $\begin{cases} 10x + 20y \ge 40 \\ 10x + 20y \le 60 \end{cases}$

Let x = # of T-shirts

y = # of pants

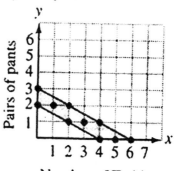

Number of T-shirts

1.

$(5, 3) \begin{cases} 3x + 2y = 21 \\ x + y = 8 \end{cases}$

substitute (5, 3) into

$3x + 2y = 21$

$3(5) + 2(3) \overset{?}{=} 21$

$15 + 6 \overset{?}{=} 21$

$\boxed{21 = 21}$ yes

substitute (5, 3) into

$x + y = 8$

$5 + 3 \overset{?}{=} 8$

$\boxed{8 = 8}$ yes

(5, 3) is a solution.

2.

$(-2, -1) \begin{cases} 4x + y = -9 \\ 2x - 3y = -7 \end{cases}$

substitute (-2, -1) into

$4x + y = -9$

$4(-2) + (-1) \overset{?}{=} -9$

$-8 - 1 \overset{?}{=} -9$

$\boxed{-9 = -9}$ yes

substitute (-2, -1) into

$2x - 3y = -7$

$2(-2) - 3(-1) \overset{?}{=} -7$

$-4 + 3 \overset{?}{=} -7$

$\boxed{-1 \neq -7}$ no

(-2, -1) is not a solution.

3.

a) A **solution** of a system of linear equations is an ordered pair that satisfies each equation.

b) A system of equations that has at least one solution is called a **consistent** system.

c) A system of equations that has no solution is called an **inconsistent** system.

d) Equations with different graphs are called **independent** equations.

e) A system of **dependent** equations has an infinite number of solutions

4.

$\begin{cases} y = 2x - 1 \\ x - 2y = -4 \end{cases}$

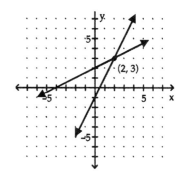

(2, 3)

5.

$\begin{cases} x + y = 5 \\ y = -x \end{cases}$

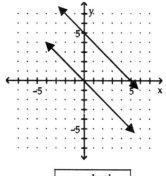

$\boxed{\text{no solution}}$

6. (30, 3,000); if 30 items are sold, the salesperson gets paid the same by both plans, $3,000.

7. Plan 1

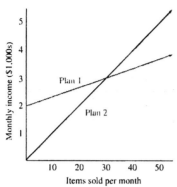

8.
$$\begin{cases} y = x - 1 \\ 2x + y = -7 \end{cases}$$
$$2x + y = -7$$
$$2x + (x - 1) = -7$$
$$3x - 1 = -7$$
$$3x - 1 + 1 = -7 + 1$$
$$3x = -6$$
$$\frac{3x}{3} = \frac{-6}{3}$$
$$\boxed{x = -2}$$
$$y = x - 1$$
$$y = -2 - 1$$
$$\boxed{y = -3}$$
solution is (-2, -3)

9.
$$\begin{cases} 3x + 6y = -15 \\ x + 2y = -5 \end{cases}$$
$$\begin{cases} 3x + 6y = -15 \\ x = -5 - 2y \end{cases}$$
$$3x + 6y = -15$$
$$(-5 - 2y) + 6y = -15$$
$$-15 - 6y + 6y = -15$$
$$-15 = -15 \text{ true}$$
infinitely many solutions

10.
$$\begin{cases} 3a + 4b = -7 \\ 2b - a = -1 \end{cases}$$
$$\begin{cases} 3a + 4b = -7 \\ 2b + 1 = a \end{cases}$$
$$3a + 4b = -7$$
$$3(2b + 1) + 4b = -7$$
$$6b + 3 + 4b = -7$$
$$10b + 3 - 3 = -7 - 3$$
$$10b = -10$$
$$\frac{10b}{10} = \frac{-10}{10}$$
$$\boxed{b = -1}$$
$$a = 2b + 1$$
$$a = 2(-1) + 1$$
$$a = -2 + 1$$
$$\boxed{a = -1}$$
solution is (-1, -1)

11.
$$\begin{cases} 3x - y = 2 \\ 2x + y = 8 \end{cases}$$
eliminate the y
$$3x - y = 2$$
$$\underline{2x + y = 8}$$
$$5x = 10$$
$$\frac{5x}{5} = \frac{10}{5}$$
$$\boxed{x = 2}$$
$$2x + y = 8 \quad 2^{nd} \text{ equation}$$
$$2(2) + y = 8$$
$$4 + y - 4 = 8 - 4$$
$$\boxed{y = 4}$$
solution is (2, 4)

12.
$$\begin{cases} 4x + 3y = \text{-}3 \\ \text{-}3x = \text{-}4y + 21 \end{cases}$$

put 2nd equation in standard form

$$\begin{cases} 4x + 3y = \text{-}3 \\ \text{-}3x + 4y = 21 \end{cases}$$

multiply 1st equation by 3

multiply 2nd equation by 4

$$\begin{cases} 12x + 9y = \text{-}9 \\ \text{-}12x + 16y = 84 \end{cases}$$

eliminate the x

$$\begin{aligned} 12x + 9y &= \text{-}9 \\ \underline{\text{-}12x + 16y} &= \underline{84} \\ 25y &= 75 \\ \frac{25y}{25} &= \frac{75}{25} \\ \boxed{y = 3} \end{aligned}$$

$$4x + 3y = \text{-}3 \quad 1^{st} \text{ equation}$$
$$4x + 3(3) = \text{-}3$$
$$4x + 9 - 9 = \text{-}3 - 9$$
$$4x = \text{-}12$$
$$\frac{4x}{4} = \frac{\text{-}12}{4}$$
$$\boxed{x = \text{-}3}$$

solution is (-3, 3)

13.
$$\begin{cases} 3x - 5y - 16 = 0 \\ \dfrac{x}{2} - \dfrac{5}{6}y = \dfrac{1}{3} \end{cases}$$

put 1st equation in standard form

remove fractions from 2nd equation

$$\begin{cases} 3x - 5y - 16 + 16 = 0 + 16 \\ 6\left(\dfrac{x}{2}\right) - 6\left(\dfrac{5}{6}y\right) = 6\left(\dfrac{1}{3}\right) \end{cases}$$

$$\begin{cases} 3x - 5y = 16 \\ 3x - 5y = 2 \end{cases}$$

multiply 2nd equation by -1

eliminate the x

$$\begin{aligned} 3x - 5y &= 16 \\ \underline{\text{-}3x + 5y} &= \underline{\text{-}2} \\ 0 &= 14 \quad \text{false} \end{aligned}$$

no solution

14. Elimination method; the terms involving y can be eliminated easily.

15. CHILD CARE

Let x = first part of commute

y = second part of commute

total commute to work: 22 miles

1st part is 6 miles less than 2nd part

$$\begin{cases} x + y = 22 \\ x = y - 6 \end{cases}$$

put 2nd equation in standard form

$$\begin{cases} x + y = 22 \\ x - y = -6 \end{cases}$$

eliminate the y

$x + y = 22$

$\underline{x - y = -6}$

$2x = 16$

$\dfrac{2x}{2} = \dfrac{16}{2}$

$\boxed{x = 8}$

$x + y = 22$

$8 + y = 22$

$8 + y - 8 = 22 - 8$

$\boxed{y = 14}$

1st part is 8 miles.

2nd part is 14 miles.

16. VACATIONING

Let x = number of adult tickets

y = number of child tickets

an adult ticket cost $21

a child ticket cost $14

# of adult and # of children = 7

total cost $119

$$\begin{cases} x + y = 7 \\ 21x + 14y = 119 \end{cases}$$

multiply 1st equation by -21

$$\begin{cases} -21x - 21y = -147 \\ 21x + 14y = 119 \end{cases}$$

eliminate the x

$-21x - 21y = -147$

$\underline{21x + 14y = 119}$

$-7y = -28$

$\dfrac{-7y}{-7} = \dfrac{-28}{-7}$

$\boxed{y = 4}$

$x + y = 7$

$x + 4 = 7$

$x + 4 - 4 = 7 - 4$

$\boxed{x = 3}$

3 adult tickets were bought.

4 child tickets were bought.

17. FINANCIAL PLANNING

Let x = amt put into 8% account

y = amt put into 9% acount

principal • rate • time = interest

total investment: $10,000

first year combined interest: $840

$$\begin{cases} x + y = 10,000 \\ 0.08x + 0.09y = 840 \end{cases}$$

$$\begin{cases} x = 10,000 - y \\ 0.08x + 0.09y = 840 \end{cases}$$

$$0.08x + 0.09y = 840$$

$$0.08(10,000 - y) + 0.09y = 840$$

$$800 - 0.08y + 0.09y = 840$$

$$800 + 0.01y - 800 = 840 - 800$$

$$0.01y = 40$$

$$\frac{0.01y}{0.01} = \frac{40}{0.01}$$

$$\boxed{y = 4,000}$$

$$x + y = 10,000$$

$$x + 4,000 = 10,000$$

$$x + 4,000 - 4,000 = 10,000 - 4,000$$

$$\boxed{x = 6,000}$$

$6,000 is put into 1st account.

$4,000 is put into 2nd account.

18. KAYAKING

Let x = speed of kayak in miles/hour

c = speed of current in miles/hour

$(x - c)$ mph

19. TETHER BALL

Let x = top $\angle$

y = bottom $\angle$

sum of 2 complementary $\angle$'s = 90

$$x + y = 90$$

20.

$$\begin{cases} 2x + 3y \le 6 \\ x > 2 \end{cases}$$

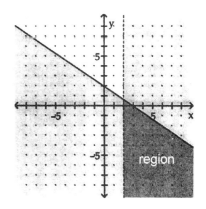

21. CLOTHES SHOPPING

$$\begin{cases} 20x + 20y \ge 80 \\ 20x + 40y \le 120 \end{cases}$$

Let x = # of $20 shirts

y = # of $40 pants

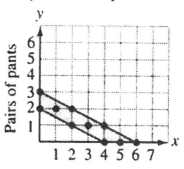

Number of   shirts

1 shirt and 2 pairs of pants

2 shirts and 2 pairs of pants

3 shirts and 1 pair of pants

There are other answers.

CHAPTER 7 CUMULATIVE REVIEW

1. CANDY SALES

   total candy sales: $6,300,000,000

   rate • base = amount

   Halloween sales: 32%

   $$a = rb$$
   $$a = (32\%)(6,300,000,000)$$
   $$a = (0.32)(6,300,000,000)$$
   $$a = 2,016,000,000$$

   $2,016,000,000 in Halloween sales.

2. $\{\ldots, -3, -2, -1, 0, 1, 2, 3, \ldots\}$

3. $$3 - 4[-10 - 4(-5)] = 3 - 4[-10 + 20]$$
   $$= 3 - 4[10]$$
   $$= 3 - 40$$
   $$= -37$$

4. $$\frac{|-45| - 2(-5) + 1^5}{2 \cdot 9 - 2^4} = \frac{|-45| - 2(-5) + 1}{2 \cdot 9 - 2^4}$$
   $$= \frac{45 + 10 + 1}{18 - 16}$$
   $$= \frac{56}{2}$$
   $$= 28$$

5. AIR CONDITIONING

   diameter: 6 inches

   radius: 3 inches

   3 inches $= \dfrac{3}{12}$ ft $= 0.25$ ft

   length or height: 6 ft

   $$V = \pi r^2 h$$
   $$V = (3.14)(0.25 \text{ ft})^2 (6 \text{ ft})$$
   $$V = (3.14)(0.25 \text{ ft})(0.25 \text{ ft})(6 \text{ ft})$$
   $$V = 1.1775 \text{ ft}^3$$

   $\boxed{V \approx 1.2 \text{ ft}^3}$

6. $$3x^2 + 2x^2 - 5x^2 = 5x^2 - 5x^2$$
   $$= 0$$

7. $$2 - (4x + 7) = 3 + 2(x + 2)$$
   $$2 - 4x - 7 = 3 + 2x + 4$$
   $$-4x - 5 = 2x + 7$$
   $$-4x - 5 + 5 = 2x + 7 + 5$$
   $$-4x = 2x + 12$$
   $$-4x - 2x = 2x + 12 - 2x$$
   $$-6x = 12$$
   $$\frac{-6x}{-6} = \frac{12}{-6}$$
   $$\boxed{x = -2}$$

8. $$\frac{2}{5}y + 3 = 9$$
   $$\frac{2}{5}y + 3 - 3 = 9 - 3$$
   $$\frac{2}{5}y = 6$$
   $$\frac{5}{2}\left(\frac{2}{5}y\right) = \frac{5}{2}\left(\frac{6}{1}\right)$$
   $$\boxed{y = 15}$$

9. $$-4x + 6 > 17$$
   $$-4x + 6 - 6 > 17 - 6$$
   $$-4x > 11$$
   $$\frac{-4x}{-4} < \frac{11}{-4}$$
   $$\boxed{x < -\frac{11}{4}}$$

   $$\left(-\infty, -\frac{11}{4}\right)$$

10. ANGLE OF ELEVATION

a right angle measures 90°

sum of measures of angles of

a triangle is 180°

Let $x°$ = measure of one angle

$2x°$ = measure of other angle

$x + 2x + 90 = 180$

$3x + 90 - 90 = 180 - 90$

$3x = 90$

$\dfrac{3x}{3} = \dfrac{90}{3}$

$\boxed{x = 30}$

$30°$ = measure of one angle

11. STOCK MARKET

total principal: $45,000

interest earned after 1 year: $4,300

principal • rate • time = interest

Let x = amt invested at 12% (mutual)

$45,000 - x$ = amt invested at 6.5% (bonds)

$0.12x + 0.065(45,000 - x) = 4,300$

$0.12x + 2,925 - 0.065x = 4,300$

$0.055x + 2,925 = 4,300$

subtract 2,925 from both sides

$-2,925 = -2,925$

$0.055x = 1,375$

$\dfrac{0.055x}{0.055} = \dfrac{1,375}{0.055}$

$\boxed{x = 25,000}$

$45,000 - x = 45,000 - 25,000$

$= 20,000$

$25,000 invested in mutual funds.

$20,000 invested in bonds.

12. Give the formula.

a) the perimeter of a rectangle

$P = 2l + 2w$

b) the area of a rectangle

$A = lw$

c) the area of a circle

$A = \pi r^2$

d) the distance traveled

$d = rt$

13. Graph: $x = -2$

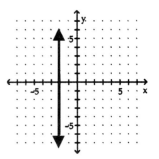

14. Graph: $y = x^2 + 1$

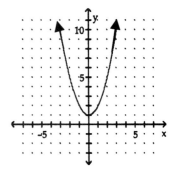

15. Find slope and y-intercept.

$m = 3$

$y - \text{intercept} = (0, -2)$

Write the equation of the line.

$y = mx + b$

$y = 3x - 2$

16. Find the slope.

(6, -2) and (-3, 2)

$(6_{x_1}, -2_{y_1})$ and $(-3_{x_2}, 2_{y_2})$

$$m = \frac{y_2 - y_1}{x_2 - x_1}$$

$$= \frac{2 - (-2)}{-3 - 6}$$

$$= \frac{2 + 2}{-3 - 6}$$

$$= -\frac{4}{9}$$

17. $(x^5)^2(x^7)^3 = (x^{5(2)})(x^{7(3)})$

$$= (x^{10})(x^{21})$$

$$= (x^{10+21})$$

$$= x^{31}$$

18. $\dfrac{16(aa^2)^3}{2a^2a^3} = \dfrac{16(a^{1+2})^3}{2a^{2+3}}$

$$= \frac{16(a^3)^3}{2a^5}$$

$$= \frac{16(a^{3(3)})}{2a^5}$$

$$= \frac{16(a^9)}{2a^5}$$

$$= 8(a^{9-5})$$

$$= 8a^4$$

19. $\dfrac{2^{-4}}{3^{-1}} = \dfrac{3^1}{2^4}$

$$= \frac{3}{16}$$

20. $(2x)^0 = 1$

21. $(5x - 8y) - (-2x + 5y) = 5x - 8y + 2x - 5y$

$$= 7x - 13y$$

22. $2x^2(3x^2 + 4x - 7) = 6x^{2+2} + 8x^{2+1} - 14x^{2+0}$

$$= 6x^4 + 8x^3 - 14x^2$$

23. $(c + 16)^2 = (c + 16)(c + 16)$

$$= c^2 + 16c + 16c + 256$$

$$= c^2 + 32c + 256$$

24. $(x + 3)(2x - 3) = 2x^2 - 3x + 6x - 9$

$$= 2x^2 + 3x - 9$$

25. $\dfrac{2x - 32}{16x} = \dfrac{2x}{16x} - \dfrac{32}{16x}$

$$= \frac{1}{8} - \frac{2}{x}$$

26.

$$3x + 1 \enclose{longdiv}{9x^2 + 6x + 1} \quad \dfrac{3x}{}$$

$$9x^2 + 3x$$

$$\dfrac{3x}{3x+1\,\overline{)9x^2 + 6x + 1}}$$

$$\underline{-9x^2 - 3x}$$

$$3x + 1$$

$$\dfrac{3x+1}{3x+1\,\overline{)9x^2 + 6x + 1}}$$

$$\underline{-9x^2 - 3x}$$

$$3x + 1$$

$$3x + 1$$

$$\boxed{3x+1}$$

$$3x + 1\,\overline{)9x^2 + 6x + 1}$$

$$\underline{-9x^2 - 3x}$$

$$3x + 1$$

$$\underline{-3x - 1}$$

$$0$$

27. $12r^2 - 3rs + 9r^2s^2 = 3r(4r - s + 3rs^2)$

28. $u^2 - 18u + 81 = (u - 9)(u - 9)$

$$= (u - 9)^2$$

29. $2y^2 - 7y + 3 = (2y - 1)(y - 3)$

30. $x^4 - 81 = (x^2 + 9)(x^2 - 9)$
$= (x^2 + 9)(x + 3)(x - 3)$

31. $(t^3 - v^3) = (t - v)(t^2 + tv + v^2)$

32. $xy - ty + xs - ts = (xy - ty) + (xs - ts)$
$= y(x - t) + s(x - t)$
$= (x - t)(y + s)$

33. $8s^2 - 16s = 0$
$8s(s - 2) = 0$
$\boxed{s = 0}$

$s - 2 = 0$
$s - 2 + 2 = 0 + 2$
$\boxed{s = 2}$
$\boxed{s = 0, 2}$

34. $x^2 + 2x - 15 = 0$
$(x + 5)(x - 3) = 0$
$x + 5 = 0$
$x + 5 - 5 = 0 - 5$
$\boxed{x = -5}$

$x - 3 = 0$
$x - 3 + 3 = 0 + 3$
$\boxed{x = 3}$
$\boxed{x = 3, -5}$

35. $\dfrac{x^2 - 25}{5x + 25} = \dfrac{(x + 5)(x - 5)}{5(x + 5)}$
$= \dfrac{x - 5}{5}$

Do not attempt to reduce the "5's".
The top "5" is term of a binomial.
The bottom "5" is a monominal.

36. $\dfrac{x^2 - x - 2}{x^2 + x} \div \dfrac{2 - x}{x} = \dfrac{(x - 2)(x + 1)}{x(x + 1)} \cdot \dfrac{x}{2 - x}$
$= \dfrac{(x - 2)}{1} \cdot \dfrac{1}{2 - x}$
$= \dfrac{(x - 2)}{-(-2 + x)}$
$= -1$

37. $\dfrac{x + 5}{xy} - \dfrac{x - 1}{x^2 y} = \dfrac{x + 5}{xy}\left(\dfrac{x}{x}\right) - \dfrac{x - 1}{x^2 y}$
$= \dfrac{x(x + 5)}{x^2 y} - \dfrac{x - 1}{x^2 y}$
$= \dfrac{x(x + 5) - (x - 1)}{x^2 y}$
$= \dfrac{x^2 + 5x - x + 1}{x^2 y}$
$= \dfrac{x^2 + 4x + 1}{x^2 y}$

38. $\dfrac{\dfrac{y}{x} + 3y}{y + \dfrac{2y}{x}} = \dfrac{\dfrac{y}{x} + 3y\left(\dfrac{x}{x}\right)}{y\left(\dfrac{x}{x}\right) + \dfrac{2y}{x}}$
$= \dfrac{\dfrac{y}{x} + \dfrac{3xy}{x}}{\dfrac{xy}{x} + \dfrac{2y}{x}}$
$= \dfrac{\dfrac{y + 3xy}{x}}{\dfrac{xy + 2y}{x}}$
$= \dfrac{y + 3xy}{x} \cdot \dfrac{x}{xy + 2y}$
$= \dfrac{y(1 + 3x)}{x} \cdot \dfrac{x}{y(x + 2)}$
$= \dfrac{3x + 1}{x + 2}$

39.

$$\frac{3}{x+1} - \frac{x-2}{2} = \frac{x-2}{x+1}$$

Find all restrictions of x.

$$x + 1 \neq 0$$

$$x + 1 - 1 \neq 0 - 1$$

$$\boxed{x \neq -1}$$

$$\frac{3}{x+1} - \frac{x-2}{2} = \frac{x-2}{x+1}$$

Multiply each fraction by $2(x+1)$
to remove all the denominators.

$$6 - (x+1)(x-2) = 2(x-2)$$

$$6 - (x^2 - x - 2) = 2x - 4$$

$$6 - x^2 + x + 2 = 2x - 4$$

$$-x^2 + x + 8 = 2x - 4$$

Add $(x^2 - x - 8)$ to both sides.

$$\underline{x^2 - x - 8 = x^2 - x - 8}$$

$$0 = x^2 + x - 12$$

$$0 = (x+4)(x-3)$$

$$x + 4 = 0$$

$$x + 4 - 4 = 0 - 4$$

$$\boxed{x = -4}$$

$$x - 3 = 0$$

$$x - 3 + 3 = 0 + 3$$

$$\boxed{x = 3}$$

$$\boxed{x = -4, 3}$$

40.

$$\frac{a}{20} = \frac{12}{15}$$

$$\frac{20}{1}\left(\frac{a}{20}\right) = \frac{20}{1}\left(\frac{12}{15}\right)$$

$$\boxed{a = 16}$$

$$\frac{b}{20} = \frac{6}{15}$$

$$\frac{20}{1}\left(\frac{b}{20}\right) = \frac{20}{1}\left(\frac{6}{15}\right)$$

$$\boxed{b = 8}$$

41.

$$\begin{cases} x + 4y = -2 \\ y = -x - 5 \end{cases}$$

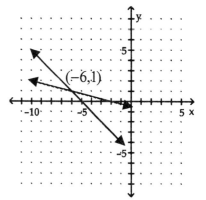

$(-6,1)$

42.

$$\begin{cases} 2x - 3y < 0 \\ y > x - 1 \end{cases}$$

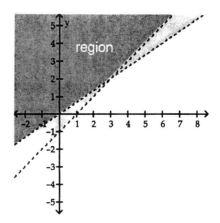

region

- 367 -

Chapter 7 Cumulative Review

43.
$$\begin{cases} x - 2y = 2 \\ 2x + 3y = 11 \end{cases}$$

$$\begin{cases} x = 2y + 2 \\ 2x + 3y = 11 \end{cases}$$

$2x + 3y = 11$ 2nd equation

$2(2y + 2) + 3y = 11$

$4y + 4 + 3y = 11$

$7y + 4 = 11$

$7y + 4 - 4 = 11 - 4$

$7y = 7$

$\dfrac{7y}{7} = \dfrac{7}{7}$

$\boxed{y = 1}$

$x = 2y + 2$

$x = 2(1) + 2$

$\boxed{x = 4}$

The solution is (1, 4)

44. Protein in 1 serving of egg noodles: 5 g

Protein in 1 serving of rice pilaf: 4 g

Wants to consume only 22 g of protein.

Fat in 1 serving of egg noodles: 3 g

Fat in 1 serving of rice pilaf: 5 g

Wants to consume only 21 g of fat.

Let x = # of servings of egg noodles

y = # of servings of rice pilaf

$$\begin{cases} 5x + 4y = 22 \\ 3x + 5y = 21 \end{cases}$$

Multiply 1st equation by -5

Multiply 2nd equation by 4

Eliminate the y

$-25x - 20y = -110$

$\underline{12x + 20y = 84}$

$-13x = -26$

$\dfrac{-13x}{-13} = \dfrac{-26}{-13}$

$\boxed{x = 2}$

$5x + 4y = 22$ 1st equation

$5(2) + 4y = 22$

$10 + 4y - 10 = 22 - 10$

$4y = 12$

$\dfrac{4y}{4} = \dfrac{12}{4}$

$\boxed{y = 3}$

2 servings of noodles.

3 servings of rice.

VOCABULARY

1. The number b is a **square** root of the number a if $b^2 = a$.

3. The symbol $-\sqrt{}$ is used to represent the **negative** square root of a number.

5. A number such as 25, 49, or $\dfrac{4}{81}$, that is the square of some rational number, is called a **perfect** square.

7. The **Pythagorean** Theorem is a formula that relates the lengths of the sides of a right triangle.

CONCEPTS

9. Every positive number has **two** square roots, one positive and one negative.

11. The positive square root of 25 is **5** and the negative square root of 25 is **-5**. In

    symbols, we write $\sqrt{25} = \underline{\mathbf{5}}$ and $-\sqrt{25} = \underline{\mathbf{-5}}$

13. 2 is a square root of 4, because $(\, \mathbf{2}\,)^2 = 4$ and -2 is a square root of 4, because $(\, \mathbf{-2}\,)^2 = 4$.

15. The distance formula is:

    $$d = \sqrt{(x_2 - \boxed{x_1})^2 + (\boxed{y_2} - y_1)^2}$$

17. irrational

19. irrational

21. 45

23. $16x^2$

25. $-\sqrt{43} = -1\sqrt{43}$

    $-1\sqrt{43} = -1\sqrt{43}$ true

PRACTICE

27. 7

29. 6

31. 10

33. 20

35. –9

37. 1.1

39. 13

41. $\dfrac{3}{5}$

43. $-\dfrac{1}{8}$

45. –0.2

47. –17

49. 50

51. 1.732

53. 3.317

55. 9.747

57. 20.688

59. 3.464

61. m

63. t^2

65. c^5

67. $2y$

69. $8b^2$

71. $a = 4$, $b = 3$, $c = ?$

    $$a^2 + b^2 = c^2$$
    $$(4)^2 + (3)^2 = c^2$$
    $$16 + 9 = c^2$$
    $$25 = c^2$$
    $$\sqrt{25} = \sqrt{c^2}$$
    $$\boxed{5 = c}$$

73.　$a = ?$, $b = 16$, $c = 34$

$$a^2 + b^2 = c^2$$
$$a^2 + (16)^2 = (34)^2$$
$$a^2 + 256 = 1,156$$
$$a^2 + 256 - 256 = 1,156 - 256$$
$$a^2 = 900$$
$$\sqrt{a^2} = \sqrt{900}$$
$$\boxed{a = 30}$$

75.　$a = 44$, $b = ?$, $c = 125$

$$a^2 + b^2 = c^2$$
$$(44)^2 + b^2 = (125)^2$$
$$1,936 + b^2 = 15,625$$
$$1,936 + b^2 - 1,936 = 15,625 - 1,936$$
$$b^2 = 13,689$$
$$\sqrt{b^2} = \sqrt{13,689}$$
$$\boxed{b = 117}$$

77.　$a = x$, $b = x$, $c = 2$ inches

$$a^2 + b^2 = c^2$$
$$x^2 + x^2 = (2)^2$$
$$2x^2 = 4$$
$$\frac{2x^2}{2} = \frac{4}{2}$$
$$x^2 = 2$$
$$\sqrt{x^2} = \sqrt{2}$$
$$\boxed{x = \sqrt{2} \text{ in.}}$$

or

$$x \approx 1.414$$
$$\boxed{x \approx 1.41 \text{ in.}}$$

79.　$(4, 6)$　and　$(1, 2)$

(x_1, y_1) and (x_2, y_2)
$$d = \sqrt{(x_2 - x_1)^2 + (y_2 - y_1)^2}$$
$$d = \sqrt{(1-4)^2 + (2-6)^2}$$
$$d = \sqrt{(-3)^2 + (-4)^2}$$
$$d = \sqrt{9 + 16}$$
$$d = \sqrt{25}$$
$$\boxed{d = 5}$$

81.　$(-2, -8)$　and　$(3, 4)$

(x_1, y_1)　　and (x_2, y_2)
$$d = \sqrt{(x_2 - x_1)^2 + (y_2 - y_1)^2}$$
$$d = \sqrt{[3-(-2)]^2 + [4-(-8)]^2}$$
$$d = \sqrt{(3+2)^2 + (4+8)^2}$$
$$d = \sqrt{(5)^2 + (12)^2}$$
$$d = \sqrt{25 + 144}$$
$$d = \sqrt{169}$$
$$\boxed{d = 13}$$

83.　$(6, -5)$　and　$(0, -4)$

(x_1, y_1)　　and (x_2, y_2)
$$d = \sqrt{(x_2 - x_1)^2 + (y_2 - y_1)^2}$$
$$d = \sqrt{(0-6)^2 + [-4-(-5)]^2}$$
$$d = \sqrt{(-6)^2 + (-4+5)^2}$$
$$d = \sqrt{(-6)^2 + (1)^2}$$
$$d = \sqrt{36 + 1}$$
$$\boxed{d = \sqrt{37}}$$
$$d \approx 6.082$$
$$\boxed{d \approx 6.08}$$

85.

$$(9, -1) \text{ and } (7, -6)$$
$$(x_1, y_1) \quad \text{and} \quad (x_2, y_2)$$
$$d = \sqrt{(x_2 - x_1)^2 + (y_2 - y_1)^2}$$
$$d = \sqrt{(7-9)^2 + [-6-(-1)]^2}$$
$$d = \sqrt{(-2)^2 + (-6+1)^2}$$
$$d = \sqrt{(-2)^2 + (-5)^2}$$
$$d = \sqrt{4 + 25}$$
$$\boxed{d = \sqrt{29}}$$
$$d \approx 5.385$$
$$\boxed{d \approx 5.39}$$

APPLICATIONS

87. HIGHWAY SAFETY

s is speed in mph

d is skid mark in feet

skid mark is 64 feet

$$s = 4.5\sqrt{d}$$
$$s = 4.5\sqrt{64}$$
$$s = 4.5(8)$$
$$\boxed{s = 36}$$

Speed of car is 36 mph.

89. DRAFTING

a)   leg is 6 inches

hypotenuse is ? inches

hypotenuse is leg • $\sqrt{2}$

$$c = \text{leg}\sqrt{2}$$
$$\boxed{c = 6\sqrt{2} \text{ in.}}$$
$$c \approx 6(1.4142135)$$
$$c \approx 8.485$$
$$\boxed{c \approx 8.49 \text{ in.}}$$

continued

89. continued

b)   hypotenuse is 10 inches

long leg is ? inchs

long leg is $\dfrac{1}{2}$ • hypotenuse • $\sqrt{3}$

$$b = \frac{1}{2}(10)\sqrt{3}$$
$$\boxed{b = 5\sqrt{3} \text{ in.}}$$
$$b \approx 5(1.7320508)$$
$$b \approx 8.660$$
$$\boxed{b \approx 8.66 \text{ in.}}$$

91. BASEBALL

baseball diamond is a square

one side (leg) is 90 feet

home plate to 2nd base is ? feet

hypotenuse is leg • $\sqrt{2}$

$$c = s\sqrt{2}$$
$$\boxed{c = 90\sqrt{2} \text{ ft}}$$
$$c \approx 90(1.4142135)$$
$$c \approx 127.279$$
$$\boxed{c \approx 127.28 \text{ ft}}$$

or

$$a = 90, \ b = 90, \ c = ?$$
$$a^2 + b^2 = c^2$$
$$(90)^2 + (90)^2 = c^2$$
$$8{,}100 + 8{,}100 = c^2$$
$$16{,}200 = c^2$$
$$\sqrt{16{,}200} = \sqrt{c^2}$$
$$\sqrt{16{,}200} = c$$
$$127.279 \approx c$$
$$\boxed{127.28 \approx c}$$

The diagonal is 127.28 feet.

93. QUALITY CONTROL

a) short side (leg) is 16 inches

long side (leg) is 30 inches

b) hypotenuse (diagonal) is ? inches

$a = 16$, $b = 30$, $c = ?$

$a^2 + b^2 = c^2$

$(16)^2 + (30)^2 = c^2$

$256 + 900 = c^2$

$1{,}156 = c^2$

$\sqrt{1{,}156} = \sqrt{c^2}$

$\boxed{34 = c}$

The diagonal is 34 inches.

95. FOOTBALL

a) hypotenuse (diagonal) is 6 yards

one side is ? yards

$a = x$, $b = x$, $c = 6$ yards

$a^2 + b^2 = c^2$

$x^2 + x^2 = (6)^2$

$2x^2 = 36$

$\dfrac{2x^2}{2} = \dfrac{36}{2}$

$x^2 = 18$

$\sqrt{x^2} = \sqrt{18}$

$x \approx 4.242$

$\boxed{x \approx 4.24}$

The side is 4.24 yds.

b) yes, 6 yds + 4.24 yds > 10 yds

WRITING

97. Answers will vary.

99. Answers will vary.

101. Answers will vary.

REVIEW

103.

$(3s^2 - 3s - 2) + (3s^2 + 4s - 3)$

$= 6s^2 + s - 5$

105.

$(3x - 2)(x + 4) = 3x^2 + 12x - 2x - 8$

$= 3x^2 + 10x - 8$

CHALLENGE PROBLEMS

107. SHORTCUTS

width (short leg) is 52 feet

length (long leg) is 165 feet

hypotenuse (diagonal) is ? feet

$a = 52$, $b = 165$, $c = ?$

$a^2 + b^2 = c^2$

$(52)^2 + (165)^2 = c^2$

$2{,}704 + 27{,}225 = c^2$

$29{,}929 = c^2$

$\sqrt{29{,}929} = \sqrt{c^2}$

$\boxed{173 = c}$

The diagonal is 173 feet.

165 ft. + 52 ft. = 217 ft.

217 ft. − 173 ft. = 44 ft.

$\boxed{\text{Using the diagonal saves 44 ft.}}$

109. $\sqrt{16x^{16n}} = \sqrt{16}\sqrt{x^{16n}}$

$= \sqrt{4^2}\sqrt{x^{8n}x^{8n}}$

$= 4x^{8n}$

SECTION 8.2

VOCABULARY

1. The first four natural number perfect **squares** are 1, 4, 9, and 16.

3. "To **simplify** $\sqrt{20}$ " means to write it as $2\sqrt{5}$.

5. The word *product* is associated with the operation of **multiplication** and the word *quotient* with **division**.

7. The expression $\sqrt{4 \cdot 3}$ is a square root of a **product** and $\sqrt{4}\sqrt{3}$ is a product of **square** roots.

CONCEPTS

9. a) The square root of the product of two positive numbers is equal to the **product** of their square roots. In symbols,

$$\sqrt{a \cdot b} = \boxed{\sqrt{a}\sqrt{b}}$$

   b) The square root of the quotient of two positive numbers is equal to the **quotient** of their square roots. In symbols,

$$\sqrt{\frac{a}{b}} = \boxed{\frac{\sqrt{a}}{\sqrt{b}}}$$

   where $b \neq 0$.

11. a) 4
    b) 49
    c) 9
    d) 16

13. $\sqrt{4 \cdot 10}$

15. a) $\sqrt{81 \cdot 2} = \sqrt{81}\ \sqrt{2}$
    b) $\sqrt{m^4 \cdot m} = \sqrt{m^4}\ \sqrt{m}$

17. a) $\sqrt{100} = 10$
    b) $\sqrt{t^2} = t$
    c) $\sqrt{a^4} = a^2$
    d) $\sqrt{144y^2} = 12y$

NOTATION

19. $\sqrt{32} = \sqrt{\boxed{16} \cdot 2}$
$$= \sqrt{\boxed{16}}\ \sqrt{2}$$
$$= \boxed{4}\sqrt{2}$$

21. a) $3 \cdot 2\sqrt{15} = 6\sqrt{15}$
    b) $5x \cdot 4x\sqrt{14} = 20x^2\sqrt{14}$

PRACTICE

23. $\sqrt{20} = \sqrt{4}\sqrt{5}$
$$= 2\sqrt{5}$$

25. $\sqrt{50} = \sqrt{25}\sqrt{2}$
$$= 5\sqrt{2}$$

27. $\sqrt{27} = \sqrt{9}\sqrt{3}$
$$= 3\sqrt{3}$$

29. $\sqrt{98} = \sqrt{49}\sqrt{2}$
$$= 7\sqrt{2}$$

31. $\sqrt{48} = \sqrt{16}\sqrt{3}$
$$= 4\sqrt{3}$$

33. $\sqrt{44} = \sqrt{4}\sqrt{11}$
$$= 2\sqrt{11}$$

35. $\sqrt{15} = \sqrt{15}$

37. $\sqrt{192} = \sqrt{64}\sqrt{3}$
$$= 8\sqrt{3}$$

39. $\sqrt{250} = \sqrt{25}\sqrt{10}$
$= 5\sqrt{10}$

41. $-\sqrt{700} = -\sqrt{100}\sqrt{7}$
$= -10\sqrt{7}$

43. $\sqrt{162} = \sqrt{81}\sqrt{2}$
$= 9\sqrt{2}$

45. $3\sqrt{32} = 3\sqrt{16}\sqrt{2}$
$= 3 \cdot 4\sqrt{2}$
$= 12\sqrt{2}$

47. $-3\sqrt{72} = -3\sqrt{36}\sqrt{2}$
$= -3 \cdot 6\sqrt{2}$
$= -18\sqrt{2}$

49. $\sqrt{x^5} = \sqrt{x^4}\sqrt{x}$
$= x^2\sqrt{x}$

51. $\sqrt{n^9} = \sqrt{n^8}\sqrt{n}$
$= n^4\sqrt{n}$

53. $2\sqrt{24} = 2\sqrt{4}\sqrt{6}$
$= 2 \cdot 2\sqrt{6}$
$= 4\sqrt{6}$

55. $\sqrt{4k} = \sqrt{4}\sqrt{k}$
$= 2\sqrt{k}$

57. $\sqrt{12x} = \sqrt{4}\sqrt{3x}$
$= 2\sqrt{3x}$

59. $6\sqrt{75t} = 6\sqrt{25}\sqrt{3t}$
$= 6 \cdot 5\sqrt{3t}$
$= 30\sqrt{3t}$

61. $\sqrt{25x^3} = \sqrt{25}\sqrt{x^2}\sqrt{x}$
$= 5x\sqrt{x}$

63. $\sqrt{10h} = \sqrt{10h}$

65. $\sqrt{9x^4 y} = \sqrt{9}\sqrt{x^4}\sqrt{y}$
$= 3x^2\sqrt{y}$

67. $\sqrt{2c^7 d^{11}} = \sqrt{c^6}\sqrt{d^{10}}\sqrt{2cd}$
$= c^3 d^5\sqrt{2cd}$

69. $-12x\sqrt{16x^2 y^3} = -12x\sqrt{16x^2 y^2}\sqrt{y}$
$= -12x \cdot 4xy\sqrt{y}$
$= -48x^2 y\sqrt{y}$

71. $\dfrac{1}{5}x^2 y\sqrt{50x^2 y^2} = \dfrac{1}{5}x^2 y\sqrt{25x^2 y^2}\sqrt{2}$
$= \dfrac{1}{5}x^2 y \cdot 5xy\sqrt{2}$
$= x^3 y^2\sqrt{2}$

73. $\sqrt{\dfrac{25}{9}} = \dfrac{\sqrt{25}}{\sqrt{9}}$
$= \dfrac{5}{3}$

75. $\sqrt{\dfrac{81}{64}} = \dfrac{\sqrt{81}}{\sqrt{64}}$
$= \dfrac{9}{8}$

77. $\sqrt{\dfrac{6}{121}} = \dfrac{\sqrt{6}}{\sqrt{121}}$
$= \dfrac{\sqrt{6}}{11}$

79.
$$\sqrt{\frac{26}{25}} = \frac{\sqrt{26}}{\sqrt{25}}$$
$$= \frac{\sqrt{26}}{5}$$

81.
$$-\sqrt{\frac{20}{49}} = -\frac{\sqrt{20}}{\sqrt{49}}$$
$$= -\frac{\sqrt{4}\sqrt{5}}{\sqrt{49}}$$
$$= -\frac{2\sqrt{5}}{7}$$

83.
$$\sqrt{\frac{48}{81}} = \frac{\sqrt{48}}{\sqrt{81}}$$
$$= \frac{\sqrt{16}\sqrt{3}}{\sqrt{81}}$$
$$= \frac{4\sqrt{3}}{9}$$

85.
$$\sqrt{\frac{32}{25}} = \frac{\sqrt{32}}{\sqrt{25}}$$
$$= \frac{\sqrt{16}\sqrt{2}}{\sqrt{25}}$$
$$= \frac{4\sqrt{2}}{5}$$

87.
$$\sqrt{\frac{72x^3}{y^2}} = \frac{\sqrt{72x^3}}{\sqrt{y^2}}$$
$$= \frac{\sqrt{36x^2}\sqrt{2x}}{\sqrt{y^2}}$$
$$= \frac{6x\sqrt{2x}}{y}$$

89.
$$\sqrt{\frac{125n^5}{64n}} = \sqrt{\frac{125n^4}{64}}$$
$$= \frac{\sqrt{25n^4}\sqrt{5}}{\sqrt{64}}$$
$$= \frac{5n^2\sqrt{5}}{8}$$

91.
$$\sqrt{\frac{128m^3n^5}{81mn^7}} = \sqrt{\frac{128m^2}{81n^2}}$$
$$= \frac{\sqrt{64m^2}\sqrt{2}}{\sqrt{81n^2}}$$
$$= \frac{8m\sqrt{2}}{9n}$$

93.
$$\sqrt{\frac{12r^7s^7}{r^5s^2}} = \sqrt{12r^2s^5}$$
$$= \sqrt{4r^2s^4}\sqrt{3s}$$
$$= 2rs^2\sqrt{3s}$$

APPLICATIONS

95. LADDERS

short leg is 5 feet

long leg is 10 feet

ladder (hypotenuse) is ? feet

$a = 5$, $b = 10$, $c = ?$
$$a^2 + b^2 = c^2$$
$$5^2 + 10^2 = c^2$$
$$25 + 100 = c^2$$
$$125 = c^2$$
$$\sqrt{125} = \sqrt{c^2}$$
$$\sqrt{25}\sqrt{5} = c$$
$$\boxed{5\sqrt{5} = c}$$

The diagonal is $5\sqrt{5}$ feet.

97. CROSSWORD PUZZLE

a) $\sqrt{28} = \sqrt{4}\sqrt{7}$

$\qquad = 2\sqrt{7}$

The length is $2\sqrt{7}$ in.

b) $\sqrt{28} \approx 5.29$

$\qquad \approx 5.3$

The length is approximately 5.3 in.

WRITING

99. Answers will vary.

101. Answers will vary.

REVIEW

103. $\begin{cases} y = 2x - 6 \\ 2x + y = 6 \end{cases}$

$2x + y = 6$

$2x + (2x - 6) = 6$

$4x - 6 = 6$

$4x - 6 + 6 = 6 + 6$

$4x = 12$

$\dfrac{4x}{4} = \dfrac{12}{4}$

$\boxed{x = 3}$

$y = 2x - 6$

$y = 2(3) - 6$

$y = 6 - 6$

$\boxed{y = 0}$

Solution is (3, 0).

105. $\begin{cases} 3x + 4y = -7 \\ 2x - y = -1 \end{cases}$

Multiply 2nd equation by 4.

Eliminate the x.

$3x + 4y = -7$

$\underline{8x - 4y = -4}$

$11x = -11$

$\dfrac{11x}{11} = \dfrac{-11}{11}$

$\boxed{x = -1}$

$2x - y = -1$

$2(-1) - y = -1$

$-2 - y + 2 = -1 + 2$

$-y = 1$

$\dfrac{-y}{-1} = \dfrac{1}{-1}$

$\boxed{y = -1}$

Solution is (-1, -1).

CHALLENGE PROBLEMS

107. Show that $\sqrt{a + b} \neq \sqrt{a} + \sqrt{b}$.

Let $a = 9$, $b = 16$

$\sqrt{a + b} \neq \sqrt{a} + \sqrt{b}$

$\sqrt{9 + 16} \neq \sqrt{9} + \sqrt{16}$

$\sqrt{25} \neq \sqrt{9} + \sqrt{16}$

$5 \neq 3 + 4$

$\boxed{5 \neq 7}$

There are many other variations.

Section 8.3

VOCABULARY

1. Square root radicals such as $\sqrt{2}$ and $5\sqrt{2}$, that have the same radicand, are called like **radicals**.

3. Combining like radicals is similar to combining like **terms**.

CONCEPTS

5. yes

7. no

9. a) $\sqrt{12} = 2\sqrt{3}$

   b) $\sqrt{4x^3} = 2x\sqrt{x}$

NOTATION

11. $9\sqrt{5} - 3\sqrt{20} = 9\sqrt{5} - 3\sqrt{\boxed{4} \cdot 5}$

    $= 9\sqrt{5} - 3\sqrt{\boxed{4}}\sqrt{5}$

    $= 9\sqrt{5} - 3 \cdot \boxed{2\sqrt{5}}$

    $= 9\sqrt{5} - \boxed{6\sqrt{5}}$

    $= \boxed{3\sqrt{5}}$

PRACTICE

13. $5\sqrt{7} + 4\sqrt{7} = 9\sqrt{7}$

15. $14\sqrt{21} - 4\sqrt{21} = 10\sqrt{21}$

17. $8\sqrt{n} + \sqrt{n} = 9\sqrt{n}$

19. $\sqrt{6} - \sqrt{6} = 0$

21. $2\sqrt{5} + 2\sqrt{5} = 4\sqrt{5}$

23. $-9\sqrt{21} + 6\sqrt{21} = -3\sqrt{21}$

25. $\sqrt{3} + \sqrt{15} = \sqrt{3} + \sqrt{15}$

27. $\sqrt{x} - 4\sqrt{x} = -3\sqrt{x}$

29. $\sqrt{2} + \sqrt{x} = \sqrt{2} + \sqrt{x}$

31. $2\sqrt{11} + 3\sqrt{11} + 5\sqrt{11} = 10\sqrt{11}$

33. $4\sqrt{2} + 4\sqrt{2} - 4\sqrt{2} = 4\sqrt{2}$

35. $5 + 3\sqrt{3} + 3\sqrt{3} = 5 + 6\sqrt{3}$

37. $-1 + 2\sqrt{r} - 3\sqrt{r} = -1 - \sqrt{r}$

39. $\sqrt{12} + \sqrt{27} = \sqrt{4}\sqrt{3} + \sqrt{9}\sqrt{3}$

    $= 2\sqrt{3} + 3\sqrt{3}$

    $= 5\sqrt{3}$

41. $\sqrt{18} - \sqrt{8} = \sqrt{9}\sqrt{2} - \sqrt{4}\sqrt{2}$

    $= 3\sqrt{2} - 2\sqrt{2}$

    $= \sqrt{2}$

43. $2\sqrt{45} + 2\sqrt{80} = 2\sqrt{9}\sqrt{5} + 2\sqrt{16}\sqrt{5}$

    $= 2 \cdot 3\sqrt{5} + 2 \cdot 4\sqrt{5}$

    $= 6\sqrt{5} + 8\sqrt{5}$

    $= 14\sqrt{5}$

45. $2\sqrt{80} - 3\sqrt{125} = 2\sqrt{16}\sqrt{5} - 3\sqrt{25}\sqrt{5}$

    $= 2 \cdot 4\sqrt{5} - 3 \cdot 5\sqrt{5}$

    $= 8\sqrt{5} - 15\sqrt{5}$

    $= -7\sqrt{5}$

47. $\sqrt{12} - \sqrt{48} = \sqrt{4}\sqrt{3} - \sqrt{16}\sqrt{3}$

    $= 2\sqrt{3} - 4\sqrt{3}$

    $= -2\sqrt{3}$

49. $\sqrt{288} - 3\sqrt{200} = \sqrt{144}\sqrt{2} - 3\sqrt{100}\sqrt{2}$

    $= 12\sqrt{2} - 3 \cdot 10\sqrt{2}$

    $= 12\sqrt{2} - 30\sqrt{2}$

    $= -18\sqrt{2}$

51. $2\sqrt{28} + 2\sqrt{112} = 2\sqrt{4}\sqrt{7} + 2\sqrt{16}\sqrt{7}$

$\qquad\qquad = 2 \cdot 2\sqrt{7} + 2 \cdot 4\sqrt{7}$

$\qquad\qquad = 4\sqrt{7} + 8\sqrt{7}$

$\qquad\qquad = 12\sqrt{7}$

53. $\sqrt{20} + \sqrt{45} + \sqrt{80}$

$\qquad = \sqrt{4}\sqrt{5} + \sqrt{9}\sqrt{5} + \sqrt{16}\sqrt{5}$

$\qquad = 2\sqrt{5} + 3\sqrt{5} + 4\sqrt{5}$

$\qquad = 9\sqrt{5}$

55. $\sqrt{200} - \sqrt{75} + \sqrt{48}$

$\qquad = \sqrt{100}\sqrt{2} - \sqrt{25}\sqrt{3} + \sqrt{16}\sqrt{3}$

$\qquad = 10\sqrt{2} - 5\sqrt{3} + 4\sqrt{3}$

$\qquad = 10\sqrt{2} - \sqrt{3}$

57. $8\sqrt{6} - 5\sqrt{2} - 3\sqrt{6} = 5\sqrt{6} - 5\sqrt{2}$

59. $\sqrt{24} + \sqrt{150} + \sqrt{240}$

$\qquad = \sqrt{4}\sqrt{6} + \sqrt{25}\sqrt{6} + \sqrt{16}\sqrt{15}$

$\qquad = 2\sqrt{6} + 5\sqrt{6} + 4\sqrt{15}$

$\qquad = 7\sqrt{6} + 4\sqrt{15}$

61. $\sqrt{48} - \sqrt{8} + \sqrt{27} - \sqrt{32}$

$\qquad = \sqrt{16}\sqrt{3} - \sqrt{4}\sqrt{2} + \sqrt{9}\sqrt{3} - \sqrt{16}\sqrt{2}$

$\qquad = 4\sqrt{3} - 2\sqrt{2} + 3\sqrt{3} - 4\sqrt{2}$

$\qquad = 7\sqrt{3} - 6\sqrt{2}$

63. $\sqrt{72a} - \sqrt{98a} = \sqrt{36}\sqrt{2a} - \sqrt{49}\sqrt{2a}$

$\qquad\qquad\quad = 6\sqrt{2a} - 7\sqrt{2a}$

$\qquad\qquad\quad = -\sqrt{2a}$

65. $\sqrt{2x^2} + \sqrt{8x^2} = \sqrt{x^2}\sqrt{2} + \sqrt{4x^2}\sqrt{2}$

$\qquad\qquad\quad = x\sqrt{2} + 2x\sqrt{2}$

$\qquad\qquad\quad = 3x\sqrt{2}$

67. $\sqrt{2d^3} + \sqrt{8d^3} = \sqrt{d^2}\sqrt{2d} + \sqrt{4d^2}\sqrt{2d}$

$\qquad\qquad\quad = d\sqrt{2d} + 2d\sqrt{2d}$

$\qquad\qquad\quad = 3d\sqrt{2d}$

69. $\sqrt{18x^2 y} - \sqrt{27x^2 y}$

$\qquad = \sqrt{9x^2}\sqrt{2xy} - \sqrt{9x^2}\sqrt{3xy}$

$\qquad = 3x\sqrt{2xy} - 3x\sqrt{3xy}$

71. $\sqrt{32x^5} - \sqrt{18x^5}$

$\qquad = \sqrt{16x^4}\sqrt{2x} - \sqrt{9x^4}\sqrt{2x}$

$\qquad = 4x^2\sqrt{2x} - 3x^2\sqrt{2x}$

$\qquad = x^2\sqrt{2x}$

73. $3\sqrt{54b^2} + 5\sqrt{24b^2}$

$\qquad = 3\sqrt{9b^2}\sqrt{6} + 5\sqrt{4b^2}\sqrt{6}$

$\qquad = 3 \cdot 3b\sqrt{6} + 5 \cdot 2b\sqrt{6}$

$\qquad = 9b\sqrt{6} + 10b\sqrt{6}$

$\qquad = 19b\sqrt{6}$

75. $y\sqrt{490y} - 2\sqrt{360y^3}$

$\qquad = y\sqrt{49}\sqrt{10y} - 2\sqrt{36y^2}\sqrt{10y}$

$\qquad = y \cdot 7\sqrt{10y} - 2 \cdot 6y\sqrt{10y}$

$\qquad = 7y\sqrt{10y} - 12y\sqrt{10y}$

$\qquad = -5y\sqrt{10y}$

77. $\sqrt{20x^3 y} + \sqrt{45x^5 y^3} - \sqrt{80x^7 y^5}$

$\qquad = \sqrt{4x^2}\sqrt{5xy} + \sqrt{9x^4 y^2}\sqrt{5xy} - \sqrt{16x^6 y^4}\sqrt{5xy}$

$\qquad = 2x\sqrt{5xy} + 3x^2 y\sqrt{5xy} - 4x^3 y^2\sqrt{5xy}$

APPLICATION

79. ANATOMY

$(2\sqrt{48} + 5\sqrt{12})$ in
$$= (2\sqrt{16}\sqrt{3} + 5\sqrt{4}\sqrt{3}) \text{ in}$$
$$= (2 \cdot 4\sqrt{3} + 5 \cdot 2\sqrt{3}) \text{ in}$$
$$= (8\sqrt{3} + 10\sqrt{3}) \text{ in}$$
$$= 18\sqrt{3} \text{ in}$$

81. HARDWARE

$(\sqrt{147} - \sqrt{27})$ in $= (\sqrt{49}\sqrt{3} - \sqrt{9}\sqrt{3})$ in
$$= (7\sqrt{3} - 3\sqrt{3}) \text{ in}$$
$$= 4\sqrt{3} \text{ in}$$

83. CAMPING

$t = 0.5s\sqrt{3}$

Parent's tent side: 6 ft.

Children's tent side: 4 ft.

$t = 2(0.5s\sqrt{3}) + 2(0.5s\sqrt{3})$ ft
$$= [2(0.5 \cdot 6\sqrt{3}) + 2(0.5 \cdot 4\sqrt{3})] \text{ ft}$$
$$= [2(3\sqrt{3}) + 2(2\sqrt{3})] \text{ ft}$$
$$= (6\sqrt{3} + 4\sqrt{3}) \text{ ft}$$
$$= 10\sqrt{3} \text{ ft}$$

WRITING

85. Answers will vary.

87. Answers will vary.

REVIEW

89. $3^{-2} = \dfrac{1}{3^2}$

$\qquad = \dfrac{1}{9}$

91. $-3^2 = -(3)(3)$
$\qquad = 9$

93. $x^{-3} = \dfrac{1}{x^3}$

95. $3^0 = 1$

CHALLENGE PROBLEMS

97. $2y\sqrt{175y} - \sqrt{\boxed{}} = 0$

$2y\sqrt{175y} - \sqrt{\boxed{4y^2(175y)}} = 0$

$2y\sqrt{175y} - \sqrt{\boxed{700y^3}} = 0$

VOCABULARY

1. The denominator of $\dfrac{1}{\sqrt{3}}$ is an **irrational** number.

3. To **rationalize** the denominator of $\dfrac{4}{\sqrt{5}}$, we multiply the fraction by $\dfrac{\sqrt{5}}{\sqrt{5}}$.

CONCEPTS

5. $\sqrt{a} \cdot \sqrt{b} = \sqrt{\boxed{a \cdot b}}$

7. $(\sqrt{a})^2 = \sqrt{a^2}$
$= \boxed{a}$

9. $\dfrac{\sqrt{6}}{\sqrt{6}} = \boxed{1}$

11. $7\sqrt{2} \cdot 9\sqrt{6} = 7 \cdot \boxed{9} \cdot \sqrt{2} \cdot \boxed{\sqrt{6}}$

13. To rationalize the denominator of $\dfrac{4}{\sqrt{3}}$, we multiply the fraction by $\dfrac{\sqrt{3}}{\sqrt{3}}$, which is a form of $\boxed{1}$.

15. a) $\sqrt{\dfrac{3}{4}}$, The radicand is a fraction.

b) $\dfrac{1}{\sqrt{10}}$, There is a radical in the denominator.

17. $\sqrt{2} \cdot \sqrt{3} = \sqrt{2 \cdot 3}$
$= \sqrt{6}$

19. $\dfrac{\sqrt{2}}{\sqrt{3}} = \dfrac{\sqrt{2}}{\sqrt{3}} \cdot \dfrac{\sqrt{3}}{\sqrt{3}}$
$= \dfrac{\sqrt{6}}{3}$

PRACTICE

21. $\sqrt{3} \cdot \sqrt{5} = \sqrt{3 \cdot 5}$
$= \sqrt{15}$

23. $\sqrt{2x}\sqrt{11} = \sqrt{2x \cdot 11}$
$= \sqrt{22x}$

25. $\sqrt{7} \cdot \sqrt{7} = \sqrt{7 \cdot 7}$
$= \sqrt{49}$
$= 7$

27. $\sqrt{5} \cdot \sqrt{10} = \sqrt{5 \cdot 10}$
$= \sqrt{50}$
$= \sqrt{25}\sqrt{2}$
$= 5\sqrt{2}$

29. $\sqrt{2}(\sqrt{27}) = \sqrt{2}(\sqrt{9}\sqrt{3})$
$= \sqrt{9}\sqrt{2}\sqrt{3}$
$= 3\sqrt{2 \cdot 3}$
$= 3\sqrt{6}$

31. $\sqrt{5} \cdot \sqrt{5d} = \sqrt{5} \cdot \sqrt{5}\sqrt{d}$
$= (\sqrt{5}\sqrt{5})\sqrt{d}$
$= \sqrt{5 \cdot 5}\sqrt{d}$
$= 5\sqrt{d}$

33. $\sqrt{2}\sqrt{8} = \sqrt{2}(\sqrt{2}\sqrt{4})$
$= (\sqrt{2}\sqrt{2})\sqrt{4}$
$= 2(2)$
$= 4$

35. $\sqrt{8}\sqrt{7} = (\sqrt{4}\sqrt{2})\sqrt{7}$
$= \sqrt{4}(\sqrt{2}\sqrt{7})$
$= 2\sqrt{14}$

37. $3\sqrt{2}\sqrt{x} = 3\sqrt{2x}$

39. $\sqrt{x^3}\sqrt{x^5} = \sqrt{x^3 x^5}$
$= \sqrt{x^{3+5}}$
$= \sqrt{x^8}$
$= x^4$

41. $(5\sqrt{6})(4\sqrt{3}) = 5(4)\sqrt{6(3)}$
$= 20\sqrt{18}$
$= 20\sqrt{9}\sqrt{2}$
$= 20(3)\sqrt{2}$
$= 60\sqrt{2}$

43. $5\sqrt{3} \cdot 2\sqrt{5} = 5 \cdot 2\sqrt{3 \cdot 5}$
$= 10\sqrt{15}$

45. $3\sqrt{y}(15\sqrt{y}) = 3(15)\sqrt{yy}$
$= 45y$

47. $\sqrt{6n^4} \cdot \sqrt{10n^5} = \sqrt{6(10)n^4 n^5}$
$= \sqrt{60n^9}$
$= \sqrt{4n^8}\sqrt{15n}$
$= 2n^4\sqrt{15n}$

49. $\sqrt{3s^6} \cdot \sqrt{3s^5} = \sqrt{3(3)s^6 s^5}$
$= \sqrt{9s^{11}}$
$= \sqrt{9s^{10}}\sqrt{s}$
$= 3s^5\sqrt{s}$

51. $\sqrt{2}(\sqrt{2} + 1) = \sqrt{2}\sqrt{2} + \sqrt{2}(1)$
$= 2 + \sqrt{2}$

53. $3\sqrt{3}(\sqrt{27} - \sqrt{2}) = 3\sqrt{3}\sqrt{27} - 3\sqrt{3}\sqrt{2}$
$= 3\sqrt{81} - 3\sqrt{6}$
$= 3(9) - 3\sqrt{6}$
$= 27 - 3\sqrt{6}$

55. $\sqrt{x}(\sqrt{3x} - 2) = \sqrt{x}\sqrt{3x} - 2\sqrt{x}$
$= \sqrt{3xx} - 2\sqrt{x}$
$= x\sqrt{3} - 2\sqrt{x}$

57. $2\sqrt{x}(\sqrt{9x} + \sqrt{x}) = 2\sqrt{x(9x)} + 2\sqrt{xx}$
$= 2\sqrt{9xx} + 2\sqrt{xx}$
$= 2(3x) + 2x$
$= 6x + 2x$
$= 8x$

59. $(\sqrt{2} + 1)(\sqrt{2} - 1) = \sqrt{4} - \sqrt{2} + \sqrt{2} - 1$
$= 2 - 1$
$= 1$

61. $(\sqrt{2} - \sqrt{3})(\sqrt{3} + \sqrt{5})$
$= \sqrt{6} + \sqrt{10} - \sqrt{9} - \sqrt{15}$
$= \sqrt{6} + \sqrt{10} - 3 - \sqrt{15}$

63. $(\sqrt{2x} + 3)(\sqrt{8x} - 6)$
$= \sqrt{16x^2} - 6\sqrt{2x} + 3\sqrt{8x} - 18$
$= 4x - 6\sqrt{2x} + 3\sqrt{4}\sqrt{2x} - 18$
$= 4x - 6\sqrt{2x} + 6\sqrt{2x} - 18$
$= 4x - 18$

65. $(2\sqrt{7} - x)(3\sqrt{2} + x)$
$= 6\sqrt{14} + 2x\sqrt{7} - 3x\sqrt{2} - x^2$

67. $(\sqrt{3x} + 2)(\sqrt{27x} - 6)$
$= \sqrt{81x^2} - 6\sqrt{3x} + 2\sqrt{27x} - 12$
$= 9x - 6\sqrt{3x} + 2\sqrt{9}\sqrt{3x} - 12$
$= 9x - 6\sqrt{3x} + 6\sqrt{3x} - 12$
$= 9x - 12$

69. $(\sqrt{6}+\sqrt{3})(\sqrt{6}-\sqrt{3})$
$\quad = \sqrt{36}-\sqrt{18}+\sqrt{18}-\sqrt{9}$
$\quad = 6-3$
$\quad = 3$

71. $(\sqrt{7}-3)(\sqrt{7}+3)$
$\quad = \sqrt{49}+3\sqrt{7}-3\sqrt{7}-9$
$\quad = 7-9$
$\quad = -2$

73. 6

75. $4 \cdot 3 = 12$

77. $100 \cdot x = 100x$

79. $y+2$

81. $\dfrac{\sqrt{12x^3}}{\sqrt{27x}} = \dfrac{\sqrt{4x^2}\sqrt{3x}}{\sqrt{9}\sqrt{3x}}$
$\quad = \dfrac{2x}{3}$

83. $\dfrac{\sqrt{18x}}{\sqrt{25x}} = \dfrac{\sqrt{9}\sqrt{2}\sqrt{x}}{\sqrt{25}\sqrt{x}}$
$\quad = \dfrac{3\sqrt{2}}{5}$

85. $\dfrac{\sqrt{196x}}{\sqrt{49x^3}} = \dfrac{\sqrt{196}\sqrt{x}}{\sqrt{49x^2}\sqrt{x}}$
$\quad = \dfrac{14\sqrt{x}}{7x\sqrt{x}}$
$\quad = \dfrac{2}{x}$

87. $\dfrac{1}{\sqrt{3}} = \dfrac{1}{\sqrt{3}} \cdot \dfrac{\sqrt{3}}{\sqrt{3}}$
$\quad = \dfrac{\sqrt{3}}{3}$

89. $\sqrt{\dfrac{13}{7}} = \dfrac{\sqrt{13}}{\sqrt{7}}$
$\quad = \dfrac{\sqrt{13}}{\sqrt{7}} \cdot \dfrac{\sqrt{7}}{\sqrt{7}}$
$\quad = \dfrac{\sqrt{91}}{7}$

91. $\dfrac{9}{\sqrt{27}} = \dfrac{9}{\sqrt{9}\sqrt{3}}$
$\quad = \dfrac{9}{3\sqrt{3}}$
$\quad = \dfrac{3}{\sqrt{3}} \cdot \dfrac{\sqrt{3}}{\sqrt{3}}$
$\quad = \dfrac{3\sqrt{3}}{3}$
$\quad = \sqrt{3}$

93. $\dfrac{3}{\sqrt{32}} = \dfrac{3}{\sqrt{16}\sqrt{2}}$
$\quad = \dfrac{3}{4\sqrt{2}}$
$\quad = \dfrac{3}{4\sqrt{2}} \cdot \dfrac{\sqrt{2}}{\sqrt{2}}$
$\quad = \dfrac{3\sqrt{2}}{4(2)}$
$\quad = \dfrac{3\sqrt{2}}{8}$

95. $\sqrt{\dfrac{12}{5}} = \dfrac{\sqrt{12}}{\sqrt{5}}$
$\quad = \dfrac{\sqrt{4}\sqrt{3}}{\sqrt{5}} \cdot \dfrac{\sqrt{5}}{\sqrt{5}}$
$\quad = \dfrac{2\sqrt{15}}{5}$

97.
$$\frac{10}{\sqrt{x}} = \frac{10}{\sqrt{x}} \cdot \frac{\sqrt{x}}{\sqrt{x}}$$
$$= \frac{10\sqrt{x}}{x}$$

99.
$$\frac{\sqrt{9y}}{\sqrt{2x}} = \frac{\sqrt{9}\sqrt{y}}{\sqrt{2x}} \cdot \frac{\sqrt{2x}}{\sqrt{2x}}$$
$$= \frac{3\sqrt{2xy}}{2x}$$

101.
$$\frac{3}{\sqrt{3}-1} = \frac{3}{\sqrt{3}-1} \cdot \frac{\sqrt{3}+1}{\sqrt{3}+1}$$
$$= \frac{3(\sqrt{3}+1)}{3-1}$$
$$= \frac{3\sqrt{3}+3}{2}$$

103.
$$\frac{3}{\sqrt{7}+2} = \frac{3}{\sqrt{7}+2} \cdot \frac{\sqrt{7}-2}{\sqrt{7}-2}$$
$$= \frac{3(\sqrt{7}-2)}{7-4}$$
$$= \frac{3(\sqrt{7}-2)}{3}$$
$$= \sqrt{7}-2$$

105.
$$\frac{12}{3-\sqrt{3}} = \frac{12}{3-\sqrt{3}} \cdot \frac{3+\sqrt{3}}{3+\sqrt{3}}$$
$$= \frac{12(3+\sqrt{3})}{9-3}$$
$$= \frac{12(3+\sqrt{3})}{6}$$
$$= 2(3+\sqrt{3})$$
$$= 6+2\sqrt{3}$$

107.
$$\frac{5}{\sqrt{3}+\sqrt{2}} = \frac{5}{\sqrt{3}+\sqrt{2}} \cdot \frac{\sqrt{3}-\sqrt{2}}{\sqrt{3}-\sqrt{2}}$$
$$= \frac{5(\sqrt{3}-\sqrt{2})}{3-2}$$
$$= 5\sqrt{3}-5\sqrt{2}$$

APPLICATIONS

109. AIR HOCKEY

width is $20\sqrt{3}$ in.

length is $30\sqrt{6}$ in.

$$A = lw$$
$$= (30\sqrt{6})(20\sqrt{3})$$
$$= 600\sqrt{18}$$
$$= 600(\sqrt{9}\sqrt{2})$$
$$= 600(3\sqrt{2})$$
$$= 1,800\sqrt{2} \text{ in.}^2$$

$\boxed{\text{Area is } 1,800\sqrt{2} \text{ in.}^2}$

111. LAWNMOWERS

length is $6\sqrt{3}$ in.

Area $= \pi r^2$

$$A = \pi r^2$$
$$= \pi(6\sqrt{3})(6\sqrt{3})$$
$$= \pi(36\sqrt{9})$$
$$= \pi(36 \cdot 3)$$
$$= 108\pi$$

$\boxed{\text{Area is } 108\pi \text{ in.}^2}$

WRITING

113. Answers will vary.

115. Answers will vary.

REVIEW

117. Is $x = -2$ a solution?

$$3x - 7 = 5x + 1$$
$$3(-2) - 7 \stackrel{?}{=} 5(-2) + 1$$
$$-6 - 7 \stackrel{?}{=} -10 + 1$$
$$-13 \neq -9 \text{ no}$$

119. positive

CHALLENGE PROBLEMS

121.

$$\frac{\sqrt{x} - 3}{\sqrt{x} + 3} = \frac{\sqrt{x} - 3}{\sqrt{x} + 3} \cdot \frac{\sqrt{x} - 3}{\sqrt{x} - 3}$$

$$= \frac{\sqrt{x^2} - 3\sqrt{x} - 3\sqrt{x} + 9}{\sqrt{x^2} - 9}$$

$$= \frac{x - 6\sqrt{x} + 9}{x - 9}$$

VOCABULARY

1. A **radical** equation is an equation that contains a variable in a radicand.

3. Squaring both sides of an equation can lead to possible solutions that do not satisfy the original equation. We call such numbers **extraneous** solutions

CONCEPTS

5. The squaring property of equality states that if $a = b$, then $a^2 = \boxed{b^2}$.

7. a) $\sqrt{x-4} - 1 = 2$

$\sqrt{x-4} - 1 + 1 = 2 + 1$

$\sqrt{x-4} = 3$

b) $8 = \sqrt{x} - x$

$8 + x = \sqrt{x} - x + x$

$x + 8 = \sqrt{x}$

9. a) $x^2 - 6x + 8 = 15$

$x^2 - 6x + 8 - 15 = 15 - 15$

$x^2 - 6x - 7 = 0$

b) $x^2 = x^2 - 3x + 12$

$x^2 - x^2 = x^2 - 3x + 12 - x^2$

$0 = -3x + 12$

NOTATION

11. $\sqrt{x-3} = 5$

$(\sqrt{x-3})^{\boxed{2}} = 5^{\boxed{2}}$

$\boxed{x-3} = 25$

$x = \boxed{28}$

13. $\sqrt{x} = 3$

$(\sqrt{x})^2 = 3^2$

$\boxed{x = 9}$

It should be noted that all answers of even index radical equations should be checked for extraneous solutions.

15. $\sqrt{y} = 12$

$(\sqrt{y})^2 = 12^2$

$\boxed{y = 144}$

17. $\sqrt{2a} = 4$

$(\sqrt{2a})^2 = 4^2$

$2a = 16$

$\dfrac{2a}{2} = \dfrac{16}{2}$

$\boxed{a = 8}$

19. $\sqrt{4n} = 6$

$(\sqrt{4n})^2 = 6^2$

$4n = 36$

$\dfrac{4n}{4} = \dfrac{36}{4}$

$\boxed{n = 9}$

21. $\sqrt{x} = -6$

$\boxed{\text{no solution}}$

23. $\sqrt{r} + 4 = 0$

$\sqrt{r} + 4 - 4 = 0 - 4$

$\sqrt{r} = -4$

$\boxed{\text{no solution}}$

25. $\sqrt{a} + 2 = 9$

$\sqrt{a} + 2 - 2 = 9 - 2$

$\sqrt{a} = 7$

$(\sqrt{a})^2 = 7^2$

$\boxed{a = 49}$

27. $\sqrt{x} + 1 = 7$

$\sqrt{x} + 1 - 1 = 7 - 1$

$\sqrt{x} = 6$

$(\sqrt{x})^2 = 6^2$

$\boxed{x = 36}$

29. $5\sqrt{x} - 11 = 9$

$5\sqrt{x} - 11 + 11 = 9 + 11$

$5\sqrt{x} = 20$

$\dfrac{5\sqrt{x}}{5} = \dfrac{20}{5}$

$(\sqrt{x})^2 = 4^2$

$\boxed{x = 16}$

31. $-2 = 2\sqrt{x} - 12$

$-2 + 12 = 2\sqrt{x} - 12 + 12$

$10 = 2\sqrt{x}$

$\dfrac{10}{2} = \dfrac{2\sqrt{x}}{2}$

$5^2 = (\sqrt{x})^2$

$\boxed{25 = x}$

33. $-\sqrt{x} = -2$

$\dfrac{-\sqrt{x}}{-1} = \dfrac{-2}{-1}$

$\sqrt{x} = 2$

$(\sqrt{x})^2 = 2^2$

$\boxed{x = 4}$

35. $10 - \sqrt{s} = 7$

$-10 + \sqrt{s} = -7$

$-10 + \sqrt{s} + 10 = -7 + 10$

$\sqrt{s} = 3$

$(\sqrt{s})^2 = 3^2$

$\boxed{s = 9}$

37. $\sqrt{x + 3} = 2$

$(\sqrt{x + 3})^2 = 2^2$

$x + 3 = 4$

$x + 3 - 3 = 4 - 3$

$\boxed{x = 1}$

39. $\sqrt{5 - T} = 10$

$(\sqrt{5 - T})^2 = 10^2$

$5 - T = 100$

$5 - T - 5 = 100 - 5$

$-T = 95$

$\dfrac{-T}{-1} = \dfrac{95}{-1}$

$\boxed{T = -95}$

41. $\sqrt{6x + 19} - 7 = 0$

$\sqrt{6x + 19} - 7 + 7 = 0 + 7$

$\sqrt{6x + 19} = 7$

$(\sqrt{6x + 19})^2 = 7^2$

$6x + 19 = 49$

$6x + 19 - 19 = 49 - 19$

$6x = 30$

$\dfrac{6x}{6} = \dfrac{30}{6}$

$\boxed{x = 5}$

43.
$$\sqrt{x-5}-3=4$$
$$\sqrt{x-5}-3+3=4+3$$
$$(\sqrt{x-5})^2=7^2$$
$$x-5=49$$
$$x-5+5=49+5$$
$$\boxed{x=54}$$

45.
$$\sqrt{7+2x}+4=1$$
$$\sqrt{7+2x}+4-4=1-4$$
$$\sqrt{7+2x}=-3$$
$$\boxed{\text{no solution}}$$

47.
$$x=\sqrt{x^2-2x+16}$$
$$x^2=(\sqrt{x^2-2x+16})^2$$
$$x^2=x^2-2x+16$$
$$x^2-x^2=x^2-2x+16-x^2$$
$$0=-2x+16$$
$$0-16=-2x+16-16$$
$$-16=-2x$$
$$\frac{-16}{-2}=\frac{-2x}{-2}$$
$$\boxed{8=x}$$

49.
$$\sqrt{4m^2+6m+6}=-2m$$
$$(\sqrt{4m^2+6m+6})^2=(-2m)^2$$
$$4m^2+6m+6=4m^2$$
$$4m^2+6m+6-4m^2=4m^2-4m^2$$
$$6m+6=0$$
$$6m+6-6=0-6$$
$$6m=-6$$
$$\frac{6m}{6}=\frac{-6}{6}$$
$$\boxed{m=-1}$$

51.
$$x-3=\sqrt{x^2-15}$$
$$(x-3)^2=(\sqrt{x^2-15})^2$$
$$x^2-6x+9=x^2-15$$
$$x^2-6x+9-x^2=x^2-15-x^2$$
$$-6x+9=-15$$
$$-6x+9-9=-15-9$$
$$-6x=-24$$
$$\frac{-6x}{-6}=\frac{-24}{-6}$$
$$\boxed{x=4}$$

53.
$$x+1=\sqrt{x^2+x+4}$$
$$(x+1)^2=(\sqrt{x^2+x+4})^2$$
$$x^2+2x+1=x^2+x+4$$
$$x^2+2x+1-x^2=x^2+x+4-x^2$$
$$2x+1=x+4$$
$$2x+1-1=x+4-1$$
$$2x=x+3$$
$$2x-x=x+3-x$$
$$\boxed{x=3}$$

55.
$$y-9=\sqrt{y-3}$$
$$(y-9)^2=(\sqrt{y-3})^2$$
$$y^2-18y+81=y-3$$
$$y^2-18y+81-y+3=y-3-y+3$$
$$y^2-19y+84=0$$
$$(y-7)(y-12)=0$$
$$\boxed{y=7,12}$$

Check for extraneous solutions.

$y=7$	$y=12$
$y-9=\sqrt{y-3}$	$y-9=\sqrt{y-3}$
$7-9\stackrel{?}{=}\sqrt{7-3}$	$12-9\stackrel{?}{=}\sqrt{12-3}$
$-2\stackrel{?}{=}\sqrt{4}$	$3\stackrel{?}{=}\sqrt{9}$
$-2=2$ false	$3=3$ true

$y=7$ is an extraneous solution.

Therefore $y=12$ is the only solution.

Section 8.5

57.
$$\sqrt{15-3t} = t-5$$
$$(\sqrt{15-3t})^2 = (t-5)^2$$
$$15-3t = t^2-10t+25$$
$$t^2-10t+25 = 15-3t$$
$$t^2-10t+25-15+3t = 15-3t-15+3t$$
$$t^2-7t+10 = 0$$
$$(t-2)(t-5) = 0$$
$$\boxed{t = 2,5}$$

$t = 2$ is an extraneous solution.

Therefore $t = 5$ is the only solution.

59.
$$b = \sqrt{2b-2}+1$$
$$b-1 = \sqrt{2b-2}$$
$$(b-1)^2 = (\sqrt{2b-2})^2$$
$$b^2-2b+1 = 2b-2$$
$$b^2-2b+1-2b+2 = 2b-2-2b+2$$
$$b^2-4b+3 = 0$$
$$(b-1)(b-3) = 0$$
$$\boxed{b = 1,3}$$

61.
$$\sqrt{24+10n}-n = 4$$
$$\sqrt{24+10n} = 4+n$$
$$(\sqrt{24+10n})^2 = (n+4)^2$$
$$24+10n = n^2+8n+16$$
$$n^2+8n+16-24-10n = 24+10n-24-10n$$
$$n^2-2n-8 = 0$$
$$(n+2)(n-4) = 0$$
$$\boxed{n = -2,4}$$

63.
$$6+\sqrt{m}-m = 0$$
$$\sqrt{m} = m-6$$
$$(\sqrt{m})^2 = (m-6)^2$$
$$m = m^2-12m+36$$
$$m^2-12m+36-m = m-m$$
$$m^2-13m+36 = 0$$
$$(m-4)(m-9) = 0$$
$$\boxed{m = 4,9}$$

$m = 4$ is an extraneous solution.

Therefore $m = 9$ is the only solution.

65.
$$\sqrt{3t-9} = \sqrt{t+1}$$
$$(\sqrt{3t-9})^2 = (\sqrt{t+1})^2$$
$$3t-9 = t+1$$
$$3t-9+9 = t+1+9$$
$$3t-t = t+10-t$$
$$2t = 10$$
$$\frac{2t}{2} = \frac{10}{2}$$
$$\boxed{t = 5}$$

67.
$$\sqrt{10-3x} = \sqrt{2x+20}$$
$$(\sqrt{10-3x})^2 = (\sqrt{2x+20})^2$$
$$10-3x = 2x+20$$
$$10-3x-10 = 2x+20-10$$
$$-3x-2x = 2x+10-2x$$
$$-5x = 10$$
$$\frac{-5x}{-5} = \frac{10}{-5}$$
$$\boxed{x = -2}$$

69.

$$\sqrt{3c-8} = \sqrt{c}$$
$$(\sqrt{3c-8})^2 = (\sqrt{c})^2$$
$$3c - 8 = c$$
$$3c - 8 - 3c = c - 3c$$
$$-8 = -2c$$
$$\frac{-8}{-2} = \frac{-2c}{-2}$$
$$\boxed{4 = c}$$

71.

$$2\sqrt{x+8} = \sqrt{x+2}$$
$$(2\sqrt{x+8})^2 = (\sqrt{x+2})^2$$
$$4(x+8) = x+2$$
$$4x + 32 = x + 2$$
$$4x + 32 - x = x + 2 - x$$
$$3x + 32 - 32 = 2 - 32$$
$$\frac{3x}{3} = \frac{-30}{3}$$
$$\boxed{x = -10}$$

The right side of the equation becomes the $\sqrt{-8}$, therefore there is no solution.

73.

$$2\sqrt{3x+4} = \sqrt{5x+9}$$
$$(2\sqrt{3x+4})^2 = (\sqrt{5x+9})^2$$
$$4(3x+4) = 5x+9$$
$$12x + 16 = 5x + 9$$
$$12x + 16 - 16 = 5x + 9 - 16$$
$$12x = 5x - 7$$
$$12x - 5x = 5x - 7 - 5x$$
$$7x = -7$$
$$\frac{7x}{7} = \frac{-7}{7}$$
$$\boxed{x = -1}$$

APPLICATIONS

75. NIAGARA FALLS

t is time it takes an object to fall d ft.

$t = 3.45$ seconds , d (ht of falls) = ?

$$t = \frac{\sqrt{d}}{4}$$
$$3.25 = \frac{\sqrt{d}}{4}$$
$$4(3.25) = \sqrt{d}$$
$$(13)^2 = (\sqrt{d})^2$$
$$\boxed{169 = d}$$

The height of the falls is 169 ft.

77. PENDULUMS

t is time it takes a pendulum of length L to swing through one full cycle

$t = 8.91$ seconds , L (one cycle) = ?

$$t = 1.11\sqrt{L}$$
$$8.91 = 1.11\sqrt{L}$$
$$\frac{8.91}{1.11} = \sqrt{L}$$
$$\left(\frac{8.91}{1.11}\right)^2 = (\sqrt{L})^2$$
$$\boxed{64.43 = L}$$

The pendulum is about 64.4 ft.

79. ROAD SAFETY (WET PAYMENT)

s is the speed in mph of car
d is the distance of the skid in feet
k is the constant for wet payment
$s = 55$ mph , $d = ?$, $k = 3.24$

$$s = k\sqrt{d}$$
$$55 = 3.24\sqrt{d}$$
$$\frac{55}{3.24} = \sqrt{d}$$
$$\left(\frac{55}{3.24}\right)^2 = (\sqrt{d})^2$$
$$\boxed{288.1 = d}$$

The skid is about 288 ft.

81. HIGHWAY DESIGN

s is the speed in mph
r is the radius of the curve in feet
$s = 65$ mph , $r = ?$

$$s = 1.45\sqrt{r}$$

$$65 = 1.45\sqrt{r}$$

$$\frac{65}{1.45} = \sqrt{r}$$

$$\left(\frac{65}{1.45}\right)^2 = (\sqrt{r})^2$$

$$\boxed{2{,}009.5 = r}$$

The radius is about 2,010 ft.

WRITING

83. Answers will vary.

85. Answers will vary.

REVIEW

87.

$$\frac{1}{2} + \frac{x}{5} = \frac{3}{4}$$

$$20\left(\frac{1}{2}\right) + 20\left(\frac{x}{5}\right) = 20\left(\frac{3}{4}\right)$$

$$10 + 4x = 15$$

$$10 + 4x - 10 = 15 - 10$$

$$4x = 5$$

$$\frac{4x}{4} = \frac{5}{4}$$

$$\boxed{x = \frac{5}{4}}$$

89.

$$\frac{2}{5}x + 1 = \frac{1}{3} + x$$

$$15\left(\frac{2}{5}x\right) + 15(1) = 15\left(\frac{1}{3}\right) + 15(x)$$

$$6x + 15 = 5 + 15x$$

$$6x + 15 - 15 = 5 + 15x - 15$$

$$6x = 15x - 10$$

$$6x - 15x = 15x - 10 - 15x$$

$$-9x = -10$$

$$\frac{-9x}{-9} = \frac{-10}{-9}$$

$$\boxed{x = \frac{10}{9}}$$

CHALLENGE PROBLEMS

91.

$$\sqrt{\sqrt{x+2}} = 3$$

$$(\sqrt{\sqrt{x+2}})^2 = (3)^2$$

$$\sqrt{x+2} = 9$$

$$(\sqrt{x+2})^2 = (9)^2$$

$$x + 2 = 81$$

$$x + 2 - 2 = 81 - 2$$

$$\boxed{x = 79}$$

SECTION 8.6

VOCABULARY

1. We read $\sqrt[3]{8}$ as the **cube** root of 8 and $\sqrt[4]{16}$ as the **fourth** root of 16.

3. "To **simplify** $\sqrt[3]{54}$" means to write it as $3\sqrt[3]{2}$.

5. The exponential expression $25^{3/2}$ has a **rational** (fractional) exponent.

CONCEPTS

7. a) $\sqrt[3]{64} = 4$ because $(\boxed{4})^3 = 64$.

   b) $\sqrt[4]{16x^4} = 2x$ because $(\boxed{2x})^4 = 16x^4$.

   c) $\sqrt[5]{-32} = -2$ because $(-2)^5 = \boxed{-32}$.

9. a) $\sqrt[n]{a \cdot b} = \boxed{\sqrt[n]{a}\ \sqrt[n]{b}}$

   b) $\sqrt[n]{\dfrac{a}{b}} = \dfrac{\sqrt[n]{a}}{\sqrt[n]{b}}$

11. $\sqrt[3]{24} = \sqrt[3]{8 \cdot 3}$

13. a) $\sqrt[3]{8}\sqrt[3]{2} = 2\sqrt[3]{2}$

    b) $\sqrt[4]{16}\sqrt[4]{3} = 2\sqrt[4]{3}$

    c) $\dfrac{\sqrt[3]{125}}{\sqrt[3]{27}} = \dfrac{5}{3}$

    d) $\dfrac{\sqrt[4]{256}}{\sqrt[4]{81}} = \dfrac{4}{3}$

15. a) $36^{-1/2} = \dfrac{1}{\sqrt{36}}$

    b) $16^{-5/4} = \dfrac{1}{(\sqrt[4]{16})^5}$

NOTATION

17. 2 , 2

PRACTICE

19. 2

21. 0

23. –2

25. –4

27. $\dfrac{1}{5}$

29. 1

31. –4

33. 9

35. 10

37. $\dfrac{2}{3}$

39. 2

41. 4

43. $-\dfrac{1}{3}$

45. not a real number

47. 2

49. –1

51. 2

53. m

55. $3a^2$

57. y

59. $2b^3$

61. x

63. s^2

65. $\sqrt[3]{24} = \sqrt[3]{8 \cdot 3}$
    $= 2\sqrt[3]{3}$

67. $\sqrt[3]{-128} = \sqrt[3]{-64 \cdot 2}$

$= -4\sqrt[3]{2}$

69. $\sqrt[3]{72} = \sqrt[3]{8 \cdot 9}$

$= 2\sqrt[3]{9}$

71. $\sqrt[3]{\dfrac{250}{27}} = \dfrac{\sqrt[3]{125 \cdot 2}}{\sqrt[3]{27}}$

$= \dfrac{5\sqrt[3]{2}}{3}$

73. $\sqrt[4]{162} = \sqrt[4]{81 \cdot 2}$

$= 3\sqrt[4]{2}$

75. $\sqrt[5]{64} = \sqrt[5]{32 \cdot 2}$

$= 2\sqrt[5]{2}$

77. $81^{1/2} = \sqrt{81}$

$= 9$

79. $16^{1/4} = \sqrt[4]{16}$

$= 2$

81. $-144^{1/2} = -\sqrt{144}$

$= -12$

83. $(-125)^{1/3} = \sqrt[3]{-125}$

$= -5$

85. $1^{1/2} = \sqrt{1}$

$= 1$

87. $\left(\dfrac{9}{16}\right)^{1/2} = \sqrt{\dfrac{9}{16}}$

$= \dfrac{3}{8}$

89. $8^{1/3} = \sqrt[3]{8}$

$= 2$

91. $16^{3/2} = (\sqrt{16})^3$

$= 4^3$

$= 64$

93. $27^{2/3} = (\sqrt[3]{27})^2$

$= 3^2$

$= 9$

95. $8^{-1/3} = \dfrac{1}{8^{1/3}}$

$= \dfrac{1}{\sqrt[3]{8}}$

$= \dfrac{1}{2}$

97. $25^{-3/2} = \dfrac{1}{25^{3/2}}$

$= \dfrac{1}{(\sqrt{25})^3}$

$= \dfrac{1}{5^3}$

$= \dfrac{1}{125}$

99. $32^{-3/5} = \dfrac{1}{32^{3/5}}$

$= \dfrac{1}{(\sqrt[5]{32})^3}$

$= \dfrac{1}{2^3}$

$= \dfrac{1}{8}$

APPLICATIONS

101. WINDMILL

S is the speed (mph) of the wind
P is the power (watts)
$P = 20$ watts

$$S = \sqrt[3]{\frac{P}{0.02}}$$

$$= \sqrt[3]{\frac{20}{0.02}}$$

$$= \sqrt[3]{1,000}$$

$$= 10$$

The speed of the wind is 10 mph.

103. HOLIDAY DECORATING

s is the length of each string of lights
h is the height of the tree
r is radius of the tree
$s = ?$, $h = 24$ ft. , $r = 10$ ft.

$$s = (r^2 + h^2)^{\frac{1}{2}}$$

$$= [(10)^2 + (24)^2]^{\frac{1}{2}}$$

$$= [100 + 576]^{\frac{1}{2}}$$

$$= [676]^{\frac{1}{2}}$$

$$= \sqrt{676}$$

$$= 26$$

The length of the lights is 26 ft.

105. SPEAKERS

A is the area of 1 face in inches
V is the volume in cubic inches
$A = ?$, $V = 216$

$$A = V^{\frac{2}{3}}$$

$$= 216^{\frac{2}{3}}$$

$$= (\sqrt[3]{216})^2$$

$$= (6)^2$$

$$= 36$$

The area of one face is 36 in.²

WRITING

107. Answers will vary.

109. Answers will vary.

REVIEW

111.

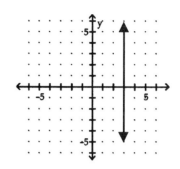

113.

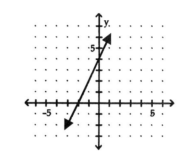

CHALLENGE PROBLEMS

115.

x	-27	-8	-1	0	1	8	27
y	-3	-2	-1	0	1	2	3

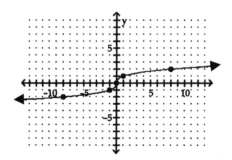

CHAPTER 8 KEY CONCEPTS

The Rules for Radicals

1. The square root of the **product** of two nonnegative numbers is equal to the product of their **square roots**. Also, the product of the **square roots** of two nonnegative numbers is equal to the square root of the **product** of the two numbers.

2. Use the product rule.

   a) $\sqrt{24x^2y^3} = \sqrt{4x^2y^2}\sqrt{6y}$
   $$= 2xy\sqrt{6y}$$

   b) $\sqrt{8} + \sqrt{18} = \sqrt{4}\sqrt{2} + \sqrt{9}\sqrt{2}$
   $$= 2\sqrt{2} + 3\sqrt{2}$$
   $$= 5\sqrt{2}$$

   c) $\sqrt{72a} - \sqrt{98a} = \sqrt{36}\sqrt{2a} - \sqrt{49}\sqrt{2a}$
   $$= 6\sqrt{2a} - 7\sqrt{2a}$$
   $$= -\sqrt{2a}$$

   d) $(2\sqrt{3n})(3\sqrt{5}) = 6\sqrt{15n}$

   e) $\dfrac{\sqrt{2}}{\sqrt{5}} = \dfrac{\sqrt{2}}{\sqrt{5}} \cdot \dfrac{\sqrt{5}}{\sqrt{5}}$
   $$= \dfrac{\sqrt{10}}{5}$$

3. The square root of the **quotient** of two positive numbers is equal to the quotient of their **square roots**. Also, the quotient of the **square roots** of two positive numbers is equal to the square root of the **quotient** of the two numbers.

4. Use the quotient rule.

   a) $\sqrt{\dfrac{36}{25}} = \dfrac{\sqrt{36}}{\sqrt{25}}$
   $$= \dfrac{6}{5}$$

   b) $\sqrt{\dfrac{37}{121}} = \dfrac{\sqrt{37}}{\sqrt{121}}$
   $$= \dfrac{\sqrt{37}}{11}$$

   c) $\sqrt{\dfrac{5}{3}} = \dfrac{\sqrt{5}}{\sqrt{3}} \cdot \dfrac{\sqrt{3}}{\sqrt{3}}$
   $$= \dfrac{\sqrt{15}}{3}$$

   d) $\dfrac{\sqrt{18x^5}}{\sqrt{2x^3}} = \sqrt{\dfrac{18x^5}{2x^3}}$
   $$= \sqrt{9x^2}$$
   $$= 3x$$

CHAPTER 8 REVIEW

SECTION 8.1
An Introduction to Square Roots

1. 6 is the square root of 36 because
    $(\boxed{6})^2 = \boxed{36}$.

2. 5

3. 7

4. -12

5. $\dfrac{4}{9}$

6. 0.8

7. 1

8. 4.583

9. -3.873

10. 5.292

11.
    $\sqrt{-2}$, not a real number

    $\sqrt{68}$, irrational

    $\sqrt{81}$, rational

    $\sqrt{3}$, irrational

12. x

13. $2b^2$

14. $-y^6$

15. ROAD SIGNS

    v is the velocity in mph

    r is the radius of the curve in ft

    $r = 360$ ft , $v = ?$

    $$v = \sqrt{2.5r}$$
    $$= \sqrt{2.5(360)}$$
    $$= \sqrt{900}$$
    $$= 30$$

    The sign should be labeled 30 mph.

16. short leg is 6

    long leg is 8

    hypotenuse is ?

    $a = 6$, $b = 8$, $c = ?$

    $$a^2 + b^2 = c^2$$
    $$(6)^2 + (8)^2 = c^2$$
    $$36 + 64 = c^2$$
    $$100 = c^2$$
    $$\sqrt{100} = \sqrt{c^2}$$
    $$\boxed{10 = c}$$

    The hypotenuse is 10 units.

17. short leg is 8

    long leg is ?

    hypotenuse is 17

    $a = 8$, $b = ?$, $c = 17$

    $$a^2 + b^2 = c^2$$
    $$(8)^2 + b^2 = (17)^2$$
    $$64 + b^2 = 289$$
    $$64 + b^2 - 64 = 289 - 64$$
    $$b^2 = 225$$
    $$\sqrt{b^2} = \sqrt{225}$$
    $$\boxed{b = 15}$$

    The long leg is 15 units.

18. THEATER SEATING

    short leg is ?

    long leg is 12 ft.

    hypotenuse is 13 ft.

    $a = ?$, $b = 12$, $c = 13$

    $$a^2 + b^2 = c^2$$
    $$a^2 + (12)^2 = (13)^2$$
    $$a^2 + 144 = 169$$
    $$a^2 + 144 - 144 = 169 - 144$$
    $$a^2 = 25$$
    $$\sqrt{a^2} = \sqrt{25}$$
    $$\boxed{a = 5}$$

    It is 5 ft higher.

19.

$(-7, 12)$ and $(-4, 8)$

(x_1, y_1) and (x_2, y_2)

$d = \sqrt{(x_2 - x_1)^2 + (y_2 - y_1)^2}$

$d = \sqrt{[-4 - (-7)]^2 + (8 - 12)^2}$

$d = \sqrt{(3)^2 + (-4)^2}$

$d = \sqrt{9 + 16}$

$d = \sqrt{25}$

$\boxed{d = 5}$

20.

$(-15, -3)$ and $(-10, -16)$

(x_1, y_1) and (x_2, y_2)

$d = \sqrt{(x_2 - x_1)^2 + (y_2 - y_1)^2}$

$d = \sqrt{[-10 - (-15)]^2 + [-16 - (-3)]^2}$

$d = \sqrt{(5)^2 + (-13)^2}$

$d = \sqrt{25 + 169}$

$d = \sqrt{194}$

$d \approx 13.928$

$\boxed{d \approx 13.93}$

SECTION 8.2
Simplifying Square Roots

21. $\sqrt{32} = \sqrt{16}\sqrt{2}$

$= 4\sqrt{2}$

22. $\sqrt{x^5} = \sqrt{x^4}\sqrt{x}$

$= x^2\sqrt{x}$

23. $\sqrt{80x^2} = \sqrt{16x^2}\sqrt{5}$

$= 4x\sqrt{5}$

24. $-2\sqrt{63} = -2\sqrt{9}\sqrt{7}$

$= -2(3)\sqrt{7}$

$= -6\sqrt{7}$

25. $\sqrt{250t^3} = \sqrt{25t^2}\sqrt{10t}$

$= 5t\sqrt{10t}$

26. $\sqrt{200x^2 y} = \sqrt{100x^2}\sqrt{2y}$

$= 10x\sqrt{2y}$

27. $\sqrt{\dfrac{16}{25}} = \dfrac{\sqrt{16}}{\sqrt{25}}$

$= \dfrac{4}{5}$

28. $\sqrt{\dfrac{60}{49}} = \dfrac{\sqrt{4}\sqrt{15}}{\sqrt{49}}$

$= \dfrac{2\sqrt{15}}{7}$

29. $\sqrt{\dfrac{242x^4}{169x^2}} = \dfrac{\sqrt{121x^2}\sqrt{2}}{\sqrt{169}}$

$= \dfrac{11x\sqrt{2}}{13}$

30. FITNESS EQUIPMENT

short leg is 2 ft.

long leg is 6 ft.

hypotenuse is ?

$a = 2$, $b = 6$, $c = ?$

$a^2 + b^2 = c^2$

$(2)^2 + (6)^2 = c^2$

$4 + 36 = c^2$

$40 = c^2$

$\sqrt{40} = \sqrt{c^2}$

$\boxed{2\sqrt{10} = c}$

The hypotenuse is $2\sqrt{10}$ ft.

or

The hypotenuse is 6.3 ft.

SECTION 8.3
Adding and Subtracting Radical Expressions

31. $\sqrt{10} + \sqrt{10} = 2\sqrt{10}$

32. $6\sqrt{x} - \sqrt{x} = 5\sqrt{x}$

33. $\sqrt{2} + \sqrt{8} - \sqrt{18}$
$$= \sqrt{2} + \sqrt{4}\sqrt{2} - \sqrt{9}\sqrt{2}$$
$$= \sqrt{2} + 2\sqrt{2} - 3\sqrt{2}$$
$$= 0$$

34. $\sqrt{3} + 4 + \sqrt{27} - 7$
$$= \sqrt{3} + 4 + \sqrt{9}\sqrt{3} - 7$$
$$= \sqrt{3} + 3\sqrt{3} - 3$$
$$= -3 + 4\sqrt{3}$$

35. $5\sqrt{28} - 3\sqrt{63} = 5\sqrt{4}\sqrt{7} - 3\sqrt{9}\sqrt{7}$
$$= 10\sqrt{7} - 9\sqrt{7}$$
$$= \sqrt{7}$$

36. $3y\sqrt{5xy^3} - y^2\sqrt{20xy}$
$$= 3y\sqrt{y^2}\sqrt{5xy} - y^2\sqrt{4}\sqrt{5xy}$$
$$= 3y^2\sqrt{5xy} - 2y^2\sqrt{5xy}$$
$$= y^2\sqrt{5xy}$$

37. The radicands are different.

38. GARDENING
$$7\sqrt{45} - 4\sqrt{20} = 7\sqrt{9}\sqrt{5} - 4\sqrt{4}\sqrt{5}$$
$$= 21\sqrt{5} - 8\sqrt{5}$$
$$= 13\sqrt{5}$$

The difference is $13\sqrt{5}$ in.

SECTION 8.4
Multiplying and Dividing Radical Expressions

39. $\sqrt{2}\sqrt{3} = \sqrt{6}$

40. $(-5\sqrt{5})^2 = (-5\sqrt{5})(-5\sqrt{5})$
$$= 25(5)$$
$$= 125$$

41. $(3\sqrt{3x^3})(4\sqrt{6x^2}) = 12\sqrt{18x^5}$
$$= 12\sqrt{9x^4}\sqrt{2x}$$
$$= 36x^2\sqrt{2x}$$

42. $(\sqrt{15} + 3x)^2 = (\sqrt{15} + 3x)(\sqrt{15} + 3x)$
$$= 15 + 3x\sqrt{15} + 3x\sqrt{15} + 9x^2$$
$$= 15 + 6x\sqrt{15} + 9x^2$$

43. $\sqrt{2}(\sqrt{8} - \sqrt{18}) = \sqrt{16} - \sqrt{36}$
$$= 4 - 6$$
$$= -2$$

44. $(\sqrt{3} + \sqrt{5})(\sqrt{3} - \sqrt{5}) = 3 - 5$
$$= -2$$

45. $(\sqrt{x})^2 = x$

46. $(\sqrt{e-1})^2 = e - 1$

47. VACUUM CLEANERS

width is $2\sqrt{6}$ in

length is $5\sqrt{3}$ in

area is length x width

$$A = lw$$
$$= (2\sqrt{6})(5\sqrt{3})$$
$$= 10\sqrt{18}$$
$$= 10\sqrt{9}\sqrt{2}$$
$$= 30\sqrt{2}$$

Area is $30\sqrt{2}$ in².

or

Area is approximately 42.4 in².

48.
$$\frac{1}{\sqrt{7}} = \frac{1}{\sqrt{7}} \cdot \frac{\sqrt{7}}{\sqrt{7}}$$
$$= \frac{\sqrt{7}}{7}$$

49.
$$\sqrt{\frac{3}{a}} = \frac{\sqrt{3}}{\sqrt{a}} \cdot \frac{\sqrt{a}}{\sqrt{a}}$$
$$= \frac{\sqrt{3a}}{a}$$

50.
$$\frac{\sqrt{27x^3}}{\sqrt{75x}} = \frac{\sqrt{27x^3}}{\sqrt{75x}} \cdot \frac{\sqrt{75x}}{\sqrt{75x}}$$
$$= \frac{\sqrt{2,025x^4}}{75x}$$
$$= \frac{45x^2}{75x}$$
$$= \frac{3x}{5}$$

51.
$$\frac{7}{\sqrt{2}+1} = \frac{7}{\sqrt{2}+1} \cdot \frac{\sqrt{2}-1}{\sqrt{2}-1}$$
$$= \frac{7(\sqrt{2}-1)}{2-1}$$
$$= 7\sqrt{2} - 7$$

SECTION 8.5
Solving Radical Equations

52.
$$\sqrt{x} = 9$$
$$(\sqrt{x})^2 = 9^2$$
$$\boxed{x = 81}$$

Remember that all even index radical equations should be checked for extraneous solutions.

53.
$$\sqrt{2x+10} = 10$$
$$(\sqrt{2x+10})^2 = 10^2$$
$$2x + 10 = 100$$
$$2x + 10 - 10 = 100 - 10$$
$$2x = 90$$
$$\frac{2x}{2} = \frac{90}{2}$$
$$\boxed{x = 45}$$

54.
$$\sqrt{3x+4} + 5 = 3$$
$$\sqrt{3x+4} + 5 - 5 = 3 - 5$$
$$\sqrt{3x+4} = -2$$

There is no solution because the principal square root is always positive.

55.
$$\sqrt{b+12} = 3\sqrt{b+4}$$
$$(\sqrt{b+12})^2 = (3\sqrt{b+4})^2$$
$$b + 12 = 9(b+4)$$
$$b + 12 = 9b + 36$$
$$b + 12 - 36 = 9b + 36 - 36$$
$$b - 24 = 9b$$
$$b - 24 - b = 9b - b$$
$$-24 = 8b$$
$$\frac{-24}{8} = \frac{8b}{8}$$
$$\boxed{-3 = b}$$

56.

$$\sqrt{p^2-3}=p+3$$
$$(\sqrt{p^2-3})^2=(p+3)^2$$
$$p^2-3=p^2+6p+9$$
$$p^2-3-p^2=p^2+6p+9-p^2$$
$$-3=6p+9$$
$$-3-9=6p+9-9$$
$$-12=6p$$
$$\frac{-12}{6}=\frac{6p}{6}$$
$$\boxed{-2=p}$$

57.

$$\sqrt{24+10y}-y=4$$
$$\sqrt{24+10y}-y+y=4+y$$
$$(\sqrt{24+10y})^2=(y+4)^2$$
$$24+10y=y^2+8y+16$$
$$y^2+8y+16-24=10y+24-24$$
$$y^2+8y-8-10y=10y-10y$$
$$y^2-2y-8=0$$
$$(y-4)(y+2)=0$$
$$\boxed{y=-2,4}$$

58. FERRIS WHEELS

t is time it takes an object to fall d ft.

$t = 2$ seconds , ht of ferris wheel = ?

$$t=\sqrt{\frac{d}{16}}$$
$$2=\sqrt{\frac{d}{16}}$$
$$(2)^2=\left(\sqrt{\frac{d}{16}}\right)^2$$
$$4=\frac{d}{16}$$
$$16(4)=d$$
$$\boxed{64=d}$$

The height of the wheel is 64 ft.

SECTION 8.6
Higher-Order Roots; Rational Exponents

59. $\sqrt[3]{125}=5,$ because $\boxed{5^3}=125;$ 5 is called the $\boxed{\text{cube}}$ root of 125.

60. -3

61. -5

62. 3

63. 2

64. 0

65. -1

66. $\dfrac{1}{4}$

67. 1

68. x

69. $3y^2$

70. $2a^3$

71. b^4

72. $\sqrt[3]{54}=\sqrt[3]{27}\sqrt[3]{2}$
$\phantom{\sqrt[3]{54}}=3\sqrt[3]{2}$

73. $\sqrt[4]{80}=\sqrt[4]{16}\sqrt[4]{5}$
$\phantom{\sqrt[4]{80}}=2\sqrt[4]{5}$

74. $\sqrt[3]{\dfrac{56}{125}}=\dfrac{\sqrt[3]{56}}{\sqrt[3]{125}}$
$\phantom{\sqrt[3]{\dfrac{56}{125}}}=\dfrac{\sqrt[3]{8}\sqrt[3]{7}}{\sqrt[3]{125}}$
$\phantom{\sqrt[3]{\dfrac{56}{125}}}=\dfrac{2\sqrt[3]{7}}{5}$

75. $49^{1/2}=\sqrt{49}$
$\phantom{49^{1/2}}=7$

76. $(-1,000)^{1/3} = \sqrt[3]{-1,000}$

$\qquad = -10$

77. $36^{3/2} = (\sqrt{36})^3$

$\qquad = 6^3$

$\qquad = 216$

78. $\left(\dfrac{8}{27}\right)^{2/3} = \left(\sqrt[3]{\dfrac{8}{27}}\right)^2$

$\qquad = \left(\dfrac{2}{3}\right)^2$

$\qquad = \dfrac{4}{9}$

79. $4^{-3/2} = \left(\dfrac{1}{4}\right)^{3/2}$

$\qquad = \left(\sqrt{\dfrac{1}{4}}\right)^3$

$\qquad = \left(\dfrac{1}{2}\right)^3$

$\qquad = \dfrac{1}{8}$

80. $-81^{5/4} = -(\sqrt[4]{81})^5$

$\qquad = -(3)^5$

$\qquad = -243$

81. DENTISTRY

Amount of original dose is $h^{-3/2}$

h is the # of hours after injection (16)

$A = h^{-3/2}$

$\quad = 16^{-3/2}$

$\quad = \dfrac{1}{16^{3/2}}$

$\quad = \dfrac{1}{(\sqrt{16})^3}$

$\quad = \dfrac{1}{4^3}$

$\quad = \dfrac{1}{64}$

$\dfrac{1}{64}$ of the dose remains.

82. There is no real number that, when squared, gives –4.

CHAPTER 8 TEST

1. 10

2. $-\dfrac{20}{3}$

3. 0.5

4. 1

5. ELECTRONICS

    I is the number of amperes (?)

    R is the resistance in ohms (20)

    P is the power in watts (980)

$$I = \sqrt{\dfrac{P}{R}}$$
$$= \sqrt{\dfrac{980}{20}}$$
$$= \sqrt{49}$$
$$= 7$$

    The number of amps is 7.

6.   short leg (ladder base) is ? feet

    long leg (up on wall) is 24 feet

    hypotenuse (the ladder) is 26 feet

$$a = ? , \quad b = 24 , \quad c = 26$$
$$a^2 + b^2 = c^2$$
$$a^2 + (24)^2 = (26)^2$$
$$a^2 + 576 = 676$$
$$a^2 + 576 - 576 = 676 - 576$$
$$a^2 = 100$$
$$\sqrt{a^2} = \sqrt{100}$$
$$\boxed{a = 10}$$

    The base is 10 feet.

7. $\sqrt{19}$, irrational

   $\sqrt{-16}$, imaginary

   $\sqrt{144}$, rational

8. $\sqrt{4x^2} = 2x$

9. $\sqrt{54x^3} = \sqrt{9x^2}\sqrt{6x}$
$$= 3x\sqrt{6x}$$

10. $\sqrt{\dfrac{50}{49}} = \dfrac{\sqrt{50}}{\sqrt{49}}$
$$= \dfrac{\sqrt{25}\sqrt{2}}{\sqrt{49}}$$
$$= \dfrac{5\sqrt{2}}{7}$$

11. $\sqrt{\dfrac{18x^2y^3}{2xy}} = \sqrt{9xy^2}$
$$= \sqrt{9y^2}\sqrt{x}$$
$$= 3y\sqrt{x}$$

12. ARCHAEOLOGY

    $(1, 1)$ and $(4, 4)$

    (x_1, y_1) and (x_2, y_2)

$$d = \sqrt{(x_2 - x_1)^2 + (y_2 - y_1)^2}$$
$$d = \sqrt{(4-1)^2 + (4-1)^2}$$
$$d = \sqrt{(3)^2 + (3)^2}$$
$$d = \sqrt{9+9}$$
$$d = \sqrt{18}$$
$$d = \sqrt{9}\sqrt{2}$$
$$\boxed{d = 3\sqrt{2}}$$

    The distance is $3\sqrt{2}$ ft.

13. $\sqrt{12} + \sqrt{27} = \sqrt{4}\sqrt{3} + \sqrt{9}\sqrt{3}$
$$= 2\sqrt{3} + 3\sqrt{3}$$
$$= 5\sqrt{3}$$

14. $(-2\sqrt{8x})(3\sqrt{12x^4}) = -6\sqrt{3} + \sqrt{9}\sqrt{3}$
$$= 2\sqrt{3} + 3\sqrt{3}$$
$$= 5\sqrt{3}$$

15. $\sqrt{3}(\sqrt{8}+\sqrt{6})=\sqrt{24}+\sqrt{18}$
$=\sqrt{4}\sqrt{6}+\sqrt{9}\sqrt{2}$
$=2\sqrt{6}+3\sqrt{2}$

16. $\sqrt{8x^3}-x\sqrt{18x}=\sqrt{4x^2}\sqrt{2x}-x\sqrt{9}\sqrt{2x}$
$=2x\sqrt{2x}-3x\sqrt{2x}$
$=-x\sqrt{2x}$

17. $(\sqrt{2}+\sqrt{3})(\sqrt{2}-\sqrt{3})=\sqrt{4}-\sqrt{9}$
$=2-3$
$=-1$

18.
$\left(2\sqrt{3t}\right)^2=2^2\cdot\sqrt{3t}\sqrt{3t}$
$=4\cdot 3t$
$=12$

19. $\left(5\sqrt{x}-1\right)\left(\sqrt{x}-4\right)=5\sqrt{x^2}-20\sqrt{x}-\sqrt{x}+4$
$=5x-21\sqrt{x}+4$

20. $(\sqrt{x+1})^2=x+1$

21. SEWING
$\sqrt{40}+\sqrt{32}+\sqrt{8}$
$=\sqrt{4}\sqrt{10}+\sqrt{16}\sqrt{2}+\sqrt{4}\sqrt{2}$
$=2\sqrt{10}+4\sqrt{2}+2\sqrt{2}$
$=2\sqrt{10}+6\sqrt{2}$

Total inches is $6\sqrt{2}+2\sqrt{10}$.

22. $\dfrac{2}{\sqrt{2}}=\dfrac{2}{\sqrt{2}}\cdot\dfrac{\sqrt{2}}{\sqrt{2}}$
$=\dfrac{2\sqrt{2}}{2}$
$=\sqrt{2}$

23.
$\dfrac{\sqrt{3}}{\sqrt{3}-1}=\dfrac{\sqrt{3}}{\sqrt{3}-1}\cdot\dfrac{\sqrt{3}+1}{\sqrt{3}+1}$
$=\dfrac{\sqrt{3}(\sqrt{3}+1)}{3-1}$
$=\dfrac{\sqrt{9}+\sqrt{3}}{2}$
$=\dfrac{3+\sqrt{3}}{2}$

24. $\sqrt{x}=15$
$(\sqrt{x})^2=(15)^2$
$\boxed{x=225}$

25.
$\sqrt{2-x}-2=6$
$\sqrt{2-x}-2+2=6+2$
$\sqrt{2-x}=8$
$(\sqrt{2-x})^2=(8)^2$
$2-x=64$
$2-x+x-64=64+x-64$
$\boxed{-62=x}$

26.
$\sqrt{3x+9}=2\sqrt{x+1}$
$(\sqrt{3x+9})^2=(2\sqrt{x+1})^2$
$3x+9=4(x+1)$
$3x+9=4x+4$
$3x+9-3x-4=4x+4-3x-4$
$\boxed{5=x}$

27.
$x-1=\sqrt{x-1}$
$(x-1)^2=(\sqrt{x-1})^2$
$x^2-2x+1=x-1$
$x^2-2x+1-x+1=x-1-x+1$
$x^2-3x+2=0$
$(x-1)(x-2)=0$
$\boxed{x=1,2}$

28. Is $x = 0$ a solution?

$$\sqrt{3x+1} = x - 1$$
$$\sqrt{3(0)+1} = 0 - 1$$
$$\sqrt{1} = -1$$
$$\boxed{1 = -1} \text{, false}$$

When 0 is substituted for x, the result is not true.

29. There is no real number that, when squared, gives -9.

30. CARPENTRY

short leg (wall) is 3 feet

long leg (floor) is 4 feet

hypotenuse (tape) is ? feet

$$a = 3, \quad b = 4, \quad c = ?$$
$$a^2 + b^2 = c^2$$
$$(3)^2 + (4)^2 = c^2$$
$$9 + 16 = c^2$$
$$25 = c^2$$
$$\sqrt{25} = \sqrt{c^2}$$
$$\boxed{5 = c}$$

The tape is 5 feet.

31. 5

32. -2

33. $\dfrac{1}{4}$

34. x

35.
$$\sqrt[3]{88} = \sqrt[3]{8}\sqrt[3]{11}$$
$$= 2\sqrt[3]{11}$$

36.
$$16^{1/2} = \sqrt{16}$$
$$= 4$$

37.
$$8^{2/3} = (\sqrt[3]{8})^2$$
$$= (2)^2$$
$$= 4$$

38.
$$9^{-1/2} = \dfrac{1}{9^{1/2}}$$
$$= \dfrac{1}{\sqrt{9}}$$
$$= \dfrac{1}{3}$$

1. a) true
    b) false
    c) true

2. $\dfrac{-3(3+2)^2-(-5)}{17-3|-4|} = \dfrac{-3(5)^2+5}{17-3(4)}$

    $= \dfrac{-3(25)+5}{17-12}$

    $= \dfrac{-75+5}{5}$

    $= \dfrac{-70}{5}$

    $= -14$

3. **BACKPACK**

    A is the amount to be carried (? lb)

    r is the percent (20%)

    b is the weight of girl (85 lb)

    $A = rb$

    $A = (0.20)(85)$

    $\boxed{A = 17}$

    Total weight should be no more than 17 lb.

4. $3p-6(p+z)+p = 3p-6p-6z+p$

    $\qquad\qquad\qquad\quad = -2p-6z$

5. $2-(4x+7) = 3+2(x+2)$

    $2-4x-7 = 3+2x+4$

    $-4x-5 = 2x+7$

    $-4x-5+5 = 2x+7+5$

    $-4x = 2x+12$

    $-4x-2x = 2x+12-2x$

    $-6x = 12$

    $\dfrac{-6x}{-6} = \dfrac{12}{-6}$

    $\boxed{x = -2}$

6. $3-3x \geq 6+x$

    $3-3x-3 \geq 6+x-3$

    $-3x \geq 3+x$

    $-3x-x \geq 3+x-x$

    $-4x \geq 3$

    $\dfrac{-4x}{-4} \leq \dfrac{3}{-4}$

    $\boxed{x \leq -\dfrac{3}{4}}$

    $\left(-\infty, -\dfrac{3}{4}\right)$

7. **SURFACE AREA**

    A is the total surface area (202 in^2)

    l is the length (9 in)

    w is the width (5 in)

    h is the height (? in)

    $A = 2lw+2wh+2lh$

    $202 = 2(9)(5)+2(5)h+2(9)h$

    $202 = 90+10h+18h$

    $202 = 90+28h$

    $202-90 = 90+28h-90$

    $112 = 28h$

    $\dfrac{112}{28} = \dfrac{28h}{28}$

    $\boxed{4 = h}$

    Height is 4 in.

8. SEARCH AND RESCUE

2 mph is the rate of team on foot.

4 mph is the rate of team on horse.

Total time (hr) for the search is ?

Total distance is 21 miles.

$$rt = d$$

$$\text{dis}_{\text{foot}} + \text{dis}_{\text{horse}} = \text{total dis}$$

Let t = time in hrs to cover 21 mi

$$rt_{\text{foot}} + rt_{\text{horse}} = d$$

$$2t + 4t = 21$$

$$6t = 21$$

$$\frac{6t}{6} = \frac{21}{6}$$

$$\boxed{t = 3.5}$$

Total time is 3.5 hours.

9. BLENDING COFFEE

40 lbs of $7/lb gourmet coffee

? lbs of $4/lb regular coffee

$5/lb the blend

amount • cost/lb = total value

$$\text{value}_{\text{regular}} + \text{value}_{\text{gourmet}} = \text{value}_{\text{blend}}$$

Let x = # of lbs. of regular coffee

$$4x + 7(40) = 5(x + 40)$$

$$4x + 280 = 5x + 200$$

$$4x + 280 - 200 = 5x + 200 - 200$$

$$4x + 80 = 5x$$

$$4x + 80 - 4x = 5x - 4x$$

$$\boxed{80 = x}$$

80 lbs of regular coffee needed.

10.

$(-2, 5)$ and $(4, 8)$

(x_1, y_1) and (x_2, y_2)

$$m = \frac{y_2 - y_1}{x_2 - x_1}$$

$$y - y_1 = m(x - x_1)$$

$$m = \frac{y_2 - y_1}{x_2 - x_1}$$

$$m = \frac{8 - 5}{4 - (-2)}$$

$$m = \frac{3}{6}$$

$$\boxed{m = \frac{1}{2}}$$

$m = \dfrac{1}{2}$, $(4, 8)$

$$y - y_1 = m(x - x_1)$$

$$y - 8 = \frac{1}{2}(x - 4)$$

$$2(y - 8) = x - 4$$

$$2y - 16 = x - 4$$

$$2y - 16 - 2y = x - 4 - 2y$$

$$-16 = x - 2y - 4$$

$$-16 + 4 = x - 2y - 4 + 4$$

$$-12 = x - 2y$$

$$\boxed{x - 2y = -12}$$

11.

use $y = mx + b$

a) $y = 3x - 7$

$$\boxed{m = 3}$$

b)
$$2x + 3y = -10$$

$$2x + 3y - 2x = -10 - 2x$$

$$3y = -2x - 10$$

$$\frac{3y}{3} = \frac{-2x}{3} - \frac{10}{3}$$

$$y = -\frac{2}{3}x - \frac{10}{3}$$

$$\boxed{m = -\frac{2}{3}}$$

12.

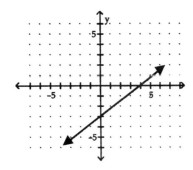

13.

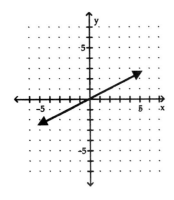

14.

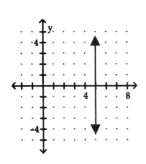

15.

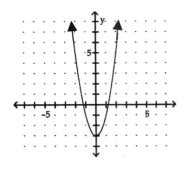

16. SHOPPING

(1994, \$819B) and (2002, \$1.235B)

(x_1, y_1) and (x_2, y_2)

$$m = \frac{y_2 - y_1}{x_2 - x_1}$$

$$m = \frac{1,235 - 819}{2002 - 1994}$$

$$m = \frac{416}{8}$$

$$\boxed{m = 52}$$

Rate of increase is \$52 billion.

17. ASTRONOMY

1 parsec is 3×10^{16} meters

1.6×10^2 parsecs

$(3 \times 10^{16})(1.6 \times 10^2) = 4.8 \times 10^{18}$

$$\boxed{\text{Distance is } 4.8 \times 10^{18} \text{ meters}}$$

18.
$$(x^5)^2(x^7)^3 = (x^{10})(x^{21})$$
$$= x^{10+21}$$
$$= x^{31}$$

19.
$$\left(\frac{a^3 b}{c^4}\right)^5 = \frac{a^{3(5)} b^5}{c^{4(5)}}$$
$$= \frac{a^{15} b^5}{c^{20}}$$

20.
$$4^{-3} \cdot 4^{-2} \cdot 4^5 = 4^{-3-2+5}$$
$$= 4^0$$
$$= 1$$

21.
$$(a^{-2} b^3)^{-4} = (a^{-2(-4)} b^{3(-4)})$$
$$= a^8 b^{-12}$$
$$= \frac{a^8}{b^{12}}$$

22.
$$(-r^4 s t^2)(2r^2 s t)(rst)$$
$$= (-2)(r^4 r^2 r)(sss)(t^2 tt)$$
$$= (-2)(r^7)(s^3)(t^4)$$
$$= -2r^7 s^3 t^4$$

23. $(-3t + 2s)(2t - 3s)$

$\qquad = -6t^2 + 9st + 4st - 6s^2$

$\qquad = -6t^2 + 13st - 6s^2$

24. $(3a^2 - 2a + 4) - (a^2 - 3a + 7)$

$\qquad = (3a^2 - 2a + 4) + (-a^2 + 3a - 7)$

$\qquad = 2a^2 + a - 3$

25.

$$
\begin{array}{r}
\boxed{2x-1} \\
x+2 \overline{) 2x^2 + 3x - 2} \\
\underline{-2x^2 - 4x} \\
-x - 2 \\
\underline{x + 2} \\
0
\end{array}
$$

26. $3x^2y - 6xy^2 = 3xy(x - 2y)$

27. $2x^2 + 2xy - 3x - 3y$

$\qquad = (2x^2 + 2xy) + (-3x - 3y)$

$\qquad = 2x(x + y) - 3(x + y)$

$\qquad = (x + y)(2x - 3)$

28. $a^3 + 8b^3 = (a)^3 + (2b)^3$

$\qquad = (a + 2b)(a^2 - 2ab + 4b^2)$

29. $6a^2 - 7a - 20 = (3a + 4)(2a - 5)$

30. $\qquad x^2 + 3x + 2 = 0$

$\qquad (x + 1)(x + 2) = 0$

$\qquad \boxed{x = -1, -2}$

31. $\qquad 5x^2 = 10x$

$\qquad 5x^2 - 10x = 10x - 10x$

$\qquad 5x^2 - 10x = 0$

$\qquad 5x(x - 2) = 0$

$\qquad \boxed{x = 0,\ 2}$

32.

$\qquad 6x^2 - x = 2$

$\qquad 6x^2 - x - 2 = 2 - 2$

$\qquad 6x^2 - x - 2 = 0$

$\qquad (2x + 1)(3x - 2) = 0$

$\qquad \boxed{x = -\dfrac{1}{2},\ \dfrac{2}{3}}$

33. $\qquad a^2 - 25 = 0$

$\qquad (a + 5)(a - 5) = 0$

$\qquad \boxed{a = -5,\ 5}$

34. CHILDREN'S STICKERS

width is 1 cm shorter than the length

lenght is ? cm

area is 20 cm^2

$lw = A$

Let l = length of rectangle in cm.

$l - 1$ = width of rectangle in cm.

$\qquad lw = A$

$\qquad l(l - 1) = 20$

$\qquad l^2 - l = 20$

$\qquad l^2 - l - 20 = 20 - 20$

$\qquad l^2 - l - 20 = 0$

$\qquad (l - 5)(l + 4) = 0$

$\qquad \boxed{l = 5, -4}$

Length of the sticker is 5 cm.

-4 is not a valid solution because

the length can not be a negative #.

35. $\dfrac{x^2 + 2x + 1}{x^2 - 1} = \dfrac{\cancel{(x+1)}(x+1)}{\cancel{(x+1)}(x-1)}$

$\qquad\qquad = \dfrac{x + 1}{x - 1}$

Chapter 8 Cumulative Review

36. $\dfrac{p^2 - p - 6}{3p - 9} \div \dfrac{p^2 + 6p + 9}{p^2 - 9}$

$\quad = \dfrac{p^2 - p - 6}{3p - 9} \cdot \dfrac{p^2 - 9}{p^2 + 6p + 9}$

$\quad = \dfrac{(p - 3)(p + 2)}{3(p - 3)} \cdot \dfrac{(p + 3)(p - 3)}{(p + 3)(p + 3)}$

$\quad = \dfrac{(p + 2)(p - 3)}{3(p + 3)}$

37. $\dfrac{x + 2}{x + 5} - \dfrac{x - 3}{x + 7}$

$\quad = \dfrac{(x + 2)(x + 7)}{(x + 5)(x + 7)} - \dfrac{(x - 3)(x + 5)}{(x + 7)(x + 5)}$

$\quad = \dfrac{(x + 2)(x + 7) - (x - 3)(x + 5)}{(x + 5)(x - 7)}$

$\quad = \dfrac{(x^2 + 9x + 14) - (x^2 + 2x - 15)}{(x + 5)(x - 7)}$

$\quad = \dfrac{(x^2 + 9x + 14) + (-x^2 - 2x + 15)}{(x + 5)(x - 7)}$

$\quad = \dfrac{7x + 29}{(x + 5)(x - 7)}$

38. $\dfrac{3a}{2b} - \dfrac{2b}{3a} = \dfrac{3a(3a)}{(2b)(3a)} - \dfrac{2b(2b)}{(3a)(2b)}$

$\quad = \dfrac{9a^2}{6ab} - \dfrac{4b^2}{6ab}$

$\quad = \dfrac{9a^2 - 4b^2}{6ab}$

39. $\dfrac{\dfrac{1}{x} + \dfrac{1}{y}}{\dfrac{1}{x} - \dfrac{1}{y}} = \dfrac{\dfrac{1(y)}{x(y)} + \dfrac{1(x)}{y(x)}}{\dfrac{1(y)}{x(y)} - \dfrac{1(x)}{y(x)}}$

$\quad = \dfrac{\dfrac{y + x}{xy}}{\dfrac{y - x}{xy}}$

$\quad = \dfrac{y + x}{xy} \cdot \dfrac{xy}{y - x}$

$\quad = \dfrac{y + x}{y - x}$

40.

$\dfrac{a + 2}{a + 3} - 1 = \dfrac{-1}{a^2 + 2a - 3}$

$\dfrac{a + 2}{a + 3} - 1 = \dfrac{-1}{(a + 3)(a - 1)}$

$\boxed{\text{restrictions are } a \neq -3, 1}$

$(a + 3)(a - 1)\left(\dfrac{a + 2}{a + 3} - 1\right) = (a + 3)(a - 1)\left(\dfrac{-1}{(a + 3)(a - 1)}\right)$

$(a + 2)(a - 1) - (a + 3)(a - 1) = -1$

$(a^2 + a - 2) - (a^2 + 2a - 3) = -1$

$(a^2 + a - 2) + (-a^2 - 2a + 3) = -1$

$-a + 1 = -1$

$-a + 1 + 1 + a = -1 + 1 + a$

$\boxed{2 = a}$

41. FILLING A POOL

one pipe can fill the pool in 5 hrs

another pipe can fill the pool in 4 hrs

$\frac{1}{5}$ of the work done in an hour

$\frac{1}{4}$ of the work done in an hour

$\frac{1}{x}$ of the work done by both in an hour

Let x = the # of hours to fill the pool

$$\frac{1}{5} + \frac{1}{4} = \frac{1}{x}$$

$$\frac{20x}{1}\left(\frac{1}{5}\right) + \frac{20x}{1}\left(\frac{1}{4}\right) = \frac{20x}{1}\left(\frac{1}{x}\right)$$

$$4x + 5x = 20$$

$$9x = 20$$

$$\frac{9x}{9} = \frac{20}{9}$$

$$\boxed{x = 2\frac{2}{9}}$$

It will take $2\frac{2}{9}$ hrs. to fill the pool.

42. ONLINE SALES

Let x = the # of sales

$$\frac{x}{360,000} = \frac{9}{500}$$

$$\frac{360,000}{1} \cdot \frac{x}{360,000} = \frac{360,000}{1} \cdot \frac{9}{500}$$

$$x = \frac{720(9)}{1}$$

$$\boxed{x = 6,480}$$

Sales transactions were 6,480.

43.

$$\begin{cases} x = y + 4 \\ 2x + y = 5 \end{cases}$$

$$\begin{cases} x - y = 4 \\ 2x + y = 5 \end{cases}$$

$$x - y = 4$$
$$\underline{2x + y = 5}$$
$$3x = 9$$
$$\frac{3x}{3} = \frac{9}{3}$$
$$\boxed{x = 3}$$

$$x = 3$$
$$x = y + 4$$
$$3 = y + 4$$
$$3 - 4 = y + 4 - 4$$
$$\boxed{-1 = y}$$

Solution is (3,-1)

44.

$$\begin{cases} \dfrac{3}{5}s + \dfrac{4}{5}t = 1 \\ -\dfrac{1}{4}s + \dfrac{3}{8}t = 1 \end{cases}$$

$$\begin{cases} 3s + 4t = 5 \\ -2s + 3t = 8 \end{cases}$$

$$\begin{cases} 2(3s + 4t) = 2(5) \\ 3(-2s + 3t) = 3(8) \end{cases}$$

$$6s + 8t = 10$$
$$\underline{-6s + 9t = 24}$$
$$17t = 34$$
$$\frac{17t}{17} = \frac{34}{17}$$
$$\boxed{t = 2}$$

$$t = 2$$
$$3s + 4t = 5$$
$$3s + 4(2) = 5$$
$$3s + 8 - 8 = 5 - 8$$
$$3s = -3$$
$$\frac{3s}{3} = \frac{-3}{3}$$
$$\boxed{s = -1}$$

Solution is (-1, 2)

45. FINANCIAL PLANNING

Let x = amt put into savings account

y = amt put into mini-mall

principal • rate • time = interest

total investment: $6,000

interest rate for savings account: 6%

interest rate for mini-mall: 12%

first year combined interest: $540

$$\begin{cases} x + y = 6,000 \\ 0.06x + 0.12y = 540 \end{cases}$$

$$\begin{cases} x = 6,000 - y \\ 0.06x + 0.12y = 540 \end{cases}$$

$$0.06x + 0.12y = 540$$
$$0.06(6,000 - y) + 0.12y = 540$$
$$360 - 0.06y + 0.12y = 540$$
$$360 + 0.06y - 360 = 540 - 360$$
$$0.06y = 180$$
$$\frac{0.06y}{0.06} = \frac{180}{0.06}$$
$$\boxed{y = 3,000}$$

$$x + y = 6,000$$
$$x + 3,000 = 6,000$$
$$x + 3,000 - 3,000 = 6,000 - 3,000$$
$$\boxed{x = 3,000}$$

$3,000 is put into savings account.

$3,000 is put into mini-mall.

46.

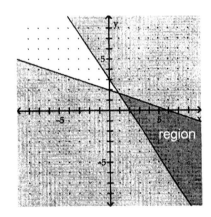

47. CARGO SPACE

width (short leg) is 48 inches

length (long leg) is 55 inches

diagonal (hypotenuse) is ?

$$a = 48, \quad b = 55, \quad c = ?$$
$$a^2 + b^2 = c^2$$
$$(48)^2 + (55)^2 = c^2$$
$$2,304 + 3,025 = c^2$$
$$5,329 = c^2$$
$$\sqrt{5,329} = \sqrt{c^2}$$
$$\boxed{73 = c}$$

A piece of plywood 73 inches wide.

48. $-12x\sqrt{16x^2y^3} = -12x\sqrt{16x^2y^2}\sqrt{y}$

$$= -12x(4xy)\sqrt{y}$$
$$= -48x^2y\sqrt{y}$$

49. $\sqrt{48} - \sqrt{8} + \sqrt{27} - \sqrt{32}$

$$= \sqrt{16}\sqrt{3} - \sqrt{4}\sqrt{2} + \sqrt{9}\sqrt{3} - \sqrt{16}\sqrt{2}$$
$$= 4\sqrt{3} - 2\sqrt{2} + 3\sqrt{3} - 4\sqrt{2}$$
$$= 7\sqrt{3} - 6\sqrt{2}$$

50. $(\sqrt{y} - 4)(\sqrt{y} - 5)$

$$= y - 5\sqrt{y} - 4\sqrt{y} + 20$$
$$= y - 9\sqrt{y} + 20$$

51. $\dfrac{4}{\sqrt{20}} = \dfrac{4}{\sqrt{4}\sqrt{5}}$

$$= \frac{4}{2\sqrt{5}} \cdot \frac{\sqrt{5}}{\sqrt{5}}$$
$$= \frac{4\sqrt{5}}{2(5)}$$
$$= \frac{2\sqrt{5}}{5}$$

52.
$$-\sqrt[3]{-27} = -(-3)$$
$$= 3$$

53.
$$\sqrt{6x+19} - 5 = 2$$
$$\sqrt{6x+19} - 5 + 5 = 2 + 5$$
$$\sqrt{6x+19} = 7$$
$$(\sqrt{6x+19})^2 = (7)^2$$
$$6x + 19 = 49$$
$$6x + 19 - 19 = 49 - 19$$
$$6x = 30$$
$$\frac{6x}{6} = \frac{30}{6}$$
$$\boxed{x = 5}$$

54.
$$16^{3/2} = (\sqrt{16})^3$$
$$= (4)^3$$
$$= 64$$

SECTION 9.1

VOCABULARY

1. $3x^2 + 2x - 1 = 0$ and $x^2 - 15 = 0$ are examples of **quadratic** equations.

3. In $x^2 - 4x + 1$, the **coefficient** of x is -4.

5. The **lead** coefficient of $5x^2 - 2x + 7 = 0$ is 5.

CONCEPTS

7. If $x^2 = 5$, then $x = \boxed{\pm\sqrt{5}}$.

9. a)
$$x^2 - 23 = 0$$
$$x^2 - 23 + 23 = 0 + 23$$
$$x^2 = 23$$

b)
$$x^2 - 1 = 4x$$
$$x^2 - 1 + 1 = 4x + 1$$
$$x^2 = 4x + 1$$
$$x^2 - 4x = 4x + 1 - 4x$$
$$x^2 - 4x = 1$$

11.
$$\frac{4x^2}{4} + \frac{5x}{4} - \frac{16}{4} = \frac{0}{4}$$
$$x^2 + \frac{5}{4}x - 4 = 0$$

13. a) $\left(\frac{6}{2}\right)^2 = \left(3^2\right)$
$$= 9$$

b) $\left(\frac{-12}{2}\right)^2 = \left(-6^2\right)$
$$= 36$$

c) $\left(\frac{3}{2}\right)^2 = \frac{3^2}{2^2}$
$$= \frac{9}{4}$$

d) $\left(\frac{-5}{2}\right)^2 = \frac{(-5)^2}{2^2}$
$$= \frac{25}{4}$$

15. a) $\dfrac{5}{1} + \dfrac{9}{4} = \dfrac{5}{1}\left(\dfrac{4}{4}\right) + \dfrac{9}{4}$
$$= \frac{20}{4} + \frac{9}{4}$$
$$= \frac{29}{4}$$

b) $\dfrac{-2}{1} + \dfrac{25}{4} = \dfrac{-2}{1}\left(\dfrac{4}{4}\right) + \dfrac{25}{4}$
$$= \frac{-8}{4} + \frac{25}{4}$$
$$= \frac{17}{4}$$

NOTATION

17. $x = \pm\sqrt{6}$

PRACTICE

19.
$$x^2 - 36 = 0$$
$$x^2 - 36 + 36 = 0 + 36$$
$$x^2 = 36$$
$$x = \sqrt{36} \quad \text{or} \quad x = -\sqrt{36}$$
$$x = 6 \qquad\qquad x = -6$$
$$\boxed{x = \pm 6}$$

21.
$$b^2 - 9 = 0$$
$$b^2 - 9 + 9 = 0 + 9$$
$$b^2 = 9$$
$$b = \sqrt{9} \quad \text{or} \quad b = -\sqrt{9}$$
$$b = 3 \qquad\qquad b = -3$$
$$\boxed{b = \pm 3}$$

23.
$$m^2 - 20 = 0$$
$$m^2 - 20 + 20 = 0 + 20$$
$$m^2 = 20$$
$$m = \sqrt{20} \quad \text{or} \quad m = -\sqrt{20}$$
$$m = \sqrt{4}\sqrt{5} \qquad m = -\sqrt{4}\sqrt{5}$$
$$m = 2\sqrt{5} \qquad\quad m = -2\sqrt{5}$$
$$\boxed{m = \pm 2\sqrt{5}}$$

25. $x^2 + 81 = 0$

   $x^2 + 81 - 81 = 0 - 81$

   $x^2 = -81$

   $\boxed{\text{no real solutions}}$

27. $3m^2 = 27$

   $\dfrac{3m^2}{3} = \dfrac{27}{3}$

   $m^2 = 9$

   $m = \sqrt{9}$ or $m = -\sqrt{9}$

   $m = 3$ $m = -3$

   $\boxed{m = \pm 3}$

29. $4x^2 = 16$

   $\dfrac{4x^2}{4} = \dfrac{16}{4}$

   $x^2 = 4$

   $x = \sqrt{4}$ or $x = -\sqrt{4}$

   $x = 2$ $x = -2$

   $x = \pm 2$

31. $x^2 = \dfrac{9}{16}$

   $x = \sqrt{\dfrac{9}{16}}$ or $x = -\sqrt{\dfrac{9}{16}}$

   $x = \dfrac{3}{4}$ $x = -\dfrac{3}{4}$

   $\boxed{x = \pm \dfrac{3}{4}}$

33. $(x + 1)^2 = 25$

   $x + 1 = \sqrt{25}$ or $x + 1 = -\sqrt{25}$

   $x + 1 = 5$ $x + 1 = -5$

   $x + 1 - 1 = 5 - 1$ $x + 1 - 1 = -5 - 1$

   $x = 4$ $x = -6$

   $\boxed{x = 4, -6}$

35. $(x + 2)^2 = 81$

   $x + 2 = \sqrt{81}$ or $x + 2 = -\sqrt{81}$

   $x + 2 = 9$ $x + 2 = -9$

   $x + 2 - 2 = 9 - 2$ $x + 2 - 2 = -9 - 2$

   $x = 7$ $x = -11$

   $\boxed{x = 7, -11}$

37. $(x - 2)^2 = 8$

   $x - 2 = \sqrt{8}$ or $x - 2 = -\sqrt{8}$

   $x - 2 = \sqrt{4}\sqrt{2}$ $x - 2 = -\sqrt{4}\sqrt{2}$

   $x - 2 = 2\sqrt{2}$ $x - 2 = -2\sqrt{2}$

   $x - 2 + 2 = 2\sqrt{2} + 2$ $x - 2 + 2 = -2\sqrt{2} + 2$

   $x = 2 + 2\sqrt{2}$ $x = 2 - 2\sqrt{2}$

   $\boxed{x = 2 \pm 2\sqrt{2}}$

39. $4(t - 7)^2 - 12 = 0$

   $4(t - 7)^2 - 12 + 12 = 0 + 12$

   $4(t - 7)^2 = 12$

   $\dfrac{4(t - 7)^2}{4} = \dfrac{12}{4}$

   $(t - 7)^2 = 3$

   $t - 7 = \sqrt{3}$ or $t - 7 = -\sqrt{3}$

   $t - 7 + 7 = \sqrt{3} + 7$ $t - 7 + 7 = -\sqrt{3} + 7$

   $t = 7 + \sqrt{3}$ $t = 7 - \sqrt{3}$

   $\boxed{t = 7 \pm \sqrt{3}}$

41. $x^2 - 14 = 0$

   $x^2 - 14 + 14 = 0 + 14$

   $x^2 = 14$

   $x = \sqrt{14}$ or $x = -\sqrt{14}$

   $x \approx 3.7416$ $x \approx -3.7416$

   $\boxed{x \approx \pm 3.74}$

43. $(y + 5)^2 - 7 = 0$

   $(y + 5)^2 - 7 + 7 = 0 + 7$

   $(y + 5)^2 = 7$

   $y + 5 = \sqrt{7}$ or $y + 5 = -\sqrt{7}$

   $y + 5 - 5 = \sqrt{7} - 5$ $y + 5 - 5 = -\sqrt{7} - 5$

   $y = -5 + \sqrt{7}$ $y = -5 - \sqrt{7}$

   $y \approx -5 + 2.65$ $y \approx -5 - 2.65$

   $\boxed{y \approx -2.35}$ $\boxed{y \approx -7.65}$

45. To complete the square, find one-half of the coefficient of x and square the result.

   $\left(\dfrac{2}{2}\right)^2 = 1^2$

   $= 1$

   Add the result to the expression and factor.

   $x^2 + 2x + 1 = (x + 1)^2$

47. To complete the square, find one-half of the coefficient of x and square the result.

$$\left(\frac{-4}{2}\right)^2 = (-2)^2$$

$$= 4$$

Add the result to the expression and factor.

$$x^2 - 4x + 4 = (x - 2)^2$$

49. To complete the square, find one-half of the coefficient of x and square the result.

$$\left(\frac{-7}{2}\right)^2 = \frac{(-7)^2}{2^2}$$

$$= \frac{49}{4}$$

Add the result to the expression and factor.

$$x^2 - 7x + \frac{49}{4} = \left(x - \frac{7}{2}\right)^2$$

51. To complete the square, find one-half of the coefficient of b and square the result.

$$\left(\frac{-\frac{2}{3}}{2}\right)^2 = \left(-\frac{2}{3} \cdot \frac{1}{2}\right)^2$$

$$= \left(-\frac{1}{3}\right)^2$$

$$= \frac{1}{9}$$

Add the result to the expression and factor.

$$b^2 - \frac{2}{3}b + \frac{1}{9} = \left(b - \frac{1}{3}\right)^2$$

53. $$k^2 - 8k + 12 = 0$$
$$k^2 - 8k + 12 - 12 = 0 - 12$$
$$k^2 - 8k = -12$$

To complete the square, find one-half of the coefficient of k and square the result.

$$\left(\frac{-8}{2}\right)^2 = (-4)^2$$

$$= 16$$

continued

53. continued

$$k^2 - 8k + 16 = -12 + 16$$
$$(k - 4)^2 = 4$$

$k - 4 = \sqrt{4}$ or $k - 4 = -\sqrt{4}$
$k - 4 = 2$ $\qquad$ $k - 4 = -2$
$k - 4 + 4 = 2 + 4$ $\quad$ $k - 4 + 4 = -2 + 4$
$\boxed{k = 6}$ $\qquad\qquad$ $\boxed{k = 2}$

55. $$x^2 + 6x + 8 = 0$$
$$x^2 + 6x + 8 - 8 = 0 - 8$$
$$x^2 + 6x = -8$$

To complete the square, find one-half of the coefficient of x and square the result.

$$\left(\frac{6}{2}\right)^2 = (3)^2$$

$$= 9$$

$$x^2 + 6x + 9 = -8 + 9$$
$$(x + 3)^2 = 1$$

$x + 3 = \sqrt{1}$ or $x + 3 = -\sqrt{1}$
$x + 3 = 1$ $\qquad$ $x + 3 = -1$
$x + 3 - 3 = 1 - 3$ $\quad$ $x + 3 - 3 = -1 - 3$
$\boxed{x = -2}$ $\qquad\qquad$ $\boxed{x = -4}$

57. $$g^2 + 5g - 6 = 0$$
$$g^2 + 5g - 6 + 6 = 0 + 6$$
$$g^2 + 5g = 6$$

$$g^2 + 5g + \left(\frac{5}{2}\right)^2 = 6 + \left(\frac{5}{2}\right)^2$$

$$g^2 + 5g + \frac{25}{4} = 6 + \frac{25}{4}$$

$$\left(g + \frac{5}{2}\right)^2 = \frac{49}{4}$$

$g + \frac{5}{2} = \sqrt{\frac{49}{4}}$ or $g + \frac{5}{2} = -\sqrt{\frac{49}{4}}$

$g + \frac{5}{2} = \frac{7}{2}$ $\qquad$ $g + \frac{5}{2} = -\frac{7}{2}$

$g + \frac{5}{2} - \frac{5}{2} = \frac{7}{2} - \frac{5}{2}$ $\quad$ $g + \frac{5}{2} - \frac{5}{2} = -\frac{7}{2} - \frac{5}{2}$

$g = \frac{2}{2}$ $\qquad\qquad$ $g = \frac{-12}{2}$

$\boxed{g = 1}$ $\qquad\qquad$ $\boxed{g = -6}$

59.
$$x^2 + 4x + 1 = 0$$
$$x^2 + 4x + 1 - 1 = 0 - 1$$
$$x^2 + 4x = -1$$
$$x^2 + 4x + \left(\frac{4}{2}\right)^2 = -1 + \left(\frac{4}{2}\right)^2$$
$$x^2 + 4x + 4 = -1 + 4$$
$$(x + 2)^2 = 3$$
$$x + 2 = \sqrt{3} \quad \text{or} \quad x + 2 = -\sqrt{3}$$
$$x + 2 - 2 = -2 + \sqrt{3} \quad x + 2 - 2 = -2 - \sqrt{3}$$
$$x = -2 + \sqrt{3} \qquad x = -2 - \sqrt{3}$$
$$\boxed{x = -2 \pm \sqrt{3}}$$

61.
$$x^2 + 8x + 6 = 0$$
$$x^2 + 8x + 6 - 6 = 0 - 6$$
$$x^2 + 8x = -6$$
$$x^2 + 8x + \left(\frac{8}{2}\right)^2 = -6 + \left(\frac{8}{2}\right)^2$$
$$x^2 + 8x + 16 = -6 + 16$$
$$(x + 4)^2 = 10$$
$$x + 4 = \sqrt{10} \quad \text{or} \quad x + 4 = -\sqrt{10}$$
$$x + 4 - 4 = \sqrt{10} - 4 \quad x + 4 - 4 = -\sqrt{10} - 4$$
$$x = -4 + \sqrt{10} \qquad x = -4 - \sqrt{10}$$
$$\boxed{x = -4 \pm \sqrt{10}}$$

63.
$$x^2 - 2x - 17 = 0$$
$$x^2 - 2x - 17 + 17 = 0 + 17$$
$$x^2 - 2x = 17$$
$$x^2 - 2x + \left(\frac{-2}{2}\right)^2 = 17 + \left(\frac{-2}{2}\right)^2$$
$$x^2 - 2x + 1 = 17 + 1$$
$$(x - 1)^2 = 18$$
$$x - 1 = \sqrt{18} \quad \text{or} \quad x - 1 = -\sqrt{18}$$
$$x - 1 = 3\sqrt{2} \qquad x - 1 = -3\sqrt{2}$$
$$x - 1 + 1 = 3\sqrt{2} + 1 \quad x - 1 + 1 = -3\sqrt{2} + 1$$
$$x = 1 + 3\sqrt{2} \qquad x = 1 - 3\sqrt{2}$$
$$\boxed{x = 1 \pm 3\sqrt{2}}$$

65.
$$n^2 - 7n - 5 = 0$$
$$n^2 - 7n - 5 + 5 = 0 + 5$$
$$n^2 - 7n = 5$$
$$n^2 - 7n + \left(\frac{-7}{2}\right)^2 = 5 + \left(\frac{-7}{2}\right)^2$$
$$n^2 - 7n + \frac{49}{4} = 5 + \frac{49}{4}$$
$$(n - \frac{7}{2})^2 = \frac{69}{4}$$
$$n - \frac{7}{2} = \sqrt{\frac{69}{4}} \quad \text{or} \quad n - \frac{7}{2} = -\sqrt{\frac{69}{4}}$$
$$n - \frac{7}{2} = \frac{\sqrt{69}}{2} \qquad n - \frac{7}{2} = -\frac{\sqrt{69}}{2}$$
$$n - \frac{7}{2} + \frac{7}{2} = \frac{\sqrt{69}}{2} + \frac{7}{2} \quad n - \frac{7}{2} + \frac{7}{2} = -\frac{\sqrt{69}}{2} + \frac{7}{2}$$
$$n = \frac{7}{2} + \frac{\sqrt{69}}{2} \qquad n = \frac{7}{2} - \frac{\sqrt{69}}{2}$$
$$\boxed{n = \frac{7}{2} \pm \frac{\sqrt{69}}{2}}$$

67.
$$x^2 = 4x + 3$$
$$x^2 - 4x = 4x + 3 - 4x$$
$$x^2 - 4x = 3$$
$$x^2 - 4x + \left(\frac{-4}{2}\right)^2 = 3 + \left(\frac{-4}{2}\right)^2$$
$$x^2 - 4x + 4 = 3 + 4$$
$$(x - 2)^2 = 7$$
$$x - 2 = \sqrt{7} \quad \text{or} \quad x - 2 = -\sqrt{7}$$
$$x - 2 + 2 = \sqrt{7} + 2 \quad x - 2 + 2 = -\sqrt{7} + 2$$
$$x = 2 + \sqrt{7} \qquad x = 2 - \sqrt{7}$$
$$\boxed{x = 2 \pm \sqrt{7}}$$

69. $a^2 - a = 3$

$$a^2 - a + \left(\frac{-1}{2}\right)^2 = 3 + \left(\frac{-1}{2}\right)^2$$

$$a^2 - a + \frac{1}{4} = 3 + \frac{1}{4}$$

$$\left(a - \frac{1}{2}\right)^2 = \frac{13}{4}$$

$$a - \frac{1}{2} = \frac{\sqrt{13}}{\sqrt{4}} \quad \text{or} \quad a - \frac{1}{2} = -\frac{\sqrt{13}}{\sqrt{4}}$$

$$a - \frac{1}{2} = \frac{\sqrt{13}}{2} \qquad a - \frac{1}{2} = -\frac{\sqrt{13}}{2}$$

$$a - \frac{1}{2} + \frac{1}{2} = \frac{\sqrt{13}}{2} + \frac{1}{2} \quad a - \frac{1}{2} + \frac{1}{2} = -\frac{\sqrt{13}}{2} + \frac{1}{2}$$

$$a = \frac{1}{2} + \frac{\sqrt{13}}{2} \qquad a = \frac{1}{2} - \frac{\sqrt{13}}{2}$$

$$\boxed{a = \frac{1}{2} \pm \frac{\sqrt{13}}{2}}$$

71.

$$3x^2 + 9x + 6 = 0$$

$$\frac{3x^2}{3} + \frac{9x}{3} + \frac{6}{3} = \frac{0}{3}$$

$$x^2 + 3x + 2 = 0$$

$$x^2 + 3x + 2 - 2 = 0 - 2$$

$$x^2 + 3x = -2$$

$$x^2 + 3x + \left(\frac{3}{2}\right)^2 = -2 + \left(\frac{3}{2}\right)^2$$

$$x^2 + 3x + \frac{9}{4} = -2 + \frac{9}{4}$$

$$\left(x + \frac{3}{2}\right)^2 = \frac{1}{4}$$

$$x + \frac{3}{2} = \frac{\sqrt{1}}{\sqrt{4}} \quad \text{or} \quad x + \frac{3}{2} = -\frac{\sqrt{1}}{\sqrt{4}}$$

$$x + \frac{3}{2} = \frac{1}{2} \qquad x + \frac{3}{2} = -\frac{1}{2}$$

$$x + \frac{3}{2} - \frac{3}{2} = \frac{1}{2} - \frac{3}{2} \quad x + \frac{3}{2} - \frac{3}{2} = -\frac{1}{2} - \frac{3}{2}$$

$$x = -\frac{3}{2} + \frac{1}{2} \qquad x = -\frac{3}{2} - \frac{1}{2}$$

$$x = -\frac{2}{2} \qquad x = -\frac{4}{2}$$

$$\boxed{x = -1} \qquad \boxed{x = -2}$$

73.

$$4x^2 + 4x + 1 = 20$$

$$4x^2 + 4x + 1 - 1 = 20 - 1$$

$$4x^2 + 4x = 19$$

$$\frac{4x^2}{4} + \frac{4x}{4} = \frac{19}{4}$$

$$x^2 + x = \frac{19}{4}$$

$$x^2 + x + \left(\frac{1}{2}\right)^2 = \frac{19}{4} + \left(\frac{1}{2}\right)^2$$

$$x^2 + x + \frac{1}{4} = \frac{19}{4} + \frac{1}{4}$$

$$\left(x + \frac{1}{2}\right)^2 = 5$$

$$x + \frac{1}{2} = \sqrt{5} \quad \text{or} \quad x + \frac{1}{2} = -\sqrt{5}$$

$$x + \frac{1}{2} - \frac{1}{2} = \sqrt{5} - \frac{1}{2} \quad x + \frac{1}{2} - \frac{1}{2} = -\sqrt{5} - \frac{1}{2}$$

$$x = -\frac{1}{2} + \sqrt{5} \qquad x = -\frac{1}{2} - \sqrt{5}$$

$$\boxed{x = -\frac{1}{2} \pm \sqrt{5}}$$

75.

$$2x^2 = 3x + 2$$

$$2x^2 - 3x = 3x + 2 - 3x$$

$$2x^2 - 3x = 2$$

$$\frac{2x^2}{2} - \frac{3x}{2} = \frac{2}{2}$$

$$x^2 - \frac{3}{2}x = 1$$

$$x^2 - \frac{3}{2}x + \left(\frac{1}{2} \bullet -\frac{3}{2}\right)^2 = 1 + \left(\frac{1}{2} \bullet -\frac{3}{2}\right)^2$$

$$x^2 - \frac{3}{2}x + \left(-\frac{3}{4}\right)^2 = 1 + \left(-\frac{3}{4}\right)^2$$

$$x^2 - \frac{3}{2}x + \frac{9}{16} = 1 + \frac{9}{16}$$

$$\left(x - \frac{3}{4}\right)^2 = \frac{25}{16}$$

$$x - \frac{3}{4} = \frac{\sqrt{25}}{\sqrt{16}} \quad \text{or} \quad x - \frac{3}{4} = -\frac{\sqrt{25}}{\sqrt{16}}$$

$$x - \frac{3}{4} = \frac{5}{4} \qquad\qquad x - \frac{3}{4} = -\frac{5}{4}$$

$$x - \frac{3}{4} + \frac{3}{4} = \frac{5}{4} + \frac{3}{4} \qquad x - \frac{3}{4} + \frac{3}{4} = -\frac{5}{4} + \frac{3}{4}$$

$$x = \frac{8}{4} \qquad\qquad x = -\frac{2}{4}$$

$$\boxed{x = 2 \qquad\qquad x = -\frac{1}{2}}$$

77.

$$2t^2 + 5t = 4$$

$$\frac{2t^2}{2} + \frac{5t}{2} = \frac{4}{2}$$

$$t^2 + \frac{5}{2}t = 2$$

$$t^2 + \frac{5}{2}t + \left(\frac{1}{2} \bullet \frac{5}{2}\right)^2 = 2 + \left(\frac{1}{2} \bullet \frac{5}{2}\right)^2$$

$$t^2 + \frac{5}{2}t + \left(\frac{5}{4}\right)^2 = 2 + \left(\frac{5}{4}\right)^2$$

$$t^2 + \frac{5}{2}t + \frac{25}{16} = 2 + \frac{25}{16}$$

$$\left(t + \frac{5}{4}\right)^2 = \frac{57}{16}$$

$$t + \frac{5}{4} = \frac{\sqrt{57}}{\sqrt{16}} \quad \text{or} \quad t + \frac{5}{4} = -\frac{\sqrt{57}}{\sqrt{16}}$$

$$t + \frac{5}{4} = \frac{\sqrt{57}}{4} \qquad\qquad t + \frac{5}{4} = -\frac{\sqrt{57}}{4}$$

$$t + \frac{5}{4} - \frac{5}{4} = \frac{\sqrt{57}}{4} - \frac{5}{4} \qquad t + \frac{5}{4} - \frac{5}{4} = -\frac{\sqrt{57}}{4} - \frac{5}{4}$$

$$t = -\frac{5}{4} + \frac{\sqrt{57}}{4} \qquad\qquad t = -\frac{5}{4} - \frac{\sqrt{57}}{4}$$

$$\boxed{t = -\frac{5}{4} \pm \frac{\sqrt{57}}{4}}$$

79.

$$x^2 - 2x - 4 = 0$$

$$x^2 - 2x - 4 + 4 = 0 + 4$$

$$x^2 - 2x = 4$$

$$x^2 - 2x + \left(-\frac{2}{2}\right)^2 = 4 + \left(-\frac{2}{2}\right)^2$$

$$x^2 - 2x + (-1)^2 = 4 + (-1)^2$$

$$(x-1)^2 = 5$$

$$x - 1 = \sqrt{5} \quad \text{or} \quad x - 1 = -\sqrt{5}$$

$$x - 1 + 1 = \sqrt{5} + 1 \quad x - 1 + 1 = -\sqrt{5} + 1$$

$$x = 1 + \sqrt{5} \qquad x = 1 - \sqrt{5}$$

$$x \approx 1 + 2.24 \qquad x \approx 1 - 2.24$$

$$\boxed{x \approx 3.24 \qquad x \approx -1.24}$$

81.

$$n^2 - 5 = 9n$$

$$n^2 - 5 + 5 = 9n + 5$$

$$n^2 = 9n + 5$$

$$n^2 - 9n = 9n + 5 - 9n$$

$$n^2 - 9n = 5$$

$$n^2 - 9n + \left(\frac{9}{2}\right)^2 = 5 + \left(\frac{9}{2}\right)^2$$

$$n^2 - 9n + \frac{81}{4} = 5 + \frac{81}{4}$$

$$\left(n - \frac{9}{2}\right)^2 = \frac{101}{4}$$

$$n - \frac{9}{2} = \sqrt{\frac{101}{4}} \quad \text{or} \quad n - \frac{9}{2} = -\sqrt{\frac{101}{4}}$$

$$n - \frac{9}{2} = \frac{\sqrt{101}}{2} \qquad n - \frac{9}{2} = -\frac{\sqrt{101}}{2}$$

$$n - \frac{9}{2} + \frac{9}{2} = \frac{\sqrt{101}}{2} + \frac{9}{2} \quad n - \frac{9}{2} + \frac{9}{2} = -\frac{\sqrt{101}}{2} + \frac{9}{2}$$

$$n = \frac{9}{2} + \frac{\sqrt{101}}{2} \qquad n = \frac{9}{2} - \frac{\sqrt{101}}{2}$$

$$n \approx \frac{9}{2} + 5.02 \qquad n \approx \frac{9}{2} - 5.02$$

$$\boxed{n \approx 9.52 \qquad n \approx -0.52}$$

83.

$$3x^2 - 6x = 1$$

$$\frac{3x^2}{3} - \frac{6x}{3} = \frac{1}{3}$$

$$x^2 - 2x = \frac{1}{3}$$

$$x^2 - 2x + \left(-\frac{2}{2}\right)^2 = \frac{1}{3} + \left(-\frac{2}{2}\right)^2$$

$$x^2 - 2x + (1)^2 = \frac{1}{3} + (1)^2$$

$$x^2 - 2x + 1 = \frac{1}{3} + 1$$

$$(x-1)^2 = \frac{4}{3}$$

$$x - 1 = \sqrt{\frac{4}{3}} \quad \text{or} \quad x - 1 = -\sqrt{\frac{4}{3}}$$

$$x - 1 = \frac{2}{\sqrt{3}} \qquad x - 1 = -\frac{2}{\sqrt{3}}$$

$$x - 1 = \frac{2\sqrt{3}}{\sqrt{3}\sqrt{3}} \qquad x - 1 = -\frac{2\sqrt{3}}{\sqrt{3}\sqrt{3}}$$

$$x - 1 = \frac{2\sqrt{3}}{3} \qquad x - 1 = -\frac{2\sqrt{3}}{3}$$

$$x - 1 + 1 = \frac{2\sqrt{3}}{3} + 1 \quad x - 1 + 1 = -\frac{2\sqrt{3}}{3} + 1$$

$$x = 1 + \frac{2\sqrt{3}}{3} \qquad x = 1 - \frac{2\sqrt{3}}{3}$$

$$x \approx 1 + 1.15 \qquad x \approx 1 - 1.15$$

$$\boxed{x \approx 2.15 \qquad x \approx -0.15}$$

85. ESCAPE VELOCITY

$$\frac{v_c^2}{2g} = R$$

$$\frac{v_e^2}{2(78.545)} = 3{,}960$$

$$\frac{v_e^2}{157.09} = 3{,}960$$

$$(157.09)\left(\frac{v_e^2}{157.09}\right) = (157.09)(3{,}960)$$

$$v_e^2 = 622{,}076.4$$

$$\sqrt{v_e^2} = \sqrt{622{,}076.4}$$

$$\boxed{v_e \approx 789}$$

87. BICYCLE SAFETY

$A = L \cdot W$
$L = x + 20$
$W = x$
$x(x + 20) = 800$
$x^2 + 20x = 800$

$$x^2 + 20x + \left(\frac{20}{2}\right)^2 = 800 + \left(\frac{20}{2}\right)^2$$

$x^2 + 20x + 100 = 800 + 100$
$(x + 10)^2 = 900$
$x + 10 = 30$ or $x + 10 = -30$
$x + 10 - 10 = 30 - 10$ $x + 10 - 10 = -30 - 10$
$x = 20$ $x = -40$
Length and width cannot be negative numbers, so x must be 20 ft.

$\boxed{\begin{array}{l} W = 20 \text{ ft} \\ L = 20 + 20 = 40 \text{ ft} \end{array}}$

WRITING

89. Answers will vary.

91. Answers will vary.

93.

$$\sqrt{5x - 6} = 2$$

$$\left(\sqrt{5x - 6}\right)^2 = (2)^2$$

$$5x - 6 = 4$$

$$5x - 6 + 6 = 4 + 6$$

$$5x = 10$$

$$\frac{5x}{5} = \frac{10}{5}$$

$$\boxed{x = 2}$$

95.

$$2\sqrt{x} = \sqrt{5x - 16}$$

$$\left(2\sqrt{x}\right)^2 = \left(\sqrt{5x - 16}\right)^2$$

$$4x = 5x - 16$$

$$4x - 5x = 5x - 16 - 5x$$

$$-x = -16$$

$$\frac{-x}{-1} = \frac{-16}{-1}$$

$$\boxed{x = 16}$$

CHALLENGE PROBLEMS

97.

$$x(x+3) - \frac{1}{2} = -2$$

$$x^2 + 3x - \frac{1}{2} = -2$$

$$x^2 + 3x - \frac{1}{2} + \frac{1}{2} = -\frac{4}{2} + \frac{1}{2}$$

$$x^2 + 3x = -\frac{3}{2}$$

$$x^2 + 3x + \left(\frac{3}{2}\right)^2 = -\frac{3}{2} + \left(\frac{3}{2}\right)^2$$

$$x^2 + 3x + \frac{9}{4} = -\frac{3}{2} + \frac{9}{4}$$

$$\left(x + \frac{3}{2}\right)^2 = \frac{3}{4}$$

$$x + \frac{3}{2} = \frac{\sqrt{3}}{\sqrt{4}} \quad \text{or} \quad x + \frac{3}{2} = -\frac{\sqrt{3}}{\sqrt{4}}$$

$$x + \frac{3}{2} = \frac{\sqrt{3}}{2} \qquad x + \frac{3}{2} = -\frac{\sqrt{3}}{2}$$

$$x + \frac{3}{2} - \frac{3}{2} = \frac{\sqrt{3}}{2} - \frac{3}{2} \qquad x + \frac{3}{2} - \frac{3}{2} = -\frac{\sqrt{3}}{2} - \frac{3}{2}$$

$$x = -\frac{3}{2} + \frac{\sqrt{3}}{2} \qquad x = -\frac{3}{2} - \frac{\sqrt{3}}{2}$$

$$\boxed{x = -\frac{3}{2} \pm \frac{\sqrt{3}}{2}}$$

SECTION 9.2

VOCABULARY

1. The general **quadratic** equation is
$ax^2 + bx + c = 0, (a \neq 0)$.

3. To **solve** a quadratic equation means to find all the values of the variable that make the equation true.

CONCEPTS

5. a)
$$x^2 + 2x = -5$$
$$x^2 + 2x + 5 = -5 + 5$$
$$x^2 + 2x + 5 = 0$$

b)
$$3x^2 = -2x + 1$$
$$3x^2 + 2x - 1 = -2x + 1 + 2x - 1$$
$$3x^2 + 2x - 1 = 0$$

7.
$$2x^2 - 4x + 8 = 0$$
$$\frac{2x^2}{2} - \frac{4x}{2} + \frac{8}{2} = \frac{0}{2}$$
$$x^2 - 2x + 4 = 0$$
$$a = 1, \ b = -2, \ c = 4$$

9. The common factor is **5**.
$$10 \pm 15\sqrt{2} = 5(2 + 3\sqrt{2})$$

11.
$$\frac{-(-4) \pm \sqrt{(-4)^2 - 4(2)(-9)}}{2(2)}$$
$$= \frac{4 \pm \sqrt{16 - (-72)}}{4}$$
$$= \frac{4 \pm \sqrt{88}}{4}$$
$$= \frac{4 \pm \sqrt{4 \cdot 22}}{4}$$
$$= \frac{4 \pm 2\sqrt{22}}{4}$$
$$= \frac{2(2 \pm \sqrt{22})}{2 \cdot 2}$$
$$= \frac{2 \pm \sqrt{22}}{2}$$

13.
$$\frac{-1 + \sqrt{5}}{2} \approx 0.62$$
$$\frac{-1 + \sqrt{5}}{2} \approx -1.62$$

15.
$$\frac{1}{2}b(b + 5) = 80$$
$$2 \cdot \frac{1}{2}b(b + 5) = 2 \cdot 80$$
$$b(b + 5) = 160$$

NOTATION

17.
$$x^2 - 5x - 6 = 0$$
$$x = \frac{-b \pm \sqrt{b^2 - 4ac}}{2a}$$
$$x = \frac{-(-5) \pm \sqrt{(-5)^2 - 4(1)(-6)}}{2(1)}$$
$$x = \frac{5 \pm \sqrt{25 + 24}}{2}$$
$$x = \frac{5 \pm \sqrt{49}}{2}$$
$$x = \frac{5 \pm 7}{2}$$
$$x = \frac{5 + 7}{2} = 6 \quad \text{or} \quad x = \frac{5 - 7}{2} = -1$$

19. In reading $x = \dfrac{-b \pm \sqrt{b^2 - 4ac}}{2a}$ we say, "the **opposite (negative)** of b, plus or **minus** the **square** root of the quanity b **squared** minus 4 **times** a times c, all **over** $2a$."

PRACTICE

21. $x^2 - 5x + 6 = 0$
        $a = 1, b = -5, c = 6$

$$x = \frac{-b \pm \sqrt{b^2 - 4ac}}{2a}$$

$$x = \frac{-(-5) \pm \sqrt{(-5)^2 - 4(1)(6)}}{2(1)}$$

$$x = \frac{5 \pm \sqrt{25 - 24}}{2}$$

$$x = \frac{5 \pm \sqrt{1}}{2}$$

$$x = \frac{5 \pm 1}{2}$$

$$x = \frac{5 + 1}{2} = \frac{6}{2} = 3 \quad \text{or} \quad x = \frac{5 - 1}{2} = \frac{4}{2} = 2$$

$$\boxed{x = 2, \ 3}$$

23. $x^2 + 7x + 12 = 0$
        $a = 1, b = 7, c = 12$

$$x = \frac{-b \pm \sqrt{b^2 - 4ac}}{2a}$$

$$x = \frac{-7 \pm \sqrt{7^2 - 4(1)(12)}}{2(1)}$$

$$x = \frac{-7 \pm \sqrt{49 - 48}}{2}$$

$$x = \frac{- \pm \sqrt{1}}{2}$$

$$x = \frac{-7 \pm 1}{2}$$

$$x = \frac{-7 + 1}{2} = \frac{-6}{2} = -3$$

$$x = \frac{-7 - 1}{2} = \frac{-8}{2} = -4$$

$$\boxed{x = -3, \ -4}$$

25. $2x^2 - x - 1 = 0$
        $a = 2, b = -1, c = -1$

$$x = \frac{-b \pm \sqrt{b^2 - 4ac}}{2a}$$

$$x = \frac{-(-1) \pm \sqrt{(-1)^2 - 4(2)(-1)}}{2(2)}$$

$$x = \frac{1 \pm \sqrt{1 - (-8)}}{4}$$

$$x = \frac{1 \pm \sqrt{9}}{4}$$

$$x = \frac{1 \pm 3}{4}$$

$$x = \frac{1 + 3}{4} = \frac{4}{4} = 1$$

$$x = \frac{1 - 3}{4} = \frac{-2}{4} = -\frac{1}{2}$$

$$\boxed{x = 1, \ -\frac{1}{2}}$$

27. $6x^2 - 13x = -6$
        $6x^2 - 13x + 6 = -6 + 6$
        $6x^2 - 13x + 6 = 0$
        $a = 6, b = -13, c = 6$

$$x = \frac{-b \pm \sqrt{b^2 - 4ac}}{2a}$$

$$x = \frac{-(-13) \pm \sqrt{(-13)^2 - 4(6)(6)}}{2(6)}$$

$$x = \frac{13 \pm \sqrt{169 - 144}}{12}$$

$$x = \frac{13 \pm \sqrt{25}}{12}$$

$$x = \frac{13 \pm 5}{12}$$

$$x = \frac{13 + 5}{12} = \frac{18}{12} = \frac{3}{2}$$

$$x = \frac{13 - 5}{12} = \frac{8}{12} = \frac{2}{3}$$

$$\boxed{x = \frac{3}{2}, \ \frac{2}{3}}$$

29.
$$4x^2 + 3x = 1$$
$$4x^2 + 3x - 1 = 1 - 1$$
$$4x^2 + 3x - 1 = 0$$
$$a = 4, b = 3, c = -1$$

$$x = \frac{-b \pm \sqrt{b^2 - 4ac}}{2a}$$

$$x = \frac{-3 \pm \sqrt{(3)^2 - 4(4)(-1)}}{2(4)}$$

$$x = \frac{-3 \pm \sqrt{9 - (-16)}}{8}$$

$$x = \frac{-3 \pm \sqrt{25}}{8}$$

$$x = \frac{-3 \pm 5}{8}$$

$$x = \frac{-3 + 5}{8} = \frac{2}{8} = \frac{1}{4}$$

$$x = \frac{-3 - 5}{8} = \frac{-8}{8} = -1$$

$$\boxed{x = \frac{1}{4}, \; -1}$$

31.
$$x^2 + 3x + 1 = 0$$
$$a = 1, b = 3, c = 1$$

$$x = \frac{-b \pm \sqrt{b^2 - 4ac}}{2a}$$

$$x = \frac{-3 \pm \sqrt{(3)^2 - 4(1)(1)}}{2(1)}$$

$$x = \frac{-3 \pm \sqrt{9 - 4}}{2}$$

$$\boxed{x = \frac{-3 \pm \sqrt{5}}{2}}$$

33.
$$3x^2 - x = 3$$
$$3x^2 - x - 3 = 3 - 3$$
$$3x^2 - x - 3 = 0$$
$$a = 3, b = -1, c = -3$$

$$x = \frac{-b \pm \sqrt{b^2 - 4ac}}{2a}$$

$$x = \frac{-(-1) \pm \sqrt{(-1)^2 - 4(3)(-3)}}{2(3)}$$

$$x = \frac{1 \pm \sqrt{1 - (-36)}}{6}$$

$$\boxed{x = \frac{1 \pm \sqrt{37}}{6}}$$

35.
$$x^2 + 5 = 2x$$
$$x^2 + 5 - 2x = 2x - 2x$$
$$x^2 - 2x + 5 = 0$$
$$a = 1, b = -2, c = 5$$

$$x = \frac{-b \pm \sqrt{b^2 - 4ac}}{2a}$$

$$x = \frac{-(-2) \pm \sqrt{(-2)^2 - 4(1)(5)}}{2(1)}$$

$$x = \frac{2 \pm \sqrt{4 - 20}}{2}$$

$$\boxed{x = \frac{2 \pm \sqrt{-16}}{2}}$$

$$\boxed{\text{No Real Solutions}}$$

Since $\sqrt{-16}$ is not a real number, there are no real solutions to this problem.

37.

$$x^2 = 1 - 2x$$
$$x^2 - 1 + 2x = 1 - 2x - 1 + 2x$$
$$x^2 + 2x - 1 = 0$$
$$a = 1, b = 2, c = -1$$

$$x = \frac{-b \pm \sqrt{b^2 - 4ac}}{2a}$$

$$x = \frac{-2 \pm \sqrt{(2)^2 - 4(1)(-1)}}{2(1)}$$

$$x = \frac{-2 \pm \sqrt{4 - (-4)}}{2}$$

$$x = \frac{-2 \pm \sqrt{8}}{2}$$

$$x = \frac{-2 \pm \sqrt{4 \bullet 2}}{2}$$

$$x = \frac{-2 \pm 2\sqrt{2}}{2}$$

$$x = \frac{2\left(-1 \pm \sqrt{2}\right)}{2}$$

$$\boxed{x = -1 \pm \sqrt{2}}$$

41.

$$x^2 + 4x = 3$$
$$x^2 + 4x - 3 = 3 - 3$$
$$x^2 + 4x - 3 = 0$$
$$a = 1, b = 4, c = -3$$

$$x = \frac{-b \pm \sqrt{b^2 \quad 4ac}}{2a}$$

$$x = \frac{-4 \pm \sqrt{(4)^2 - 4(1)(-3)}}{2(1)}$$

$$x = \frac{-4 \pm \sqrt{16 - (-12)}}{2}$$

$$x = \frac{-4 \pm \sqrt{28}}{2}$$

$$x = \frac{-4 \pm \sqrt{4 \bullet 7}}{2}$$

$$x = \frac{-4 \pm 2\sqrt{7}}{2}$$

$$x = \frac{2\left(-2 \pm \sqrt{7}\right)}{2}$$

$$\boxed{x = -2 \pm \sqrt{7}}$$

39.

$$3x^2 = 6x + 2$$
$$3x^2 - 6x - 2 = 6x + 2 - 6x - 2$$
$$3x^2 - 6x - 2 = 0$$
$$a = 3, b = -6, c = -2$$

$$x = \frac{-b \pm \sqrt{b^2 - 4ac}}{2a}$$

$$x = \frac{-(-6) \pm \sqrt{(-6)^2 - 4(3)(-2)}}{2(3)}$$

$$x = \frac{6 \pm \sqrt{36 - (-24)}}{6}$$

$$x = \frac{6 \pm \sqrt{60}}{6}$$

$$x = \frac{6 \pm \sqrt{4 \bullet 15}}{6}$$

$$x = \frac{6 \pm 2\sqrt{15}}{6}$$

$$x = \frac{2\left(3 \pm \sqrt{15}\right)}{2 \bullet 3}$$

$$\boxed{x = \frac{3 \pm \sqrt{15}}{3}}$$

43.

$$2x^2 + 10x + 11 = 0$$
$$a = 2, b = 10, c = 11$$

$$x = \frac{-b \pm \sqrt{b^2 - 4ac}}{2a}$$

$$x = \frac{-10 \pm \sqrt{(10)^2 - 4(2)(11)}}{2(2)}$$

$$x = \frac{-10 \pm \sqrt{100 - 88}}{4}$$

$$x = \frac{-10 \pm \sqrt{12}}{4}$$

$$x = \frac{-10 \pm \sqrt{4 \bullet 3}}{4}$$

$$x = \frac{-10 \pm 2\sqrt{3}}{4}$$

$$x = \frac{2\left(-5 \pm \sqrt{3}\right)}{2 \bullet 2}$$

$$\boxed{x = \frac{-5 \pm \sqrt{3}}{2}}$$

45.

$$7x^2 - x = -8$$
$$7x^2 - x + 8 = -8 + 8$$
$$7x^2 - x + 8 = 0$$
$$a = 7, b = -1, c = 8$$

$$x = \frac{-b \pm \sqrt{b^2 - 4ac}}{2a}$$

$$x = \frac{-(-1) \pm \sqrt{(-1)^2 - 4(7)(8)}}{2(7)}$$

$$x = \frac{1 \pm \sqrt{1 - 224}}{14}$$

$$\boxed{x = \frac{1 \pm \sqrt{-223}}{14}}$$

$$\boxed{\text{No Real Solutions}}$$

Since $\sqrt{-223}$ is not a real number, there are no real solutions to this problem.

47.

$$-8x - 5 = 2x^2$$
$$-8x - 5 - 2x^2 = 2x^2 - 2x^2$$
$$-2x^2 - 8x - 5 = 0$$
$$a = -2, b = -8, c = -5$$

$$x = \frac{-b \pm \sqrt{b^2 - 4ac}}{2a}$$

$$x = \frac{-(-8) \pm \sqrt{(-8)^2 - 4(-2)(-5)}}{2(-2)}$$

$$x = \frac{8 \pm \sqrt{64 - 40}}{-4}$$

$$x = \frac{8 \pm \sqrt{24}}{-4}$$

$$x = \frac{8 \pm \sqrt{4 \bullet 6}}{-4}$$

$$x = \frac{8 \pm 2\sqrt{6}}{-4}$$

$$x = \frac{2(4 \pm \sqrt{6})}{-2 \bullet 2}$$

$$x = \frac{4 \pm \sqrt{6}}{-2}$$

$$\boxed{x = \frac{-4 \pm \sqrt{6}}{2}}$$

49. Use the Square Root Method

$$(2y - 1)^2 = 25$$

$$2y - 1 = 5 \quad \text{or} \quad 2y - 1 = -5$$
$$2y = 6 \qquad\qquad\quad 2y = -4$$
$$\frac{2y}{2} = \frac{6}{2} \qquad\qquad \frac{2y}{2} = \frac{-4}{2}$$
$$y = 3 \qquad\qquad\qquad y = -2$$

$$\boxed{y = 3, -2}$$

51. Use the Quadratic Formula

$$2x^2 + x = 5$$
$$2x^2 + x - 5 = 5 - 5$$
$$2x^2 + x - 5 = 0$$
$$a = 2, b = 1, c = -5$$

$$x = \frac{-b \pm \sqrt{b^2 - 4ac}}{2a}$$

$$x = \frac{-1 \pm \sqrt{(1)^2 - 4(2)(-5)}}{2(2)}$$

$$x = \frac{-1 \pm \sqrt{1 - (-40)}}{4}$$

$$x = \frac{-1 \pm \sqrt{41}}{4}$$

$$x = \frac{-1 + \sqrt{41}}{4} \qquad x = \frac{-1 - \sqrt{41}}{4}$$

$$x \approx \frac{-1 + 6.40}{4} \qquad x \approx \frac{-1 - 6.40}{4}$$

$$x \approx \frac{5.40}{4} \qquad\quad x \approx \frac{-7.40}{4}$$

$$\boxed{x \approx 1.35} \qquad\qquad \boxed{x \approx -1.85}$$

53. Use the Quadratic Formula

$$x^2 - 2x - 1 = 0$$
$$a = 1, b = -2, c = -1$$

$$x = \frac{-b \pm \sqrt{b^2 - 4ac}}{2a}$$

$$x = \frac{-(-2) \pm \sqrt{(-2)^2 - 4(1)(-1)}}{2(1)}$$

$$x = \frac{2 \pm \sqrt{4 - (-4)}}{2}$$

$$x = \frac{2 \pm \sqrt{8}}{2}$$

$$x = \frac{2 \pm \sqrt{4 \cdot 2}}{2}$$

$$x = \frac{2 \pm 2\sqrt{2}}{2}$$

$$x = \frac{2\left(1 \pm \sqrt{2}\right)}{2}$$

$$x = 1 \pm \sqrt{2}$$

$$x \approx 1 + \sqrt{2} \qquad x \approx 1 - \sqrt{2}$$

$$x \approx 1 + 1.41 \qquad x \approx 1 - 1.41$$

$$\boxed{x \approx 2.41} \qquad \boxed{x \approx -0.41}$$

55. Use the Factoring Method

$$x^2 - 2x - 35 = 0$$
$$(x - 7)(x + 5) = 0$$
$$x - 7 = 0 \quad \text{or} \quad x + 5 = 0$$
$$x - 7 + 7 = 0 + 7 \quad x + 5 - 5 = 0 - 5$$
$$x = 7 \quad \text{or} \quad x = -5$$
$$\boxed{x = 7, -5}$$

57. Use the Quadratic Formula

$$x^2 + 2x + 7 = 0$$
$$a = 1, b = 2, c = 7$$

$$x = \frac{-b \pm \sqrt{b^2 - 4ac}}{2a}$$

$$x = \frac{-2 \pm \sqrt{(2)^2 - 4(1)(7)}}{2(1)}$$

$$x = \frac{-2 \pm \sqrt{4 - 28}}{2}$$

continued

57. continued

$$x = \frac{-2 \pm \sqrt{-22}}{2}$$

$$\boxed{\text{No Real Solutions}}$$

Since $\sqrt{-22}$ is not a real number, there are no real solutions to this problem.

59. Use the Factoring Method

$$4c^2 + 16c = 0$$
$$4c(c + 4) = 0$$
$$4c = 0 \qquad \text{or} \qquad c + 4 = 0$$
$$\frac{4c}{4} = \frac{0}{4} \qquad\qquad c + 4 - 4 = 0 - 4$$
$$c = 0 \qquad\qquad\qquad c = -4$$
$$\boxed{c = 0, -4}$$

61. Use the Square Root Method

$$18 = 3y^2$$
$$\frac{18}{3} = \frac{3y^2}{3}$$
$$6 = y^2$$
$$\boxed{y = \sqrt{6}} \quad \text{or} \quad \boxed{y = -\sqrt{6}}$$
$$y \approx 2.45 \qquad\qquad y \approx -2.45$$
$$\boxed{y \approx \pm 2.45}$$

63. Use the Quadratic Formula

$$2.4x^2 - 9.5x + 6.2 = 0$$
$$a = 2.4, b = -9.5, c = 6.2$$

$$x = \frac{-b \pm \sqrt{b^2 - 4ac}}{2a}$$

$$x = \frac{-(-9.5) \pm \sqrt{(-9.5)^2 - 4(2.4)(6.2)}}{2(2.4)}$$

$$x \approx \frac{9.5 \pm \sqrt{90.25 - 59.52}}{4.8}$$

$$x \approx \frac{9.5 \pm \sqrt{30.73}}{4.8}$$

$$x \approx \frac{9.5 + \sqrt{30.73}}{4.8} \qquad x \approx \frac{9.5 - \sqrt{30.73}}{4.8}$$

$$x \approx \frac{9.5 + 5.54}{4.8} \qquad x \approx \frac{9.5 - 5.54}{4.8}$$

$$x \approx \frac{15.04}{4.8} \qquad x \approx \frac{3.96}{4.8}$$

$$\boxed{x \approx 3.1} \qquad \boxed{x \approx 0.8}$$

65. HEIGHT OF TRIANGLE

$$A = \frac{1}{2}bh$$

$$\frac{1}{2}(x+4)(x) = 30$$

$$2\left[\frac{1}{2}(x+4)(x)\right] = 2(30)$$

$$(x+4)(x) = 60$$

$$x^2 + 4x = 60$$

$$x^2 + 4x - 60 = 0$$

Use the Quadratic Formula.
$a = 1, b = 4, c = -60$

$$x = \frac{-b \pm \sqrt{b^2 - 4ac}}{2a}$$

$$x = \frac{-4 \pm \sqrt{(4)^2 - 4(1)(-60)}}{2(1)}$$

$$x = \frac{-4 \pm \sqrt{16 + 240}}{2}$$

$$x = \frac{-4 \pm \sqrt{256}}{2}$$

$$x = \frac{-4 \pm 16}{2}$$

$$x = \frac{-4 + 16}{2} \qquad x = \frac{-4 - 16}{2}$$

$$x = \frac{12}{2} \qquad\qquad x = \frac{-20}{2}$$

$$\boxed{x = 6} \qquad\qquad x = -10$$

Since the height cannot be negative, the height of the triangle is 6 in.

67. FLAGS

$$A = L \cdot W$$
$$A = 3{,}102, \ L = x + 19, \ W = x$$
$$x(x + 19) = 3{,}102$$
$$x^2 + 19x = 3{,}102$$
$$x^2 + 19x - 3{,}102 = 3{,}102 - 3{,}102$$
$$x^2 + 19x - 3{,}102 = 0 - 3{,}102$$

Use the Quadratic Formula.
$a = 1, b = 19, c = -3{,}102$

$$x = \frac{-b \pm \sqrt{b^2 - 4ac}}{2a}$$

$$x = \frac{-19 \pm \sqrt{(19)^2 - 4(1)(-3{,}102)}}{2(1)}$$

$$x = \frac{-19 \pm \sqrt{361 - (-12{,}408)}}{2}$$

$$x = \frac{-19 \pm \sqrt{12{,}769}}{2}$$

$$x = \frac{-19 \pm 113}{2}$$

$$x = \frac{-19 + 113}{2} \qquad x = \frac{-19 - 113}{2}$$

$$x = \frac{94}{2} \qquad\qquad x = \frac{-132}{2}$$

$$\boxed{x = 47} \qquad\qquad x = -66$$

Since the width cannot be negative, the width of the flag is 47 ft.

$W = x$	$L = x + 19$
$= 47$ ft	$= 47 + 19$
	$= 66$ ft

69. DAREDEVILS

$$s = 16t^2$$
$$200 = 16t^2$$
$$\frac{200}{16} = \frac{16t^2}{16}$$
$$12.5 = t^2$$
$$t = \sqrt{12.5} \quad \text{or} \quad t = -\sqrt{12.5}$$
$$t \approx 3.53 \quad \text{or} \quad t \approx -3.53$$
$$\boxed{t \approx 3.53 \text{ sec.}}$$

71. EARTHQUAKES

Use the Pythagorean Theorem.
$$50^2 = L^2 + (L - 10)^2$$
$$2{,}500 = L^2 + L^2 - 20L + 100$$
$$2{,}500 - 2{,}500 = L^2 + L^2 - 20L + 100 - 2{,}500$$
$$0 = 2L^2 - 20L - 2{,}400$$
$$\frac{0}{2} = \frac{2L^2}{2} - \frac{20L}{2} - \frac{2{,}400}{2}$$
$$0 = L^2 - 10L - 1{,}200$$
$$0 = (L - 40)(L + 30)$$
$$L - 40 = 0 \quad \text{or} \quad L + 30 = 0$$
$$L - 40 + 40 = 0 + 40 \quad L + 30 - 30 = 0 - 30$$
$$\boxed{L = 40} \qquad\qquad L = -30$$

Since the length cannot be negative, the length of the window is 40.

$$\boxed{\begin{aligned} L = 40 \text{ in.} \quad H &= L - 10 \\ &= 40 - 10 \\ &= 30 \text{ in.} \end{aligned}}$$

73. INVESTING

$$A = P(1 + r)^2$$
$$5{,}724.50 = 5{,}000(1 + r)^2$$
$$\frac{5{,}724.50}{5{,}000} = \frac{5{,}000(1 + r)^2}{5{,}000}$$
$$1.1449 = (1 + r)^2$$
$$(1 + r)^2 = 1.1449$$
$$1 + r = 1.07 \quad \text{or} \quad 1 + r = -1.07$$
$$1 + r - 1 = 1.07 - 1 \quad 1 + r - 1 = -1.07 - 1$$
$$r = 1.07 - 1 \qquad r = -1.07 - 1$$
$$r = 0.07 \qquad\qquad r = -2.07$$
$$\boxed{r = 7\%}$$

WRITING

75. Answers will vary.

77. Answers will vary.

REVIEW

79.
$$A = p + prt$$
$$A - p = p + prt - p$$
$$A - p = prt$$
$$\frac{A - p}{pt} = \frac{prt}{pt}$$
$$\boxed{\frac{A - p}{pt} = r}$$

81.
$$\frac{1}{r} = \frac{1}{r_1} + \frac{1}{r_2}$$
$$rr_1r_2\left(\frac{1}{r}\right) = rr_1r_2\left(\frac{1}{r_1} + \frac{1}{r_2}\right)$$
$$r_1r_2 = rr_2 + rr_1$$
$$r_1r_2 = r(r_2 + r_1)$$
$$\frac{r_1r_2}{r_2 + r_1} = \frac{r(r_2 + r_1)}{r_2 + r_1}$$
$$\boxed{\frac{r_1r_2}{r_2 + r_1} = r}$$

CHALLENGE PROBLEMS

83. DECKING

Length $= 24 + x + x$
$\qquad = 2x + 24$
Width $= 14 + x + x$
$\qquad = 2x + 14$

Total area − area of pool = area of deck
$(2x + 24)(2x + 14) - (24)(14) = 368$
$4x^2 + 28x + 48x + 336 - 336 = 368$
$\qquad\qquad\qquad 4x^2 + 76x = 368$
$\qquad\qquad 4x^2 + 76x - 368 = 368 - 368$
$\qquad\qquad 4x^2 + 76x - 368 = 0$

$$x = \frac{-b \pm \sqrt{b^2 - 4ac}}{2a}$$

$$x = \frac{-76 \pm \sqrt{(76)^2 - 4(4)(-368)}}{2(4)}$$

$$x = \frac{-76 \pm \sqrt{5,776 - (-5,888)}}{8}$$

$$x = \frac{-76 \pm \sqrt{11,664}}{8}$$

$$x = \frac{-76 \pm 108}{8}$$

$$x = \frac{-76 + 108}{8} \qquad x = \frac{-76 - 108}{8}$$

$$x = \frac{32}{8} \qquad\qquad x = \frac{-184}{8}$$

$$\boxed{x = 4} \qquad\qquad x = -23$$

Since the width cannot be negative, the
width of the deck would be 4 ft.

VOCABULARY

1. A **complex** number is any number that can be written in the form $a + bi$.

3. The complex numbers $6 + 3i$ and $6 - 3i$ are called complex **conjugates**.

CONCEPTS

5. a) $i = \sqrt{-1}$
   b) $i^2 = -1$

7. $\sqrt{-25} = \sqrt{-1 \cdot 25} = \sqrt{-1}\ \sqrt{25} = 5\ i$

9. $i \cdot i = i^2 = -1$

11. To simplify $\dfrac{2-3i}{6-i}$, we multiply it by

$\dfrac{6+i}{6+i}$.

13. a) True
    b) True
    c) False
    d) True

15. a) $2 + 9i$
    b) $-8 + i$
    c) $0 + 4i$
    d) $0 - 11i$

NOTATION

17. a) $\sqrt{7}i = i\sqrt{7}$
    b) $2\sqrt{3}i = 2i\sqrt{3}$

PRACTICE

19.
$$\sqrt{-9} = \sqrt{-1 \cdot 9}$$
$$= \sqrt{9}\sqrt{-1}$$
$$= 3i$$

21.
$$\sqrt{-7} = \sqrt{-1 \cdot 7}$$
$$= \sqrt{7}\sqrt{-1}$$
$$= i\sqrt{7}$$

23.
$$\sqrt{-24} = \sqrt{-1 \cdot 4 \cdot 6}$$
$$= \sqrt{-1 \cdot 4}\sqrt{6}$$
$$= 2i\sqrt{6}$$

25.
$$-\sqrt{-24} = -\sqrt{-1 \cdot 4 \cdot 6}$$
$$= -\sqrt{-1 \cdot 4}\sqrt{6}$$
$$= -2i\sqrt{6}$$

27.
$$5\sqrt{-81} = 5\sqrt{-1 \cdot 81}$$
$$= 5 \cdot 9i$$
$$= 45i$$

29.
$$\sqrt{-\frac{25}{9}} = \sqrt{-1} \cdot \sqrt{\frac{25}{9}}$$
$$= i \cdot \frac{5}{3}$$
$$= \frac{5}{3}i$$

31.
$$x^2 + 9 = 0$$
$$x^2 + 9 - 9 = 0 - 9$$
$$x^2 = -9$$
$$x = \sqrt{-9} \quad \text{or} \quad x = -\sqrt{-9}$$
$$x = 3i \quad \text{or} \quad x = -3i$$
$$\boxed{x = 0 \pm 3i}$$

33. $x^2 = -36$
$$x = \sqrt{-36} \quad \text{or} \quad x = -\sqrt{-36}$$
$$x = 6i \quad \text{or} \quad x = -6i$$
$$\boxed{x = 0 \pm 6i}$$

35. $x^2 + 8 = 0$
$$x^2 + 8 - 8 = 0 - 8$$
$$x^2 = -8$$
$$x = \sqrt{-8} \quad \text{or} \quad x = -\sqrt{-8}$$
$$x = \sqrt{-1 \cdot 4 \cdot 2} \quad \text{or} \quad x = -\sqrt{-1 \cdot 4 \cdot 2}$$
$$x = 2i\sqrt{2} \quad \text{or} \quad x = -2i\sqrt{2}$$
$$\boxed{x = 0 \pm 2i\sqrt{2}}$$

37. $x^2 = -\dfrac{16}{9}$

$\quad x = \sqrt{-\dfrac{16}{9}} \qquad$ or $\qquad x = -\sqrt{-\dfrac{16}{9}}$

$\quad x = \dfrac{4}{3}i \qquad\qquad$ or $\qquad x = -\dfrac{4}{3}i$

$\boxed{x = 0 \pm \dfrac{4}{3}i}$

39. $x^2 - 3x + 4 = 0$
$a = 1,\ b = -3,\ c = 4$

$x = \dfrac{-b \pm \sqrt{b^2 - 4ac}}{2a}$

$\quad = \dfrac{-(-3) \pm \sqrt{(-3)^2 - 4(1)(4)}}{2(1)}$

$\quad = \dfrac{3 \pm \sqrt{9 - 16}}{2}$

$\quad = \dfrac{3 \pm \sqrt{-7}}{2}$

$\quad = \dfrac{3 \pm i\sqrt{7}}{2}$

$\boxed{x = \dfrac{3}{2} \pm \dfrac{\sqrt{7}}{2}i}$

41. $2x^2 + x + 1 = 0$
$a = 2,\ b = 1,\ c = 1$

$x = \dfrac{-b \pm \sqrt{b^2 - 4ac}}{2a}$

$\quad = \dfrac{-1 \pm \sqrt{(1)^2 - 4(2)(1)}}{2(2)}$

$\quad = \dfrac{-1 \pm \sqrt{1 - 8}}{4}$

$\quad = \dfrac{-1 \pm \sqrt{-7}}{4}$

$\quad = \dfrac{-1 \pm i\sqrt{7}}{4}$

$\boxed{x = -\dfrac{1}{4} \pm \dfrac{\sqrt{7}}{4}i}$

43. $x^2 + 2x + 2 = 0$
$a = 1,\ b = 2,\ c = 2$

$x = \dfrac{-b \pm \sqrt{b^2 - 4ac}}{2a}$

$\quad = \dfrac{-2 \pm \sqrt{(2)^2 - 4(1)(2)}}{2(1)}$

$\quad = \dfrac{-2 \pm \sqrt{4 - 8}}{2}$

$\quad = \dfrac{-2 \pm \sqrt{-4}}{2}$

$\quad = \dfrac{-2 \pm 2i}{2}$

$x = \dfrac{-2}{2} \pm \dfrac{2}{2}i$

$\boxed{x = -1 \pm i}$

45. $3x^2 + 2x + 1 = 0$
$a = 3,\ b = 2,\ c = 1$

$x = \dfrac{-b \pm \sqrt{b^2 - 4ac}}{2a}$

$\quad = \dfrac{-2 \pm \sqrt{(2)^2 - 4(3)(1)}}{2(3)}$

$\quad = \dfrac{-2 \pm \sqrt{4 - 12}}{6}$

$\quad = \dfrac{-2 \pm \sqrt{-8}}{6}$

$\quad = \dfrac{-2 \pm \sqrt{-1 \bullet 4 \bullet 2}}{6}$

$\quad = \dfrac{-2 \pm 2i\sqrt{2}}{6}$

$x = -\dfrac{2}{6} \pm \dfrac{2\sqrt{2}}{6}i$

$\boxed{x = -\dfrac{1}{3} \pm \dfrac{\sqrt{2}}{3}i}$

47. $3x^2 - 2x = -3$

$3x^2 - 2x + 3 = -3 + 3$

$3x^2 - 2x + 3 = 0$

$a = 3, b = -2, c = 3$

$x = \dfrac{b \pm \sqrt{b^2 - 4ac}}{2a}$

$= \dfrac{-(-2) \pm \sqrt{(-2)^2 - 4(3)(3)}}{2(3)}$

$= \dfrac{2 \pm \sqrt{4 - 36}}{6}$

$= \dfrac{2 \pm \sqrt{-32}}{6}$

$= \dfrac{2 \pm \sqrt{-1 \bullet 16 \bullet 2}}{6}$

$= \dfrac{2 \pm 4i\sqrt{2}}{6}$

$x = \dfrac{2}{6} \pm \dfrac{4\sqrt{2}}{6} i$

$\boxed{x = \dfrac{1}{3} \pm \dfrac{2\sqrt{2}}{3} i}$

49. $(3 + 4i) + (5 - 6i) = (3 + 5) + (4i - 6i)$

$= 8 - 2i$

51. $(7 - 3i) - (4 + 2i) = 7 - 3i - 4 - 2i$

$= (7 - 4) + (-3i - 2i)$

$= 3 - 5i$

53. $(6 - i) + (9 + 3i) = (6 + 9) + (-1i + 3i)$

$= 15 + 2i$

55. $(-3 - 8i) - (-3 - 9i) = -3 - 8i + 3 + 9i$

$= (-3 + 3) + (-8i + 9i)$

$= 0 + i$

57. $(14 - 4i) - 9i = 14 - 4i - 9i$

$= 14 - 13i$

59. $15 + (-3 - 9i) = 15 - 3 - 9i$

$= 12 - 9i$

61. $3(2 - i) = 3(2) - 3(i)$

$= 6 - 3i$

63. $-4(3 + 4i) = -4(3) - 4(4i)$

$= -12 - 16i$

65. $-5i(5 - 5i) = -5i(5) - 5i(-5i)$

$= -25i + 25i^2$

$= -25i + 25(-1)$

$= -25i - 25$

$= -25 - 25i$

67. $(2 + i)(3 - i) = 6 - 2i + 3i - i^2$

$= 6 + i - (-1)$

$= 6 + i + 1$

$= 7 + i$

69. $(3 - 2i)(2 + 3i) = 6 + 9i - 4i - 6i^2$

$= 6 + 5i - 6(-1)$

$= 6 + 5i + 6$

$= 12 + 5i$

71. $(4 + i)(3 - i) = 12 - 4i + 3i - i^2$

$= 12 - i - (-1)$

$= 12 - i + 1$

$= 13 - i$

73. $(2 + i)^2 = (2 + i)(2 + i)$

$= 4 + 2i + 2i + i^2$

$= 4 + 4i - 1$

$= 3 + 4i$

75.

$\dfrac{5}{2 - i} = \dfrac{5}{2 - i} \bullet \dfrac{2 + i}{2 + i}$

$= \dfrac{10 + 5i}{4 - i^2}$

$= \dfrac{10 + 5i}{4 - (-1)}$

$= \dfrac{10 + 5i}{5}$

$= \dfrac{10}{5} + \dfrac{5i}{5}$

$= 2 + i$

77.

$\dfrac{3}{5 + i} = \dfrac{3}{5 + i} \bullet \dfrac{5 - i}{5 - i}$

$= \dfrac{15 - 3i}{25 - i^2}$

$= \dfrac{15 - 3i}{25 - (-1)}$

$= \dfrac{15 - 3i}{26}$

$= \dfrac{15}{26} - \dfrac{3}{26} i$

79.

$$\frac{5i}{6+2i} = \frac{5i}{6+2i} \cdot \frac{6-2i}{6-2i}$$

$$= \frac{30i - 10i^2}{36 - 4i^2}$$

$$= \frac{30i - 10(-1)}{36 - 4(-1)}$$

$$= \frac{30i + 10}{36 + 4}$$

$$= \frac{10 + 30i}{40}$$

$$= \frac{10}{40} + \frac{30}{40}i$$

$$= \frac{1}{4} + \frac{3}{4}i$$

81.

$$\frac{2+3i}{2-3i} = \frac{2+3i}{2-3i} \cdot \frac{2+3i}{2+3i}$$

$$= \frac{4 + 6i + 6i + 9i^2}{4 - 9i^2}$$

$$= \frac{4 + 12i + 9(-1)}{4 - 9(-1)}$$

$$= \frac{4 + 12i - 9}{4 + 9}$$

$$= \frac{-5 + 12i}{13}$$

$$= -\frac{5}{13} + \frac{12}{13}i$$

83.

$$\frac{3+2i}{3+i} = \frac{3+2i}{3+i} \cdot \frac{3-i}{3-i}$$

$$= \frac{9 - 3i + 6i - 2i^2}{9 - i^2}$$

$$= \frac{9 + 3i - 2(-1)}{9 - (-1)}$$

$$= \frac{9 + 3i + 2}{9 + 1}$$

$$= \frac{11 + 3i}{10}$$

$$= \frac{11}{10} + \frac{3}{10}i$$

APPLICATIONS

85. ELECTRICITY

$$V = (10.31 - 5.97i) + (8.14 + 3.79i)$$
$$V = (10.31 + 8.14) + (-5.97i + 3.79i)$$
$$V = 18.45 - 2.18i$$

WRITING

87. Answers will vary.

89. Answers will vary.

REVIEW

91.

$$\frac{7}{x+3} + \frac{4}{x+6} = \frac{7}{x+3}\left(\frac{x+6}{x+6}\right) + \frac{4x}{x+6}\left(\frac{x+3}{x+3}\right)$$

$$= \frac{7x + 42}{(x+3)(x+6)} + \frac{4x^2 + 12x}{(x+3)(x+6)}$$

$$= \frac{7x + 42 + 4x^2 + 12x}{(x+3)(x+6)}$$

$$= \frac{4x^2 + 19x + 42}{(x+3)(x+6)}$$

93.

$$\frac{x^2 - 5x + 6}{x - 3} = \frac{(x-2)(x-3)}{(x-3)}$$

$$= x - 2$$

CHALLENGE PROBLEMS

95.

$$\left(2\sqrt{2} + i\sqrt{2}\right)\left(3\sqrt{2} - i\sqrt{2}\right)$$

$$= 6\sqrt{4} - 2i\sqrt{4} + 3i\sqrt{4} - i^2\sqrt{4}$$

$$= 6(2) - 2i(2) + 3i(2) - (-1)(2)$$

$$= 12 - 4i + 6i + 2$$

$$= 14 + 2i$$

SECTION 9.4

VOCABULARY

1. $y = 3x^2 + 5x - 1$ is a **quadratic** equation in two variables.

3. The lowest point on a parabola that opens upward, and the highest point on a parabola that opens downward, is called the **vertex** of the parabola.

5. The point where a parabola intersects the y-axis is called the y-**intercept** of the graph.

CONCEPTS

7. a) The graph of $y = ax^2 + bx + c$ opens downward when $a \boxed{<} 0$.
    b) The graph of $y = ax^2 + bx + c$ opens upward when $a > \boxed{0}$.

9. a) To find the y-intercepts of a graph, substitute $\boxed{0}$ for x in the given equation and solve for y.
    b) To find the x-intercepts of a graph, substitute 0 for y in the given equation and solve for $\boxed{x}$.

11. a) parabola
    b) (1, 0) and (3, 0)
    c) (0, -3)
    d) (2, 1)
    e) $x = 2$

13. Let $x = 3$
    $y = -(3)^2 + 6(3) - 7$
    $y = -9 + 18 - 7$
    $y = 2$

15.
    zero x-intercepts:

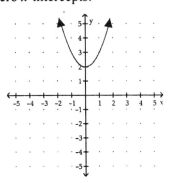

one x-intercept:

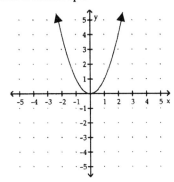

two x-intercepts:

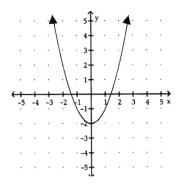

NOTATION

17. a) $a = 2, b = 4, c = -8$

    b) $-\dfrac{b}{2a} = -\dfrac{4}{2(2)}$

    $= -\dfrac{4}{4}$

    $= -1$

19. two solutions because there are two x-intercepts; $x = -3, x = 1$

PRACTICE

21. $a = 2 > 0$; upward

23. $a = -6 < 0$; downward

25.
$$x = -\frac{b}{2a}$$
$$= -\frac{-4}{2(2)}$$
$$= -\frac{-4}{4}$$
$$= 1$$

$$y = 2(1)^2 - 4(1) + 1$$
$$= 2 - 4 + 1$$
$$= -1$$

vertex = (1, -1)

27.
$$x = -\frac{b}{2a}$$
$$= -\frac{6}{2(-1)}$$
$$= -\frac{6}{-2}$$
$$= 3$$

$$y = -(3)^2 + 6(3) - 8$$
$$= -9 + 18 - 8$$
$$= 1$$

vertex = (3, 1)

29. To find the x-intercepts, let $y = 0$ and solve the equation for x.

$$0 = x^2 - 2x + 1$$
$$0 = (x - 1)(x - 1)$$

$$x - 1 = 0$$
$$x - 1 + 1 = 0 + 1$$
$$x = 1$$

(1, 0)

To find the y-intercepts, let $x = 0$ and solve the equation for y.
$$y = 0^2 - 2(0) + 1$$
$$y = 0 - 0 + 1$$
$$y = 1$$
(0, 1)

x-intercept: (1, 0)
y-intercept: (0, 1)

31. To find the x-intercepts, let $y = 0$ and solve the equation for x.

$$0 = -x^2 - 10x - 21$$
$$0 = -1(x^2 + 10x + 21)$$
$$0 = -1(x + 3)(x + 7)$$
$$x + 7 = 0 \quad \text{or} \quad x + 3 = 0$$
$$x + 7 - 7 = 0 - 7 \quad\quad x + 3 - 3 = 0 - 3$$
$$x = -7 \quad\quad\quad\quad x = -3$$

(-7, 0)   (-3, 0)

To find the y-intercepts, let $x = 0$ and solve the equation for y.

$$y = -(0)^2 - 10(0) - 21$$
$$y = -(0) - 10(0) - 21$$
$$y = 0 - 0 - 21$$
$$y = -21$$
(0, -21)

x-intercepts: (-7, 0) and (-3, 0);
y-intercept: (0, -21)

33. $y = x^2 + 2x - 3$

Step 1: Since $a = 1 > 0$, the parabola opens upward.

Step 2: Find the vertex.

$x = -\dfrac{b}{2a}$ $y = (-1)^2 + 2(-1) - 3$

$= -\dfrac{2}{2(1)}$ $= 1 - 2 - 3$

                          $= -4$

$= -\dfrac{2}{2}$

$= -1$

vertex: (-1, -4)

Step 3: Find the x- and y-intercepts.
Since $c = -3$, the y-intercept is (0, -3).
To find the x-intercepts, let $y = 0$ and solve the equation for x.

$0 = x^2 + 2x - 3$
$0 = (x + 3)(x - 1)$
$x + 3 = 0$ or $x - 1 = 0$
$x + 3 - 3 = 0 - 3$ $x - 1 + 1 = 0 + 1$
$x = -3$ $x = 1$
(-3, 0)   (1, 0)
x-intercepts: (-3, 0) and (1, 0);
y-intercept: (0, -3)

Step 4: Find another point(s).
Let $x = -2$.
$y = (-2)^2 + 2(-2) - 3 = 4 - 4 - 3 = -3$
(-2, -3)

Step 5: Plots the points and draw the parabola.

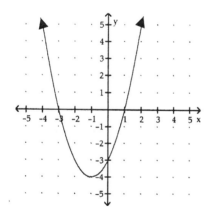

35. $y = 2x^2 + 8x + 6$

Step 1: Since $a = 2 > 0$, the parabola opens upward.

Step 2: Find the vertex.

$x = -\dfrac{b}{2a}$ $y = 2(-2)^2 + 8(-2) + 6$

$= -\dfrac{8}{2(2)}$ $= 8 - 16 + 6$

                          $= -2$

$= -\dfrac{8}{4}$

$= -2$

vertex: (-2, -2)

Step 3: Find the x- and y-intercepts.
Since $c = 6$, the y-intercept is (0, 6).
To find the x-intercepts, let $y = 0$ and solve the equation for x.

$0 = 2x^2 + 8x + 6$
$0 = 2(x^2 + 4x + 3)$
$0 = (x + 3)(x + 1)$
$x + 3 = 0$ or $x + 1 = 0$
$x + 3 - 3 = 0 - 3$ $x + 1 - 1 = 0 - 1$
$x = -3$ $x = -1$
(-3, 0)   (-1, 0)
x-intercepts: (-3, 0) and (-1, 0);
y-intercept: (0, 6)

Step 4: Find another point(s).
Let $x = -4$.
$y = 2(-4)^2 + 8(-4) + 6 = 32 - 32 + 6 = 6$
(-4, 6)

Step 5: Plots the points and draw the parabola.

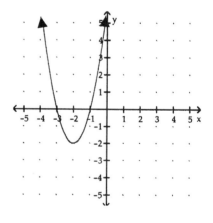

37. $y = -x^2 + 2x + 3$

Step 1: Since $a = -1 < 0$, the parabola opens downward.

Step 2: Find the vertex.

$x = -\dfrac{b}{2a}$
$= -\dfrac{2}{2(-1)}$
$= -\dfrac{2}{-2}$
$= 1$

$y = -(1)^2 + 2(1) + 3$
$= -1 + 2 + 3$
$= 4$

vertex: $(1, 4)$

Step 3: Find the x- and y-intercepts.
Since $c = 3$, the y-intercept is $(0, 3)$.
To find the x-intercepts, let $y = 0$ and solve the equation for x.

$0 = -x^2 + 2x + 3$
$0 = -1(x^2 - 2x - 3)$
$0 = (x - 3)(x + 1)$

$x - 3 = 0$ or $x + 1 = 0$
$x - 3 = 0 + 3$ $\quad$ $x + 1 - 1 = 0 - 1$
$x = 3$ or $x = -1$
$(3, 0)$ or $(-1, 0)$

x-intercepts: $(3, 0)$ and $(1, 0)$;
y-intercept: $(0, 9)$

Step 4: Find another point(s).
Let $x = 2$.
$y = -(2)^2 + 2(2) + 3 = -4 + 4 + 3 = 3$
$(2, 3)$

Step 5: Plots the points and draw the parabola.

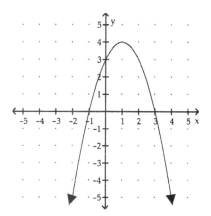

39. $y = -x^2 - 2x - 1$

Step 1: Since $a = -1 < 0$, the parabola opens downward.

Step 2: Find the vertex.

$x = -\dfrac{b}{2a}$
$= -\dfrac{-2}{2(-1)}$
$= -\dfrac{-2}{-2}$
$= -1$

$y = -(-1)^2 - 2(-1) - 1$
$= -1 + 2 - 1$
$= 0$

vertex: $(-1, 0)$

Step 3: Find the x- and y-intercepts.
Since $c = -1$, the y-intercept is $(0, -1)$.
To find the x-intercepts, let $y = 0$ and solve the equation for x.

$0 = -x^2 - 2x - 1$
$0 = -1(x^2 + 2x + 1)$
$0 = (x + 1)(x + 1)$

$x + 1 = 0$ or $x + 1 = 0$
$x + 1 - 1 = 0 - 1$ $\quad$ $x + 1 - 1 = 0 - 1$
$x = -1$ or $x = -1$
$(-1, 0)$ or $(-1, 0)$

x-intercept: $(-1, 0)$;
y-intercept: $(0, -1)$

Step 4: Find another point(s).
Let $x = 1$.
$y = -(1)^2 - 2(1) - 1 = -1 - 2 - 1 = -4$
$(1, -4)$
Use symmetry to find the points
$(-2, -1)$ and $(-3, -4)$

Step 5: Plots the points and draw the parabola.

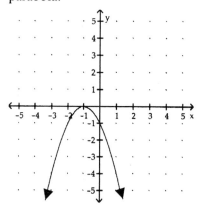

41. $y = x^2 - 2x$

Step 1: Since $a = 1 > 0$, the parabola opens upward.

Step 2: Find the vertex.

$$x = -\frac{b}{2a} \qquad y = (1)^2 - 2(1)$$
$$= -\frac{-2}{2(1)} \qquad = 1 - 2$$
$$= -\frac{-2}{2} \qquad = -1$$
$$= 1$$

vertex: (1, -1)

Step 3: Find the x- and y-intercepts.
Since $c = 0$, the y-intercept is (0, 0).
To find the x-intercepts, let $y = 0$ and solve the equation for x.

$$0 = x^2 - 2x$$
$$0 = x(x - 2)$$

$x = 0$ or $x - 2 = 0$
                    $x - 2 + 2 = 0 + 2$
                    $x = 2$
(0, 0)   or   (2, 0)
x-intercepts: (0, 0) and (2, 0);
y-intercept: (0, 0)

Step 4: Find another point(s).
Let $x = 3$.
$y = 3^2 - 2(3) = 9 - 6 = 3$
(3, 3)
Use symmetry to find the point (-1, 3).

Step 5: Plots the points and draw the parabola.

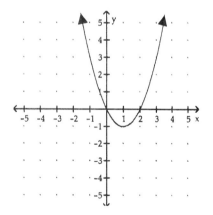

43. $y = x^2 + 4x + 4$

Step 1: Since $a = 1 > 0$, the parabola opens upward.

Step 2: Find the vertex.

$$x = -\frac{b}{2a} \qquad y = (-2)^2 + 4(-2) + 4$$
$$= -\frac{4}{2(1)} \qquad = 4 - 8 + 4$$
$$= -\frac{4}{2} \qquad = 0$$
$$= -2$$

vertex: (-2, 0)

Step 3: Find the x- and y-intercepts.
Since $c = 4$, the y-intercept is (0, 4).
To find the x-intercepts, let $y = 0$ and solve the equation for x.

$$0 = x^2 + 4x + 4$$
$$0 = (x + 2)(x + 2)$$

$x + 2 = 0$ or $x + 2 = 0$
$x + 2 - 2 = 0 - 2$ $x + 2 - 2 = 0 - 2$
$x = -2$ or $x = -2$
(-2, 0)   or   (-2, 0)
x-intercept: (-2, 0);
y-intercept: (0, 4)

Step 4: Find another point(s).
Let $x = -1$.
$y = (-1)^2 + 4(-1) + 4 = 1 - 4 + 4 = 1$
(-1, 1)
Use symmetry to find the points (-3, 1) and (-4, 4).

Step 5: Plots the points and draw the parabola.

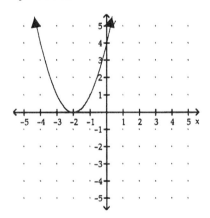

45. $y = -x^2 + 2x$

Step 1: Since $a = -1 < 0$, the parabola opens downward.

Step 2: Find the vertex.

$$x = -\frac{b}{2a} \qquad y = -(1)2 + 2(1)$$
$$= -\frac{2}{2(-1)} \qquad = -1 + 2$$
$$\qquad\qquad\quad = 1$$
$$= -\frac{2}{-2}$$
$$= 1$$

vertex: $(1, 1)$

Step 3: Find the x- and y-intercepts.
Since $c = 0$, the y-intercept is $(0, 0)$.
To find the x-intercepts, let $y = 0$ and solve the equation for x.
$$0 = -x^2 + 2x$$
$$0 = -x(x - 2)$$
$$-x = 0 \qquad \text{or} \qquad x - 2 = 0$$
$$\underline{-x = 0} \qquad \text{or} \qquad x - 2 + 2 = 0 + 2$$
$$-1 \quad -1$$
$$x = 0 \qquad \text{or} \qquad x = 2$$
$$(0, 0) \qquad \text{or} \qquad (2, 0)$$
x-intercepts: $(0, 0)$ and $(2, 0)$;
y-intercept: $(0, 0)$

Step 4: Find another point(s).
Let $x = 3$.
$y = -(3)^2 + 2(3) = -9 + 6 = -3$
$(3, -3)$
Use symmetry to find the point $(-1, -3)$.

Step 5: Plots the points and draw the parabola.

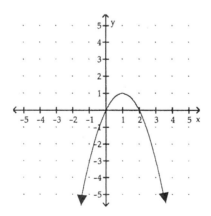

47. $y = 2x^2 + 3x - 2$

Step 1: Since $a = 2 > 0$, the parabola opens upward.

Step 2: Find the vertex.

$$x = -\frac{b}{2a} \qquad y = 2\left(-\frac{3}{4}\right)^2 + 3\left(-\frac{3}{4}\right) - 2$$
$$= -\frac{3}{2(2)} \qquad = 2\left(\frac{9}{16}\right) + 3\left(-\frac{3}{4}\right) - 2$$
$$= -\frac{3}{4} \qquad = \frac{9}{8} - \frac{9}{4} - 2$$
$$\qquad\qquad = \frac{9}{8} - \frac{18}{8} - \frac{16}{8}$$
$$\qquad\qquad = -\frac{25}{8}$$
$$\qquad\qquad = -3\frac{1}{8}$$

vertex: $\left(-\frac{3}{4}, -3\frac{1}{8}\right)$

Step 3: Find the x- and y-intercepts.
Since $c = -2$, the y-intercept is $(0, -2)$.
To find the x-intercepts, let $y = 0$ and solve the equation for x.
$$0 = 2x^2 + 3x - 2$$
$$0 = (2x - 1)(x + 2)$$
$$2x - 1 = 0 \quad \text{or} \qquad x + 2 = 0$$
$$2x - 1 + 1 = 0 + 1 \qquad x + 2 - 2 = 0 - 2$$
$$2x = 1$$
$$x = \frac{1}{2} \qquad \text{or} \qquad x = -2$$
$$\left(\frac{1}{2}, 0\right) \qquad \text{or} \qquad (-2, 0)$$
x-intercepts: $\left(\frac{1}{2}, 0\right)$ and $(-2, 0)$

y-intercept: $(0, -2)$

Step 4: Find another point(s).
Let $x = 1$.
$y = 2(1)^2 + 3(1) - 2 = 2 + 3 - 2 = 3$
$(1, 3)$
Let $x = -1$.
$y = 2(-1)^2 + 3(-1) - 2 = 2 - 3 - 2 = -3$
$(-1, -3)$

Continued

47. continued

Step 5: Plots the points and draw the parabola.

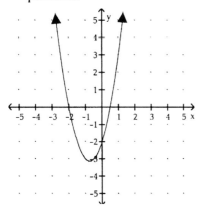

49. $y = 4x^2 - 12x + 9$

Step 1: Since $a = 4 > 0$, the parabola opens upward.

Step 2: Find the vertex.

$$x = -\frac{b}{2a} \qquad y = 4\left(\frac{3}{2}\right)^2 - 12\left(\frac{3}{2}\right) + 9$$

$$= -\frac{-12}{2(4)} \qquad = 4\left(\frac{9}{4}\right) - 12\left(\frac{3}{2}\right) + 9$$

$$= \frac{12}{8} \qquad\qquad = 9 - 18 + 9$$

$$= \frac{3}{2} \qquad\qquad = 0$$

vertex: $\left(\frac{3}{2},\ 0\right)$

Step 3: Find the x- and y-intercepts. Since $c = 9$, the y-intercept is $(0, 9)$. To find the x-intercepts, let $y = 0$ and solve the equation for x.

$$0 = 4x^2 - 12x + 9$$
$$0 = (2x - 3)(2x - 3)$$

$2x - 3 = 0$ or	$2x - 3 = 0$
$2x - 3 + 3 = 0 + 3$	$2x - 3 + 3 = 0 + 3$
$2x = 3$ or	$2x = 3$
$x = \dfrac{3}{2}$ or	$x = \dfrac{3}{2}$
$\left(1\dfrac{1}{2},\ 0\right)$ or	$\left(1\dfrac{1}{2},\ 0\right)$

x-intercept: $\left(1\dfrac{1}{2},\ 0\right)$

y-intercept: $(0, 9)$

continued

49. continued

Step 4: Find another point(s).
Let $x = 1$.
$y = 4(1)^2 - 12(1) + 9 = 4 - 12 + 9 = 1$
$(1, 1)$
Use symmetry to find the points $(2, 1)$ and $(3, 9)$

Step 5: Plots the points and draw the parabola.

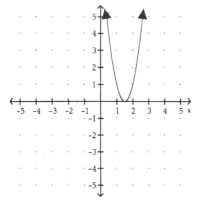

APPLICATIONS

51. BIOLOGY

53. COST ANALYSIS

The ordered pair $(30, 100)$ indicates that the cost to manufacture a carburetor is lowest ($\$100$) for a production run of 30 units.

55. TRAMPOLINE

a) The ordered pair $(0.5, 14)$ indicates that she is **14 feet** above the ground ½ second after bounding upward.
b) The ordered pairs $(0.25, 9)$ and $(1.75, 9)$ indicate that she is 9 feet above the ground at **0.25 seconds** and **1.75 seconds**.
c) The vertex $(1, 18)$ indicates the maximum number of feet above the ground she gets—**18 feet above the ground after 1 second.**

57. BRIDGES

x	-80	-60	-40	-20	0
y	32	18	8	2	0

x	20	40	60	80
y	2	8	18	32

$x = -80$
$y = 0.005(-80)^2$
$\quad = 0.005(6,400)$
$\quad = 32$

$x = -60$
$y = 0.005(-60)^2$
$\quad = 0.005(3,600)$
$\quad = 18$

$x = -40$
$y = 0.005(-40)^2$
$\quad = 0.005(1,600)$
$\quad = 8$

$x = -20$
$y = 0.005(-20)^2$
$\quad = 0.005(400)$
$\quad = 2$

$x = 0$
$y = 0.005(0)^2$
$\quad = 0.005(0)$
$\quad = 0$

$x = 20$
$y = 0.005(20)^2$
$\quad = 0.005(400)$
$\quad = 2$

$x = 40$
$y = 0.005(40)^2$
$\quad = 0.005(1,600)$
$\quad = 8$

$x = 60$
$y = 0.005(60)^2$
$\quad = 0.005(3,600)$
$\quad = 18$

$x = 80$
$y = 0.005(80)^2$
$\quad = 0.005(6,400)$
$\quad = 32$

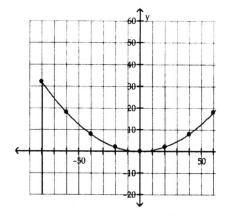

WRITING

59. Answers will vary.

61. Answers will vary.

REVIEW

63.
$$\sqrt{8} - \sqrt{50} + \sqrt{72} = \sqrt{4}\sqrt{2} - \sqrt{25}\sqrt{2} + \sqrt{36}\sqrt{2}$$
$$= 2\sqrt{2} - 5\sqrt{2} + 6\sqrt{2}$$
$$= 3\sqrt{2}$$

65.
$$3\sqrt{z}\left(\sqrt{4z} - \sqrt{z}\right) = 3\sqrt{z} \bullet \sqrt{4z} - 3\sqrt{z} \bullet \sqrt{z}$$
$$= 3\sqrt{4z^2} - 3\sqrt{z^2}$$
$$= 3 \bullet 2z - 3 \bullet z$$
$$= 6z - 3z$$
$$= 3z$$

CHALLENGE PROBLEMS

67. The x-coordinates of the x-intercepts are the solutions: **(-1, 0) and (2, 0).**

SECTION 9.5

VOCABULARY

1. A **function** is a rule that assigns to each value of one variable (called the independent variable) exactly one value of another variable (called the dependent variable).

3. We can think of a function as a machine that takes some **input** x and turns it into some **output** $f(x)$.

5. The graph of a **linear** function is a straight line.

CONCEPTS

7. FEDERAL MINIMUM WAGE

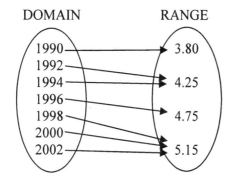

9. $f(-5) = (-5)^2 + 8$
$f(-5) = 25 + 8$
$f(-5) = 33$

11. If a **vertical** line intersects a graph in more than one point, the graph is not the graph of a function.

NOTATION

13. We read $f(x) = 5x - 6$ as "f **of** x is $5x$ minus 6."

15. The notation $f(4) = 5$ indicates that when the x-value **4** is input into a function rule, the output is **5**. This fact can be shown graphically by plotting the ordered pair **(4, 5)**.

PRACTICE

17. Yes

19. No. 4 is assigned to 2, 4, and 6..
(4, 2), (4, 4) and (4, 6).

21. Yes

23. No. -1 is assigned to 0 and 2.
(-1, 0) and (-1, 2)

25. No. 3 is assigned to 4 and -4 and 4 is assigned to 3 and -3.
(3, 4), (3, -4), (4, 3) and (4, -3)

32. $f(x) = 4x - 1$

a) $f(1) = 4(1) - 1$
$f(1) = 4 - 1$
$f(1) = 3$

b) $f(-2) = 4(-2) - 1$
$f(-2) = -8 - 1$
$f(-2) = -9$

c) $f\left(\dfrac{1}{4}\right) = 4\left(\dfrac{1}{4}\right) - 1$
$f\left(\dfrac{1}{4}\right) = 1 - 1$
$f\left(\dfrac{1}{4}\right) = 0$

d) $f(50) = 4(50) - 1$
$f(50) = 200 - 1$
$f(50) = 199$

29. $f(x) = 2x^2$

    a) $f(0.4) = 2(0.4)^2$
       $f(0.4) = 2(0.16)$
       $f(0.4) = 0.32$

    b) $f(-3) = 2(-3)^2$
       $f(-3) = 2(9)$
       $f(-3) = 18$

    c) $f(1,000) = 2(1,000)^2$
       $f(1,000) = 2(1,000,000)$
       $f(1,000) = 2,000,000$

    d) $f\left(\dfrac{1}{8}\right) = 2\left(\dfrac{1}{8}\right)^2$
       $f\left(\dfrac{1}{8}\right) = 2\left(\dfrac{1}{64}\right)$
       $f\left(\dfrac{1}{8}\right) = \dfrac{1}{32}$

31. $h(x) = |x - 7|$

    a) $h(0) = |0 - 7|$
       $h(0) = |-7|$
       $h(0) = 7$

    b) $h(-7) = |-7 - 7|$
       $h(-7) = |-14|$
       $h(-7) = 14$

    c) $h(7) = |7 - 7|$
       $h(7) = |0|$
       $h(7) = 0$

    d) $h(8) = |8 - 7|$
       $h(8) = |1|$
       $h(8) = 1$

33. $g(x) = x^3 - x$

    a) $g(1) = 1^3 - 1$
       $g(1) = 1 - 1$
       $g(1) = 0$

    b) $g(10) = 10^3 - 10$
       $g(10) = 1,000 - 10$
       $g(10) = 990$

    c) $g(-3) = (-3)^3 - (-3)$
       $g(-3) = -27 + 3$
       $g(-3) = -24$

    d) $g(6) = 6^3 - 6$
       $g(6) = 216 - 6$
       $g(6) = 210$

35. $f(-0.3) = 3.4(-0.3)^2 - 1.2(-0.3) + 0.5$
    $f(-0.3) = 3.4(-0.09) - 1.2(-0.3) + 0.5$
    $f(-0.3) = 0.306 + 0.36 + 0.5$
    $f(-0.3) = 1.166$

37. Yes, it is a function.

39. No, it is not a function. Answers will vary. Possible values of x assigned to more than one value of y are (3, 4) and (3, -1).

41. No, it is not a function. Answers will vary. Possible values of x assigned to more than one value of y are (0, 2) and (0, -4).

43. $f(x) = -2 - 3x$

x	$f(x)$
0	$-2 - 3(0) = -2 - 0$ $= \mathbf{-2}$
1	$-2 - 3(1) = -2 - 3$ $= \mathbf{-5}$
-1	$-2 - 3(-1) = -2 + 3$ $= \mathbf{1}$
-2	$-2 - 3(-2) = -2 + 6$ $= \mathbf{4}$

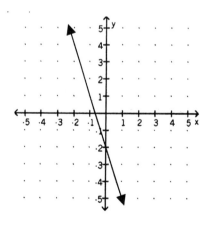

Section 9.5

45. $f(x) = \dfrac{1}{2}x - 2$

x	$f(x)$
0	$\dfrac{1}{2}(0) - 2 = 0 - 2$ $= -2$
2	$\dfrac{1}{2}(2) - 2 = 1 - 2$ $= -1$
4	$\dfrac{1}{2}(4) - 2 = 2 - 2$ $= 0$
-2	$\dfrac{1}{2}(-2) - 2 = -1 - 2$ $= -3$

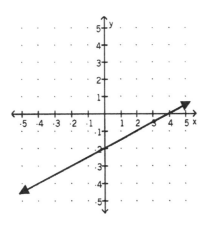

47. $g(x) = 2 - x^2$

x	$g(x)$
0	$2 - 0^2 = 2 - 0$ $= \mathbf{2}$
1	$2 - 1^2 = 2 - 1$ $= \mathbf{1}$
2	$2 - 2^2 = 2 - 4$ $= \mathbf{-2}$
-1	$2 - (-1)^2 = 2 - 1$ $= \mathbf{1}$
-2	$2 - (-2)^2 = 2 - 4$ $= \mathbf{-2}$

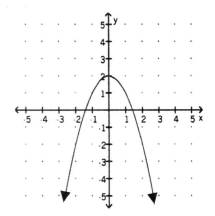

APPLICATIONS

49. REFLECTIONS

$f(x) = |x|$

51. LIGHTING

$D(5) = \dfrac{5}{5}$

$D(5) = 1$

If it is 5 seconds between seeing the lightning and hearing the thunder, the lightning strike was 1 mile away.

53. LAWN SPRINKERS

$A(5) \approx (3.14159)(5)^2$
$A(5) \approx (3.14159)(25)$
$A(5) \approx 78.5 \text{ ft}^2$

$A(20) \approx (3.14159)(20)^2$
$A(20) \approx (3.14159)(400)$
$A(20) \approx 1,256.6 \text{ ft}^2$

WRITING

55. Answers will vary.

57. Answers will vary.

REVIEW

59. COFFEE BLENDS

Let x = the number of pounds of
regular coffee he uses

	regular coffee	gourmet coffee	blend
# of pounds	x	40	$x + 40$
value	$4x$	$7(40)$	$5(x + 40)$

$$4x + 7(40) = 5(x + 40)$$
$$4x + 280 = 5x + 200$$
$$4x + 280 - 4x = 5x + 200 - 4x$$
$$280 = x + 200$$
$$280 - 200 = x + 200 - 200$$
$$80 = x$$

80 lbs of regular coffee is used.

CHALLENGE PROBLEMS

61. No. The graph contains many values of x
that are assigned more than one value of y,
such as (2, 1) and (2, 2)

VOCABULARY

1. The equation $y = kx$ defines **direct** variation.

3. In $y = kx$, the **constant** of variation is k.

5. The equation $y = kxz$ represents **joint** variation.

CONCEPTS

7. direction variation

9. inverse variation

11.

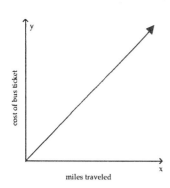

13. $d = kn$

15. $l = \dfrac{k}{w}$

17. $A = kr^2$

19. $d = kst$

21. $I = \dfrac{kV}{R}$

NOTATION

23. a) yes
    b) no
    c) no
    d) yes

25. $y = kx$;
    Find k:
    $10 = k(2)$
    $10 = 2k$
    $\dfrac{10}{2} = \dfrac{2k}{2}$
    $5 = k$

    Find y if $x = 7$ and $k = 5$.
    $y = 5 \cdot 7$
    $\boxed{y = 35}$

27. $r = ks$;
    Find k:
    $21 = k(6)$
    $21 = 6k$
    $\dfrac{21}{6} = \dfrac{6k}{6}$
    $\dfrac{7}{2} = k$

    Find r if $s = 12$ and $k = \dfrac{7}{2}$.

    $r = \dfrac{7}{2} \cdot 12$
    $\boxed{r = 42}$

29. $s = kt^2$;
    Find k:
    $12 = k(4)^2$
    $12 = k \cdot 16$
    $\dfrac{12}{16} = \dfrac{16k}{16}$
    $\dfrac{3}{4} = k$

    Find s if $t = 30$ and $k = \dfrac{3}{4}$.

    $s = \dfrac{3}{4} \cdot (30)^2$
    $s = \dfrac{3}{4} \cdot 900$
    $\boxed{s = 675}$

31. $y = \dfrac{k}{x}$

Find k:

$$8 = \frac{k}{1}$$

$$1 \bullet 8 = \frac{k}{1} \bullet 1$$

$$8 = k$$

Find y if $x = 8$ and $k = 8$.

$$y = \frac{8}{8}$$

$$\boxed{y = 1}$$

33. $r = \dfrac{k}{s}$

Find k:

$$40 = \frac{k}{10}$$

$$10 \bullet 40 = \frac{k}{10} \bullet 10$$

$$400 = k$$

Find r if $s = 15$ and $k = 400$.

$$r = \frac{400}{15}$$

$$\boxed{r = \frac{80}{3}}$$

35. $y = \dfrac{k}{x^2}$

Find k:

$$6 = \frac{k}{4^2}$$

$$6 = \frac{k}{16}$$

$$16 \bullet 6 = \frac{k}{16} \bullet 16$$

$$96 = k$$

Find y if $x = 2$ and $k = 96$.

$$y = \frac{96}{2^2}$$

$$y = \frac{96}{4}$$

$$\boxed{y = 24}$$

37. $y = krs$

Find k:

$$4 = k(2)(6)$$

$$4 = 12k$$

$$\frac{4}{12} = \frac{12k}{12}$$

$$\frac{1}{3} = k$$

Find y if $r = 3$, $s = 4$ and $k = \dfrac{1}{3}$.

$$y = \frac{1}{3}(3)(4)$$

$$y = 1(4)$$

$$\boxed{y = 4}$$

39. $D = kpq$

Find k:

$$20 = k(5)(5)$$

$$20 = 25k$$

$$\frac{20}{25} = \frac{25k}{25}$$

$$\frac{4}{5} = k$$

Find D if $p = 10$, $q = 10$ and $k = \dfrac{4}{5}$.

$$D = \frac{4}{5}(10)(10)$$

$$D = 8(10)$$

$$\boxed{D = 80}$$

41. $y = \dfrac{ka}{b}$

Find k.

$$1 = \frac{k(2)}{10}$$

$$1 = \frac{1}{5}k$$

$$\frac{5}{1} \bullet 1 = \frac{1}{5}k \bullet \frac{5}{1}$$

$$5 = k$$

Find y if $a = 7$, $b = 14$, and $k = 5$.

$$y = \frac{5(7)}{14}$$

$$y = \frac{35}{14}$$

$$\boxed{y = \frac{5}{2}}$$

APPLICATIONS

43. COMMUTING DISTANCE

Let d = distance and g = gallons of gas
Then, $d = kg$
Find k.

$360 = 15k$

$\dfrac{360}{15} = \dfrac{15k}{15}$

$24 = k$

Find d if $g = 7$ and $k = 24$.

$d = 24(7)$

$\boxed{d = 168 \text{ miles}}$

45. DOSAGE

Let d = dosage and w = weight
Then, $d = kw$
Find k.

$18 = 30k$

$\dfrac{18}{30} = \dfrac{30k}{30}$

$\dfrac{3}{5} = k$

Find d if $w = 45$ and $k = \dfrac{3}{5}$.

$d = \dfrac{3}{5}(45)$

$\boxed{d = 27 \text{ mg}}$

47. CIDER

Let i = inches of stick cinnamon and
s = number of servings.
Then, $i = ks$
Find k.

$6 = 8k$

$\dfrac{6}{8} = \dfrac{8k}{8}$

$\dfrac{3}{4} = k$

Find i if $s = 30$ and $k = \dfrac{3}{4}$.

$i = \dfrac{3}{4}(30)$

$\boxed{i = 22.5 \text{ inches}}$

49. COMMUTING TIME

Let t = time and r = rate
Then, $t = \dfrac{k}{r}$
Find k.

$3 = \dfrac{k}{50}$

$50 \bullet 3 = \dfrac{k}{50} \bullet 50$

$150 = k$

Find t if $r = 60$ and $k = 150$.

$t = \dfrac{150}{60}$

$\boxed{t = 2.5 \text{ hours}}$

51. ELECTRICTY

Let c = current and r = resistance.
Then, $c = \dfrac{k}{r}$
Find k.

$30 = \dfrac{k}{4}$

$4 \bullet 30 = \dfrac{k}{4} \bullet 4$

$120 = k$

Find c if $r = 15$ and $k = 120$.

$c = \dfrac{120}{15}$

$\boxed{c = 8 \text{ amps}}$

53. COMPUTING PRESSURES

Let v = volume and p = pressure

Then, $v = \dfrac{k}{p}$

Find k.

$$40 = \frac{k}{8}$$

$$8 \cdot 40 = \frac{k}{8} \cdot 8$$

$$320 = k$$

Find v if $p = 6$ and $k = 320$.

$$v = \frac{320}{6}$$

$$\boxed{v = 55\frac{1}{3}\,\text{m}^3}$$

55. COMPUTING INTEREST

Let I = interest earned, a = annual interest rate, and t = time left on deposit.

$I = kat$

Find k.

$$120 = k(0.08)(2)$$

$$120 = k(0.16)$$

$$\frac{120}{0.16} = \frac{0.16k}{0.16}$$

$$750 = k$$

Find I if $a = 0.12$, $t = 3$, and $k = 750$

$$I = (750)(0.12)(3)$$

$$I = 90(3)$$

$$\boxed{I = \$270}$$

57. ELECTRONICS

Let c = current, v = voltage, and r = resistance.

$$c = \frac{kv}{r}$$

Find k.

$$4 = \frac{k \cdot 36}{9}$$

$$4 = 4k$$

$$\frac{4}{4} = \frac{4k}{4}$$

$$1 = k$$

Find c if $v = 42$, $r = 11$, and $k = 1$.

$$c = \frac{1 \cdot 42}{11}$$

$$c = \frac{42}{11}$$

$$\boxed{c = 3\frac{9}{11}\,\text{amps}}$$

WRITING

59. Answers will vary.

61. Answers will vary.

REVIEW

63. Solve by factoring.

$$x^2 - 5x - 6 = 0$$

$$(x - 6)(x + 1) = 0$$

$x - 6 = 0$ or $x + 1 = 0$

$x - 6 + 6 = 0 + 6$ $x + 1 - 1 = 0 - 1$

$\boxed{x = 6 \qquad\text{or}\qquad x = \text{-}1}$

65. Solve by factoring.

$$2(y - 4) = \text{-}y^2$$

$$2y - 8 = \text{-}y^2$$

$$2y - 8 + y^2 = \text{-}y^2 + y^2$$

$$y^2 + 2y - 8 = 0$$

$$(y + 4)(y - 2) = 0$$

$y + 4 = 0$ or $y - 2 = 0$

$y + 4 - 4 = 0 - 4$ $y - 2 + 2 = 0 + 2$

$y = \text{-}4$ or $y = 2$

67. Solve by factoring.

$$5a^3 - 125a = 0$$
$$5a(a^2 - 25) = 0$$
$$5a(a - 5)(a + 5) = 0$$

$5a = 0$ or $a - 5 = 0$ or $a + 5 = 0$

$$\frac{5a}{5} = \frac{0}{5} \qquad a - 5 + 5 = 0 + 5 \qquad a + 5 - 5 = 0 - 5$$

$a = 0$	$a = 5$	$a = -5$

CHALLENGE PROBLEMS

69. GRAVITY

F = earth's force, d = distance from center of the earth

$$F = \frac{k}{d^2}$$

$$80 = \frac{k}{4,000^2}$$

$$80 = \frac{k}{16,000,000}$$

$$16,000,000 \bullet 80 = \frac{k}{16,000,000} \bullet 16,000,000$$

$$1,280,000,000 = k$$

Find F if $d = 5,000$ and $k = 1,280,000,000$

$$F = \frac{1,280,000,000}{5,000^2}$$

$$F = \frac{1,280,000,000}{25,000,000}$$

$$\boxed{F = 51.2 \text{ pounds}}$$

1. $4x^2 - x = 0$
    $x(4x - 1) = 0$
    $x = 0$ or $4x - 1 = 0$
                              $4x = 1$
                              $x = \dfrac{1}{4}$

    $x = 0, \dfrac{1}{4}$

2. $a^2 - a - 56 = 0$
    $(a - 8)(a + 7) = 0$
    $a - 8 = 0$ or $a + 7 = 0$
    $a = 8$ $a = \text{-}7$

    $a = 8, \text{-}7$

3. $x^2 = 27$
    $x = \sqrt{27}$ or $x = -\sqrt{27}$
    $x = \sqrt{9}\sqrt{3}$ or $x = -\sqrt{9}\sqrt{3}$
    $x = 3\sqrt{3}$ or $x = -3\sqrt{3}$
    $x = \pm\, 3\sqrt{3}$

4. $(x + 3)^2 = 16$
    $x + 3 = \sqrt{16}$ or $x + 3 = -\sqrt{16}$
    $x + 3 = 4$ $x + 3 = \text{-}4$
    $x = 4 - 3$ $x = \text{-}4 - 3$
    $x = 1$ $x = \text{-}7$
    $x = 1, \text{-}7$

5. $x^2 + 4x + 1 = 0$
    $x^2 + 4x + 1 - 1 = 0 - 1$
    $x^2 + 4x = \text{-}1$
    $x^2 + 4x + \left(\dfrac{4}{2}\right)^2 = \text{-}1 + \left(\dfrac{4}{2}\right)^2$
    $x^2 + 4x + 4 = \text{-}1 + 4$
    $(x + 2)^2 = 3$
    $x + 2 = \sqrt{3}$ or $x + 2 = -\sqrt{3}$
    $x = \text{-}2 + \sqrt{3}$ $x = \text{-}2 - \sqrt{3}$
    $x = \text{-}2 \pm \sqrt{3}$

6. $x^2 + 3x + 1 = 0$
    $a = 1, b = 3, c = 1$
    $x = \dfrac{-b \pm \sqrt{b^2 - 4ac}}{2a}$
    $= \dfrac{-3 \pm \sqrt{3^2 - 4(1)(1)}}{2(1)}$
    $= \dfrac{-3 \pm \sqrt{9 - 4}}{2}$
    $x = \dfrac{-3 \pm \sqrt{5}}{2}$

7. A $\pm$ sign should be written in front of $\sqrt{20}$ and $2\sqrt{5}$.
    $x^2 = 20$
    $x = \sqrt{20}$ or $x = -\sqrt{20}$
    $x = 2\sqrt{5}$ or $x = -2\sqrt{5}$
    $x = \pm 2\sqrt{5}$

8. Write $x^2 - 3x = 10$ as $x^2 - 3x - 10 = 0$.
    $a = 1, b = 3, c = 1$
    $x = \dfrac{-b \pm \sqrt{b^2 - 4ac}}{2a}$
    $= \dfrac{-(-3) \pm \sqrt{(-3)^2 - 4(1)(-10)}}{2(1)}$
    $= \dfrac{3 \pm \sqrt{9 - (-40)}}{2}$
    $= \dfrac{3 \pm \sqrt{49}}{2}$
    $= \dfrac{3 \pm 7}{2}$
    $x = \dfrac{3 + 7}{2} = \dfrac{10}{2} = 5$
    $x = \dfrac{3 - 7}{2} = \dfrac{-4}{2} = \text{-}2$
    $x = 5, \text{-}2$

9. 9 should be added to both sides.

$x^2 + 6x - 10 = 0$

$x^2 + 6x - 10 + 10 = 0 + 10$

$x^2 + 6x = 10$

$x^2 + 6x + \left(\dfrac{6}{2}\right)^2 = 10 + \left(\dfrac{6}{2}\right)^2$

$x^2 + 6x + 9 = 10 + 9$

$(x + 3)^2 = 19$

$x + 3 = \sqrt{19}$ or $x + 3 = -\sqrt{19}$

$x = -3 + \sqrt{19}$ $x = -3 - \sqrt{19}$

$x = -3 \pm \sqrt{19}$

10. The fraction bar should be extended.

$x^2 + 5x + 1 = 0$

$a = 1, b = 5, c = 1$

$x = \dfrac{-b \pm \sqrt{b^2 - 4ac}}{2a}$

$= \dfrac{-5 \pm \sqrt{5^2 - 4(1)(1)}}{2(1)}$

$= \dfrac{-5 \pm \sqrt{25 - 4}}{2}$

$x = \dfrac{-5 \pm \sqrt{21}}{2}$

SECTION 9.1
The Square Root Property; Completing the Square

1. $x^2 = 25$
   $x = \sqrt{25}$ or $x = -\sqrt{25}$
   $x = 5$ $\qquad x = -5$
   $\boxed{x = \pm 5}$

2. $t^2 = 8$
   $t = \sqrt{8}$ or $t = -\sqrt{8}$
   $t = \sqrt{4}\sqrt{2}$ $\qquad t = -\sqrt{4}\sqrt{2}$
   $t = 2\sqrt{2}$ $\qquad t = -2\sqrt{2}$
   $\boxed{t = \pm 2\sqrt{2}}$

3. $2x^2 - 1 = 149$
   $2x^2 - 1 + 1 = 149 + 1$
   $2x^2 = 150$
   $\dfrac{2x^2}{2} = \dfrac{150}{2}$
   $x^2 = 75$
   $x = \sqrt{75}$ or $x = -\sqrt{75}$
   $x = \sqrt{25}\sqrt{3}$ $\qquad x = -\sqrt{25}\sqrt{3}$
   $x = 5\sqrt{3}$ $\qquad x = -5\sqrt{3}$
   $\boxed{x = \pm 5\sqrt{3}}$

4. $(x - 1)^2 = 25$
   $x - 1 = \sqrt{25}$ or $x - 1 = -\sqrt{25}$
   $x - 1 = 5$ $\qquad x - 1 = -5$
   $x - 1 + 1 = 5 + 1$ $\quad x - 1 + 1 = -5 + 1$
   $x = 6$ $\qquad x = -4$
   $\boxed{x = 6, -4}$

5. $4(x - 2)^2 = 9$
   $\dfrac{4(x - 2)^2}{4} = \dfrac{9}{4}$
   $(x - 2)^2 = \dfrac{9}{4}$
   $x - 2 = \sqrt{\dfrac{9}{4}}$ or $x - 2 = -\sqrt{\dfrac{9}{4}}$
   $x - 2 = \dfrac{3}{2}$ $\qquad x - 2 = -\dfrac{3}{2}$
   $x - 2 + 2 = \dfrac{3}{2} + 2$ $\qquad x - 2 + 2 = -\dfrac{3}{2} + 2$
   $x = 2 + \dfrac{3}{2}$ $\qquad x = 2 - \dfrac{3}{2}$
   $x = \dfrac{4}{2} + \dfrac{3}{2}$ $\qquad x = \dfrac{4}{2} - \dfrac{3}{2}$
   $x = \dfrac{7}{2}$ $\qquad x = \dfrac{1}{2}$
   $\boxed{x = \dfrac{7}{2}, \dfrac{1}{2}}$

6. $(x - 8)^2 = 40$
   $x - 8 = \sqrt{40}$ or $x - 8 = -\sqrt{40}$
   $x - 8 = \sqrt{4}\sqrt{10}$ $\qquad x - 8 = -\sqrt{4}\sqrt{10}$
   $x - 8 = 2\sqrt{10}$ $\qquad x - 8 = -2\sqrt{10}$
   $x - 8 + 8 = 2\sqrt{10} + 8$ $\quad x - 8 + 8 = -2\sqrt{10} + 8$
   $x = 8 + 2\sqrt{10}$ $\qquad x = 8 - 2\sqrt{10}$
   $\boxed{x = 8 \pm 2\sqrt{10}}$

7. $x^2 = 12$
   $x = \sqrt{12}$ or $x = -\sqrt{12}$
   $x \approx 3.46$ $\qquad x \approx -3.46$
   $\boxed{x \approx \pm 3.46}$

8. $(x - 1)^2 = 55$
   $x - 1 = \sqrt{55}$ or $x - 1 = -\sqrt{55}$
   $x - 1 \approx 7.42$ $\qquad x - 1 \approx -7.42$
   $x - 1 + 1 \approx 7.42 + 1$ $\quad x - 1 + 1 \approx -7.42 + 1$
   $x \approx 1 + 7.42$ $\qquad x \approx 1 - 7.42$
   $x \approx 8.42$ $\qquad x \approx -6.42$
   $\boxed{x \approx 8.42, -6.42}$

9. $x^2 + 4x$

$$x^2 + 4x + \left(\frac{4}{2}\right)^2 = x^2 + 4x + 4$$
$$= (x + 2)^2$$

10. $t^2 - 5t$

$$t^2 - 5t + \left(\frac{5}{2}\right)^2 = t^2 - 5t + \frac{25}{4}$$
$$= \left(t - \frac{5}{2}\right)^2$$

11. $x^2 - 8x + 15 = 0$

$$x^2 - 8x + 15 - 15 = 0 - 15$$
$$x^2 - 8x = -15$$
$$x^2 - 8x + \left(\frac{8}{2}\right)^2 = -15 + \left(\frac{8}{2}\right)^2$$
$$x^2 - 8x + 16 = -15 + 16$$
$$(x - 4)^2 = 1$$

$x - 4 = \sqrt{1}$ or $x - 4 = -\sqrt{1}$

$x - 4 = 1$ $x - 4 = -1$

$x - 4 + 4 = 1 + 4$ $x - 4 + 4 = -1 + 4$

$x = 1 + 4$ $x = -1 + 4$

$x = 5$ $x = 3$

$\boxed{x = 5, 3}$

12. $x^2 = -5x + 14$

$$x^2 + 5x = -5x + 14 + 5x$$
$$x^2 + 5x = 14$$
$$x^2 + 5x + \left(\frac{5}{2}\right)^2 = 14 + \left(\frac{5}{2}\right)^2$$
$$x^2 + 5x + \frac{25}{4} = 14 + \frac{25}{4}$$
$$\left(x + \frac{5}{2}\right)^2 = \frac{81}{4}$$

$x + \dfrac{5}{2} = \sqrt{\dfrac{81}{4}}$ or $x + \dfrac{5}{2} = -\sqrt{\dfrac{81}{4}}$

$x + \dfrac{5}{2} = \dfrac{9}{2}$ $x + \dfrac{5}{2} = -\dfrac{9}{2}$

$x + \dfrac{5}{2} - \dfrac{5}{2} = \dfrac{9}{2} - \dfrac{5}{2}$ $x + \dfrac{5}{2} - \dfrac{5}{2} = -\dfrac{9}{2} - \dfrac{5}{2}$

$x = -\dfrac{5}{2} + \dfrac{9}{2}$ $x = -\dfrac{5}{2} - \dfrac{9}{2}$

$x = \dfrac{4}{2}$ $x = -\dfrac{14}{2}$

$x = 2$ $x = -7$

$\boxed{x = 2, -7}$

13. $2x^2 + 5x = 3$

$$\frac{2x^2}{2} + \frac{5x}{2} = \frac{3}{2}$$
$$x^2 + \frac{5}{2}x - \frac{3}{2}$$
$$x^2 + \frac{5}{2}x + \left(\frac{1}{2} \bullet \frac{5}{2}\right)^2 = \frac{3}{2} + \left(\frac{1}{2} \bullet \frac{5}{2}\right)^2$$
$$x^2 + \frac{5}{2}x + \left(\frac{5}{4}\right)^2 = \frac{3}{2} + \left(\frac{5}{4}\right)^2$$
$$x^2 + \frac{5}{2}x + \frac{25}{16} = \frac{3}{2} + \frac{25}{16}$$
$$\left(x + \frac{5}{4}\right)^2 = \frac{49}{16}$$

$x + \dfrac{5}{4} = \sqrt{\dfrac{49}{16}}$ or $x + \dfrac{5}{4} = -\sqrt{\dfrac{49}{16}}$

$x + \dfrac{5}{4} = \dfrac{7}{4}$ $x + \dfrac{5}{4} = -\dfrac{7}{4}$

$x + \dfrac{5}{4} - \dfrac{5}{4} = \dfrac{7}{4} - \dfrac{5}{4}$ $x + \dfrac{5}{4} - \dfrac{5}{4} = -\dfrac{7}{4} - \dfrac{5}{4}$

$x = -\dfrac{5}{4} + \dfrac{7}{4}$ $x = -\dfrac{5}{4} - \dfrac{7}{4}$

$x = \dfrac{2}{4}$ $x = -\dfrac{12}{4}$

$\boxed{x = \dfrac{1}{2}, \quad -3}$

14.
$$x^2 - 2x - 1 = 0$$
$$2x^2 - 2x - 1 + 1 = 0 + 1$$
$$2x^2 - 2x = 1$$
$$\frac{2x^2}{2} - \frac{2x}{2} = \frac{1}{2}$$
$$x^2 - x = \frac{1}{2}$$
$$x^2 - x + \left(\frac{1}{2}\right)^2 = \frac{1}{2} + \left(\frac{1}{2}\right)^2$$
$$x^2 - x + \frac{1}{4} = \frac{1}{2} + \frac{1}{4}$$
$$\left(x - \frac{1}{2}\right)^2 = \frac{3}{4}$$

$$x - \frac{1}{2} = \sqrt{\frac{3}{4}} \quad \text{or} \quad x - \frac{1}{2} = -\sqrt{\frac{3}{4}}$$

$$x - \frac{1}{2} = \frac{\sqrt{3}}{2} \qquad\qquad x - \frac{1}{2} = -\frac{\sqrt{3}}{2}$$

$$x - \frac{1}{2} + \frac{1}{2} = \frac{\sqrt{3}}{2} + \frac{1}{2} \quad x - \frac{1}{2} + \frac{1}{2} = -\frac{\sqrt{3}}{2} + \frac{1}{2}$$

$$x = \frac{1}{2} + \frac{\sqrt{3}}{2} \qquad\qquad x = \frac{1}{2} - \frac{\sqrt{3}}{2}$$

$$x = \frac{1}{2} \pm \frac{\sqrt{3}}{2}$$

$$\boxed{x = \frac{1 \pm \sqrt{3}}{2}}$$

15.
$$x^2 + 4x + 1 = 0$$
$$x^2 + 4x + 1 - 1 = 0 - 1$$
$$x^2 + 4x = -1$$
$$x^2 + 4x + \left(\frac{4}{2}\right)^2 = -1 + \left(\frac{4}{2}\right)^2$$
$$x^2 + 4x + 4 = -1 + 4$$
$$(x + 2)^2 = 3$$

$$x + 2 = \sqrt{3} \quad \text{or} \quad x + 2 = -\sqrt{3}$$
$$x + 2 - 2 = \sqrt{3} \qquad x + 2 - 2 = -\sqrt{3} - 2$$
$$x = -2 + \sqrt{3} \qquad\quad x = -2 - \sqrt{3}$$
$$x \approx -2 + 1.73 \qquad\quad x \approx -2 - 1.73$$
$$x \approx -0.27 \qquad\qquad x \approx -3.73$$
$$\boxed{x \approx -0.27, -3.73}$$

16. PLAYGROUND EQUIPMENT

$$A = \pi r^2$$
$$28.3 \approx 3.14 r^2$$
$$\frac{28.3}{3.14} \approx \frac{3.14 r^2}{3.14}$$
$$9.01 \approx r^2$$
$$r \approx \sqrt{9.01} \qquad \text{or} \qquad r = -\sqrt{9.01}$$
$$r \approx 3.0 \qquad\quad \text{or} \qquad r \approx -3.0$$
The radius cannot be negative.
$$\boxed{r \approx 3.0 \text{ ft}}$$

SECTION 9.2
The Quadratic Formula

17.
$$x^2 + 2x = -5$$
$$x^2 + 2x + 5 = -5 + 5$$
$$x^2 + 2x + 5 = 0$$
$$a = 1, b = 2, c = 5$$

18.
$$6x^2 = 2x + 1$$
$$6x^2 - 2x - 1 = 2x + 1 - 2x - 1$$
$$6x^2 - 2x - 1 = 0$$
$$a = 6, b = -2, c = -1$$

19. $x^2 - 2x - 15 = 0$
$$a = 1, b = -2, c = -15$$
$$x = \frac{-b \pm \sqrt{b^2 - 4ac}}{2a}$$
$$= \frac{-(-2) \pm \sqrt{(-2)^2 - 4(1)(-15)}}{2(1)}$$
$$= \frac{2 \pm \sqrt{4 - (-60)}}{2}$$
$$= \frac{2 \pm \sqrt{64}}{2}$$
$$= \frac{2 \pm 8}{2}$$
$$x = \frac{2 + 8}{2} = \frac{10}{2} = 5$$
$$x = \frac{2 - 8}{2} = \frac{-6}{2} = -3$$
$$\boxed{x = 5, -3}$$

20. $x^2 - 6x = 7$
$x^2 - 6x - 7 = 7 - 7$
$x^2 - 6x - 7 = 0$
$a = 1, b = -6, c = -7$

$$x = \frac{-b \pm \sqrt{b^2 - 4ac}}{2a}$$

$$= \frac{-(-6) \pm \sqrt{(-6)^2 - 4(1)(-7)}}{2(1)}$$

$$= \frac{6 \pm \sqrt{36 - (-28)}}{2}$$

$$= \frac{6 \pm \sqrt{64}}{2}$$

$$= \frac{6 \pm 8}{2}$$

$$x = \frac{6 + 8}{2} = \frac{14}{2} = 7$$

$$x = \frac{6 - 8}{2} = \frac{-2}{2} = -1$$

$$\boxed{x = 7, \, -1}$$

21. $6x^2 = 7x + 3$
$6x^2 - 7x - 3 = 7x + 3 - 7x - 3$
$6x^2 - 7x - 3 = 0$
$a = 6, b = -7, c = -3$

$$x = \frac{-b \pm \sqrt{b^2 - 4ac}}{2a}$$

$$= \frac{-(-7) \pm \sqrt{(-7)^2 - 4(6)(-3)}}{2(6)}$$

$$= \frac{7 \pm \sqrt{49 - (-72)}}{12}$$

$$= \frac{7 \pm \sqrt{121}}{12}$$

$$= \frac{7 \pm 11}{12}$$

$$x = \frac{7 + 11}{12} = \frac{18}{12} = \frac{3}{2}$$

$$x = \frac{7 - 11}{12} = \frac{-4}{12} = -\frac{1}{3}$$

$$\boxed{x = \frac{3}{2}, \, -\frac{1}{3}}$$

22. $x^2 - 6x + 7 = 0$
$a = 1, b = -6, c = 7$

$$x = \frac{-b \pm \sqrt{b^2 - 4ac}}{2a}$$

$$= \frac{-(-6) \pm \sqrt{(-6)^2 - 4(1)(7)}}{2(1)}$$

$$= \frac{6 \pm \sqrt{36 - 28}}{2}$$

$$= \frac{6 \pm \sqrt{8}}{2}$$

$$= \frac{6 \pm \sqrt{4}\sqrt{2}}{2}$$

$$= \frac{6 \pm 2\sqrt{2}}{2}$$

$$= \frac{2(3 \pm \sqrt{2})}{2}$$

$$\boxed{x = 3 \pm \sqrt{2}}$$

23. $3x^2 + 2x - 2 = 0$
$a = 3, b = 2, c = -2$

$$x = \frac{-b \pm \sqrt{b^2 - 4ac}}{2a}$$

$$= \frac{-2 \pm \sqrt{(2)^2 - 4(3)(-2)}}{2(3)}$$

$$= \frac{-2 \pm \sqrt{4 - (-24)}}{6}$$

$$= \frac{-2 \pm \sqrt{28}}{6}$$

$$= \frac{-2 \pm \sqrt{4}\sqrt{7}}{6}$$

$$= \frac{-2 \pm 2\sqrt{7}}{6}$$

$$= \frac{2(-1 \pm \sqrt{7})}{2 \cdot 3}$$

$$x = \frac{-1 \pm \sqrt{7}}{3}$$

$$x \approx \frac{-1 + 2.65}{3} \approx \frac{1.65}{3} \approx 0.55$$

$$x \approx \frac{-1 - 2.65}{3} \approx \frac{-3.65}{3} \approx -1.22$$

$$\boxed{x \approx -1.22, \, 0.55}$$

24. SECURITY GATE

Use the Pythagorean Theorem.
$$w^2 + (w + 14)^2 = 26^2$$
$$w^2 + w^2 + 28w + 196 = 676$$
$$2w^2 + 28w + 196 = 676$$
$$2w^2 + 28w + 196 - 676 = 676 - 676$$
$$2w^2 + 28w - 480 = 0$$
$$a = 2, b = 28, c = -480$$
$$x = \frac{-b \pm \sqrt{b^2 - 4ac}}{2a}$$
$$= \frac{-28 \pm \sqrt{(28)^2 - 4(2)(-480)}}{2(2)}$$
$$= \frac{-28 \pm \sqrt{784 - (-3,840)}}{4}$$
$$= \frac{-28 \pm \sqrt{4,624}}{4}$$
$$= \frac{-28 \pm 68}{4}$$
$$x = \frac{-28 + 68}{4} = \frac{40}{4} = 10$$
$$x = \frac{-28 - 68}{4} = \frac{-96}{4} = -24$$

Since the width cannot be negative, $w = 10$
and $w + 14 = 10 + 14 = 24$
$$\boxed{10 \text{ ft and } 24 \text{ ft}}$$

25. MILITARY

Let $h = 0$.
$$0 = 3,000 + 40t - 16t^2$$
$$0 - 3,000 - 40t + 16t^2 = 3,000 + 40t - 16t^2 - 3,000 - 40t + 16t^2$$
$$16t^2 - 40t - 3,000 = 0$$
$$a = 16, b = -40, c = -3,000$$
$$t = \frac{-b \pm \sqrt{b^2 - 4ac}}{2a}$$
$$= \frac{-(-40) \pm \sqrt{(-40)^2 - 4(16)(-3,000)}}{2(16)}$$
$$= \frac{40 \pm \sqrt{1,600 - (-192,000)}}{32}$$
$$= \frac{40 \pm \sqrt{193,600}}{32}$$
$$= \frac{40 \pm 440}{32}$$
$$t = \frac{40 + 440}{32} = \frac{480}{32} = 15$$
$$t = \frac{40 - 440}{32} = \frac{-400}{32} = -\frac{25}{2}$$

Since t represents the time after the rocket
is released, t cannot be negative.
$$\boxed{t = 15 \text{ seconds}}$$

26.

$$x^2 + 6x + 2 = 0$$
$$a = 1, b = 6, c = 2$$
$$x = \frac{-b \pm \sqrt{b^2 - 4ac}}{2a}$$
$$= \frac{-6 \pm \sqrt{(6)^2 - 4(1)(2)}}{2(1)}$$
$$= \frac{-6 \pm \sqrt{36 - 8}}{2}$$
$$= \frac{-6 \pm \sqrt{28}}{2}$$
$$= \frac{-6 \pm \sqrt{4}\sqrt{7}}{2}$$
$$= \frac{-6 \pm 2\sqrt{7}}{2}$$
$$= \frac{2(-3 \pm \sqrt{7})}{2}$$
$$\boxed{x = -3 \pm \sqrt{7}}$$

27.

$$(y + 3)^2 = 16$$

$y + 3 = \sqrt{16}$ or $y + 3 = -\sqrt{16}$
$y + 3 = 4$ $y + 3 = -4$
$y + 3 - 3 = 4 - 3$ $y + 3 - 3 = -4 - 3$
$y = 4 - 3$ $y = -4 - 3$
$y = 1$ $y = -7$
$$\boxed{y = 1, -7}$$

28.

$$x^2 + 5x = 0$$
$$x(x + 5) = 0$$

$x = 0$ or $x + 5 = 0$
                        $x + 5 - 5 = 0 - 5$
                        $x = -5$
$$\boxed{x = 0, -5}$$

29.

$$2x^2 + x = 5$$
$$2x^2 + x - 5 = 5 - 5$$
$$2x^2 + x - 5 = 0$$
$$a = 2, b = 1, c = -5$$
$$x = \frac{-b \pm \sqrt{b^2 - 4ac}}{2a}$$
$$= \frac{-1 \pm \sqrt{(1)^2 - 4(2)(-5)}}{2(2)}$$
$$= \frac{-1 \pm \sqrt{1 - (-40)}}{4}$$
$$\boxed{x = \frac{-1 \pm \sqrt{41}}{4}}$$

30.
$$g^2 - 20 = 0$$
$$g^2 - 20 + 20 = 0 + 20$$
$$g^2 = 20$$

$g = \sqrt{20}$ or $g = -\sqrt{20}$

$g = \sqrt{4}\sqrt{5}$ or $g = -\sqrt{4}\sqrt{5}$

$g = 2\sqrt{5}$ or $g = -2\sqrt{5}$

$$\boxed{g = \pm 2\sqrt{5}}$$

31.
$$a^2 = 4a - 4$$
$$a^2 - 4a + 4 = 4a - 4 - 4a + 4$$
$$a^2 - 4a + 4 = 0$$
$$(a-2)(a-2) = 0$$

$a - 2 = 0$ or $a - 2 = 0$

$a - 2 + 2 = 0 + 2$ $a - 2 + 2 = 0 + 2$

$a = 2$ $a = 2$

$$\boxed{a = 2, 2}$$

32. $a^2 - 2a + 5 = 0$

$a = 1, b = -2, c = 5$

$$x = \frac{-b \pm \sqrt{b^2 - 4ac}}{2a}$$

$$= \frac{-(-2) \pm \sqrt{(-2)^2 - 4(1)(5)}}{2(1)}$$

$$= \frac{2 \pm \sqrt{4 - 20}}{2}$$

$$= \frac{2 \pm \sqrt{-16}}{2}$$

No Real Solutions since $\sqrt{-16}$ is not real.

SECTION 9.3
Complex Numbers

33.
$$\sqrt{-25} = \sqrt{25}\sqrt{-1}$$
$$= 5i$$

34.
$$\sqrt{-18} = \sqrt{-1}\sqrt{9}\sqrt{2}$$
$$= 3i\sqrt{2}$$

35.
$$-\sqrt{-49} = -\sqrt{49}\sqrt{-1}$$
$$= -7i$$

36.
$$\sqrt{-\frac{9}{64}} = \sqrt{\frac{9}{64}}\sqrt{-1}$$
$$= \frac{3}{8}i$$

37.

Complex Numbers	
Real numbers	Imaginary numbers

38. a) True
b) True
c) False
d) False

39. $x^2 = -9$

$x = \sqrt{-9}$ or $x = -\sqrt{-9}$

$x = 3i$ or $x = -3i$

$$\boxed{x = 0 \pm 3i}$$

40. $3x^2 = -16$

$$\frac{3x^2}{3} = \frac{-16}{3}$$

$$x^2 = -\frac{16}{3}$$

$$x = \sqrt{-\frac{16}{3}} \qquad\qquad x = -\sqrt{-\frac{16}{3}}$$

$$x = \frac{4i}{\sqrt{3}} \qquad\qquad x = -\frac{4i}{\sqrt{3}}$$

$$x = \frac{4i}{\sqrt{3}} \cdot \frac{\sqrt{3}}{\sqrt{3}} \qquad\qquad x = -\frac{4i}{\sqrt{3}} \cdot \frac{\sqrt{3}}{\sqrt{3}}$$

$$x = \frac{4\sqrt{3}}{3}i \qquad\qquad x = -\frac{4\sqrt{3}}{3}i$$

$$\boxed{x = 0 \pm \frac{4\sqrt{3}}{3}i}$$

41. $2x^2 - 3x + 2 = 0$

$a = 2, b = -3, c = 2$

$$x = \frac{-b \pm \sqrt{b^2 - 4ac}}{2a}$$

$$= \frac{-(-3) \pm \sqrt{(-3)^2 - 4(2)(2)}}{2(2)}$$

$$= \frac{3 \pm \sqrt{9 - 16}}{4}$$

$$= \frac{3 \pm \sqrt{-7}}{4}$$

$$x = \frac{3 \pm i\sqrt{7}}{4}$$

$$\boxed{x = \frac{3}{4} \pm \frac{\sqrt{7}}{4}i}$$

42. $x^2 + 2x = -2$

$x^2 + 2x + 2 = -2 + 2$

$x^2 + 2x + 2 = 0$

$a = 1, b = 2, c = 2$

$$x = \frac{-b \pm \sqrt{b^2 - 4ac}}{2a}$$

$$= \frac{-2 \pm \sqrt{(2)^2 - 4(1)(2)}}{2(1)}$$

$$= \frac{-2 \pm \sqrt{4 - 8}}{2}$$

$$= \frac{-2 \pm \sqrt{-4}}{2}$$

$$= \frac{-2 \pm 2i}{2}$$

$$= \frac{2(-1 \pm i)}{2}$$

$$\boxed{x = -1 \pm i}$$

43. $3 - 6i$

44. $-1 + 7i$

45. $0 - 19i$

46. $0 + i$

47. $(3 + 4i) + (5 - 6i) = (3 + 5) + (4i - 6i)$

$= 8 - 2i$

48. $(7 - 3i) - (4 + 2i) = 7 - 3i - 4 - 2i$

$= (7 - 4) + (-3i - 2i)$

$= 3 - 5i$

49. $3i(2 - i) = 3i(2) - 3i(i)$

$= 6i - 3i^2$

$= 6i - 3(-1)$

$= 6i + 3$

$= 3 + 6i$

50. $(2 + 3i)(3 - i) = 6 - 2i + 9i - 3i^2$

$= 6 + 7i - 3(-1)$

$= 6 + 7i + 3$

$= 9 + 7i$

51.

$$\frac{3}{5 + i} = \frac{3}{5 + i} \bullet \frac{5 - i}{5 - i}$$

$$= \frac{15 - 3i}{25 - i^2}$$

$$= \frac{15 - 3i}{25 - (-1)}$$

$$= \frac{15 - 3i}{26}$$

$$= \frac{15}{26} - \frac{3}{26}i$$

52.

$$\frac{2 + 3i}{2 - 3i} = \frac{2 + 3i}{2 - 3i} \bullet \frac{2 + 3i}{2 + 3i}$$

$$= \frac{4 + 6i + 6i + 9i^2}{4 - 9i^2}$$

$$= \frac{4 + 12i + 9(-1)}{4 - 9(-1)}$$

$$= \frac{4 + 12i - 9}{4 + 9}$$

$$= \frac{-5 + 12i}{13}$$

$$= -\frac{5}{13} + \frac{12}{13}i$$

SECTION 9.4
Graphing Quadratic Equations

53. a) (-3, 0) and (1, 0)—The points where
the parabola crosses the   x-axis.
    b) (0, -3)—The point where the parabola
crosses the y-axis.
    c) (-1, -4)
    d) x = -1

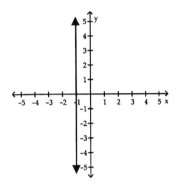

54. (400, 16,000) - The maximum profit of
$16,000 is obtained from the sale of 400
units.

55. upward

$$x = -\frac{b}{2a}$$

$$= -\frac{-4}{2(2)}$$

$$= -\frac{-4}{4}$$

$$= 1$$

$y = 2(1)^2 - 4(1) + 7$
$y = 2(1) - 4(1) + 7$
$y = 2 - 4 + 7$
$y = 5$
vertex: (1, 5)

56. downward

$$x = -\frac{b}{2a}$$

$$= -\frac{18}{2(-3)}$$

$$= -\frac{18}{-6}$$

$$= 3$$

$y = -3(3)^2 + 18(3) - 11$
$y = -3(9) + 18(3) - 11$
$y = -27 + 54 - 11$
$y = 16$
vertex: (3, 16)

57. To find the x-intercepts, let y = 0 and solve
for x.
$\quad 0 = x^2 + 6x + 5$
$\quad 0 = (x + 1)(x + 5)$
$\quad x + 1 = 0 \qquad$ or $\qquad x + 5 = 0$
$\quad x + 1 - 1 = 0 - 1 \qquad x + 5 - 5 = 0 - 5$
$\quad x = -1 \qquad\qquad\quad x = -5$
$\quad (-1, 0) \qquad$ and $\qquad (-5, 0)$

To find the y-intercept, let x = 0.
$\quad y = 0^2 + 6(0) + 5 = 5$
$\quad (0, 5)$
x-intercepts: (-1, 0) and (-5, 0)
y-intercept: (0, 5)

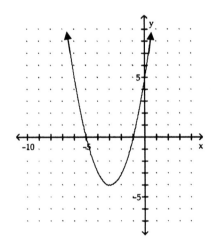

58. $y = x^2 + 2x - 3$

Step 1: Since $a = 1 > 0$, the parabola opens upward.

Step 2: Find the vertex.
$$x = -\frac{b}{2a} = -\frac{2}{2(1)} = -\frac{2}{2} = -1$$
$$y = (-1)^2 + 2(-1) - 3 = 1 - 2 - 3 = -4$$
vertex: $(-1, -4)$

Step 3: Find the x- and y-intercepts.
Since $c = -3$, the y-intercept is $(0, -3)$.
To find the x-intercepts, let $y = 0$ and solve the equation for x.
$$0 = x^2 + 2x - 3$$
$$0 = (x + 3)(x - 1)$$
$$x + 3 = 0 \quad \text{or} \quad x - 1 = 0$$
$$x + 3 - 3 = 0 - 3 \quad x - 1 + 1 = 0 + 1$$
$$x = -3 \qquad\qquad x = 1$$
$$(-3, 0) \qquad\qquad (1, 0)$$
x-intercepts: $(-3, 0)$ and $(1, 0)$;
y-intercept: $(0, -3)$

Step 4: Find another point(s).
Let $x = -2$.
$$y = (-2)^2 + 2(-2) - 3 = 4 - 4 - 3 = -3$$
$(-2, -3)$

Step 5: Plots the points and draw the parabola.

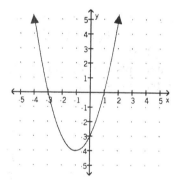

59. $y = -2x^2 + 4x - 2$

Step 1: Since $a = -2 < 0$, the parabola opens downward.

Step 2: Find the vertex.
$$x = -\frac{b}{2a} = -\frac{4}{2(-2)} = -\frac{4}{-4} = 1$$
$$y = -2(1)^2 + 4(1) - 2 = -2 + 4 - 2 = 0$$
vertex: $(1, 0)$

Step 3: Find the x- and y-intercepts.
Since $c = -2$, the y-intercept is $(0, -2)$.
To find the x-intercepts, let $y = 0$ and solve the equation for x.
$$0 = -2x^2 + 4x - 2$$
$$0 = -2(x^2 - 2x + 1)$$
$$0 = -2(x - 1)(x - 1)$$

$$x - 1 = 0 \quad \text{or} \quad x - 1 = 0$$
$$x - 1 + 1 = 0 + 1 \quad x - 1 + 1 = 0 + 1$$
$$x = 1 \qquad\qquad x = 1$$
x-intercept: $(1, 0)$
y-intercept: $(0, -2)$

Step 4: Find another point(s).
Let $x = -1$.
$$y = -2(-1)^2 + 4(-1) - 2 = -2 - 4 - 2 = -8$$
$(-1, -8)$
Use symmetry to find the points $(3, -8)$ and $(2, -2)$.

Step 5: Plots the points and draw the parabola.

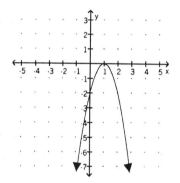

60. a) $x^2 + x - 2 = 0$ has **2** real number solution(s). The solutions are **-2, 1**.

b) $2x^2 + 12x + 18 = 0$ has **1** real number solution(s). The solution is **-3**.

c) $-x^2 + 4x - 5 = 0$ has **no** real number solution(s). The solutions are **none.**

SECTION 9.5
An Introduction to Functions

61. Yes.

62. Yes.

63. Yes.

64. No. $(-1, 2)$ and $(-1, 4)$

65. domain, input, output, range
$$f(2) = 2^3 - 4$$
$$= 8 - 4$$
$$= 4$$

66. Since $y = \boxed{f(x)}$, the equations $y = 2x - 8$ and $f(x) = 2x - 8$ are equivalent.

Chapter 9 Review

67. $f(1) = 1^2 - 4(1)$
$\quad\quad = 1 - 4$
$\quad\quad = \text{-}3$

68. $f(0) = 0^2 - 4(0)$
$\quad\quad = 0 - 4$
$\quad\quad = \text{-}4$

69. $f(\text{-}3) = (\text{-}3)^2 - 4(\text{-}3)$
$\quad\quad = 9 + 12$
$\quad\quad = 21$

70.
$$f\left(\frac{1}{2}\right) = \left(\frac{1}{2}\right)^2 - 4\left(\frac{1}{2}\right)$$
$$= \frac{1}{4} - 2$$
$$= \frac{1}{4} - \frac{8}{4}$$
$$= -\frac{7}{4}$$

71. $g(1) = 1 - 6(1)$
$\quad\quad = 1 - 6$
$\quad\quad = \text{-}5$

72. $g(\text{-}6) = 1 - 6(\text{-}6)$
$\quad\quad = 1 + 36$
$\quad\quad = 37$

73. $g(0.5) = 1 - 6(0.5)$
$\quad\quad = 1 - 3$
$\quad\quad = \text{-}2$

74.
$$g\left(\frac{3}{2}\right) = 1 - 6\left(\frac{3}{2}\right)$$
$$= 1 - 9$$
$$= -8$$

75. No. Answers may vary. Possible answers are (1, 4) and (1, 0.5).

76. Yes, it is a function.

77.

x	$f(x) = 1 -$ $\lvert$ x $\rvert$
0	$1 - \lvert \mathbf{0} \rvert = 1 - 0$ $= 1$
1	$1 - \lvert \mathbf{1} \rvert = 1 - 1$ $= 0$
2	$1 - \lvert \mathbf{2} \rvert = 1 - 2$ $= \text{-}1$
-1	$1 - \lvert \mathbf{\text{-}1} \rvert = 1 - 1$ $= 0$
-2	$1 - \lvert \mathbf{\text{-}2} \rvert = 1 - 2$ $= \text{-}1$
-3	$1 - \lvert \mathbf{\text{-}3} \rvert = 1 - 3$ $= \text{-}2$

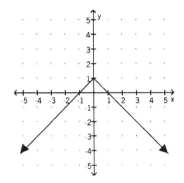

78. $f(x) = 15.7(8)^2$
$\quad\quad = 15.7(64)$
$\quad\quad = 1,004.8 \text{ in}^3$

SECTION 9.6
Variation

79. $\quad y = kx$
$\quad 500 = k \cdot 750$
$$\frac{500}{750} = k$$
$$\frac{2}{3} = k$$

Find y if $x = 1,250$ and $k = \dfrac{2}{3}$

$$y = \frac{2}{3} \bullet 1,250$$
$$\boxed{y = \$833.33}$$

80.

$$l = \frac{k}{w}$$

$$30 = \frac{k}{20}$$

$$20 \bullet 30 = \frac{k}{20} \bullet 20$$

$$\boxed{600 = k}$$

81. ELECTRICTY

$$c = \frac{k}{r}$$

$$2.5 = \frac{k}{150}$$

$$150 \bullet 2.5 = \frac{k}{150} \bullet 150$$

$$375 = k$$

Find c if $k = 375$ and $r = 150$

$$c = \frac{375}{2(150)}$$

$$c = \frac{375}{300}$$

$$c = 1.25$$

82. inverse variation

83. $R = kbc$

$$72 = k(4)(24)$$

$$72 = 96k$$

$$\frac{72}{96} = \frac{96k}{96}$$

$$k = \frac{72}{96}$$

$$k = \frac{3}{4}$$

Find R if $k = \frac{3}{4}, b = 6$, and $c = 18$.

$$R = \frac{3}{4}(6)(18)$$

$$\boxed{R = 81}$$

84.

$$s = \frac{wk}{m^2}$$

$$\frac{7}{4} = \frac{4k}{4^2}$$

$$\frac{7}{4} = \frac{4k}{16}$$

$$\frac{7}{4} = \frac{k}{4}$$

$$4 \bullet \frac{7}{4} = \frac{k}{4} \bullet 4$$

$$7 = k$$

Find s if $k = 7$, $w = 5$, and $m = 7$.

$$s = \frac{5(7)}{7^2}$$

$$s = \frac{35}{49}$$

$$\boxed{s = \frac{5}{7}}$$

CHAPTER 9 TEST

1. $x^2 = 17$

   $x = \sqrt{17}$ or $x = -\sqrt{17}$

   $\boxed{x = \pm\sqrt{17}}$

2. $u^2 = 24$

   $u = \sqrt{24}$ or $u = -\sqrt{24}$

   $u = \sqrt{4}\sqrt{6}$ or $u = -\sqrt{4}\sqrt{6}$

   $u = 2\sqrt{6}$ or $u = -2\sqrt{6}$

   $\boxed{u = \pm 2\sqrt{6}}$

3. $4y^2 = 25$

   $\dfrac{4y^2}{4} = \dfrac{25}{4}$

   $y = \sqrt{\dfrac{25}{4}}$ or $y = -\sqrt{\dfrac{25}{4}}$

   $y = \dfrac{5}{2}$ or $y = -\dfrac{5}{2}$

   $\boxed{y = \pm\dfrac{5}{2}}$

4. $(x - 2)^2 = 3$

   $x - 2 = \sqrt{3}$ or $x - 2 = -\sqrt{3}$

   $x - 2 + 2 = \sqrt{3} + 2$ $x - 2 + 2 = -\sqrt{3} + 2$

   $x = 2 + \sqrt{3}$ $x = 2 - \sqrt{3}$

   $\boxed{x = 2 \pm \sqrt{3}}$

5. $A = \pi r^2$

   $5{,}026.5 \approx 3.14 r^2$

   $\dfrac{5{,}026.5}{3.14} \approx \dfrac{3.14 r^2}{3.14}$

   $1{,}600.80 \approx r^2$

   $r \approx \sqrt{1{,}600.80}$ or $r \approx -\sqrt{1{,}600.80}$

   $r \approx 40$ $r \approx -40$

   $\boxed{40 \text{ cm.}}$

6. $\left(\dfrac{-14}{2}\right)^2 = (-7)^2 = 49$

7. $a^2 + 2a - 4 = 0$

   $a^2 + 2a - 4 + 4 = 0 + 4$

   $a^2 + 2a = 4$

   $a^2 + 2a + \left(\dfrac{2}{2}\right)^2 = 4 + \left(\dfrac{2}{2}\right)^2$

   $a^2 + 2a + 1 = 4 + 1$

   $(a + 1)^2 = 5$

   $a + 1 = \sqrt{5}$ or $a + 1 = -\sqrt{5}$

   $a + 1 - 1 = \sqrt{5} - 1$ $a + 1 - 1 = -\sqrt{5} - 1$

   $a = -1 + \sqrt{5}$ $a = -1 - \sqrt{5}$

   $\boxed{a = -1 \pm \sqrt{5}}$

   $a \approx -1 + 2.24 \approx 1.24$

   $a \approx -1 - 2.24 \approx -3.24$

8. $2x^2 = 3x + 2$

   $2x^2 - 3x = 3x + 2 - 3x$

   $2x^2 - 3x = 2$

   $\dfrac{2x^2}{2} - \dfrac{3x}{2} = \dfrac{2}{2}$

   $x^2 - \dfrac{3}{2}x = 1$

   $x^2 - \dfrac{3}{2}x + \left(\dfrac{3}{2} \bullet \dfrac{1}{2}\right)^2 = 1 + \left(\dfrac{3}{2} \bullet \dfrac{1}{2}\right)^2$

   $x^2 - \dfrac{3}{2}x + \left(\dfrac{3}{4}\right)^2 = 1 + \left(\dfrac{3}{4}\right)^2$

   $x^2 - \dfrac{3}{2}x + \dfrac{9}{16} = \dfrac{16}{16} + \dfrac{9}{16}$

   $\left(x - \dfrac{3}{4}\right)^2 = \dfrac{25}{16}$

   $x - \dfrac{3}{4} = \sqrt{\dfrac{25}{16}}$ or $x - \dfrac{3}{4} = -\sqrt{\dfrac{25}{16}}$

   $x - \dfrac{3}{4} = \dfrac{5}{4}$ or $x - \dfrac{3}{4} = -\dfrac{5}{4}$

   $x - \dfrac{3}{4} + \dfrac{3}{4} = \dfrac{3}{4} + \dfrac{5}{4}$ $x - \dfrac{3}{4} + \dfrac{3}{4} = \dfrac{3}{4} - \dfrac{5}{4}$

   $x = \dfrac{8}{4}$ or $x = \dfrac{-2}{4}$

   $\boxed{x = 2 \quad \text{or} \quad x = -\dfrac{1}{2}}$

9. $2x^2 - 5x = 12$

$2x^2 - 5x - 12 = 12 - 12$

$2x^2 - 5x - 12 = 0$

$a = 2, b = -5, c = -12$

$$x = \frac{-b \pm \sqrt{b^2 - 4ac}}{2a}$$

$$= \frac{-(-5) \pm \sqrt{(-5)^2 - 4(2)(-12)}}{2(2)}$$

$$= \frac{5 \pm \sqrt{25 - (-96)}}{4}$$

$$= \frac{5 \pm \sqrt{121}}{4}$$

$$= \frac{5 \pm 11}{4}$$

$$x = \frac{5 + 11}{4} = \frac{16}{4} = 4$$

$$x = \frac{5 - 11}{4} = \frac{-6}{4} = -\frac{3}{2}$$

$$\boxed{x = 4, -\frac{3}{2}}$$

10. $x^2 = 4x - 2$

$x^2 - 4x + 2 = 4x - 2 - 4x + 2$

$x^2 - 4x + 2 = 0$

$a = 1, b = -4, c = 2$

$$x = \frac{-b \pm \sqrt{b^2 - 4ac}}{2a}$$

$$= \frac{-(-4) \pm \sqrt{(-4)^2 - 4(1)(2)}}{2(1)}$$

$$= \frac{4 \pm \sqrt{16 - 8}}{2}$$

$$= \frac{4 \pm \sqrt{8}}{2}$$

$$= \frac{4 \pm \sqrt{4}\sqrt{2}}{2}$$

$$= \frac{4 \pm 2\sqrt{2}}{2}$$

$$= \frac{2(2 \pm \sqrt{2})}{2}$$

$$\boxed{x = 2 \pm \sqrt{2}}$$

11. $3x^2 - x - 1 = 0$

$a = 3, b = -1, c = -1$

$$x = \frac{-b \pm \sqrt{b^2 - 4ac}}{2a}$$

$$= \frac{-(-1) \pm \sqrt{(-1)^2 - 4(3)(-1)}}{2(3)}$$

$$= \frac{1 \pm \sqrt{1 - (-12)}}{6}$$

$$x = \frac{1 \pm \sqrt{13}}{6}$$

$$x \approx \frac{1 + 3.61}{6} \approx \frac{4.61}{6} \approx 0.77$$

$$x \approx \frac{1 - 3.61}{6} \approx \frac{-2.61}{6} \approx -0.43$$

$$\boxed{x \approx -0.43, 0.77}$$

12. $A = 128{,}775, \ L = 2x - 5, \ W = x$

$x(2x - 5) = 128{,}775$

$2x^2 - 5x = 128{,}775$

$2x^2 - 5x - 128{,}775 = 128{,}775 - 128{,}775$

$2x^2 - 5x - 128{,}775 = 0$

$a = 2, b = -5, c = -128{,}775$

$$x = \frac{-b \pm \sqrt{b^2 - 4ac}}{2a}$$

$$= \frac{-(-5) \pm \sqrt{(-5)^2 - 4(2)(-128{,}775)}}{2(2)}$$

$$= \frac{5 \pm \sqrt{25 - (-1{,}030{,}200)}}{4}$$

$$= \frac{5 \pm \sqrt{1{,}030{,}225}}{4}$$

$$= \frac{5 \pm 1{,}015}{4}$$

$$x = \frac{5 + 1{,}015}{4} = \frac{1{,}020}{4} = 255$$

$$x = \frac{5 - 1{,}015}{4} = \frac{-1{,}010}{4} = -\frac{505}{2}$$

$$\boxed{\begin{array}{l} W = 255 \text{ ft} \\ L = 2(255) - 5 = 510 - 5 = 505 \text{ ft} \end{array}}$$

13.

$$\sqrt{-100} = \sqrt{100}\sqrt{-1}$$
$$= 10i$$

14.

$$-\sqrt{-18} = -\sqrt{-1}\sqrt{9}\sqrt{2}$$
$$= -3i\sqrt{2}$$

15. A complex number is any number that can be written in the form $a + bi$, where a and b are real numbers and $i = \sqrt{-1}$.
Example: $2 + 8i$

16. $3x^2 + 2x + 1 = 0$
$a = 3, b = 2, c = 1$

$$x = \frac{-b \pm \sqrt{b^2 - 4ac}}{2a}$$

$$= \frac{-2 \pm \sqrt{(2)^2 - 4(3)(1)}}{2(3)}$$

$$= \frac{-2 \pm \sqrt{4 - 12}}{6}$$

$$= \frac{-2 \pm \sqrt{-8}}{6}$$

$$= \frac{-2 \pm \sqrt{-1}\sqrt{4}\sqrt{2}}{6}$$

$$= \frac{-2 \pm 2i\sqrt{2}}{6}$$

$$= \frac{2\left(-1 \pm i\sqrt{2}\right)}{2 \bullet 3}$$

$$x = \frac{-1 \pm i\sqrt{2}}{3}$$

$$\boxed{x = -\frac{1}{3} \pm \frac{\sqrt{2}}{3}i}$$

17. $(8 + 3i) + (-7 - 2i) = (8 + -7) + (3i - 2i)$
$\qquad\qquad\qquad\qquad = 1 + i$

18. $(5 + 3i) - (6 - 9i) = 5 + 3i - 6 + 9i$
$\qquad\qquad\qquad\qquad = (5 - 6) + (3i + 9i)$
$\qquad\qquad\qquad\qquad = -1 + 12i$

19. $(2 - 4i)(3 + 2i) = 6 + 4i - 12i - 8i^2$
$\qquad\qquad\qquad\quad = 6 - 8i - 8(-1)$
$\qquad\qquad\qquad\quad = 6 - 8i + 8$
$\qquad\qquad\qquad\quad = 14 - 8i$

20.

$$\frac{3 - 2i}{3 + 2i} = \frac{3 - 2i}{3 + 2i} \bullet \frac{3 - 2i}{3 - 2i}$$

$$= \frac{9 - 6i - 6i + 4i^2}{9 - 4i^2}$$

$$= \frac{9 - 12i + 4(-1)}{9 - 4(-1)}$$

$$= \frac{9 - 12i - 4}{9 + 4}$$

$$= \frac{5 - 12i}{13}$$

$$= \frac{5}{13} - \frac{12}{13}i$$

21. (3, 18)—The most air conditioners sold in a week (18) occurred when 3 ads were run.

22. $y = x^2 + 6x + 5$

Step 1: Since $a = 1 > 0$, the parabola opens upward.

Step 2: Find the vertex.
$$x = -\frac{b}{2a} = -\frac{6}{2(1)} = -\frac{6}{2} = -3$$
$$y = (-3)^2 + 6(-3) + 5 = 9 - 18 + 5 = -4$$
vertex: $(-3, -4)$

Step 3: Find the x- and y-intercepts.
Since $c = 5$, the y-intercept is $(0, 5)$.
To find the x-intercepts, let $y = 0$ and solve the equation for x.
$$0 = x^2 + 6x + 5$$
$$0 = (x + 5)(x + 1)$$

$x + 5 = 0$ or	$x + 1 = 0$
$x + 5 - 5 = 0 - 5$	$x + 1 - 1 = 0 - 1$
$x = -5$	$x = -1$
$(-5, 0)$	$(-1, 0)$

x-intercepts: $(-5, 0)$ and $(-1, 0)$;
y-intercept: $(0, 5)$

Step 4: Find another point(s).
Let $x = -2$.
$$y = (-2)^2 + 6(-2) + 5 = 4 - 12 + 5 = -3$$
$(-2, -3)$
Use symmetry to find $(-4, -3)$.

Step 5: Plots the points and draw the parabola.

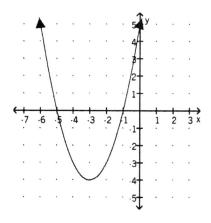

23. The domain of a function is the set of all possible values that can be used for the independent variable (inputs). The range is the set of all values of the dependent variable (outputs).

24. Yes.

25. No. (2, 3.5) and (2, -3.5)

26. Yes.

$f(-3) = 2(-3) - 7$
$\quad\quad = -6 - 7$
$\quad\quad = -13$

27. $g(6) = 3.5(6)^3$
$\quad\quad = 3.5(216)$
$\quad\quad = 756$

28. $C(45) = 0.30(45) + 15 = 13.5 + 15 = 28.50$
It costs $28.50 to make 45 calls.

29. $f(x) = |x| - 1$

x	$f(x) = \|x\| - 1$
0	$\|0\| - 1 = 0 - 1$ $= -1$
1	$\|1\| - 1 = 1 - 1$ $= 0$
2	$\|2\| - 1 = 2 - 1$ $= 1$
-1	$\|-1\| - 1 = 1 - 1$ $= 0$
-2	$\|-2\| - 1 = 2 - 1$ $= 1$
-3	$\|-3\| - 1 = 3 - 1$ $= 2$

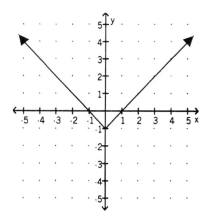

30. $f = ks$
$130 = k \cdot 6.5$
$130 = 6.5k$
$$\frac{130}{6.5} = \frac{6.5k}{6.5}$$
$20 = k$

Find f if $k = 20$ and $s = 5$.
$f = 20(5)$
$f = 100$ pounds

31.

$$r = \frac{k}{s}$$

$$40 = \frac{k}{10}$$

$$10 \bullet 40 = \frac{k}{10} \bullet 10$$

$$400 = k$$

Find $r = k = 400$ and $s = 15$.

$$r = \frac{400}{15}$$

$$\boxed{r = \frac{80}{3}}$$

1. a) true
   b) true
   c) true

2. $-4 + 2[-7 - 3(-9)] \; = -4 + 2[-7 + 27]$
   $$= -4 + 2[20]$$
   $$= -4 + 40$$
   $$= 36$$

3. DRIVING SAFETY

   2 hours before the salt is spread

4. EMPLOYMENT

   5.3% of what is 5,200?
   $0.053 \cdot x = 5,200$
   $$\frac{0.053x}{0.053} = \frac{5,200}{0.053}$$
   $$\boxed{x \approx 98,113}$$

5. $3p - 6(p + z) + p = 3p - 6p - 6z + p$
   $$= (3p - 6p + p) - 6z$$
   $$= -2p - 6z$$

6.
   $$\frac{5}{6}k = 10$$
   $$\frac{6}{5} \bullet \frac{5}{6}k = 10 \bullet \frac{6}{5}$$
   $$k = \frac{60}{5}$$
   $$\boxed{k = 12}$$

7. $-(3a + 1) + a = 2$
   $$-3a - 1 + a = 2$$
   $$-2a - 1 = 2$$
   $$-2a - 1 + 1 = 2 + 1$$
   $$-2a = 3$$
   $$\boxed{a = -\frac{3}{2}}$$

8. $\quad 5x + 7 < 2x + 1$
   $$5x + 7 - 2x < 2x + 1 - 2x$$
   $$3x + 7 < 1$$
   $$3x + 7 - 7 < 1 - 7$$
   $$3x < -6$$
   $$\frac{3x}{3} < \frac{-6}{3}$$
   $$x < -2$$
   $$(-\infty, 2)$$

   2

9. ENTREPRENEURS

   Let x = amount lent at 7%.
   Then $(28,000 - x)$ = amount lent at 10%.

   $0.07x + 0.10(28,000 - x) = 2,560$
   $$0.07x + 2,800 - 0.10x = 2,560$$
   $$-0.03x + 2,800 = 2,560$$
   $$-0.03x + 2,800 - 2,800 = 2,560 - 2,800$$
   $$-0.03x = -240$$
   $$\frac{-0.03x}{-0.03} = \frac{-240}{-0.03}$$
   $$\boxed{x = 8,000}$$

   $28,000 - x = 28,000 - 8,000$
   $$= 20,000$$

   $8,000 at 7% and $20,000 at 10%

10. $(-2 - 5)^2 + (1 - (-3))^2 = (-7)^2 + (4)^2$
    $$= 49 + 16$$
    $$= 65$$

11.
    $$\left|\frac{4}{5} \bullet 10 - 12\right| = \left|\frac{40}{5} - 12\right|$$
    $$= |8 - 12|$$
    $$= |-4|$$
    $$= 4$$

12.
$$m = \frac{y_2 - y_1}{x_2 - x_1}$$
$$= \frac{-8 - (-2)}{-12 - (-2)}$$
$$= \frac{-6}{-10}$$
$$= \frac{3}{5}$$

13.
$$2y - 2x = 6$$
$$2y - 2x + 2x = 6 + 2x$$
$$2y = 2x + 6$$
$$\frac{2y}{2} = \frac{2x}{2} + \frac{6}{2}$$
$$y = x + 3$$

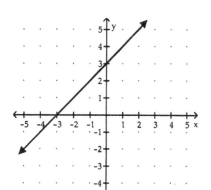

14. $y = -3$

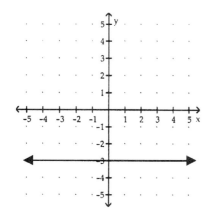

15. $y = -x + 2$

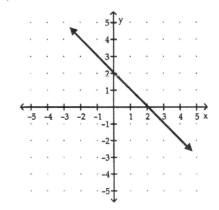

16. $y < 3x$

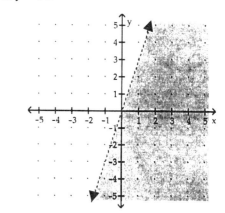

17.

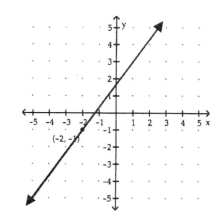

(-2, -1)

18. $y = x^3 - 2$

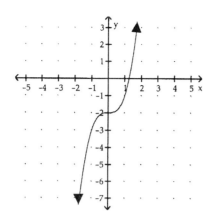

19. $y = mx + b$
$y = -2x + 1$

20. $\quad y - y_1 = m\left(x - x_1\right)$

$y - 1 = \dfrac{1}{4}(x - 8)$

$y - 1 = \dfrac{1}{4}(x) - \dfrac{1}{4}(8)$

$y - 1 = \dfrac{1}{4}x - 2$

$y - 1 + 1 = \dfrac{1}{4}x - 2 + 1$

$$\boxed{y = \dfrac{1}{4}x - 1}$$

21. $\quad 4x + 5y = 6$

$4x + 5y - 4x = 6 - 4x$

$5y = -4x + 6$

$\dfrac{5y}{5} = \dfrac{-4x}{5} + \dfrac{6}{5}$

$y = -\dfrac{4}{5}x + \dfrac{6}{5}$

$$\boxed{m = -\dfrac{4}{5}}$$

22. INSURANCE

$\text{rate of change} = \dfrac{y_2 - y_1}{x_2 - x_1}$

$= \dfrac{530 - 418}{2002 - 1995}$

$= \dfrac{112}{7}$

$= 16$

An increase of \$16 per year.

23.
$$y^3\left(y^2 y^4\right) = y^3\left(y^{2+4}\right)$$
$$= y^3\left(y^6\right)$$
$$= y^{3+6}$$
$$= y^9$$

24.
$$\left(\dfrac{b^2}{3a}\right)^3 = \dfrac{b^{2 \cdot 3}}{3^3 \, a^{1 \cdot 3}}$$
$$= \dfrac{b^6}{27a^3}$$

25.
$$\dfrac{10a^4 a^{-2}}{5a^2 a^0} = \dfrac{10a^{4+(-2)}}{5a^{2+0}}$$
$$= \dfrac{10a^2}{5a^2}$$
$$= 2a^{2-2}$$
$$= 2a^0$$
$$= 2(1)$$
$$= 2$$

26.
$$\left(\dfrac{21x^{-2}y^2 z^{-2}}{7x^3 y^{-1}}\right)^{-2} = \left(3x^{-2-3}y^{2-(-1)}z^{-2}\right)^{-2}$$
$$= \left(3x^{-5}y^3 z^{-2}\right)^{-2}$$
$$= 3^{-2} x^{-5 \cdot -2} y^{3 \cdot -2} z^{-2 \cdot -2}$$
$$= \dfrac{1}{9}x^{10} y^{-6} z^4$$
$$= \dfrac{x^{10} z^4}{9y^6}$$

27. FIVE-CARD POKER

2,600,000 to 1

28. 7.3×10^{-4}

29. $4(4x^3 + 2x^2 - 3x - 8) - 5(2x^3 - 3x + 8)$
$= 16x^3 + 8x^2 - 12x - 32 - 10x^3 + 15x - 40$
$= 6x^3 + 8x^2 + 3x - 72$

30. $(-2a^3)(3a^2) = (-2 \cdot 3)a^{3+2}$
$= -6a^5$

Chapter 9 Cumulative Review

31. $(2b - 1)(3b + 4) = 6b^2 + 8b - 3b - 4$
$= 6b^2 + 5b - 4$

32. $(2x + 5y)^2 = (2x + 5y)(2x + 5y)$
$= 4x^2 + 10xy + 10xy + 25y^2$
$= 4x^2 + 20xy + 25y^2$

33. $(3x + y)(2x^2 - 3xy + y^2)$
$= 3x(2x^2 - 3xy + y^2) + y(2x^2 - 3xy + y^2)$
$= 6x^3 - 9x^2y + 3xy^2 + 2x^2y - 3xy^2 + y^3$
$= 6x^3 - 7x^2y + y^3$

34.

$$
\begin{array}{r}
2x + 1 \\
x - 3 \overline{\smash{\big)}\, 2x^2 - 5x - 3} \\
\underline{2x^2 - 6x} \\
x - 3 \\
\underline{x - 3} \\
0
\end{array}
$$

35. $6a^2 - 12a^3b + 36ab = 6a(a - 2a^2b + 6b)$

36. $2x + 2y + ax + ay = (2x + 2y) + (ax + ay)$
$= 2(x + y) + a(x + y)$
$= (x + y)(2 + a)$

37. $b^3 + 125 = (b + 5)(b^2 - b(5) + 5^2)$
$= (b + 5)(b^2 - 5b + 25)$

38. $t^4 - 16 = (t^2 + 4)(t^2 - 4)$
$= (t^2 + 4)(t + 2)(t - 2)$

39. $3x^2 + 8x = 0$
$x(3x + 8) = 0$
$x = 0$ or $3x + 8 = 0$
$3x = -8$
$x = -\dfrac{8}{3}$

$$\boxed{x = 0, \; -\dfrac{8}{3}}$$

40. $15x^2 - 2 = 7x$
$15x^2 - 2 - 7x = 7x - 7x$
$15x^2 - 7x - 2 = 0$
$(5x + 1)(3x - 2) = 0$
$5x + 1 = 0$ or $3x - 2 = 0$
$5x + 1-1 = 0-1$ $3x - 2+2 = 0+2$
$5x = -1$ $3x = 2$

$$\boxed{x = -\dfrac{1}{5} \qquad x = \dfrac{2}{3}}$$

41. $2(2x^3 - x) + 2(x^3 + 3x)$
$= 4x^3 - 2x + 2x^3 + 6x$
$= 6x^3 + 4x$

42. HEIGHT OF A TRIANGLE

$$\frac{1}{2}(x)(x + 4) = 22.5$$

$$2 \bullet \frac{1}{2}(x)(x + 4) = 2 \bullet 22.5$$

$$x(x + 4) = 45$$

$$x^2 + 4x = 45$$

$$x^2 + 4x - 45 = 45 - 45$$

$$x^2 + 4x - 45 = 0$$

$$(x + 9)(x - 5) = 0$$

$x + 9 = 0$ or $x - 5 = 0$
$x + 9-9 = 0-9$ $x - 5+5 = 0+5$
$x = -9$ $x = 5$

The height cannot be negative.

$$\boxed{\text{The height is 5 in.}}$$

43. $x + 8 = 0$
$x + 8 - 8 = 0 - 8$
$x = -8$ makes the fraction undefined.

44.

$$\frac{3x^2 - 27}{x^2 + 3x - 18} = \frac{3(x^2 - 9)}{(x + 6)(x - 3)}$$

$$= \frac{3(x - 3)(x + 3)}{(x + 6)(x - 3)}$$

$$= \frac{3(x + 3)}{(x + 6)}$$

45.

$$\frac{x^2 - x - 6}{2x^2 + 9x + 10} \div \frac{x^2 - 25}{2x^2 + 15x + 25}$$

$$= \frac{x^2 - x - 6}{2x^2 + 9x + 10} \bullet \frac{2x^2 + 15x + 25}{x^2 - 25}$$

$$= \frac{(x - 3)(x + 2)}{(2x + 5)(x + 2)} \bullet \frac{(2x + 5)(x + 5)}{(x + 5)(x - 5)}$$

$$= \frac{x - 3}{x - 5}$$

46.

$$\frac{x+5}{xy} - \frac{x-1}{x^2 y} = \frac{x+5}{xy}\left(\frac{x}{x}\right) - \frac{x-1}{x^2 y}$$

$$= \frac{x^2 + 5x}{x^2 y} - \frac{x-1}{x^2 y}$$

$$= \frac{x^2 + 5x - (x-1)}{x^2 y}$$

$$= \frac{x^2 + 5x - x + 1}{x^2 y}$$

$$= \frac{x^2 + 4x + 1}{x^2 y}$$

47.

$$\frac{x}{x-2} + \frac{3x}{x^2 - 4} = \frac{x}{x-2} + \frac{3x}{(x+2)(x-2)}$$

$$= \frac{x}{x-2}\left(\frac{x+2}{x+2}\right) + \frac{3x}{(x+2)(x-2)}$$

$$= \frac{x^2 + 2x}{(x-2)(x+2)} + \frac{3x}{(x+2)(x-2)}$$

$$= \frac{x^2 + 2x + 3x}{(x-2)(x+2)}$$

$$= \frac{x^2 + 5x}{(x-2)(x+2)}$$

48.

$$\frac{\dfrac{5}{y} + \dfrac{4}{y+1}}{\dfrac{4}{y} - \dfrac{5}{y+1}} = \frac{\dfrac{5}{y} + \dfrac{4}{y+1}}{\dfrac{4}{y} - \dfrac{5}{y+1}} \bullet \frac{y(y+1)}{y(y+1)}$$

$$= \frac{\dfrac{5}{y} \bullet y(y+1) + \dfrac{4}{y+1} \bullet y(y+1)}{\dfrac{4}{y} \bullet y(y+1) - \dfrac{5}{y+1} \bullet y(y+1)}$$

$$= \frac{5(y+1) + 4y}{4(y+1) - 5y}$$

$$= \frac{5y + 5 + 4y}{4y + 4 - 5y}$$

$$= \frac{9y + 5}{-y + 4}$$

$$= \frac{9y + 5}{4 - y}$$

49.

$$\frac{2p}{3} - \frac{1}{p} = \frac{2p-1}{3}$$

$$3p\left(\frac{2p}{3} - \frac{1}{p}\right) = 3p\left(\frac{2p-1}{3}\right)$$

$$p(2p) - 3(1) = p(2p-1)$$

$$2p^2 - 3 = 2p^2 - p$$

$$2p^2 - 3 - 2p^2 = 2p^2 - p - 2p^2$$

$$-3 = -p$$

$$\frac{-3}{-1} = \frac{-p}{-1}$$

$$\boxed{3 = p}$$

50.

$$\frac{7}{q^2 - q - 2} + \frac{1}{q+1} = \frac{3}{q-2}$$

$$\frac{7}{(q+1)(q-2)} + \frac{1}{q+1} = \frac{3}{q-2}$$

$$(q+1)(q-2)\left(\frac{7}{(q+1)(q-2)} + \frac{1}{q+1}\right) = (q+1)(q-2)\left(\frac{3}{q-2}\right)$$

$$7 + q - 2 = 3(q+1)$$

$$q + 5 = 3q + 3$$

$$q + 5 - q = 3q + 3 - q$$

$$5 = 2q + 3$$

$$5 - 3 = 2q + 3 - 3$$

$$2 = 2q$$

$$\frac{2}{2} = \frac{2q}{2}$$

$$\boxed{1 = q}$$

Chapter 9 Cumulative Review

51.

$$\frac{1}{a} + \frac{1}{b} = 1$$

$$ab\left(\frac{1}{a} + \frac{1}{b}\right) = ab(1)$$

$$b + a = ab$$

$$b + a - a = ab - a$$

$$b = ab - a$$

$$b = a(b - 1)$$

$$\frac{b}{b-1} = \frac{a(b-1)}{b-1}$$

$$\boxed{\frac{b}{b-1} = a}$$

52. ROOFING

$$\frac{1}{7} + \frac{1}{4} = \frac{1}{x}$$

$$28x\left(\frac{1}{7} + \frac{1}{4}\right) = 28x\left(\frac{1}{x}\right)$$

$$4x + 7x = 28$$

$$11x = 28$$

$$\frac{11x}{11} = \frac{28}{11}$$

$$\boxed{x = 2\frac{6}{11}}$$

It will take both of them $2\frac{6}{11}$ days.

53. LOSING WEIGHT

$$\frac{350}{10} = \frac{x}{25}$$

$$350(25) = 10x$$

$$8{,}750 = 10x$$

$$\frac{8{,}750}{10} = \frac{10x}{10}$$

$$\boxed{875 = x}$$

It will take 875 days.

54.

$$\frac{6}{x} = \frac{4}{26}$$

$$6(26) = x \cdot 4$$

$$156 = 4x$$

$$\frac{156}{4} = \frac{4x}{4}$$

$$\boxed{39 = x}$$

55.   (-2, 3)

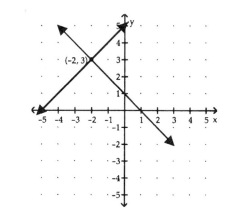

56.
$$y = 2x + 5$$
$$x + 2y = -5$$

Substitute $(2x + 5)$ for y into the second equation.

$$x + 2(\mathbf{2x + 5}) = -5$$
$$x + 4x + 10 = -5$$
$$5x + 10 = -5$$
$$5x + 10 - 10 = -5 - 10$$
$$5x = -15$$
$$\frac{5x}{5} = \frac{-15}{5}$$
$$x = -3$$

Substitute -3 for x into the first equation.
$$y = 2x + 5$$
$$= 2(-3) + 5$$
$$= -6 + 5$$
$$= -1$$

$$\boxed{(-3, -1)}$$

57.

$$\frac{3}{5}s + \frac{4}{5}t = 1$$

$$-\frac{1}{4}s + \frac{3}{8}t = 1$$

Multiply by a LCD

$$5\left(\frac{3}{5}s + \frac{4}{5}t = 1\right)$$

$$8\left(-\frac{1}{4}s + \frac{3}{8}t = 1\right)$$

$3s + 4t = 5$ Multiply 1st equation by 2

$-2s + 3t = 8$ Multiply 2nd equation by 3

$$2(3s + 4t = 5)$$

$$3(-2s + 3t = 8)$$

$6s + 8t = 10$

$-6s + 9t = 24$

$17t = 34$

$$\frac{17t}{17} = \frac{34}{17}$$

$$\boxed{t = 2}$$

Substitute 2 in 1st equation

$3s + 4(\mathbf{2}) = 5$

$3s + 8 = 5$

$3s = -3$

$$\frac{3s}{3} = \frac{-3}{3}$$

$$\boxed{s = -1}$$

$$\boxed{\text{Solution point is } (-1, 2).}$$

58. AVIATION

Let r = rate of plane in still air

w = rate of wind.

$d = rt$

$3,000 = 5(r + w)$

$3,000 = 6(r - w)$

$3,000 = 5r + 5w$

$3,000 = 6r - 6w$

$6(3,000 = 5r + 5w)$

$5(3,000 = 6r - 6w)$

$18,000 = 30r + 30w$

$15,000 = 30r - 30w$

$33,000 = 60r$

$$\frac{33,000}{60} = \frac{60r}{60}$$

$$\boxed{550 = r}$$

The plane's speed is 550 mph.

59. MIXING CANDY

Let x = pounds of hard candy

y = pounds of soft candy.

$x + y = 48$

$3x + 6y = 4(48)$

$x + y = 48$

$3x + 6y = 192$

$-3(x + y = 48)$

$3x + 6y = 192$

$-3x - 3y = -144$

$3x + 6y = 192$

$3y = 48$

$$\frac{3y}{3} = \frac{48}{3}$$

$$\boxed{y = 16}$$

Substitute 16 in 1st equation

$x + y = 48$

$x + 16 = 48$

$x + 16 - 16 = 48 - 16$

$$\boxed{x = 32}$$

32 pounds of hard candy.

16 pounds of soft candy.

60.

$$\begin{cases} 3x + 4y \geq -7 \\ 2x - 3y \geq 1 \end{cases}$$

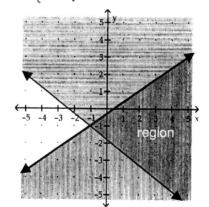

     Chapter 9 Cumulative Review

61.

$$\sqrt{50x^2} = \sqrt{25x^2}\sqrt{2}$$
$$= 5x\sqrt{2}$$

62. $-\sqrt{100a^6b^4} = -10a^3b^2$

63.

$$3\sqrt{24} + \sqrt{54} = 3\sqrt{4}\sqrt{6} + \sqrt{9}\sqrt{6}$$
$$= 3 \bullet 2\sqrt{6} + 3\sqrt{6}$$
$$= 6\sqrt{6} + 3\sqrt{6}$$
$$= 9\sqrt{6}$$

64.

$$\left(\sqrt{2}+1\right)\left(\sqrt{2}-3\right) = \sqrt{4} - 3\sqrt{2} + 1\sqrt{2} - 3$$
$$= 2 - 2\sqrt{2} - 3$$
$$= -1 - 2\sqrt{2}$$

65.

$$\sqrt{\frac{72x^3}{y^2}} = \frac{\sqrt{36x^2}\sqrt{2x}}{\sqrt{y^2}}$$
$$= \frac{6x\sqrt{2x}}{y}$$

66.

$$\frac{8}{\sqrt{10}} = \frac{8}{\sqrt{10}} \bullet \frac{\sqrt{10}}{\sqrt{10}}$$
$$= \frac{8\sqrt{10}}{\sqrt{100}}$$
$$= \frac{8\sqrt{10}}{10}$$
$$= \frac{4\sqrt{10}}{5}$$

67. $\sqrt[3]{\frac{27m^3}{8n^6}} = \frac{3m}{2n^2}$

68. $\sqrt[4]{16} = 2$

69.

$$25^{3/2} = \left(\sqrt{25}\right)^3$$
$$= 5^3$$
$$= 125$$

70.

$$(-8)^{-4/3} = \frac{1}{(-8)^{4/3}}$$
$$= \frac{1}{\left(\sqrt[3]{-8}\right)^4}$$
$$= \frac{1}{(-2)^4}$$
$$= \frac{1}{16}$$

71.

$$\sqrt{6x+1} + 2 = 7$$
$$\sqrt{6x+1} + 2 - 2 = 7 - 2$$
$$\sqrt{6x+1} = 5$$
$$\left(\sqrt{6x+1}\right)^2 = (5)^2$$
$$6x + 1 = 25$$
$$6x + 1 - 1 = 25 - 1$$
$$6x = 24$$
$$\frac{6x}{6} = \frac{24}{6}$$
$$\boxed{x = 4}$$

72. $x^2 + 8x + 12 = 0$
$$x^2 + 8x + 12 - 12 = 0 - 12$$
$$x^2 + 8x = -12$$
$$x^2 + 8x + \left(\frac{8}{2}\right)^2 = -12 + \left(\frac{8}{2}\right)^2$$
$$x^2 + 8x + 16 = -12 + 16$$
$$(x+4)^2 = 4$$
$$x + 4 = \sqrt{4} \quad \text{or} \quad x + 4 = -\sqrt{4}$$
$$x + 4 = 2 \qquad\qquad x + 4 = -2$$
$$x + 4 - 4 = 2 - 4 \qquad x + 4 - 4 = -2 - 4$$
$$\boxed{x = -2 \qquad\qquad x = -6}$$

73.

$$t^2 = 75$$
$$t = \sqrt{75} \qquad\qquad t = -\sqrt{75}$$
$$t = \sqrt{25}\sqrt{3} \qquad\quad t = -\sqrt{25}\sqrt{3}$$
$$t = 5\sqrt{3} \qquad\qquad t = -5\sqrt{3}$$
$$\boxed{t = \pm 5\sqrt{3}}$$

74. STORAGE CUBES

Use the Pythagorean Theorem
$$x^2 + x^2 = 15^2$$
$$2x^2 = 225$$
$$x^2 = \frac{225}{2}$$
$$x = \sqrt{\frac{225}{2}}$$
$$x \approx 10.6$$
$$2x \approx 2(10.6)$$
$$= 21.2$$

The height of the cube is 21.2 in.

75. $3x^2 - x - 1 = 0$
$a = 3, b = -1, c = -1$

$$x = \frac{-b \pm \sqrt{b^2 - 4ac}}{2a}$$

$$x = \frac{-(-1) \pm \sqrt{(-1)^2 - 4(3)(-1)}}{2(3)}$$

$$x = \frac{1 \pm \sqrt{1 - (-12)}}{6}$$

$$x = \frac{1 \pm \sqrt{13}}{6}$$

$$x \approx \frac{1 + 3.61}{6} \qquad x \approx \frac{1 - 3.61}{6}$$

$$x \approx \frac{4.61}{6} \qquad x \approx \frac{-2.61}{6}$$

$$x \approx 0.77 \qquad x \approx -0.44$$

$$\boxed{x \approx 0.77, \quad -0.44}$$

76. QUILTS

Let $W = x$ and $L = 2x - 11$.
$$x(2x - 11) = 12{,}865$$
$$2x^2 - 11x = 12{,}865$$
$$2x^2 - 11x - 12{,}865 = 12{,}865 - 12{,}865$$
$$2x^2 - 11x - 12{,}865 = 0$$

$a = 2, b = -11, c = -12{,}865$

$$x = \frac{-b \pm \sqrt{b^2 - 4ac}}{2a}$$

$$x = \frac{-(-11) \pm \sqrt{(-11)^2 - 4(2)(-12{,}865)}}{2(2)}$$

$$x = \frac{11 \pm \sqrt{121 - (-102{,}920)}}{4}$$

$$x = \frac{11 \pm \sqrt{103{,}041}}{4}$$

$$x = \frac{11 \pm 321}{4}$$

$$x = \frac{11 + 321}{4} \qquad x = \frac{11 - 321}{4}$$

$$x = \frac{332}{4} \qquad x = \frac{-310}{4}$$

$$\boxed{x = 83 \qquad x = -\frac{155}{2}}$$

Since the length and width cannot be negative, $x = 83$.

$$\text{Width} = 83$$
$$\text{Length} = 2(\mathbf{83}) - 11$$
$$= 166 - 11$$
$$= 155$$

The dimensions are 83 ft × 155 ft.

77.
$$\sqrt{-49} = \sqrt{49}\sqrt{-1}$$
$$= 7i$$

78.
$$\sqrt{-54} = \sqrt{-1}\sqrt{9}\sqrt{6}$$
$$= 3i\sqrt{6}$$

79. $(2 + 3i) - (1 - 2i) = 2 + 3i - 1 + 2i$
$$= 1 + 5i$$

80. $(7 - 4i) + (9 + 2i) = (7 + 9) + (-4i + 2i)$
$$= 16 - 2i$$

81. $(3 - 2i)(4 - 3i) = 12 - 9i - 8i + 6i^2$
$$= 12 - 17i + 6(-1)$$
$$= 12 - 17i - 6$$
$$= 6 - 17i$$

82.

$$\frac{3-i}{2+i} = \frac{3-i}{2+i} \cdot \frac{2-i}{2-i}$$

$$= \frac{6-3i-2i+i^2}{4-i^2}$$

$$= \frac{6-5i+(-1)}{4-(-1)}$$

$$= \frac{5-5i}{5}$$

$$= \frac{5(1-i)}{5}$$

$$= 1-i$$

83.

$$x^2 + 16 = 0$$
$$x^2 + 16 - 16 = 0 - 16$$
$$x^2 = -16$$
$$x = \sqrt{-16} \quad \text{or} \quad x = -\sqrt{-16}$$
$$x = 4i \qquad\qquad x = -4i$$
$$\boxed{x = 0 \pm 4i}$$

84.

$$x^2 - 4x = -5$$
$$x^2 - 4x + 5 = -5 + 5$$
$$x^2 - 4x + 5 = 0$$

$$a = 1, b = -4, c = 5$$

$$x = \frac{-b \pm \sqrt{b^2 - 4ac}}{2a}$$

$$= \frac{-(-4) \pm \sqrt{(-4)^2 - 4(1)(5)}}{2(1)}$$

$$= \frac{4 \pm \sqrt{16 - 20}}{2}$$

$$= \frac{4 \pm \sqrt{-4}}{2}$$

$$= \frac{4 \pm 2i}{2}$$

$$= \frac{2(2 \pm i)}{2}$$

$$x = 2 \pm i$$

85. $y = 2x^2 + 8x + 6$

Step 1: Since $a = 2 > 0$, the parabola opens upward.

Step 2: Find the vertex.

$$x = -\frac{b}{2a} = -\frac{8}{2(2)} = -\frac{8}{4} = -2$$

$$y = 2(-2)^2 + 8(-2) + 6 = 8 - 16 + 6 = -2$$

vertex: $(-2, -2)$

Step 3: Find the x- and y-intercepts.
Since $c = 6$, the y-intercept is $(0, 6)$.
To find the x-intercepts, let $y = 0$ and solve the equation for x.

$$0 = 2x^2 + 8x + 6$$
$$0 = 2(x^2 + 4x + 3)$$
$$0 = (x + 3)(x + 1)$$
$$x + 3 = 0 \quad \text{or} \quad x + 1 = 0$$
$$x + 3 - 3 = 0 \qquad x + 1 - 1 = 0 - 1$$
$$x = -3 \qquad\qquad x = -1$$
$$(-3, 0) \qquad\qquad (-1, 0)$$

x-intercepts: $(-3, 0)$ and $(-1, 0)$;
y-intercept: $(0, 6)$

Step 4: Find another point(s).
Let $x = -4$.

$$y = 2(-4)^2 + 8(-4) + 6 = 32 - 32 + 6 = 6$$
$$(-4, 6)$$

Step 5: Plots the points and draw the parabola.

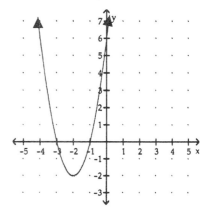

86. POWER OUTPUT

4,000 rpm

87.
$$f(-1) = 3(-1)^2 + 3(-1) - 8$$
$$f(-1) = 3(1) + 3(-1) - 8$$
$$f(-1) = 3 - 3 - 8$$
$$f(-1) = -8$$

88. input

89. yes

90. GEAR

$$s = \frac{k}{t}$$

$$3 = \frac{k}{10}$$

$$10 \bullet 3 = \frac{k}{10} \bullet 10$$

$$30 = k$$

Find s if $k = 30$ and $t = 25$.

$$s = \frac{30}{25}$$

$$s = \frac{6}{5}$$

$$\boxed{s = 1.2}$$

APPENDIX 1

STATISTICS

PRACTICE

1. $\text{mean} = \dfrac{7+5+9+10+8+6+6+7+9+12+9}{11}$

    $= \dfrac{88}{11}$

    $= 8$

2. 5, 6, 6, 7, 7, 8, 9, 9, 9, 10, 12
    median = 8 since 8 is the middle value

3. mode = 9 since 9 occurs most frequently

4. 8, 8, 10, 12, 12, 12, 14, 16, 16, 23, 23, 26

    $\text{median} = \dfrac{12+14}{2}$

    $= \dfrac{26}{2}$

    $= 13$

5. mode = 12 since 12 occurs most frequently

6. $\text{mean} = \dfrac{8+12+23+12+10+16+26+12+14+8+16+23}{12}$

    $= \dfrac{180}{12}$

    $= 15$

7. $\text{mean} = \dfrac{24+27+30+27+31+30+27}{7}$

    $= \dfrac{196}{7}$

    $= 28$

    24, 27, 27, 27, 30, 30, 31
    median = 27
    mode = 27

8. $\text{mean} = \dfrac{85+87+88+82+85+91+88+88}{8}$

    $= \dfrac{694}{8}$

    $= 86.75$

    82, 85, 85, 87, 88, 88, 88, 91

    $\text{median} = \dfrac{87+88}{2}$

    $= \dfrac{175}{2}$

    $= 87.5$

    mode = 88

APPLICATIONS

9. FOOTBALL

    $\text{means} = \dfrac{-8+2+(-6)+6+4+(-7)+(-5)}{7}$

    $= \dfrac{-14}{7}$

    $= -2 \text{ yd}$

10. SALES

    $\text{means} = \dfrac{1,525+785+1,628+1,214+917+1,197}{7}$

    $= \dfrac{7,266}{6}$

    $= \$1,211$

11. VIRUSES

    $\text{means} = \dfrac{2.5+105.1+74.9+137.4+52.6}{5}$

    $= \dfrac{372.5}{5}$

    $= 74.5$

12. SALARIES

$$\text{mean} = \frac{\text{sum of all salaries}}{10}$$
$$= \frac{26,065}{10}$$
$$= \$2,606.50$$

13. JOB TESTING

$$85 = \frac{78 + 91 + 87 + x}{4}$$
$$4 \bullet 85 = \frac{78 + 91 + 87 + x}{4} \bullet 4$$
$$340 = 78 + 91 + 87 + x$$
$$340 = 256 + x$$
$$340 - 256 = 256 + x - 256$$
$$\boxed{84 = x}$$

14. GAS MILEAGE

$$20.8 = \frac{20.3 + 14.1 + 28.2 + 16.9 + x}{5}$$
$$20.8 = \frac{79.5 + x}{5}$$
$$5 \bullet 20.8 = \frac{79.5 + x}{5} \bullet 5$$
$$104 = 79.5 + x$$
$$104 - 79.5 = 79.5 + x - 79.5$$
$$\boxed{24.5 = x}$$

15. SPORT FISHING

3, 3, 4, 4, 4, 4, 4, 5, 5, 5, 6, 6, 7, 8, 9, 9, 13
median = 5
mode = 4

16. SALARIES

median = \$2,575
mode = \$2,415

17. FUEL EFFICIENCY

28, 29, 29, 29, 29, 30, 31, 33, 35, 39
$$\text{median} = \frac{29 + 30}{2} = \frac{59}{2} = 29.5$$
mode = 29

18. FUEL EFFICIENCY

35, 37, 38, 38, 38, 38, 40, 40, 41, 43
$$\text{median} = \frac{38 + 38}{2} = \frac{76}{2} = 38$$
mode = 38

WRITING

19. Answers may vary.

20. Answers may vary.

21. Answers may vary.

22. Answers may vary.

Appendix - Statistics